LINEAR DIFFERENTIAL OPERATORS

CORNELIUS LANCZOS

DOVER PUBLICATIONS, INC.
Mineola, New York

Bibliographical Note

This Dover edition, first published in 1997, is an unabridged
and unaltered republication of the work first published by
D. Van Nostrand Company Ltd., London and New York, in
1961.

Library of Congress Cataloging-in-Publication Data

Lanczos, Cornelius, 1893–
    Linear differential operators / Cornelius Lanczos.
        p.   cm.
    Originally published: London ; New York : Van Nos-
trand, 1961.
    Includes Bibliographical references and index.
    ISBN 0-486-68035-5 (pbk.)
    1. Calculus, Operational. 2. Differential equations,
Linear. I. Title.
QA432.L3   1997
515'.7242—dc21   97-26069
                                               CIP

Manufactured in the United States of America
Dover Publications, Inc., 31 East 2nd Street, Mineola, N.Y. 11501

A l'apôtre de l'humanité universelle, le Père
Pire, dont la charité ne connaît pas de limites.

PREFACE

In one of the (unfortunately lost) comedies of Aristophanes the Voice of the Mathematician appeared, as it descended from a snow-capped mountain peak, pronouncing in a ponderous sing-song—and words which to the audience sounded like complete gibberish—his eternal Theorems, Lemmas, and Corollaries. The laughter of the listeners was enhanced by the implication that in fifty years' time another Candidate of Eternity would pronounce from the same snow-capped mountain peak exactly the same theorems, although in a modified but scarcely less ponderous and incomprehensible language.

Since the days of antiquity it has been the privilege of the mathematician to engrave his conclusions, expressed in a rarefied and esoteric language, upon the rocks of eternity. While this method is excellent for the codification of mathematical results, it is not so acceptable to the many addicts of mathematics, for whom the science of mathematics is not a logical game, but the language in which the physical universe speaks to us, and whose mastery is inevitable for the comprehension of natural phenomena.

In his previous books the author endeavoured to establish a more discursive manner of presentation in which the esoteric shorthand formulation of mathematical deductions and results was replaced by a more philosophic exposition, putting the emphasis on ideas and concepts and their mutual interrelations, rather than on the mere manipulation of formulae. Our symbolic mechanism is eminently useful and powerful, but the danger is ever-present that we become drowned in a language which has its well-defined grammatical rules but eventually loses all content and becomes a nebulous sham. Hence the author's constant desire to penetrate below the manipulative surface and comprehend the hidden springs of mathematical equations.

To the author's surprise this method (which, of course, is not his monopoly) was well received and made many friends and few enemies. It is thus his hope that the present book, which is devoted to the fundamental aspects of the theory of Linear Differential Operators, will likewise find its adherents. The book is written at advanced level but does not require any specific knowledge which goes beyond the boundaries of the customary introductory courses, since the necessary tools of the subject are developed as the narration proceeds.

Indeed, the first three chapters are of an introductory nature, exploring some of the technical tools which will be required in the later treatment. Since the algebraic viewpoint will be the principal beacon throughout our journey, the problem of obtaining a function from a discrete set of values, that is the problem of *Interpolation*, is the first station on our itinerary. We investigate the properties of the Gregory-Newton and Stirling type of interpolations and their limitations, encountering more than one strange phenomenon which is usually left unheeded. The second station is *Harmonic Analysis*. Here we have a chance to study at close range the remarkable manner in which a series of orthogonal functions, terminated after a finite number of terms, approximates a function. We can hardly find a better introduction to those "orthogonal expansions", which will play such a vital role in our later studies, than by studying the nature of the Fourier series. The third station is *Matrix Calculus*. Here we encounter for the first time that fundamental "decomposition theorem" which in proper re-interpretation will become the theme-song of our later explorations.

Through the concept of the *Function Space* we establish the link between matrices and the continuous domain, and we proceed to the central problem of the *Green's Function* which—being the inverse operator—plays such a central role in the solution of differential equations. Certain elementary aspects of *Communication Problems* provide a number of interesting applications of the Green's function method in engineering problems, whence we proceed to the *Sturm-Liouville Problems* which played such a remarkable role in the historical development of mathematical physics. In the chapter on *Boundary Value Problems* we get acquainted with the classical examples of the solution method known as the "separation of variables", but we add some highly non-traditional types of boundary value problems which bring the peculiar "parasitic spectrum" in appearance. The book comes to a close with a brief chapter on the *Numerical Solution of Trajectory Problems*.

One may well ask why it was necessary to add another treatise on differential equations to the many excellent textbooks which are already on the market. It is the author's contention, however, that neither the treatment nor the selection of the material duplicates any of the standard treatises. Points are stressed which often find scanty attention, while large fields are neglected which take a prominent part in other treatments. The emphasis is constantly on the one question: *what are the basic and characteristic properties of linear differential operators?* Manipulative skill is relegated to a more secondary place—although it is the author's conviction that the student who works seriously on the 350 "Problems" posed (and solved) in the course of discussions, will in fact develop a "feel" for the peculiarities of differential equations which will enable him to try his hands on more specialised and technically more involved problems, encountered in physical and industrial research. (The designation "Problem" instead of "Exercise" may be resented as too pretentious. However, the author does not feel himself in the role of a teacher, who hands out "home-work" to the student, in order to prepare him for examinations. These "Problems" arise naturally from the proceedings in the form of questions or puzzles which deserve an answer. They often complement the text on insufficiently treated details and induce the student to ask questions of his own, "flying off at a tangent", if necessary. At the same time they force him to develop those manipulative skills, without which the successful study of mathematics is not conceivable.)

It is the author's hope that his book will stimulate discussions and research at the graduate level. Although the scope of the book is restricted to certain fundamental aspects of the theory of linear differential operators, the thorough and comprehensive study of these aspects seemed to him well worth pursuing. By a peculiar quirk of historical development the brilliant researches of Fredholm and Hilbert in the field of integral equations overshadowed the importance of differential operators, and the tendency is widespread to transform a given differential equation immediately into an integral equation, and particularly an integral equation of the Fredholm type which in algebraic language is automatically equivalent to the $n \times n$ type of matrices. This is a tendency which completely overlooks the true nature of partial differential operators. The present book departs sharply from the preconceived notion of "well-posed" problems and puts the general—that is arbitrarily over-determined or under-determined—case in the focus of interest. The properties of differential operators are thus examined on an unbiased basis and a theory is developed which submerges the "well-posed" type of problems in a much more comprehensive framework.

The author apologises to the purist and the modernist that his language is that of classical mathematics to which he is bound by tradition and conviction. In his opinion the classical methods can go a long way in the investigation of the fundamental problems which arise in the field of differential operators. This is not meant, however, as a slight on those who with more powerful tools may reach much more sweeping results. Yet, there was still another viewpoint which militated against an overly "modernistic" treatment. This book is written primarily for the natural scientist and engineer to whom a problem in ordinary or partial differential equations is not a problem of logical acrobatism, but a problem in the exploration of the physical universe. To get an explicit solution of a given boundary value problem is in this age of large electronic computers no longer a basic question. The problem can be coded for the machine and the numerical answer obtained. But of what value is the numerical answer if the scientist does not understand the peculiar analytical properties and idiosyncrasies of the given operator? The author hopes that this book will help him in this task by telling him something about the manifold aspects of a fascinating field which is still far from being properly explored.

Acknowledgements. In the Winter Semester 1957–58 the author had the privilege to give a course on "Selected Topics of Applied Analysis" in the Graduate Seminar of Professor A. Lonseth, Oregon State College, Corvallis, Oregon. The lecture notes of that course form the basic core from which the present book took its start.

By the generous invitation of Professor R. E. Langer the excellent research facilities and stimulating associations of the Mathematics Research Center of the U.S. Army in Madison, Wis., were opened to the author, in the winter of 1959–60.   He is likewise indebted to Professor John W. Carr III, Director of the Computation Center, University of North Carolina, Chapel Hill, N.C., for the memorable time spent with him and his graduate group.

These opportunities "far from the home base" were happily complemented by the stimulating daily "tea-time" discussions with the junior and senior staff of the School of Theoretical Physics, Dublin Institute for Advanced Studies, in particular with Professor John L. Synge, Director of the School, whose animated inquiries brought to the surface and elucidated many hidden corners of the subject.

Finally the author wishes to express his heartfelt thanks to his publishers, the Van Nostrand Company, for their unfailing courtesy and understanding.

Dublin, November 1960   C. L.

CONTENTS

xi

BIBLIOGRAPHY

The following textbooks, written in the English language, and selected from a very extensive literature, contain material which in parts parallel the discussions of the present volume, and which can be recommended for collateral or more advanced reading. Additional sources are listed at the end of each chapter. References in braces { } refer to the books of the general Bibliography, those in brackets [ ] to the books of the chapter bibliographies.

{1} Courant, R. and D. Hilbert, *Methods of Mathematical Physics*, Vol. 1 (Interscience Publishers, New York, 1953)

{2} Duff, G. F. D., *Partial Differential Equations* (University of Toronto Press, 1956)

{3} Friedman, B., *Principles and Techniques of Applied Mathematics* (John Wiley & Sons, 1957)

{4} Ince, E. L., *Ordinary Differential Equations* (Dover, New York, 1944)

{5} Jeffreys, H. and B. S. Jeffreys, *Mathematical Methods of Physics* (Cambridge University Press, 1956)

{6} Margenau, H. and G. M. Murphy, *The Mathematics of Physics and Chemistry*, 2nd Ed. (Van Nostrand, 1956)

{7} Morse, P. M. and H. Feshbach, *Methods of Theoretical Physics* (McGraw-Hill, 1953)

{8} Page, C. H., *Physical Mathematics* (Van Nostrand, 1955)

{9} Sneddon, I. N., *Elements of Partial Differential Equations* (McGraw-Hill, 1957)

{10} Sommerfeld, A., *Partial Differential Equations of Physics* (Academic Press, New York, 1949)

{11} Webster, A. G., *Partial Differential Equations of Mathematical Physics* (Hafner, New York, 1941)

{12} Whittaker, E. T., and G. N. Watson, *A Course of Modern Analysis* (Cambridge University Press, 1940)

CHAPTER 1

INTERPOLATION

Synopsis. We investigate the two types of equidistant interpolation procedures, corresponding to the Gregory-Newton and the Stirling formulae. We get acquainted with the Laguerre polynomials and their intimate relation to equidistant interpolation. We learn that only a very restricted class of functions, characterised by a certain integral transform, allows the Gregory-Newton type of interpolation, while another integral transform is characteristic for the Stirling type of interpolation.

1.1. Introduction

The art of interpolation goes back to the early Hindu algebraists. The idea of "linear interpolation" was in fact known by the early Egyptians and Babylonians and belongs to the earliest arithmetic experiences of mankind. But the science of interpolation in its more intricate forms starts with the time of Newton and Wallis. The art of table-making brought into the foreground the idea of obtaining some *intermediate* values of the tabulated function in terms of the calculated tabular values, and the aim was to achieve an accuracy which could match the accuracy of the basic values. Since these values were often obtained with a large number of significant figures, the art of interpolation had to be explored with great circumspection. And thus we see the contemporaries of Newton, particularly Gregory, Stirling, and Newton himself, developing the fundamental tools of the calculus of interpolation.

The unsettled question remained, to what extent can we trust the *convergence* of the various interpolation formulas. This question could not be settled without the evolution of that exact "limit concept" which came about in the beginning of the 19th century, through the efforts of Cauchy and Gauss. But the true nature of equidistant interpolation was discovered even later, around 1900, through the investigations of Runge and Borel.

Our aim in the present chapter will be to discuss some of the fundamental aspects of the theory of interpolation, in particular those features of the theory which can be put to good use in the later study of differential equations. As a general introduction to the processes of higher analysis one could hardly find a more suitable subject than the theory of interpolation.

1

1.2. The Taylor expansion

One of the most fundamental tools of higher mathematics is the well-known "Taylor expansion" which is known on the one hand as an infinite series and on the other as a finite series with a remainder term. We assume that the function $f(x)$ is "analytical" in the neighbourhood of a certain point $x = a$. This means that $f(x + iy)$, considered as a function of the complex variable $x + iy$, possesses a unique derivative $f'(x + iy)$ at that particular point $x = a$, $y = 0$. In that case the derivatives of *all* orders exist and we can consider the infinity of values

$$f(a), f'(a), f''(a), \ldots, f^{(k)}(a), \ldots \tag{1.2.1}$$

from which we can construct the infinite series

$$F(z) = f(a) + f'(a)(z - a) + \frac{f''(a)}{2!}(z - a)^2 + \ldots + \frac{f^{(k)}(a)}{k!}(z - a)^k + \ldots \tag{1.2.2}$$

Although by formal differentiation on both sides we can prove that $F(z)$ coincides with $f(z)$ in all its derivatives at the point $z = a$, this does not prove that the infinite series (2)* is meaningful and that it represents $f(z)$. But it is shown in the theory of analytical functions that in fact the infinite series (2) does converge in a certain domain of the complex variable $z = x + iy$ and actually converges to $f(z)$ at every point of the domain of convergence. We can say even more and designate the domain of convergence quite accurately. It is determined by the inside of a circle whose centre is at the point $z = a$ of the complex plane and whose radius extends to the nearest "singular" point of the function, that is a point in which the analytical character of $f(z)$ ceases to exist; ($f(z)$ might become infinite for example). If it so happens that $f(z)$ is an "entire function"—which remains analytical for all finite values of z—then the radius of convergence becomes *infinite*, i.e. the Taylor series converges for *all* values of z. Generally, however, the radius of convergence is restricted to a definite value beyond which the expansion (2) diverges and loses its meaning.

Exactly *on* the circle of convergence the series may or may not converge, depending on the individuality of the function.

The remarkable feature of the expansion (2) is that the given data are taken from the infinitesimal neighbourhood of the point $x = a$, the "centre of expansion". If $f(x)$ is given between $x = a - \epsilon$ and $x = a + \epsilon$—no matter how small ϵ is chosen—we can form the successive difference coefficients of first, second, third, ... order with a Δx which is sufficiently small and which converges to zero. Hence our data do not involve more

---

* Equations encountered in the current section are quoted by the last digit only; hence (2), encountered in Section 2, refers to equation (1.2.2). Equations quoted by two digits refer to another section of the same chapter; e.g. (4.1) refers to equation (1.4.1) of the present chapter, while the same equation, if encountered in a later chapter, would be quoted as (1.4.1). "Chapter 5.18" refers to section 18 of Chapter 5. The author's book *Applied Analysis* (Prentice-Hall, 1957) is quoted by A. A.

than an *infinitesimal* element of the function and yet we can *predict* what the value of $f(x)$ will be outside of the point $x = a$, within a circle of the complex plane whose centre is at $x = a$. The Taylor series is thus not an interpolating but an *extrapolating* series.

Problem 1. Given the following function:

$$f(x) = (x^3 - 3\pi x^2 + 2\pi^2 x)e^{ix}(\sin x)^{-1} \qquad (1.2.3)$$

Find the convergence radius of the Taylor series if the centre of expansion is at $x = \pi$.

[Answer: $r = 2\pi$]

Problem 2. The mere existence of all the derivatives of $f^{(n)}(a)$ on the *real* axis is not sufficient for the existence of the Taylor series. Show that the function

$$f(x) = e^{-1/x^2} \qquad (1.2.4)$$

possesses derivatives of all order at $x = 0$ (if $f(x)$ is considered as function of the real variable x), and that all these derivatives vanish. The corresponding Taylor series vanishes identically and does not converge to $f(x)$, except at the single point $x = 0$. Show that $f(x + iy)$ is *not* analytical at the point $x = 0$, $y = 0$.

Problem 3. Find the radius of convergence of the Taylor expansion of (4), if the centre of expansion is at the point $x = 4$.

[Answer: $r = 4$]

1.3. The finite Taylor series with the remainder term

In the early days of calculus an "infinite series" was taken literally, viz. the actual sum of an infinity of terms. The exact limit theory developed by Gauss and Cauchy attached a more precise meaning to the expression "infinity of terms". It is obviously impossible to add up an infinity of terms and what we actually mean is that we add more and more terms and thus hope to approach $f(x)$ more and more. Under no circumstances can an infinite series be conceived to be more than a *never ending approximation process*. No matter how many terms of the series we have added, we still have not obtained $f(x)$ *exactly*. However, the "convergence" of an infinite series permits us to make the remaining difference between the sum $f_n(x)$ and $f(x)$ itself *as small as we wish*, although we cannot make it *zero*. This is the meaning of the statement that the infinite series gives us *in the limit* $f(x)$:

$$f(x) = \lim_{n \to \infty} f_n(x) \qquad (1.3.1)$$

The unfortunate feature of this symbolism is that the equality sign is used for an infinite process in which in fact equality never occurs.

Instead of operating with the infinite Taylor series we may prefer the use of the *finite* series

$$f_n(x) = f(a) + f'(a)(x - a) + f''(a)\frac{(x - a)^2}{2!} + \ldots + f^{(n-1)}(a)\frac{(x - a)^{n-1}}{(n - 1)!} \qquad (1.3.2)$$

together with an estimation of the "remainder" of the series, defined by

$$\eta_n(x) = f(x) - f_n(x) \tag{1.3.3}$$

We shall see later (cf. Chapter 5.18) that on the basis of the general theory of differential operators we can derive a very definite expression for $\eta_n(x)$ in the form of the following *definite integral*:

$$\eta_n(x) = \frac{1}{(n-1)!} \int_a^x f^{(n)}(\xi)(x - \xi)^{n-1} d\xi \tag{1.3.4}$$

which we may put in the frequently more convenient form

$$\eta_n(a + t) = \frac{1}{(n-1)!} \int_0^t f^{(n)}(a + \xi)(t - \xi)^{n-1} d\xi \tag{1.3.5}$$

Now the remainder of an approximation process need not be known with full accuracy. What we want is merely an *estimation* of the error. This is possible in the case of (5), on the basis of the mean value theorem of integral calculus:

$$\int_a^b \rho(\xi)f(\xi)d\xi = f(\bar{x}) \int_a^b \rho(\xi)d\xi \tag{1.3.6}$$

which holds if $\rho(x)$ does not change its sign in the interval $[a, b]$ and $f(x)$ is continuous; $\bar{x}$ is some unknown point of the interval $[a, b]$. These conditions are satisfied in the case of (5) if we identify $f(\xi)$ with $f^{(n)}(a + \xi)$ and $\rho(\xi)$ with $(t - \xi)^{n-1}$. Hence we obtain the estimation

$$\eta_n(x) = f^{(n)}(\bar{x}) \frac{(x - a)^n}{n!} \tag{1.3.7}$$

where $\bar{x}$ is some unknown point of the interval $[a, b]$.

The finite expansion (2) with the remainder (7) gives more information than the infinite series (2.2). The analytical nature of $f(x + iy)$ is no longer assumed. It is not even demanded that the derivatives of all order exist at the point $x = a$ since derivatives of higher than n^{th} order do not appear in either $f_n(x)$ or $\eta_n(x)$. Nor is the convergence of the series (2) with increasing n demanded. It may happen that $\eta_n(x)$ decreases up to a certain point and then increases again. In fact, it is even possible that $\eta_n(x)$ increases to infinity with increasing n. We may yet obtain a very close value of $f(x)$—that is a very small error $\eta_n(x)$—if we *stop* at the proper value of n.

The difference between a convergent and a divergent series is not that the first one yields the right value of $f(x)$ while the second has to be discarded as mathematically valueless. What is true is that they are both *approximations* of $f(x)$ with an error which in the first case can be made as small as we wish, while in the second case the error cannot be reduced below a certain finite minimum.

Problem 4. Prove the truth of the following statement: "If $|f^{(n)}(x)|$ has a maximum at the centre of expansion, the error of the truncated Taylor series is smaller than the first neglected term."

Problem 5. Consider the Taylor expansion around $x = 0$ of the function

$$f(x) = (1 + x)^{3.5} \qquad (1.3.8)$$

for $x > 0$. Show that for any $n \geq 4$ the remainder of the series is smaller than the first neglected term.

Problem 6. The infinite Taylor series of the function (8) converges only up to $x = 1$. Let us assume that we want to obtain $f(2)$. How many terms of the series shall we employ for maximum accuracy, and what error bound do we obtain for it? Demonstrate by the discussion of the error term (4) that the error can be greatly diminished by adding the first neglected term *with the weight* $\frac{1}{2}$.

[Answer: $n = 9$, $\quad |\eta_9| < \dfrac{35}{128} = 0.273$

$\qquad f_9(2) = 46.8984$, corrected by $\frac{1}{2}$ of next term: $f^*_9(2) = 46.7617$

$\qquad\qquad\qquad$ correct value: $\quad f(2) = 46.7654$]

1.4. Interpolation by polynomials

The finite Taylor series of n terms can be interpreted as a polynomial approximation of $f(x)$ which has the property that the functional value and the derivatives up to the order $n - 1$ coincide at the centre of expansion $x = a$. We can equally say that we have constructed a polynomial of $n - 1^{\text{st}}$ order which has the property that it coincides with $f(x)$ at n points which are infinitely near to the point $x = a$.

An obvious generalisation of this problem can be formulated as follows: Construct a polynomial of the order $n - 1$ which shall coincide with $f(x)$ at the n *arbitrarily given* points

$$x = x_1, x_2, \ldots, x_n \qquad (1.4.1)$$

This problem was solved with great ingenuity by Lagrange, who proceeded as follows.

We construct the "fundamental polynomial" $F_n(x)$ by multiplying all the root factors:

$$F_n(x) = (x - x_1)(x - x_2) \ldots (x - x_n)$$

$$= x^n + c_1 x^{n-1} + c_2 x^{n-2} + \ldots + c_n \qquad (1.4.2)$$

Dividing synthetically by the root factors $x - x_k$ we now construct the n auxiliary polynomials

$$\frac{F_n(x)}{x - x_k} = \phi_k(x) \qquad (1.4.3)$$

and

$$p_k(x) = \frac{\phi_k(x)}{\phi_k(x_k)} = \frac{F_n(x)}{F'_n(x_k)(x - x_k)} \qquad (1.4.4)$$

These auxiliary polynomials $p_k(x)$ have the property that they give zero at all the root points x_i, except at x_k where they give the value 1:

$$p_k(x_i) = \delta_{ki} \tag{1.4.5}$$

where δ_{ki} is "Kronecker's symbol"

$$\begin{aligned} \delta_{ik} &= 1 \qquad (i = k) \\ &= 0 \qquad (i \neq k) \end{aligned} \tag{1.4.6}$$

If now we form the sum

$$P_{n-1}(x) = \sum_{k=1}^{n} f(x_k) p_k(x) \tag{1.4.7}$$

we obtain a polynomial which has the following properties: its order is $n - 1$ and it assumes the values $f(x_k)$ at the prescribed points (1). Hence it solves the Lagrangian interpolation problem from which we started. The formula (7) may also be written in the form

$$P_{n-1}(x) = F_n(x) \sum_{k=1}^{n} \frac{f(x_k)}{F'_n(x_k)} \frac{1}{x - x_k} \tag{1.4.8}$$

It is called "Lagrange's interpolation formula".

Problem 7. Construct the Lagrangian polynomials $p_k(x)$ for the following distribution of points:

$$x_k = -3, -2, -1, 0, 1, 2, 3 \tag{1.4.9}$$

Obtain the interpolating polynomial for the following function:

$$f(x) = |x| \tag{1.4.10}$$

[Answer:

$$P_6(x) = \frac{1}{60} (x^6 - 15x^4 + 74x^2) \tag{1.4.11}]$$

Problem 8. Show that if the x_k are evenly distributed around the origin (i.e. every x_k appears with $+$ and $-$ signs), the interpolating polynomial contains only even powers if $f(x)$ is an even function: $f(x) = f(-x)$ and only odd powers if $f(x)$ is an odd function: $f(x) = -f(-x)$. Show that this is not the case if the x_k are not evenly distributed.

1.5. The remainder of the Lagrangian interpolation formula

If a function $y = f(x)$ is given and we have constructed a polynomial of the order $n - 1$ which fits the functional values $y_k = f(x_k)$ at the n prescribed points (4.1), this does not mean that the interpolation will necessarily be very close at the points *between* the points of interpolation. It will be our task to find an estimation for the "remainder", or "error" of our interpolating polynomial $P_{n-1}(x)$, that is the difference

$$\eta_n(x) = f(x) - P_{n-1}(x) \tag{1.5.1}$$

For this purpose we want to assume that $f(x)$ is n times differentiable, although it is obvious that we may approximate a function by a polynomial which does not satisfy this condition (cf. for example Problem 7).

If we differentiate the equation (1) n times, the second term on the right side will drop out since the nth derivative of any polynomial of not higher than $n - 1^{\text{st}}$ order vanishes. Accordingly we obtain

$$\eta_n^{(n)}(x) = f^{(n)}(x) \tag{1.5.2}$$

We can consider this differential equation as the *defining equation* for $\eta_n(x)$, although a differential equation of nth order cannot have a unique solution without adding n "boundary conditions". These conditions are provided by the added information that $\eta_n(x)$ vanishes at the n points of interpolation $x = x_k$:

$$\eta_n(x_k) = 0, \quad (k = 1, 2, \ldots, n) \tag{1.5.3}$$

Although these are *inside conditions* rather than boundary conditions, they make our problem uniquely determined.

At this point we anticipate something that will be fully proved in Chapter 5. The solution of our problem (3) (with the given auxiliary conditions) can be obtained with the help of an auxiliary function called the "Green's function: $G(x, \xi)$" which is constructed according to definite rules. It is quite independent of the given "right side" of the differential equation (2). The solution appears in the form of a definite integral:

$$\eta_n(x) = \int_{x_1}^{x_n} f^{(n)}(\xi) G(x, \xi) d\xi \tag{1.5.4}$$

We have assumed that the points of interpolation x_k are arranged in increasing magnitude:

$$x_1 < x_2 < \ldots < x_n \tag{1.5.5}$$

and that x is some point inside the interval $[x_1, x_n]$:

$$x_1 < x < x_n \tag{1.5.6}$$

As we shall demonstrate later, the function $G(x, \xi)$, considered as a function of ξ, has the property that it *does not change its sign* throughout the interval $[x_1, x_n]$. But then we can again make use of the mean value theorem (3.6) of integral calculus and obtain

$$\eta_n(x) = f^{(n)}(\bar{x}) \int_{x_1}^{x_n} G(x, \xi) d\xi \tag{1.5.7}$$

where $\bar{x}$ is some unknown point of the interval $[x_1, x_n]$. The second factor does not depend on ξ any more, but is a pure function of x, which is independent of $f(x)$. Hence we can evaluate it by choosing *any* $f(x)$ we like.

We will choose for $f(x)$ the special function

$$f(x) = \frac{F_n(x)}{n!} \tag{1.5.8}$$

where $F_n(x)$ is the fundamental polynomial (4.2). This function has the property that it vanishes at all the points of interpolation and thus the interpolating polynomial $P_{n-1}(x)$ *vanishes identically*. Hence $\eta_n(x)$ becomes $f(x)$ itself. Moreover, this choice has the advantage that it eliminates the unknown position of $\bar{x}$ since here the nth derivative of $f(x)$ is simply 1 throughout the range. Hence we obtain from (7):

$$\frac{F_n(x)}{n!} = \int_{x_1}^{x_n} G(x, \xi)d\xi \qquad (1.5.9)$$

The second factor of the right side of (7) is now determined and we obtain the estimation

$$\eta_n(x) = \frac{f^{(n)}(\bar{x})}{n!} F_n(x) \qquad (1.5.10)$$

This is *Lagrange's form of the remainder of a polynomial interpolation*.

The result can be extended to the case of a point x which lies *outside* the realm $[x_1, x_n]$, in which case we cannot speak of *inter*polation any more but of *extra*polation. The only difference is that the point $\bar{x}$ becomes now some unknown point of the interval $[x_1, x]$ if $x > x_n$ and of the interval $[x, x_n]$ if $x < x_1$.

The disadvantage of the formula (10) from the numerical standpoint is that it demands the knowledge of a derivative of high order which is frequently difficult to evaluate.

Problem 9. Deduce the remainder (3.7) of the truncated Taylor series from the general Lagrangian remainder formula (10).

1.6. Equidistant interpolation

Particular interest is attached to the case of equidistantly placed points x_k:

$$x_{k+1} - x_k = \Delta x = \text{const.} \qquad (1.6.1)$$

If a function $f(x)$ is tabulated, we shall almost always give the values of $f(x)$ in equidistant arguments (1). Furthermore, if a function is observed by physical measurements, our measuring instruments (for example clock mechanisms) will almost exclusively provide us with functional values which belong to equidistant intervals. Hence the interpolation between equidistant arguments was from the beginning of interpolation theory treated as the most important special case of Lagrangian interpolation. Here we need not operate with the general formulae of Lagrangian interpolation (although in Chapter 2.21 we shall discover a particularly interesting property of equidistant interpolation exactly on the basis of the general formula of Lagrangian interpolation) but can develop a specific solution of our problem by a certain operational approach which uses the Taylor series as its model and translates the operational properties of this series into the realm of difference calculus.

In the calculus of finite differences it is customary to normalise the given constant interval Δx of the independent variable to 1. If originally the tabulation occurred in intervals of $\Delta x = h$, we change the original x to a new independent variable x/h and thus make the new Δx equal to 1. We will assume that this normalisation has already been accomplished.

The fundamental operation of the calculus of finite differences is the difference quotient $\Delta/\Delta x$ which takes the place of the derivative d/dx of infinitesimal calculus. But if Δx is normalised to 1, it suffices to consider the operation

$$\Delta f(x) = f(x + 1) - f(x) \qquad (1.6.2)$$

without any denominator which simplifies our formulae greatly. The operation can obviously be repeated, for example

$$\Delta^2 f(x) = f(x + 2) - 2f(x + 1) + f(x) \qquad (1.6.3)$$

and so on.

Now let us start with the truncated Taylor series, choosing the centre of expansion as the point $x = 0$:

$$f(x) = f(0) + f'(0)x + f''(0)\frac{x^2}{2!} + \ldots + f^{(n-1)}(0)\frac{x^{n-1}}{(n-1)!} + \eta_n(x) \qquad (1.6.4)$$

By differentiating on both sides and putting $x = 0$ we can prove that $\eta_n(x)$ has the property that it vanishes at the point $x = 0$, together with all of its derivatives, up to the order $n - 1$. The proof is based on the fact that the functions

$$\Phi_k(x) = \frac{x^k}{k!} \qquad (1.6.5)$$

satisfy the following functional equation

$$\frac{d}{dx}\Phi_k(x) = \Phi_{k-1}(x) \qquad (1.6.6)$$

together with the boundary condition

$$\Phi_k(0) = 0 \qquad (k \neq 0) \qquad (1.6.7)$$

while

$$\Phi_0(x) = 1 \qquad (1.6.8)$$

If now we can find a corresponding set of polynomials which satisfy the fundamental equation

$$\Delta\phi_k(x) = \phi_{k-1}(x) \qquad (1.6.9)$$

together with the same boundary conditions (7) and (8), then we can translate the Taylor series into the calculus of finite differences by putting

$$f(x) = f(0) + \Delta f(0)\phi_1(x) + \Delta^2 f(0)\phi_2(x) + \ldots + \Delta^{n-1}f(0)\phi_{n-1}(x) + \eta_n(x) \qquad (1.6.10)$$

and proving that $\eta_n(x)$ vanishes at $x = 0$, together with its first, second, $\ldots, n - 1^{\text{st}}$ differences. But this means that the polynomial

$$P_{n-1}(x) = f(0) + \Delta f(0)\phi_1(x) + \ldots + \Delta^{n-1}f(0)\phi_{n-1}(x) \qquad (1.6.11)$$

coincides with $f(x)$ in all its differences up to the order $n - 1$ and since these differences are formed in terms of the functional values

$$f(0), f(1), f(2), \ldots, f(n - 1) \qquad (1.6.12)$$

we see that $P_{n-1}(x)$ coincides with $f(x)$ at the points

$$x = 0, 1, 2, \ldots, n - 1 \qquad (1.6.13)$$

and thus solves the problem of equidistant interpolation. Our problem is thus reduced to the solution of the functional equation (9), in conjunction with the boundary conditions (7) and (8).

Now the application of the Δ operation to the Newtonian "binomial coefficients"

$$\phi_k(x) = \binom{x}{k} = \frac{x(x - 1)(x - 2) \ldots (x - k + 1)}{1 \cdot 2 \cdot 3 \ldots k} \qquad (1.6.14)$$

shows that these functions indeed satisfy the functional equation (9):

$$\Delta\binom{x}{k} = \binom{x}{k - 1} \qquad (1.6.15)$$

with the proper boundary (actually initial) conditions. Hence we have in (14) the proper auxiliary functions which take the place of the functions (5) of the Taylor series, and we can write down the "Gregory-Newton interpolation formula"

$$P_{n-1}(x) = f(0) + \Delta f(0)x + \Delta^2 f(0)\binom{x}{2} + \ldots + \Delta^{n-1}f(0)\binom{x}{n - 1} \qquad (1.6.16)$$

The successive differences $\Delta f(0)$, $\Delta^2 f(0)$, $\ldots$ are obtainable by setting up a "difference table" and reading off the values which belong to the line $x = 0$.* But they are equally obtainable by the following "binomial weighting" of the original functional values:

$$\Delta^k f(0) = f(k) - kf(k - 1) + \binom{k}{2}f(k - 2) - \ldots + (-1)^k f(0)$$

$$= \sum_{m=0}^{k} (-1)^{k+m}\binom{k}{m}f(m) \qquad (1.6.17)$$

or in symbolic notation

$$\Delta^k f(0) = [f - 1]^k \qquad (1.6.18)$$

with the understanding that in the expansion on the right side we replace f^m by $f(m)$.

* Cf. A. A., p. 308.

Problem 10. Show that the function

$$f(x) = (1 + a)^x \qquad (1.6.19)$$

satisfies the functional equation

$$\Delta^k f(x) = a^k f(x) \qquad (1.6.20)$$

Problem 11. Show that Newton's formula of the "binomial expansion"

$$(1 + a)^m = 1 + \binom{m}{1}a + \binom{m}{2}a^2 + \ldots + \binom{m}{m}a^m \qquad (1.6.21)$$

can be conceived as a special application of the general interpolation formula (16) (with $n = m + 1$).

1.7. Local and global interpolation

Let a certain continuous and differentiable function $y = f(x)$ be tabulated in equidistant intervals, normalised to $\Delta x = 1$. Let our table start with the value $x = 0$ and continue with $x = 1, 2, \ldots, n - 1$. We now want to obtain $f(x)$ at a point ξ which is between zero and 1. The simplest form of interpolation is "linear interpolation":

$$f(\xi) = f(0) + \Delta f(0)\xi \qquad (1.7.1)$$

Here we have connected the functional values $f(0)$ and $f(1)$ by a *straight line*. We may want greater accuracy, obtainable by laying a *parabola* through the points $f(0), f(1), f(2)$. We now get the "quadratic interpolation"

$$f(\xi) = f(0) + \xi \Delta f(0) + \frac{\xi(\xi - 1)}{2} \Delta^2 f(0) \qquad (1.7.2)$$

The procedure can obviously be continued to polynomials of higher and higher order by taking into account more and more terms of the interpolation formula (6.16). The analogy with the Taylor series would induce us to believe that we constantly gain in accuracy as we go to polynomials of ever-increasing order. There is, however, an essential difference between the two series. In the case of the Taylor series we are at a *constant distance* from the centre of expansion, while here we stay at a certain point ξ but the *range* in which our polynomials interpolate, increase all the time. For a linear interpolation only the two neighbouring points play a role. But if we operate with a polynomial of the order $n - 1$, we use n successive points of interpolation and lay a polynomial of high order through points which are in fact *quite far* from the point at which we want to obtain the functional value. We can generally not expect that the error oscillations will necessarily decrease by this process. That we take in more data, is an advantage. But our approximating polynomial spreads over an ever-increasing range and that may counteract the beneficial effect of more data. Indeed, in the case of the Taylor series the functions (6.5) have a strongly decreasing tendency since every new function provides the added factor x in the numerator and the factor k in the denominator. On the other hand, the functions (6.14) yield likewise the factor k in the denominator but in the numerator we now obtain $x + 1 - k$ and thus we see that for large k we have gained a factor which is not much smaller than 1. Convergence can only be expected on account of the successive differences $\Delta^k f(0)$. These differences may go down for a while but then a minimum may be reached and afterwards the terms will perhaps go up again. In this case we have to *stop* with the proper order $n - 1$ of the interpolating polynomial since the addition of further terms will increase rather than decrease the error of interpolation.

We see that under such circumstances we cannot trust the automatic functioning of the interpolation formula (6.16). We have to use our judgement in deciding how far we should go in the series to obtain the closest approximation. We stop at a term which just precedes the minimum term. We may also add the minimum term with half weight, thus giving much higher accuracy but losing the chance of estimating the committed error. The first neglected term of the series does not yield necessarily a safe error bound of our interpolation, except if the first two neglected terms are of opposite signs and $f^{(n+1)}(x)$ does not change its sign in the range of interpolation (cf. 5.10). (Safe error bounds for monotonously increasing or decreasing series are not easily available.)

A good illustration is provided by the example studied by Runge. Let the function

$$y = f(x) = \frac{100}{4 + x^2} \tag{1.7.3}$$

be given at the integer points $x = 0, \pm 1, \pm 2, \ldots, \pm 10$. We want to obtain $f(x)$ at the point $x = 9.5$.

First of all we shift the point of reference to the point $x = 10$ and count the x-values *backward*. Hence we make a table of the given y-values, starting with y_{10}, and continuing with $y_9, y_8, y_7 \ldots$. We then evaluate the successive $\Delta^k f(0)$, by setting up a difference table—or quicker by binomial weighting according to the formula (6.17)—and apply the Gregory-Newton interpolation formula (6.16) for $x = 0.5$. The successive terms and their sum is tabulated below:

$\Delta^k f(0)\binom{x}{k}$	$f_n(x)$
0.961538	0.961538
0.107466	1.069004
−0.009898	1.059106
0.002681	1.061788
−0.001251	1.060537
0.000851	1.061388
−0.000739	1.060649
0.000568	1.061217
0.000995	1.062212

$$\tag{1.7.4}$$

[correct: $f(9.5) = 1.061008$]

We observe that in the beginning the error fluctuates with alternating sign and has the tendency to decrease. After 8 steps a minimum is reached and from then on the terms *increase* again and have no tendency to converge. In fact the differences of high order become enormous. In view of the change of sign of the seventh and eighth terms we can estimate that the correct value will lie between 1.061388 and 1.060649. The arithmetic mean of these two bounds: 1.061018, approaches in fact the correct functional value 1.061008 with a high degree of accuracy. Beyond that, however, we cannot go.

Runge's discovery was that this pattern of the error behaviour cannot be remedied by adding more and more data between, thus reducing Δx to smaller and smaller values. No matter how dense our data are, the interpolation for some x-value between will show the same general character: reduction of the error to a certain finite minimum, which cannot be surpassed since afterwards the errors increase again and in fact become exceedingly large.

In the present problem our aim has been to obtain the functional value $f(x)$ at a given point x. This is the problem of *local* interpolation. We can use our judgement how far we should go with the interpolating series, that is how many of our data we should actually use for a minimisation of our error. We may have, however, quite a different problem. We may want an analytical expression which should fit the function $y = f(x)$ with reasonable accuracy in an *extended* range of x, for example in the entire range $[-10, +10]$ of our data. Here we can no longer stop with the interpolation formula at a judiciously chosen point. For example in our previous procedure, where we wanted to obtain $f(9.5)$, we decided to stop with $n = 6$ or 7. This means that we used a polynomial of 5th or 6th order which fits our data between $x = 4$ (or 5) and 10. But this polynomial would completely fail in the representation of $f(x)$ for values which are between -10 and 0. On the other hand, if we use a polynomial of the order 20 in order to include all our data, we would get for $f(9.5)$ a completely absurd value because now we would have to engage that portion of the Gregory-Newton interpolation formula which does not converge at all. We thus come to the conclusion that *interpolation in the large by means of high order polynomials is not obtainable by Lagrangian interpolation of equidistant data*. If we fit our data exactly by a Lagrangian polynomial of high order we shall generally encounter exceedingly large error oscillations around the end of the range. In order to obtain a truly well fitting polynomial of high order, we have to make systematic *errors* in the data points. We will return to this puzzling behaviour of equidistant polynomial interpolation when we can elucidate it from an entirely different angle (cf. Chapter 2.21).

1.8. Interpolation by central differences

The Gregory-Newton formula (6.16) takes into account the given data in the sequence $f(0), f(1), f(2), \ldots$. If for example we want to obtain $f(0.4)$ and we employ 5 terms of the series, we operate in fact with a polynomial of fourth order which fits the functional values given at $x = 0, 1, 2, 3, 4$. Now the point $x = 4$ is rather far from the point $x = 0.4$ at which the function is desired. We might imagine that it would be preferable to use data which are nearer to the desired point. This would require that our data proceed in *both directions* from $x = 0.4$ and we could use the functional data at $x = -2, -1, 0, 1, 2$. This interpolation procedure is associated with the name of *Stirling*. In Stirling's formula we employ the data *symmetrically* to the left and to the right and thus gain greatly in convergence.

The difference table we now set up is known as a "central difference table".* It is still the previous difference table but in new arrangement. The fundamental operation on which Stirling's formula is based, is called the "second central difference" and is traditionally denoted by δ^2. This notation is operationally misleading since δ^2 is in fact a *basic operation* which cannot be conceived as the square of the operation δ. For this reason we will deviate from the commonly accepted notation and denote the traditional δ^2 by δ:

$$\delta f(x) = f(x + 1) - 2f(x) + f(x - 1) \tag{1.8.1}$$

The even part of the function $f(x)$ can be expanded in the even Stirling series:

$$\tfrac{1}{2}[f(x) + f(-x)] = f(0) + \delta f(0)\Phi_2(x) + \delta^2 f(0)\Phi_4(x) + \cdots$$
$$+ \delta^n f(0)\Phi_{2n}(x) + \eta_{2n}(x) \tag{1.8.2}$$

where the Stirling's functions $\Phi_{2k}(x)$ are defined as follows:

$$\Phi_2(x) = \frac{x^2}{2!}, \quad \Phi_4(x) = \frac{x^2(x^2 - 1)}{4!}$$

$$\Phi_{2k}(x) = \frac{x^2(x^2 - 1) \ldots [x^2 - (k-1)^2]}{(2k)!} = \frac{1}{2}\left[\binom{x + k}{2k} + \binom{x + k - 1}{2k}\right] \tag{1.8.3}$$

The odd part of $f(x)$ can be made even by multiplication by x and thus expanding the function $xf(x)$ according to (2). The final result is expressible in the form of the following expansion:

$$f(x) = f(0) + \gamma f(0)x + \delta f(0)\frac{x^2}{2!} + \gamma \delta f(0)\frac{x(x^2 - 1)}{3!}$$

$$+ \delta^2 f(0)\frac{x^2(x^2 - 1)}{4!} + \gamma \delta^2 f(0)\frac{x(x^2 - 1)(x^2 - 4)}{5!} + \cdots$$

$$+ \delta^n f(0)\frac{x^2(x^2 - 1) \ldots (x^2 - (k-1)^2)}{(2n)!} + \eta_{2n}(x) \tag{1.8.4}$$

The operation γ has the following significance:

$$\gamma f(x) = \tfrac{1}{2}[f(x + 1) - f(x - 1)] \tag{1.8.5}$$

* See A. A., pp. 309, 310.

The formula (4) shows that the odd Stirling functions $\Phi_{2k+1}(x)$ have to be defined as follows:

$$\Phi_{2k+1}(x) = \frac{x(x^2 - 1) \ldots (x^2 - k^2)}{(2k + 1)!} = \binom{x + k}{2k + 1} \tag{1.8.6}$$

A comparison with (3) shows that

$$\Phi_{2k}(x) = \frac{x}{2k} \Phi_{2k-1}(x) \tag{1.8.7}$$

Now let $f(x)$ be *odd*. $f(-x) = -f(x)$. Then the Stirling expansion becomes

$$f(x) = \gamma f(0)\Phi_1(x) + \gamma \delta f(0)\Phi_3(x) + \ldots + \gamma \delta^n f(0)\Phi_{2n+1}(x) \tag{1.8.8}$$

but at the same time $g(x) = xf(x)$ is *even* and permits the expansion

$$g(x) = \delta g(0)\Phi_2(x) + \delta^2 g(0)\Phi_4(x) + \ldots + \delta^{n+1} g(0)\Phi_{2n+2}(x). \tag{1.8.9}$$

Dividing on both sides by x we obtain, in view of (7):

$$\gamma \delta^k f(0) = \frac{1}{2(k + 1)} \delta^{k+1} g(0) \tag{1.8.10}$$

This means that the coefficients of the odd terms are obtainable in terms of the δ operation alone if this operation is applied to the function $xf(x)$, and the final result divided by $2k + 2$.

Here again the direct construction of a central difference table can be avoided in favour of a direct weighting of the functional values, in analogy to (6.18). We now obtain

$$\delta^k f(0) = \frac{[f - 1]^{2k}}{f^k} \tag{1.8.11}$$

$$\gamma \delta^k f(0) = \frac{1}{2} \left\{ \frac{[f - 1]^{2k+1}}{f^k} + \frac{[f - 1]^{2k+1}}{f^{k+1}} \right\} \tag{1.8.12}$$

For example:

$$\delta^2 f(0) = f(2) - 4f(1) + 6f(0) - 4f(-1) + f(-2)$$
$$\gamma \delta^2 f(0) = \tfrac{1}{2}[f(3) - 4f(2) + 5f(1) - 5f(-1) + 4f(-2) - f(-3)]$$
$$\delta^3 f(0) = f(3) - 6f(2) + 15f(1) - 20f(0) + 15f(-1) - 6f(-2) + f(-3) \tag{1.8.13}$$

(The operation $\gamma \delta^2 f(0)$ is equally obtainable by applying the operation $\delta^3 f(0)$ to $xf(x)$ and dividing by $2 \cdot 3 = 6$.)

Problem 12. Show that the expansion (4) is equivalent to the application of the Gregory-Newton formula if we shift the centre of expansion to the point $-n$:

$$f(x) = f(-n) + \Delta f(-n)(n + x) + \Delta^2 f(-n)\binom{n + x}{2}$$
$$+ \ldots \Delta^{2n} f(-n)\binom{n + x}{2n} + \eta_{2n}(x) \tag{1.8.14}$$

Problem 13. The exponential function $y = e^{-x}$ is given at the following points: $f(0) = 1$, $f(1) = 0.367879$, $f(2) = 0.135335$, $f(3) = 0.049787$, $f(4) = 0.018316$ Obtain $f(0.5)$ by the Gregory-Newton formula of five terms. Then, adding the data

$$f(-1) = 2.718282 \qquad f(-2) = 7.38906$$

obtain $f(0.5)$ by the Stirling formula of five terms; (omitting $f(3)$ and $f(4)$).

[Answer: Gregory-Newton: $f(\frac{1}{2}) = 0.61197$
Stirling: $f(\frac{1}{2}) = 0.61873$
correct value: $f(\frac{1}{2}) = 0.60653$

The value obtained by the Stirling interpolation is here *less accurate* than the G.-N. value.]

1.9. Interpolation around the midpoint of the range

We will return once more to our problem (7.3), examined before in Section 7. We have assumed that our data were given in the 21 integer points between $x = -10$ and $x = 10$. We have seen that around the end of the range the Gregory-Newton formula had a "semi-convergent" behaviour: the terms decrease up to a certain point and then increase again. We approach the functional value only if we do not go too far in the series: Quite different is the behaviour of the interpolating series if we stay near to the *middle* of the range and operate with central differences. Let us investigate the convergence behaviour of the Stirling series by trying to obtain $f(0.5)$ by interpolation on the basis of central differences.

First of all we notice that the Stirling functions $\Phi_k(x)$—defined by (8.3) and (8.6)—have better convergence properties than the Gregory-Newton functions (6.14). The factor we gain as we go from $2k$ to $2k + 2$ is

$$\frac{x^2 - k^2}{(2k + 1)(2k + 2)} \tag{1.9.1}$$

and that is better than $\frac{1}{4}$. In order to obtain a comparison with the Gregory-Newton formula we should multiply the functions $\Phi_k(x)$ by 2^k and divide the central differences by 2^k. In our problem $f(x)$ is even and thus only the even differences δ^k will appear, in conjunction with the functions $\Phi_{2k}(x)$. Hence we will divide $\delta^k f(0)$ by 4^k and multiply the associated $\Phi_{2k}(x)$ by 4^k. In this fashion we keep better track of the order of magnitude of the successive terms which form the final sum (8.2).

In this instance we do not encounter that divergent behaviour of the interpolating series that we have encountered earlier in Section 7. But now we encounter another strange phenomenon, namely that the convergence is *too rapid*. The terms we obtain are all negative, with the exception of the first term. Hence we approach the final value monotonously from *above*. The successive terms diminish rapidly and there is no reason to stop before the contributions of all our data are taken into account. But the peculiar thing is that the peripheral values contribute *too little* to the sum so that we get the impression that we are much nearer to the correct value than this is actually the case. The successive convergents are given in the following table (the notation $f*(0.5)$ refers to the interpolated values, obtained on the basis of 1, 3, 5, ... 21 data, going from the centre to the periphery, and taking into account the data to the right and to the left in pairs):

$4^k \Phi_{2k}(0.5)$	$4^{-k} \delta^k f(0)$	$f*(0.5)$	
1	25	25	
0.5	-2.5	23.75	
-0.125	0.9375	23.63281	
0.0625	-0.54086	23.59901	
-0.03906	0.37860	23.58422	
0.02734	-0.29375	23.57619	(1.9.2)
-0.02051	0.24234	23.57122	
0.01611	-0.20805	23.56786	
-0.01309	0.18357	23.56546	
0.01091	-0.16521	23.56366	
-0.00927	0.15092	23.56226	

As we watch the successive convergents, we should think that the correct value can be guaranteed to at least two decimal places while in actual fact

$$f(0.5) = 23.52941 \qquad (1.9.3)$$

The great distance of $f*_{21}(0.5)$ from the correct value makes it doubtful whether the addition of the data $f(\pm 11)$, $f(\pm 12)$ ... *out to infinity*, would be able to bridge the gap. A closer analysis corroborates this impression. The series (8.2) remains convergent as n tends to infinity but *the limit does not coincide with $f(x)$ at the point* $x = 0.5$ (see Chapter 2.21).

1.10. The Laguerre polynomials

The experience of the last section brings us to the following problem. Our previous discussions were devoted to a function $y = f(x)$ which was tabulated in a *finite* interval. We have studied the behaviour of a polynomial interpolation in this interval. But let us now assume that our function is in fact tabulated in an *infinite* interval, and first we will assume that this interval is $[0, \infty]$, the function being given at the points

$$x = 0, 1, 2, 3, \ldots \qquad (1.10.1)$$

Accordingly we will operate with the Gregory-Newton formula, form the successive differences

$$f(0) = y_0, \quad \Delta f(0) = y_1 - y_0, \quad \Delta^2 f(0) = y_2 - 2y_1 + y_0, \ldots \qquad (1.10.2)$$

and construct the formal infinite series*

$$f^*(x) = \sum_{k=0}^{\infty} \Delta^k f(0) \binom{x}{k} \tag{1.10.3}$$

What can we say about the behaviour of this series? Will it converge and if so, will it converge to $f(x)$? (We will ask the corresponding problem for the interval $[-\infty, +\infty]$ and the use of central differences somewhat later, in Section 16.)

This problem is closely related to the properties of a remarkable set of polynomials, called the "Laguerre polynomials". We will thus begin our study with the exposition of the basic properties of these polynomials shaped to the aims of the interpolation theory.

We define our function $y = f(x)$ in the interval $[0, \infty]$ in the following specific manner:

$$f(x) = \frac{t^x}{x!} \tag{1.10.4}$$

We form the successive differences (2), either by setting up a difference table, or by directly taking the functional values y_k and weighting them binomially according to the formula (6.12). We thus obtain

$$L_k(t) = \Delta^k f(0) = \sum_{n=0}^{k} (-1)^{k+m} \binom{k}{m} \frac{t^m}{m!} \tag{1.10.5}$$

These are so-called "normalised Laguerre polynomials" which we will denote by $L_k(t)$.† For example:

$$\begin{aligned}
L_0(t) &= 1 \\
L_1(t) &= -1 + t \\
L_2(t) &= 1 - 2t + \frac{t^2}{2} \\
L_3(t) &= -1 + 3t + 3\frac{t^2}{2} - \frac{t^3}{6} \\
L_4(t) &= 1 - 4t + 6\frac{t^2}{2} - 4\frac{t^3}{6} + \frac{t^4}{24}
\end{aligned} \tag{1.10.6}$$

and so on. These polynomials have the remarkable property that they are *orthogonal* to each other in the interval $[0, \infty]$, with respect to the weight factor e^{-t}, while their norm is 1:

$$\int_0^{\infty} L_k(t) L_m(t) e^{-t} dt = \delta_{km} \tag{1.10.7}$$

---

* The notation $f^*(x)$ in the sense of an "approximation of $f(x)$" seems rather ill-chosen, in view of the convention that the asterisk denotes in algebra the "complex conjugate" of a complex number. An ambiguity need not be feared, however, because in all instances when this notation occurs, $f(x)$ is a *real* function of x.

† The customary notation $L_n(t)$ refers to the present $L_n(t)$, multiplied by $n!$

They thus form an "ortho-normal" set of functions. Moreover, in view of the fact that the powers of t come in succession, from t^0 to t^∞, these functions form a *complete* ortho-normal set.

Traditionally this property of the Laguerre polynomials is proved on the basis of the *differential equation* which they satisfy. But we can demonstrate this property directly on the basis of the definition (5). We form the Gregory-Newton series

$$f(x) = \sum_{n=0}^{\infty} \Delta^k f(0) \binom{x}{k} \qquad (1.10.8)$$

We do not know yet whether this series will converge or not—we will give the proof in the next section—but for any *integer* value $x = m$ the series *terminates* after $m + 1$ terms and the question of convergence does not arise. For such values the series (8) is an *algebraic identity*, no matter how the key-values $f(j)$ $(j = 0, 1, 2, \ldots, m)$ may be prescribed.

Let us apply this expansion to the function (1), obtaining

$$f(x) = \frac{t^x}{x!} = \sum_{k=0}^{\infty} \binom{x}{k} L_k(t) \qquad (1.10.9)$$

We now multiply on both sides by $\Delta^m f(0)e^{-t}$, obtaining

$$\int_0^{\infty} \frac{t^x}{x!} L_m(t)e^{-t}dt = \int_0^{\infty} \sum_{k=0}^{\infty} \binom{x}{k} L_k(t) L_m(t)e^{-t}dt \qquad (1.10.10)$$

Since integration and differencing are two independent operations, we can on the left side multiply by $f(\xi)$, perform the integration, *and then* take the mth difference at $\xi = 0$:

$$\int_0^{\infty} \frac{t^x}{x!} \frac{t^\xi}{\xi!} e^{-t}dt = \frac{1}{x!\xi!} \int_0^{\infty} t^{x+\xi}e^{-t}dt = \frac{(x+\xi)!}{x!\xi!} = \binom{x+\xi}{x} \qquad (1.10.11)$$

Now, making use of the fundamental relation (6.15)—our variable is ξ— we obtain

$$\Delta\binom{x+\xi}{x} = \binom{x+\xi}{x-1}$$

$$\Delta^m\binom{x+\xi}{x} = \binom{x+\xi}{x-m} \qquad (1.10.12)$$

and putting $\xi = 0$ we obtain for the left side of (10):

$$\int_0^{\infty} \frac{t^x}{x!} L_m(t)e^{-t}dt = \binom{x}{x-m} = \frac{x!}{m!(x-m)!} = \binom{x}{m} \qquad (1.10.13)$$

Substituting in (10), we get

$$\sum_{k=0}^{\infty} \int_0^{\infty} L_m(t)L_k(t)e^{-t}\binom{x}{k}dt = \binom{x}{m} \qquad (1.10.14)$$

which at once yields the orthogonality relation (7). (It is important to emphasise that the convergence of the infinite expansion (8) is *not* involved in this argument. What we need are only *integer* values of x for which the convergence is automatic.)

A natural generalisation of (4) offers itself. Let us define the fundamental function $f(x)$ as follows

$$f(x) = \frac{p! t^x}{(p+x)!} \tag{1.10.15}$$

where p is an arbitrary positive constant. The successive differences (2) once more define an infinite set of polynomials, called the "generalised Laguerre polynomials" $L_k{}^p(t)$. For example:

$$L_0{}^p(t) = 1$$

$$L_1{}^p(t) = -1 + \frac{t}{p+1}$$

$$L_2{}^p(t) = 1 - \frac{2t}{p+1} + \frac{t^2}{(p+1)(p+2)} \tag{1.10.16}$$

$$L_3{}^p(t) = -1 + \frac{3t}{p+1} - \frac{3t^2}{(p+1)(p+2)} + \frac{t^3}{(p+1)(p+2)(p+3)}$$

and so on. The formula (9) is now to be modified as follows:

$$\frac{p! t^x}{(p+x)!} = \sum_{k=0}^{\infty} \binom{x}{k} L_k{}^p(t) \tag{1.10.17}$$

Moreover, if we multiply on both sides by $f(\xi) t^p e^{-t}$, and integrate with respect to t between 0 and ∞, we obtain on the left side

$$\phi_p(\xi) = \frac{p!^2 (p+x+\xi)!}{(p+\xi)!(p+\xi)!} = \frac{p!^2 x!}{(p+x)!} \binom{p+x+\xi}{x} \tag{1.10.18}$$

This leads, by exactly the same reasoning as before in the case of (12):

$$\Delta^m \phi_p(0) = \frac{p!^2 x!}{(p+x)!} \binom{p+x}{x-m}$$

$$= \frac{p!^2}{(p+m)!} \frac{x!}{(x-m)!} = \frac{p!^2 m!}{(p+m)!} \binom{x}{m} \tag{1.10.19}$$

This means

$$\int_0^\infty \frac{p! t^{x+p}}{(x+p)!} e^{-t} L_m{}^p(t) dt = \frac{p!^2 m!}{(p+m)!} \binom{x}{m} \tag{1.10.20}$$

and in view of (17):

$$\int_0^\infty \sum_{k=0}^{\infty} t^p e^{-t} L_m{}^p(t) L_k{}^p(t) \binom{x}{m} dt = \frac{p!^2 m!}{(p+m)!} \binom{x}{m} \tag{1.10.21}$$

We thus obtain the orthogonality of the generalised Laguerre polynomials $L_k{}^p(x)$ in the following sense

$$\int_0^\infty t^p e^{-t} L_m{}^p(t) L_k{}^p(t) dt = \frac{p!^2 m!}{(p + m)!} \, \delta_{mk} \qquad (1.10.22)$$

Problem 15. Show that all these relations remain valid if the condition $p > 0$ is generalised to $p > -1$.

Problem 16. The hypergeometric series

$$F(\alpha, \beta, \gamma \, ; x) = 1 + \frac{\alpha\beta}{\gamma} x + \frac{\alpha(\alpha + 1)\beta(\beta + 1)}{\gamma(\gamma + 1)} \frac{x^2}{1.2} + \ldots \qquad (1.10.23)$$

is convergent for all (real or complex) $|x| < 1$. Put $x = z/\beta$ and let β go to infinity. Show that the new series, called the "confluent hypergeometric series", convergent for all z, becomes

$$F(\alpha, \gamma \, ; z) = 1 + \frac{\alpha}{\gamma} z + \frac{\alpha(\alpha + 1)}{\gamma(\gamma + 1)} \frac{z^2}{2!} + \ldots \qquad (1.10.24)$$

Show that the Laguerre polynomials $L_k{}^p(t)$ are special cases of this series, namely

$$L_k{}^p(t) = (-1)^k F(-k, p + 1 \, ; t) \qquad (1.10.25)$$

1.11. Binomial expansions

The infinite Taylor series

$$f(z) = f(0) + f'(0)z + f''(0) \frac{z^2}{2!} + \ldots$$

$$= \sum_{k=0}^\infty f^{(k)}(0) \frac{z^k}{k!} \qquad (1.11.1)$$

expands the function $f(z)$ into a power series, in terms of the value of $f(z)$ and all its derivatives at the centre of expansion $z = 0$. We obtain a counterpart of the infinite Taylor series, replacing differentiation d/dx by differencing Δ, in the form of the Gregory-Newton series

$$f(z) = f(0) + \Delta f(0)z + \Delta^2 f(0) \binom{z}{2} + \ldots$$

$$= \sum_{k=0}^\infty \Delta^k f(0) \binom{z}{k} \qquad (1.11.2)$$

which substitutes for the functions $z^k/k!$ the "binomial coefficients" $\binom{z}{k}$ and for the successive *derivatives* of $f(z)$ at $z = 0$ the successive *differences* of $f(z)$ at $z = 0$.

There is, however, a deep-seated difference between these two types of expansions. The Taylor series (1) has a *very general* validity since it holds for all analytical functions within a certain convergence radius $|z| < r$ of the complex variable z. On the other hand, the example of Section 7 has demonstrated that we cannot expect the convergence of the series (2) even under completely analytical conditions. What we can expect, however, is that there may exist a definite *class of functions* which will allow representation with the help of the infinite expansion (2).

In order to find this class, we are going to make use of the orthogonality and completeness of the Laguerre polynomials in the range [0, ∞]. Let us assume that $f(t)$ is a function which is absolutely integrable in any finite range of the interval [0, ∞] while its behaviour in infinity is such that

$$\int_0^\infty |f(t)|e^{-t/2}dt = \text{finite} \tag{1.11.3}$$

Then the function $f(t)e^{-t/2}$ can be expanded into the orthogonal Laguerre functions $L_k(t)e^{-t/2}$ which leaves us with an expansion of $f(t)$ itself into the Laguerre polynomials:

$$f(t) = \sum_{k=0}^\infty c_k L_k(t) \tag{1.11.4}$$

where

$$c_k = \int_0^\infty f(t)L_k(t)e^{-t}dt \tag{1.11.5}$$

As an example let us consider the expansion of the function

$$f(t) = \frac{t^x}{x!} \tag{1.11.6}$$

in Laguerre polynomials where x is any positive constant, or in fact any complex constant whose real part is positive or zero:

$$R(x) \geq 0 \tag{1.11.7}$$

For this purpose we need the expansion coefficients

$$c_k = \frac{1}{x!} \int_0^\infty t^x e^{-t} L_k(t)dt \tag{1.11.8}$$

which we will now evaluate. For this purpose we imagine that we replace $L_k(t)$ by the actual power series, integrating term by term:

$$\begin{aligned} c_k &= \frac{1}{x!} \int_0^\infty t^x e^{-t}(\lambda_0 + \lambda_1 t + \ldots + \lambda_k t^k)dt \\ &= \frac{1}{x!}[\lambda_0 x! + \lambda_1(x+1)! + \ldots + \lambda_k(x+k)!] \\ &= p_k(x) \end{aligned} \tag{1.11.9}$$

where $p_k(x)$ is some polynomial in x.

We have a very definite information about the *roots* of this polynomial. Let us namely assume that in the integral (8) x takes the value of any *integer* less than k. In view of the fact that any power x^j can be conceived as a certain linear combination of the Laguerre polynomials $L_\alpha(t)$ ($\alpha \le j$) and that $L_k(t)$ is orthogonal to all the polynomials $L_j(t)$ of lower order, we observe that all these integrals must *vanish*. Hence $p_k(x)$ is zero for $x = 0, 1, 2, \ldots, (k - 1)$. This identifies $p_k(x)$ to

$$p_k(x) = Cx(x - 1)(x - 2) \ldots (x - k + 1) \tag{1.11.10}$$

and the only remaining uncertainty is the constant C. But we know from the definition of $L_k(t)$ that the coefficient of t^k is $1/(k!)$. Therefore, if we let x go to infinity, the coefficient of x^k must become $1/k!$. This determines the constant C and we obtain

$$\frac{1}{x!} \int_0^\infty t^x e^{-t} L_k(t) dt = c_k = p_k(x) = \binom{x}{k} \tag{1.11.11}$$

in agreement with our earlier formula (10.13). The expansion (4) thus becomes

$$\frac{t^x}{x!} = \sum_{k=0}^\infty L_k(t) \binom{x}{k} = \sum_{k=0}^\infty \Delta^k f(0) \binom{x}{k} \tag{1.11.12}$$

We have now found a *special case* of a function which permits the infinite Gregory-Newton expansion and thus the interpolation by powers of the functional values $f(m)$, given between 0 and ∞.

We will draw two conclusions from this result. First of all, let us put $t = 1$. Then we obtain the expansion

$$\frac{1}{x!} = \sum_{k=0}^\infty \binom{x}{k} L_k(1) \tag{1.11.13}$$

The factorial $x!$ itself goes by far too rapidly to infinity to allow the Gregory-Newton type of interpolation. But the *reciprocal* of the factorial is amenable to such an interpolation. If we let x go toward zero, we obtain in the limit an interesting approximation of the celebrated "Euler's constant"

$$\gamma = \lim_{n \to \infty} \left(1 + \frac{1}{2} + \frac{1}{3} + \ldots \frac{1}{n} - \log n \right) \tag{1.11.14}$$

because the derivative of $x!$ at $x = 0$ is $-\gamma$:

$$\gamma = \sum_{k=1}^\infty \frac{(-1)^{k-1}}{k} L_k(1) \tag{1.11.15}$$

The convergence of this series is very slow and of no practical significance. But a similar method, applied to the series (13) at some integer point $x = m$ instead of $x = 0$, yields expansions of much quicker convergence.

Another conclusion can be drawn concerning the interpolation of the exponential function $e^{\alpha x}$. We can write

$$e^{\alpha x} = [1 + (e^\alpha - 1)]^x \tag{1.11.16}$$

which can be expanded in a binomial series according to Newton's formula [cf. (6.16)], provided that

$$e^{\alpha} < 2 \tag{1.11.17}$$

that is $\alpha < 0.69315\ldots$. If, however, we divide by $x!$ and interpolate the new function, we obtain convergence for *all* values of α.

Problem 17. The values of e^x for $x = 0, 1, 2, \ldots, 7$ are:

$$
\begin{aligned}
y_0 &= 1 \\
y_1 &= 2.71828 \\
y_2 &= 7.38906 \\
y_3 &= 20.08554 \\
y_4 &= 54.59815 \\
y_5 &= 148.41316 \\
y_6 &= 403.42879 \\
y_7 &= 1096.63316
\end{aligned}
\tag{1.11.18}
$$

Obtain an upper and lower bound for $e^{1/2}$ by Gregory-Newton interpolation, without and with the weight factor $(x!)^{-1}$. (The latter series is convergent but the convergence is very slow.)

[Answer: (a) $1.8071 > e^{1/2} > 1.4901$

(b) $1.7298 > e^{1/2} > 1.6246$] $\qquad (e^{1/2} = 1.64872)$

Problem 18. By an argument quite similar to that used in the proof of (11), but now applied to the generalised Laguerre polynomials $L_k{}^p(x)$, show the validity of the following relation

$$\frac{1}{(x+p)!} \int_0^{\infty} t^{x+p} e^{-t} L_k{}^p(t)dt = \frac{p!\,k!}{(p+k)!} \binom{x}{k} \tag{1.11.19}$$

and deduce the expansion

$$f_p(x) = \frac{p!\,t^x}{(x+p)!} = \sum_{k=0}^{\infty} L_k{}^p \binom{x}{k} = \sum_{k=0}^{\infty} \Delta_k f_p(0) \binom{x}{k} \tag{1.11.20}$$

1.12. The decisive integral transform

We will now proceed to the construction of a certain *integral transform* which is fundamental in answering our initial problem: "Find the class of functions which allow the Gregory-Newton type of expansion (11.2)."

For our construction we make use of an auxiliary function $g(t)$, defined in the range $[0, \infty]$, which satisfies the condition (11.3) and thus permits an expansion into Laguerre polynomials. We define our integral transform as follows:

$$f(x) = \frac{1}{x!} \int_0^{\infty} t^x e^{-t} g(t)dt \tag{1.12.1}$$

Let us now expand $g(t)$ in Laguerre polynomials:

$$g(t) = \sum_{k=0}^{\infty} g_k L_k(t) \tag{1.12.2}$$

If this expansion is substituted in (1) and we integrate term by term, we obtain, in view of the relation (11.11):

$$f(x) = \sum_{k=0}^{\infty} g_k \binom{x}{k} \qquad (1.12.3)$$

But

$$g_k = \int_0^{\infty} g(t)e^{-t}L_k(t)dt \qquad (1.12.4)$$

and in view of the fact that $L_k(t)$ can be conceived as the kth difference of the function $t^{\xi}/\xi!$ (at $\xi = 0$), we can replace $L_k(t)$ by this function, integrate, and then take the kth difference (considering ξ as a variable), and finally replacing ξ by 0. The result of the integration becomes

$$\frac{1}{\xi!}\int_0^{\infty} g(t)e^{-t}t^{\xi}dt = f(\xi) \qquad (1.12.5)$$

and thus

$$g_k = \Delta^k f(0) \qquad (1.12.6)$$

Substituting this value of g_k in (3) we obtain the infinite binomial expansion

$$f(x) = \sum_{k=0}^{\infty} \Delta^k f(0)\binom{x}{k} \qquad (1.12.7)$$

which shows that *the integral transform* (1) *defines a class of functions which allows the infinite binomial expansion of Gregory-Newton.*

On the other hand, let us assume that we have a Gregory-Newton expansion which is convergent:

$$f(x) = \sum_{k=0}^{\infty} g_k \binom{x}{k} \qquad (1.12.8)$$

Then we define a function $g(t)$ by the infinite sum

$$g(t) = \sum_{k=0}^{\infty} g_k L_k(t) \qquad (1.12.9)$$

and obtain $f(x)$ by constructing the integral transform (1). Hence we see that the integral transform (1) is sufficiently general to characterise the *entire class of functions which allow the Gregory-Newton type of interpolation in the infinite integral* $[0, \infty]$.

The analytical form (1) of the function $f(x)$ shows that it is in fact an analytical function of x, throughout the right complex plane $R(x) \geq 0$. Moreover, the interpolation formula (7) remains valid not only on the positive real axis but everywhere in the right complex z-plane $R(z) \geq 0$. Hence we have obtained an expansion which not only *interpolates* properly the discrete functional values $f(m)$ to the values $f(x)$ *between* the given data, but also extrapolates $f(z)$ properly at every point z of the right complex half plane.

Problem 19. Carry through the procedure with respect to the generalised integral transform

$$f_p(x) = \frac{p!}{(x+p)!} \int_0^\infty t^{x+p} e^{-t} g(t) dt \qquad (1.12.10)$$

expanding $g(t)$ in the polynomials $L_k{}^p(t)$. Show that $f_p(x)$ allows the expansion (7) throughout the right complex half plane. The expansion may also be written in the form

$$f_p(x) = \sum_{k=0}^\infty g_k{}^p \binom{x}{k}$$

with

$$g_k{}^p = \int_0^\infty t^p e^{-t} g(t) L_k{}^p(t) dt \qquad (1.12.11)$$

Problem 20. Comparing the integral transforms (1) and (10), demonstrate the following theorem: *If a function allows the infinite Gregory-Newton expansion, it allows that expansion also if the centre of expansion is shifted by an arbitrary amount to the right.*

1.13. Binomial expansions of the hypergeometric type

Some special choices of the function $g(t)$ lead to a number of interesting binomial expansions which are closely related to the hypergeometric function $F(\alpha, \beta, \gamma; x)$ [cf. (10.23)].

Problem 21. Choose the function $g(t)$ of the integral transform (12.1) to be

$$g(t) = \frac{t^\mu}{\mu!} \qquad R(\mu) \geq 0 \qquad (1.13.1)$$

and obtain the following binomial expansion:

$$\frac{(x+\mu)!}{x!\mu!} = \sum_{k=0}^\infty \binom{x}{k}\binom{\mu}{k} \qquad (1.13.2)$$

Problem 22. Employ the same function in the integral transform (10) and derive the following binomial expansion:

$$\frac{(x+\mu+p)!}{(\mu+p)!(x+p)!} = \sum_{k=0}^\infty \frac{k!}{(p+k)!}\binom{x}{k}\binom{\mu}{k} \qquad (1.13.3)$$

Problem 23. Show that the right sides of (2) and (3) are in the following relation to the hypergeometric series (10.23):

$$\sum_{k=0}^\infty \binom{x}{k}\binom{\mu}{k} = F(-x, -\mu, 1; 1) \qquad (1.13.4)$$

$$\sum_{k=0}^\infty \frac{k!}{(p+k)!}\binom{x}{k}\binom{\mu}{k} = F(-x, -\mu, 1+p; 1) \qquad (1.13.5)$$

Problem 24. In the expansion (2) substitute $\mu = \epsilon$ and make ϵ infinitesimal. Obtain in this fashion the following binomial expansion for the "logarithmic derivative of the gamma function":

$$\psi(z) = \frac{(z!)'}{z!} = \frac{\Gamma'(z+1)}{\Gamma(z+1)} \tag{1.13.6}$$

$$\psi(z) = \psi(0) + \sum_{k=1}^{\infty} \frac{(-1)^{k-1}}{k} \binom{z}{k} \tag{1.13.7}$$

Problem 25. Doing the same in the expansion (5) obtain the following generalisation of (7):

$$\psi(z+p) = \psi(p) + \sum_{k=1}^{\infty} \frac{p!\,k!}{(p+k)!} \frac{(-1)^{k-1}}{k} \binom{z}{k} \tag{1.13.8}$$

1.14. Recurrence relations

The Taylor series (11.1) has the property that the operation of "differentiation" leaves its form unchanged, merely shifting the coefficients by one unit to the left. We can ask for the operations which will leave the form of the binomial expansion

$$f(x) = \sum_{k=0}^{\infty} g_k \binom{x}{k} \tag{1.14.1}$$

invariant. First of all we have the "differencing operation" Δ [cf. (6.1)] at our disposal. If we apply this operation to the series, we obtain, in view of (6.10):

$$\Delta f(x) = \sum_{k=0}^{\infty} g_{k+1} \binom{x}{k} \tag{1.14.2}$$

Hence the operation Δ on the function has the effect that the coefficient g_k is changed to g_{k+1}. This operation can be repeated, of course, any number of times, obtaining each time a jump in the index of g_k by one. Particularly important is the operation

$$f(x+1) = (1+\Delta)f(x) \tag{1.14.3}$$

and consequently

$$f(x+m) = (1+\Delta)^m f(x) \tag{1.14.4}$$

There is a second fundamental operation which leaves the form of the series (1) unchanged. Let us namely multiply $f(x)$ by x but take $f(x)$ at the point $x - 1$. This operation shall be denoted by the symbol Γ:

$$\Gamma f(x) = x f(x-1) \tag{1.14.5}$$

If we perform this operation on the binomial coefficients, we find

$$x \binom{x-1}{k} = (k+1) \binom{x}{k+1} \tag{1.14.6}$$

Hence

$$\Gamma f(x) = \sum_{k=0}^{\infty} \binom{x}{k} k g_{k-1} \tag{1.14.7}$$

Here we find a shift of the subscript of g to the *left*, together with a multiplication by k.

The operation

$$xf(x) = [(1 + \Delta)\Gamma - 1]f(x) \tag{1.14.8}$$

is a consequence of the fundamental definitions. Accordingly

$$x^m f(x) = [(1 + \Delta)\Gamma - 1]^m f(x) \tag{1.14.9}$$

We see that by the combination of these two operations Δ and Γ we can express any linear combination of the functional values $f(x + m)$, multiplied by any polynomials of x.

If now the function $f(x)$, which can be expanded binomially, satisfies some linear functional equation between the values $f(x + m)$ whose coefficients are rational functions of x, this relation will find a counterpart in a corresponding *recurrence relation* between the coefficients g_k of its binomial expansion.

Let us consider for example the expansion (11.12):

$$\frac{t^x}{x!} = \sum_{k=0}^{\infty} L_k(t) \binom{x}{k} \tag{1.14.10}$$

Here the expansion coefficients g_k become $L_k(t)$. The function on the left satisfies the following simple functional equation:

$$\frac{(x + 1)t^{x+1}}{(x + 1)!} = t \frac{t^x}{x!} \tag{1.14.11}$$

that is:

$$(x + 1)f(x + 1) - tf(x) = 0 \tag{1.14.12}$$

According to the rules (3) and (8) we can write this equation in the form

$$\{(1 + \Delta)[(1 - \Delta)\Gamma - 1] - t\}f(x) = 0 \tag{1.14.13}$$

Translated to the coefficients g_k the corresponding relation becomes:

$$kg_{k-1} + 2(k + 1)g_k + (k + 2)g_{k+1} - (g_k + g_{k+1} + tg_k) = 0$$

$$(k + 1)g_{k+1} + (2k + 1 - t)g_k + kg_{k-1} = 0 \tag{1.14.14}$$

which yields the following recurrence relation for the Laguerre polynomials $L_k(t)$:

$$(k + 1)L_{k+1}(t) + (2k + 1 - t)L_k(t) + kL_{k-1}(t) = 0 \tag{1.14.15}$$

Problem 26. Show that the operations Δ and Γ are *commutative*: $\Gamma\Delta = \Delta\Gamma$.

Problem 27.   In the expansion (11.20) the function $f_p(x)$ satisfies the following functional equation:

$$(x + p + 1)f_p(x + 1) - tf_p(x) = 0 \qquad (1.14.16)$$

Translate this equation into the realm of the expansion coefficients and obtain the following recurrence relation for the generalised Laguerre polynomials $L_k{}^p(t)$:

$$(k + p + 1)L_{k+1}{}^p(t) + (2k + p + 1 - t)L_k{}^p(t) + kL_{k-1}{}^p(t) = 0 \qquad (1.14.17)$$

Problem 28.   The left side of the expansion (13.2) satisfies the functional equation

$$(x + 1)f(x + 1) - (x + \mu + 1)f(x) = 0 \qquad (1.14.18)$$

Find the corresponding recurrence relation for the expansion coefficients and verify its validity.

[Answer:
$$[(1 + \Delta)\Delta\Gamma - (\mu + 1 + \Delta)]f(x) = 0$$
$$(k + 1)g_{k+1} + (k - \mu)g_k = 0 \qquad (1.14.19)]$$

Problem 29.   Do the same for the binomial expansion (13.3).

[Answer:
$$[(1 + \Delta)\Delta\Gamma - (\mu + 1) + (p - 1)\Delta]f(x) = 0$$
$$(k + p + 1)g_{k+1} + (k - \mu)g_k = 0 \qquad (1.14.20)]$$

Problem 30.   The general hypergeometric series (10.23) for $F(-x, \beta, \gamma; z)$ can be conceived as a binomial expansion in x, considering β, γ, z as mere parameters. The coefficients of this expansion:

$$g_k = (-1)^k \frac{(\beta + k - 1)!(\gamma - 1)!}{(\gamma + k - 1)!(\beta - 1)!} z^k \qquad (1.14.21)$$

satisfy the following recurrence relation:

$$(\gamma + k)g_{k+1} + (\beta + k)zg_k = 0 \qquad (1.14.22)$$

Translate this relation into a functional equation for F, considered as a function of x.   Write down the resulting formula in the usual notation $F(\alpha, \beta, \gamma; z)$.

[Answer:
$$[\Delta^2\Gamma + (\gamma - 2)\Delta + z(\Delta\Gamma + \beta - 1)]f(x) = 0$$
$$(x + \gamma)f(x + 1) - [2x + \gamma - z(x + \beta)]f(x) + x(1 - z)f(x - 1) = 0$$
$$(\gamma - \alpha)F(\alpha - 1, \beta, \gamma; z) + \alpha(z - 1)F(\alpha + 1, \beta, \gamma; z)$$
$$- [\gamma - \alpha + \alpha(z - 1) - \beta z]F(\alpha, \beta, \gamma; z) = 0 \qquad (1.14.23)]$$

1.15. The Laplace transform

Let us choose the input function of the integral transform (12.1) in the form

$$g(t) = e^{(1-\alpha)t} \qquad (1.15.1)$$

In order to satisfy the condition (11.3), it is necessary and sufficient that the real part of α shall be larger than $\frac{1}{2}$:

$$R(\alpha) > \tfrac{1}{2} \qquad (1.15.2)$$

With this choice of $g(t)$ the integral transform (12.1) becomes

$$f(x) = \frac{1}{x!} \int_0^\infty t^x e^{-\alpha t} dt \tag{1.15.3}$$

We can evaluate this integral by making the substitution $\alpha t = t_1$, obtaining

$$f(x) = \frac{1}{\alpha \alpha^x} = \frac{1}{\alpha} \left[1 + \left(\frac{1}{\alpha} - 1 \right) \right]^x \tag{1.15.4}$$

If we expand the right side by Newton's binomial formula, we obtain

$$f(x) = \frac{1}{\alpha} \sum_{k=0}^\infty \left(\frac{1}{\alpha} - 1 \right)^k \binom{x}{k} \tag{1.15.5}$$

On the other hand, according to the general theory [cf. (12.3–4)]:

$$f(x) = \sum_{k=0}^\infty g_k \binom{x}{k} \tag{1.15.6}$$

with

$$g_k = \int_0^\infty g(t) e^{-t} L_k(t) dt \tag{1.15.7}$$

Hence we obtain, in view of (5):

$$\int_0^\infty e^{-\alpha t} L_k(t) dt = \frac{1}{\alpha} \left(\frac{1}{\alpha} - 1 \right)^k \tag{1.15.8}$$

The integral transform

$$f(\alpha) = \int_0^\infty e^{-\alpha t} g(t) dt \tag{1.15.9}$$

called "Laplace transform", is one of the most important transforms of applied analysis, fundamental in many problems of mathematical physics and engineering (cf. Chapter 6.10). The left side of (8) is by definition the Laplace transform of $L_k(t)$. The right side yields this transform in a remarkably simple explicit form.

The result (8) has the following important consequence. Let us expand the input function $g(t)$ of the Laplace transform into Laguerre polynomials:

$$g(t) = \sum_{k=0}^\infty g_k L_k(t) \tag{1.15.10}$$

If this sum is introduced in (9) and we integrate term by term, we get:

$$f(\alpha) = \frac{1}{\alpha} \sum_{k=0}^\infty g_k \left(\frac{1}{\alpha} - 1 \right)^k \tag{1.15.11}$$

This expansion converges for all values of the complex variable α whose real part is greater than zero.

Sometimes our aim is to obtain the input function $g(t)$ from a known Laplace transform $f(\alpha)$. In this case the expansion of $g(t)$ in Laguerre polynomials would not be feasible since this expansion goes beyond all bounds as t goes to infinity. But if it so happens that $g(t)$ is quadratically integrable without the weight factor e^{-t}; $\int_0^\infty g^2(t)dt = $ finite, then we can expand $g(t)$ into the orthonormal Laguerre *functions*, obtained by multiplying the Laguerre polynomials by $e^{-t/2}$. In this case:

$$g(t) = \sum_{k=0}^\infty g_k L_k(t) e^{-t/2} \qquad (1.15.12)$$

and

$$f(\alpha) = \frac{1}{\alpha + \frac{1}{2}} \sum_{k=0}^\infty \left(\frac{1}{\alpha + \frac{1}{2}} - 1 \right)^k g_k \qquad (1.15.13)$$

If we now introduce a new variable ξ by putting

$$\frac{1}{\alpha + \frac{1}{2}} - 1 = \xi \qquad (1.15.14)$$

and expand the function

$$G(\xi) = (\alpha + \tfrac{1}{2})f(\alpha) \qquad (1.15.15)$$

in a Taylor series around the centre 1:

$$G(\xi) = g_0 + g_1(\xi - 1) + g_2(\xi - 1)^2 + \cdots \qquad (1.15.16)$$

the coefficients of this series yield directly the coefficients of the series (12). This procedure is frequently satisfactory even from the numerical standpoint.*

Does the Laplace transform permit the Gregory-Newton type of interpolation? This is indeed the case, as we can see if we consider that the function $e^{-x\xi}$ allows a binomial expansion, on account of Newton's formula:

$$e^{-x\xi} = [1 + (e^{-\xi} - 1)]^x$$
$$= \sum_{k=0}^\infty (e^{-\xi} - 1)^k \binom{x}{k} \qquad (1.15.17)$$

If we multiply by $\phi(\xi)$ and integrate between 0 and ∞ term by term—assuming that $\phi(\xi)$ goes to infinity weaker than $e^{-\epsilon t}$, ϵ being an arbitrarily small positive constant—then our expansion remains convergent and yields the Gregory-Newton expansion of the Laplace transform

$$f(x) = \int_0^\infty e^{-x\xi} \phi(\xi) d\xi \qquad (1.15.18)$$

* Cf. A. A., p. 292.

Problem 31. Show that the condition (2) yields for the convergence of the binomial expansion (5) the condition

$$\left| \frac{1}{\alpha} - 1 \right| < 1 \tag{1.15.19}$$

Problem 32. Choose the input function of the integral transform (12.10) in the form (1) and deduce the following relation:

$$\int_0^\infty t^p L_k{}^p(t) e^{-\alpha t} dt = p! \frac{1}{\alpha^{p+1}} \left(\frac{1}{\alpha} - 1 \right)^k \tag{1.15.20}$$

This gives the Laplace transform of $t^p L_k{}^p$ in explicit form.

Problem 33. Find the input function $g(t)$ of the integral transform (12.1) which leads to the Laplace transform (18).

[Answer:

$$g(t) = \int_1^\infty e^{(1-\xi)t} \phi(\log \xi) d\xi \tag{1.15.21}]$$

Problem 34. Obtain the Gregory-Newton expansion of the Laplace transform (18) whose input function is

$$\phi(\xi) = \rho e^{-\rho \xi} \qquad R(\rho) > 0$$

[Answer:

$$\frac{\rho}{\rho + x} = 1 - \frac{x}{\rho} + \frac{x(x-1)}{\rho(\rho+1)} - \frac{x(x-1)(x-2)}{\rho(\rho+1)(\rho+2)} + \cdots$$

$$= \sum_{k=0}^\infty (-1)^k \frac{\rho! k!}{(\rho+k)!} \binom{x}{k} \tag{1.15.22}]$$

Problem 35. Show that a Laplace transform allows binomial interpolation in a still different sense namely by applying the Gregory-Newton expansion to the function $f(x)/x!$. [Hint: Assume that in the integral transform (12.1) $g(t) = 0$ for $t > 1$.]

Problem 36. Show that the binomial interpolation of the Laplace transform is possible not only with $f(m)$ as key-values but with $f(\beta m)$ as key-values where β is an arbitrary positive constant.

1.16. The Stirling expansion

The interpolation by central differences led to the Stirling type of expansion (cf. Section 8). Here again the question arises whether this method of interpolation could be used unlimitedly on tabular values of a function which is tabulated in equidistant intervals between $-\infty$ and $+\infty$. Hence the key-values are now $f(\pm m)$ where m assumes the values $0, 1, 2, \ldots$, to infinity. Once more it is clear that only a restricted class of functions will submit itself to this kind of interpolation and our aim will be to circumscribe this class.

If in our previous discussions the fundamental functions were the Laguerre polynomials—which represented a special class of hypergeometric functions—

we will once more turn to the hypergeometric series and consider two particular cases, characterised by the following choice of the parameters:

$$\alpha = -x, \quad \beta = x, \quad \gamma = \tfrac{1}{2} \qquad (1.16.1)$$

and

$$\alpha = -x + 1, \quad \beta = x + 1, \quad \gamma = \tfrac{3}{2} \qquad (1.16.2)$$

If we multiply numerator and denominator by 2^k, we get rid of the factions in the denominator and the product $\gamma(\gamma + 1) \ldots (\gamma + k - 1)$ becomes $1 \cdot 3 \cdot 5 \ldots (2k - 1)$ in the first case and $1 \cdot 3 \cdot 5 \ldots (2k + 1)$ in the second. Furthermore, the $k!$ in the denominator can be written in the form $2 \cdot 4 \ldots (2k)$, if we multiply once more numerator and denominator by 2^k. The two factors of the denominator combine into $(2k)!$ in the first case and $(2k + 1)!$ in the second. We thus obtain the two expansions

$$F(-x, x, \tfrac{1}{2}; t) = 1 - \frac{x^2}{2} 4t + \frac{x^2(x^2 - 1)}{4!} (4t)^2 - \ldots$$

$$= \sum_{k=0}^{\infty} \Phi_{2k}(x)(-4t)^k \qquad (1.16.3)$$

$$xF(-x + 1, x + 1, \tfrac{3}{2}; t) = x - \frac{x(x^2 - 1)}{3!} 4t + \frac{x(x^2 - 1)(x^2 - 4)}{5!} (4t)^2 - \ldots$$

$$= \sum_{k=0}^{\infty} \Phi_{2k+1}(x)(-4t)^k \qquad (1.16.4)$$

where $\Phi_{2k}(x)$ and $\Phi_{2k+1}(x)$ are the Stirling functions, encountered earlier in (8.3) and (8.6).

The hypergeometric functions represented by these expansions are obtainable in *closed form*. Let us consider the differential equation of Gauss which defines the hypergeometric function

$$y = F(\alpha, \beta, \gamma; t) : t(1 - t)y'' + [\gamma - (\alpha + \beta + 1)t]y' - \alpha\beta y = 0 \quad (1.16.5)$$

For the special case (1) this differential equation becomes

$$t(1 - t)y'' + (\tfrac{1}{2} - t)y' + x^2 y = 0 \qquad (1.16.6)$$

while the choice (2) yields

$$t(1 - t)y'' + (\tfrac{3}{2} - 3t)y' + (x^2 - 1)y = 0 \qquad (1.16.7)$$

Problem 37. Transform t into the new variable θ by the transformation

$$t = \frac{1 - \cos \theta}{2} = \sin^2 \frac{\theta}{2} \qquad (1.16.8)$$

Show that in the new variable the differential equation (6) becomes

$$y'' + x^2 y = 0 \qquad (1.16.9)$$

while in the case (7) we get

$$u'' + x^2 u = 0 \tag{1.16.10}$$

for

$$u(\theta) = y(\theta) \sin \theta \tag{1.16.11}$$

If we adopt the new angle variable θ for the expansions (3) and (4), we observe that the functions $F(-x, x, \tfrac{1}{2}; \sin^2 \theta/2)$ and $F(-x + 1, x + 1, \tfrac{3}{2}; \sin^2 \theta/2)$ are *even* functions of θ. Hence in the general solution of (9):

$$y = A \cos x\theta + B \sin x\theta$$

the sine-part must drop out, while the constant A must be chosen as 1, since for $\theta = 0$ the right side is reduced to 1. We thus obtain

$$F\left(-x, x, \tfrac{1}{2}; \sin^2 \frac{\theta}{2}\right) = \cos x\theta \tag{1.16.12}$$

and by a similar argument

$$xF\left(-x + 1, x + 1, \tfrac{3}{2}; \sin^2 \frac{\theta}{2}\right) = \frac{\sin x\theta}{\sin \theta} \tag{1.16.13}$$

Hence

$$\cos x\theta = \sum_{k=0}^{\infty} \Phi_{2k}(x)\left(-4 \sin^2 \frac{\theta}{2}\right)^k \tag{1.16.14}$$

$$\frac{\sin x\theta}{\sin \theta} = \sum_{k=0}^{\infty} \Phi_{2k+1}(x)\left(-4 \sin^2 \frac{\theta}{2}\right)^k \tag{1.16.15}$$

1.17. Operations with the Stirling functions

The two fundamental operations in the process of central differencing are (cf. Section 8):

$$\delta f(x) = f(x + 1) - 2f(x) + f(x - 1) \tag{1.17.1}$$

$$\gamma f(x) = \tfrac{1}{2}[f(x + 1) - f(x - 1)] \tag{1.17.2}$$

Let us perform these operations on the left sides of the series (14) and (15):

$$\delta \cos x\theta = 2(\cos \theta - 1) \cos x\theta$$

$$= -4 \sin^2 \frac{\theta}{2} \cos x\theta \tag{1.17.3}$$

$$\delta \sin x\theta = 2(\cos \theta - 1) \sin x\theta$$

$$= -4 \sin^2 \frac{\theta}{2} \sin x\theta \tag{1.17.4}$$

If we apply the same operation term by term on the right sides, we obtain the following operational equations (considering $t = \sin^2 \theta/2$ as a variable and equating powers of t):

$$\delta \Phi_{2k}(x) = \Phi_{2k-2}(x) \tag{1.17.5}$$

$$\delta \Phi_{2k+1}(x) = \Phi_{2k-1}(x) \tag{1.17.6}$$

Furthermore, the identities

$$\gamma \cos x\theta = -\sin \theta \sin x\theta$$

$$= -4 \sin^2 \frac{\theta}{2} \left(1 - \sin^2 \frac{\theta}{2}\right) \frac{\sin x\theta}{\sin \theta} \qquad (1.17.7)$$

$$\gamma \frac{\sin x\theta}{\sin \theta} = \cos x\theta \qquad (1.17.8)$$

yield the following operational relations:

$$\gamma \Phi_{2k+1}(x) = \Phi_{2k}(x) \qquad (1.17.9)$$

$$\gamma \Phi_{2k+2}(x) = -\Phi_{2k+1}(x) + \tfrac{1}{2}\Phi_{2k-1}(x) \qquad (1.17.10)$$

On the basis of these relations we see that, if we put

$$f(x) = \sum_{k=0}^{\infty} g_k \Phi_k(x) \qquad (1.17.11)$$

we must have

$$g_{2k} = \delta^k f(0) \qquad (1.17.12)$$

and

$$g_{2k+1} = \gamma \delta^k f(0) \qquad (1.17.13)$$

Problem 38.   Show that

$$\delta^k(\cos \theta x)_{x=0} = (-1)^k \left(2 \sin \frac{\theta}{2}\right)^{2k} \qquad (1.17.14)$$

and

$$\gamma \delta^k \left(\frac{\sin \theta x}{\sin \theta}\right)_{x=0} = (-1)^k \left(2 \sin \frac{\theta}{2}\right)^{2k} \qquad (1.17.15)$$

This establishes the two hypergeometric series (16.14) and (16.15) as *infinite Stirling expansions*.

1.18. An integral transform of the Fourier type

On the basis of our previous results we can now establish a particular but important class of functions which allow the infinite Stirling expansion. First of all we will combine the two series (16.14) and (16.15) in the following complex form:

$$f(x) = \cos x\theta - i \sin x\theta = e^{-ix\theta} = \sum_{k=0}^{\infty} \rho_k(\theta)\Phi_k(x) \qquad (1.18.1)$$

with

$$\rho_{2k}(\theta) = \left(2i \sin \frac{\theta}{2}\right)^{2k} \qquad (1.18.2)$$

$$\rho_{2k+1}(\theta) = -\left(2i \sin \frac{\theta}{2}\right)^{2k+1} \cos \frac{\theta}{2} \qquad (1.18.3)$$

Since the hypergeometric series converges for all $|t| = |\sin^2 \theta/2| < 1$, we can make use of this series for any θ which varies between $-\pi$ and $+\pi$. If we now multiply by an absolutely integrable function $\varphi(\theta)$ and integrate between the limits $-\pi$, $+\pi$, we obtain the following integral transform:

$$f(x) = \int_{-\pi}^{+\pi} e^{-ix\theta}\varphi(\theta)d\theta \qquad (1.18.4)$$

This $f(x)$ allows the infinite Stirling expansion which means that $f(x)$ is uniquely determined if it is given at all integer values $x = \pm m$. We can now form the successive central differences $\delta^k f(0)$ and $\gamma\delta^k f(0)$—also obtainable according to (8.11) and (8.12) by a binomial weighting of the functional values $f(\pm m)$ themselves—and expand $f(x)$ in an infinite series:

$$f(x) = \sum_{k=0}^{\infty} g_k \Phi_k(x) \qquad (1.18.5)$$

$$= \sum_{k=0}^{\infty} [\delta^k f(0)\Phi_{2k}(x) + \gamma\delta^{2k}f(0)\Phi_{2k+1}(x)] \qquad (1.18.6)$$

However, the formulae (2-4) show that the coefficients g_k are also obtainable by evaluating the following definite integrals:

$$g_{2k} = \int_{-\pi}^{+\pi} \varphi(\theta)\left(2i\sin\frac{\theta}{2}\right)^{2k}d\theta \qquad (1.18.7)$$

$$g_{2k+1} = -i\int_{-\pi}^{+\pi} \varphi(\theta)\left(2i\sin\frac{\theta}{2}\right)^{2k}\sin\theta\,d\theta \qquad (1.18.8)$$

The integral transform (4) is a special case of the so-called "Fourier transform" which is defined quite similarly to (4) but with the limits $\pm\infty$. We can conceive the transform (4) as that case of the Fourier transform for which $\varphi(\theta)$ vanishes everywhere outside the limits $\pm\pi$. The analytical form of (4) shows that it represents an *analytical function of the complex variable* x, throughout the entire complex plane. Furthermore, we know from the nature of the hypergeometric series that the series (16.14–15) remain valid for arbitrary *complex* values of x. Hence the series (6) not only *interpolates* the functional values $f(\pm m)$ on the real axis, but *extrapolates* them to any value of the complex plane.

Problem 39. Given the following data. The function $f(x) = \cos \pi x$ assumes at integer points the values $f(\pm m) = (-1)^m$. Moreover, the function allows the Stirling type of interpolation. Show that these data are sufficient for the unique determination of $\cos \pi x$ at all points x. [Hint: derive the series (16.14) (for $\theta = \pi$) from the given data by forming the successive central differences.]

Problem 40. Given the following data. The function $\sin \pi x$ vanishes at all integer points. At $x = 0$ it goes to zero like πx. It is, if divided by x, expandable into an infinite Stirling series. Show that these data are sufficient for obtaining $\sin \pi x$ at all points. [Hint: Consider the Stirling expansion of $\sin \pi x / \pi x$ and derive the following series:

$$\frac{\sin \pi x}{\pi x} = \sum_{k=0}^{\infty} \binom{x}{k}\binom{-x}{k} \qquad (1.18.9)]$$

Problem 41. Assume the input function $\varphi(\theta)$ of the integral transform (4) in the form

$$\varphi(\theta) = \frac{(-1)^m}{2\pi} e^{im\theta} \qquad (1.18.10)$$

(m = integer). Then the function $f(x)$ becomes

$$f(x) = (-1)^m \frac{\sin \pi(x - m)}{\pi(x - m)} = \frac{\sin \pi x}{\pi(x - m)} \qquad (1.18.11)$$

The values of $f(x)$ at integer points are all zero, except at $x = m$ where the function assumes the value $(-1)^m$. Hence the binomial weighting of the functional values is particularly simple. Derive the expansion

$$\frac{\sin \pi x}{\pi(x - m)} = \sum_{k=0}^{\infty} \frac{k!^2}{(k + m)!(k - m)!} \binom{x}{k}\binom{-x}{k}\left(1 + \frac{m}{x}\right) \qquad (1.18.12)$$

which, if written in the general form (17.11) possesses the following expansion coefficients:

$$g_{2k} = (-1)^k \binom{2k}{k - m}, \qquad g_{2k+1} = (-1)^{k+1} \binom{2k + 2}{k + 1 - m} \frac{m}{2k + 2} \qquad (1.18.13)$$

The same coefficients are obtainable, however, on the basis of the integrals (7) and (8). Hence obtain the following formulae:

$$\frac{1}{2\pi} \int_{-\pi}^{+\pi} \sin^{2k} \frac{\theta}{2} \cos m\theta \, d\theta = \frac{(-1)^m}{4^k} \binom{2k}{k - m} \qquad (1.18.14)$$

$$\frac{1}{2\pi} \int_{-\pi}^{+\pi} \sin^{2k} \frac{\theta}{2} \sin m\theta \sin \theta \, d\theta = \frac{(-1)^{m+1}}{4^k} \frac{m}{2k + 2} \binom{2k + 2}{k + 1 - m} \qquad (1.18.15)$$

(The second integral is reducible to the first. Show the consistency of the two expressions.)

Problem 42. Show that the first $2m - 1$ terms of the Stirling expansion (12) drop out, because their coefficients are zero.

1.19. Recurrence relations associated with the Stirling series

As in the case of the Gregory-Newton series (cf. Section 14), we can once more ask for those operations which *leave the Stirling series invariant*. We have found already *two* such operations: γ and δ. They had the property that if they operate on the functions Φ_k, they generate a linear combination of these functions, without changing the form of the series. They merely *re-arrange* the coefficients g_k. We will employ the following notation. If we write δg_k, this should mean: *the change of the g_k, due to the operation*

$\delta f(x)$. Hence, e.g. the equation $\delta g_k = g_{k+2}$ shall signify that in consequence of the operation $\delta f(x)$ the coefficient g_k of the expansion is to be replaced by g_{k+2}. With this convention we obtain from the operational equations (17.5, 6, 9, 10):

$$\delta g_k = g_{k+2} \tag{1.19.1}$$

$$\gamma g_{2k} = g_{2k+1} \tag{1.19.2}$$

$$\gamma g_{2k+1} = g_{2k+1} + \tfrac{1}{4} g_{2k+4} \tag{1.19.3}$$

Now the two operations γ and δ can be combined and repeated any number of times.

Since by definition

$$\begin{aligned}
f(x+1) + f(x-1) &= (\delta + 2)f(x) \\
f(x+1) - f(x-1) &= 2\gamma f(x)
\end{aligned} \tag{1.19.4}$$

we obtain

$$\begin{aligned}
f(x+1) &= (1 + \gamma + \tfrac{1}{2}\delta)f(x) \\
f(x-1) &= (1 - \gamma + \tfrac{1}{2}\delta)f(x)
\end{aligned} \tag{1.19.5}$$

and hence we can obtain an arbitrary $f(x \pm m)$ with the help of the two operations γ and δ. But we still need another operation we possessed in the case of simple differences, namely the *multiplication by x* (cf. 14.8). This operation is obtainable by the *differentiation* of the series (16.14) and (16.15). For this purpose we return to our original variable t (cf. 16.8), but multiplied by -4:

$$-4t = -4 \sin^2 \frac{\theta}{2} = \tau \tag{1.19.6}$$

Then the series (16.14) and (16.15) become

$$\cos x\theta = \sum_{k=0}^{\infty} \Phi_{2k}(x)\tau^k \tag{1.19.7}$$

$$\frac{\sin x\theta}{\sin \theta} = \sum_{k=0}^{\infty} \Phi_{2k+1}(x)\tau^k \tag{1.19.8}$$

and now, differentiating the first series with respect to τ and subtracting the second series, after multiplying it by $x/2$, we obtain

$$\sum_{k=0}^{\infty} \tau^{k-1}\left[k\Phi_{2k}(x) - \frac{x\tau}{2}\Phi_{2k+1}(x) \right] = 0 \tag{1.19.9}$$

which leads to the relation, encountered before (cf. 8.7):

$$x\Phi_{2k+1}(x) = (2k+2)\Phi_{2k+2}(x) \tag{1.19.10}$$

Let us now differentiate the second series with respect to τ and multiply it by $\sin^2 \theta = -\tau(1 + \tau/4)$. This gives

$$\frac{1}{2} \frac{\sin x\theta}{\sin \theta} \cos \theta - \frac{x \cos x\theta}{2} = - \sum_{k=0}^{\infty} k\Phi_{2k+1}(x)\tau^k\left(1 + \frac{\tau}{4}\right) \qquad (1.19.11)$$

and, moving over the first term to the right side:

$$-\frac{x \cos x\theta}{2} = - \sum_{k=0}^{\infty} \Phi_{2k+1}(x)\tau^k\left[k + \frac{1}{2} + (k + 1)\frac{\tau}{4}\right] \qquad (1.19.12)$$

This, in view of (9), yields the relation

$$x\Phi_{2k}(x) = (2k + 1)\Phi_{2k+1}(x) + \frac{k}{2} \Phi_{2k-1}(x) \qquad (1.19.13)$$

Accordingly, we can extend the rules (1–3) by the two additional rules:

$$xg_{2k+1} = (2k + 1)g_{2k} + \frac{k + 1}{2} g_{2k+2} \qquad (1.19.14)$$

$$xg_{2k+2} = (2k + 2)g_{2k+1} \qquad (1.19.15)$$

We see that *any linear recurrence relation which may exist between the functional values $f(x + m)$, with coefficients which are polynomials of x, can be translated into a linear recurrence relation for the coefficients of the Stirling expansion.*

Problem 43.　Show that the two operations δ and γ are *commutative*:

$$\gamma\delta = \delta\gamma \qquad (1.19.16)$$

Moreover show that the operation γ^2 is reducible to the operation δ, according to the relation

$$\gamma^2 = \delta + \tfrac{1}{4}\delta^2 \qquad (1.19.17)$$

Problem 44.　Find a recurrence relation for the expansion coefficients (18.13) on the basis of a recurrence relation for the function (18.11). Verify this relation. [Answer:

$$(x - m)f(x) = 0 \qquad (1.19.18)$$

(since $\sin \pi x$ vanishes at all integer points).

$$\begin{aligned}
(2k + 2)g_{2k+1} - mg_{2k+2} &= 0 \\
(2k + 1)g_{2k} + \frac{k + 1}{2} g_{2k+2} - mg_{2k+1} &= 0\,]
\end{aligned} \qquad (1.19.19)$$

Problem 45.　Obtain the recurrence relations corresponding to the functional equation

$$(x^2 - m^2)f(x) = 0 \qquad (1.19.20)$$

and show that both the g_{2k} (representing the even part of (18.11)), and the g_{2k+1} (representing the odd part of (18.11)) satisfy the appropriate relation.

[Answer:

$$[(k + 1)^2 - m^2]g_{2k+2} + (2k + 1)(2k + 2)g_{2k} = 0$$
$$[(k + 1)^2 - m^2]g_{2k+1} + 2k(2k + 1)g_{2k-1} = 0\,] \tag{1.19.21}$$

1.20. Interpolation of the Fourier transform

We have discussed in detail the Stirling expansion of the integral transform (18.4) which was a special example of the class of functions which permit the infinite Stirling series. It so happens, however, that the same transform can be interpolated in a still different manner, although employing once more the same key-values $f(\pm m)$. We will expand $\varphi(\theta)$ in an *infinite Fourier series* (cf. Chapter 2):*

$$\varphi(\theta) = \sum_{k=0}^{\pm\infty} c_k e^{ik\theta} \tag{1.20.1}$$

Then we obtain (cf. 18.11) the series

$$f(x) = 2\pi \sum_{k=0}^{\pm\infty} c_k \frac{\sin \pi(x - k)}{\pi(x - k)}$$

$$= 2 \sin \pi x \sum_{k=0}^{\pm\infty} \frac{(-1)^k c_k}{x - k} \tag{1.20.2}$$

The coefficients c_k of the expansion (1) are the Fourier coefficients

$$c_k = \frac{1}{2\pi} \int_{-\pi}^{+\pi} \varphi(\theta) e^{-ik\theta} d\theta = \frac{1}{2\pi} f(k) \tag{1.20.3}$$

Hence the expansion (2) becomes

$$f(x) = \frac{\sin \pi x}{\pi} \sum_{k=0}^{\pm\infty} \frac{(-1)^k f(k)}{x - k} \tag{1.20.4}$$

This series is very different from the Stirling series since the functions of interpolation are not polynomials in x but the trigonometric functions

$$\frac{\sin \pi(x - k)}{\pi(x - k)} \tag{1.20.5}$$

which are bounded by ± 1. Moreover, the functional values $f(\pm m)$ appear *in themselves*, and not in binomial weighting. The convergence of the new series is thus *much stronger* than that of the Stirling series.

---

* The notation $\pm\infty$ as summation limits means that the terms $k = \pm m$ are taken into account in succession while m assumes the values $1, 2, 3, \ldots, \ldots$ (the initial value $k = 0$ is taken only once).

If we separate the even and the odd parts of the function $f(x)$, the expansion (4) will appear in the following form:

$$f(x) = \frac{x \sin \pi x}{\pi} \sum_{k=0}^{\infty}{}' (-1)^k \frac{f(k) + f(-k)}{x^2 - k^2}$$

$$+ \frac{\sin \pi x}{\pi} \sum_{k=1}^{\infty} (-1)^k k \frac{f(k) - f(-k)}{x^2 - k^2} \qquad (1.20.6)$$

(The prime in the first sum refers to the convention that the term $k = 0$ should be taken with *half weight*.)

The function $\varphi(\theta)$ of the transform (18.4) may be chosen in the following extreme fashion: $\varphi(\theta)$ vanishes everywhere, except in the infinitesimal neighbourhood of the point $\theta = \theta_1$. With this choice of $\varphi(\theta)$ we see that $e^{-i\theta x}$ *itself* may be considered as a Fourier transform which permits the expansion (6), provided that θ is smaller than π.

Problem 46.   Obtain the following expansions:

$$\frac{1}{\cos \frac{\pi}{2} x} = \frac{4}{\pi} \left(\frac{1}{1 - x^2} - \frac{3}{9 - x^2} + \frac{5}{25 - x^2} - \cdots \right)$$

$$= \frac{4}{\pi} \sum_{k=0}^{\infty} (-1)^k \frac{2k + 1}{(2k + 1)^2 - x^2} \qquad (1.20.7)$$

$$\frac{1}{\sin \frac{\pi}{2} x} = \frac{4}{\pi} x \left(\frac{1}{2x^2} + \frac{1}{4 - x^2} - \frac{1}{16 - x^2} + \frac{1}{36 - x^2} - \cdots \right)$$

$$= \frac{4}{\pi} x \sum_{k=0}^{\infty}{}' \frac{(-1)^k}{x^2 - (2k)^2} \qquad (1.20.8)$$

Problem 47.   Obtain the following expansions:

$$\frac{1}{\cos \frac{\pi}{2} x} = 1 + \frac{4x^2}{\pi} \left(\frac{1}{1 - x^2} - \frac{1}{3(9 - x^2)} + \frac{1}{5(25 - x^2)} - \cdots \right)$$

$$= 1 - \frac{4x^2}{\pi} \sum_{k=0}^{\infty} (-1)^k \frac{1}{(2k + 1)(x^2 - (2k + 1)^2)} \qquad (1.20.9)$$

$$\tan \frac{\pi}{2} x = \frac{4}{\pi} x \left(\frac{1}{1 - x^2} + \frac{1}{9 - x^2} + \frac{1}{25 - x^2} + \cdots \right)$$

$$= \frac{4}{\pi} x \sum_{k=0}^{\infty} \frac{1}{(2k + 1)^2 - x^2} \qquad (1.20.10)$$

Problem 48. The limiting value $\theta = \pi$ is still permissible for the expansion of $\cos \theta x$. Obtain the series

$$\cos \pi x = \frac{2}{\pi} x \left(\frac{1}{2x^2} + \frac{1}{x^2 - 1} + \frac{1}{x^2 - 4} + \frac{1}{x^2 - 9} + \cdots \right)$$

$$= \frac{2}{\pi} x \sum_{k=0}^{\infty}{}' \frac{1}{x^2 - k^2} \tag{1.20.11}$$

Problem 49. Consider the Fourier transform

$$f(x) = \int_{-N}^{+N} e^{-ix\theta} \varphi(\theta) d\theta \tag{1.20.12}$$

and show that it permits an interpolation by powers and also by trigonometric functions in terms of the key values $f\left(\pm \dfrac{\pi}{N} k \right)$. Write down the infinite Stirling series associated with this function.

[Answer:

$$f(x) = \sum_{k=0}^{\infty} g_k \Phi_k \left(\frac{N}{\pi} x \right) \tag{1.20.13}]$$

Problem 50. Show that the integral transform (18.4) allows the Stirling expansion also in the key-values $x = \pm \beta m$, where β is any positive number between 0 and 1.

1.21. The general integral transform associated with the Stirling series

The two series (16.3) and (16.4) had been of great value for the derivation of the fundamental operational properties of the Stirling functions. The same series were used in the construction of the integral transform (18.4) which characterised a large class of functions which permitted the Stirling kind of interpolation (and extrapolation) in an infinite domain. We will now generalise our construction to an integral transform which shall include the *entire* class of functions to which the infinite Stirling expansion is applicable. We first consider the *even* part of the function: $\frac{1}{2}[f(x) + f(-x)]$, which can be expanded with the help of the even Stirling functions $\Phi_{2k}(x)$. Once more we use the special series (16.3), but without abandoning the original variable t which shall now be considered as a *complex* variable $-z$ whose absolute value is smaller than 1:

$$\cos [x \operatorname{arc} \cos (1 + 2z)] = \sum_{k=0}^{\infty} \Phi_{2k}(x)(4z)^k \tag{1.21.1}$$

In a similar manner as before, we multiply by an auxiliary function $g(z)$ and integrate over a certain path. However, instead of choosing as the path of integration the real axis between $z = -1$ and 0, we will now choose a closed *circle* of the complex plane:

$$z = \rho e^{i\theta} \tag{1.21.2}$$

We then obtain the integral transform

$$\frac{1}{2\pi} \oint \cos\left[x \arccos(1 + 2\rho e^{i\theta})\right] g(\theta) d\theta = \frac{1}{2\pi} \int \sum_{k=0}^{\infty} \Phi_{2k}(x) 4^k \rho^k e^{ik\theta} g(\theta) d\theta$$

(1.21.3)

We can approach the limit $\rho = 1$, without losing convergence. If we do so, we get

$$\frac{1}{2\pi} \int_{-\pi}^{+\pi} \cos\left[x \arccos(1 + 2e^{i\theta})\right] g(\theta) d\theta = \sum_{k=0}^{\infty} g_{2k} \Phi_{2k}(x)$$

(1.21.4)

with

$$g_{2k} = \frac{1}{2\pi} 4^k \int_{-\pi}^{+\pi} g(\theta) e^{ik\theta} d\theta$$

(1.21.5)

Conversely, let us assume that we have an infinite Stirling series

$$f(x) = \sum_{k=0}^{\infty} g_{2k} \Phi_{2k}(x)$$

(1.21.6)

which converges. Then we define the function $g(\theta)$ by the infinite Fourier series

$$g(\theta) = \sum_{k=0}^{\infty} \frac{1}{4^k} g_{2k} e^{-ik\theta}$$

(1.21.7)

because, in view of the orthogonality of the Fourier functions $e^{ik\theta}$, the integral (5) becomes indeed g_{2k}. The integral transform (4) is thus not only *sufficient* but even *necessary* for the characterisation of an even function which possesses an infinite convergent Stirling expansion.

The function $g(\theta)$ is closely related to a function of the complex variable z defined as follows:

$$G(z) = \sum_{k=0}^{\infty} \frac{g_{2k}}{4^k} \frac{1}{z^{k+1}}$$

(1.21.8)

Then on the unit circle we obtain $g(\theta)/e^{i\theta}$ and if the series converges at $|z| = 1$, it will certainly converge also for $|z| > 1$. Hence the integral transform (4) may also be written in the form

$$f(x) = \frac{1}{2\pi i} \oint \cos\left[x \arccos(1 + 2z)\right] G(z) dz$$

(1.21.9)

with the understanding that the range of integration is any closed curve on which $G(z)$ is analytical, and which includes all singularities of $G(z)$, but excludes the point $z = -1$ (which is the point of singularity of the function (1)). The function $G(z)$ is analytical everywhere outside the unit circle and will frequently remain analytical even *inside*, except for certain singular points.

As to the odd part $\frac{1}{2}[f(x) - f(-x)]$ of the function $f(x)$, we have seen that the Stirling expansion of an odd function is formally identical with the Stirling expansion of an even function (with the absolute term zero) divided by x (cf. Section 8). Hence the general representation of the class of functions which permits the Stirling kind of interpolation in an infinite domain, may be given in the form of the following integral transform:

$$f(x) = \frac{1}{2\pi i} \oint \cos\left[x \arccos\left(1 + 2z\right)\right]\left[G_1(z) + \frac{G_2(z)}{x}\right]dz \quad (1.21.10)$$

where $G_1(z)$, $G_2(z)$ are arbitrary functions, analytical outside and on the unit circle, and satisfying the auxiliary condition

$$\oint G_2(z)dz = 0 \quad (1.21.11)$$

Problem 51. Let the function $f(x)$ be defined as one of the Stirling functions $\Phi_{2k}(x)$, respectively $\Phi_{2k-1}(x)$. Find the corresponding generating functions $G_1(z)$, $G_2(z)$.

[Answer:

$$\Phi_{2k}(x): G_1(z) = \frac{1}{4^k}\frac{1}{z^{k+1}} \qquad G_2(z) = 0$$

$$(1.21.12)$$

$$\Phi_{2k-1}(x): G_1(z) = 0 \qquad G_2(z) = \frac{2k}{4^k z^{k+1}}\,]$$

Problem 52. Show that, if $f(x)$ is an even polynomial of the order $2n$, the generating function $G_1(z)$ is a polynomial of the order $n + 1$ in z^{-1}, while $G_2(z) = 0$. If $f(x)$ is an odd polynomial of the order $2n - 1$, the same is true of $G_2(z)$ (with the term z^{-1} missing), while $G_1(z) = 0$.

Problem 53. Find the generating functions of the functions (16.12) and (16.13).

[Answer:

$$\cos x\theta: G_1(z) = \frac{1}{z + \sin^2\dfrac{\theta}{2}} \qquad G_2(z) = 0$$

$$(1.21.13)$$

$$\frac{\sin x\theta}{\sin \theta}: G_1(z) = 0 \qquad G_2(z) = \frac{1}{2}\frac{1}{\left(z + \sin^2\dfrac{\theta}{2}\right)^2}\,]$$

Problem 54. Find the generating functions of the integral transform (18.4).

$$G_1(z) = \int_{-\pi}^{+\pi}\frac{\varphi(\theta)d\theta}{z + \sin^2\dfrac{\theta}{2}} \qquad G_2(z) = \frac{-i}{2}\int_{-\pi}^{+\pi}\frac{\varphi(\theta)\sin\theta\,d\theta}{\left(z + \sin^2\dfrac{\theta}{2}\right)^2} \quad (1.21.14)$$

In the case that $\varphi(\theta)$ is differentiable, $G_2(z)$ may be written as follows:

$$G_2(z) = \frac{i[\varphi(\pi) - \varphi(-\pi)]}{z + 1} - i\int_{-\pi}^{+\pi}\frac{\varphi'(\theta)d\theta}{z + \sin^2\dfrac{\theta}{2}} \quad (1.21.15)$$

Problem 55.   Find the generating functions of (18.11).

[Answer:

$$G_1(z) = \frac{(-1)^m}{2\pi} \int_{-\pi}^{+\pi} \frac{\cos m\theta \, d\theta}{z + \sin^2 \dfrac{\theta}{2}}$$

$$= \frac{(-1)^m}{4^m z^{m+1}} F\left(m + \frac{1}{2}, m + 1, 2m + 1; -\frac{1}{z}\right)$$

$$G_2(z) = mG_1(z) \tag{1.21.16}]$$

1.22. Interpolation of the Bessel functions

Our previous discussions have shown that the interpolation of an equidistantly tabulated function with the help of central differences is not necessarily a convergent process.   In fact, only a very limited class of entire analytical functions which allow representation in the form (1.21.10), can be interpolated in the Stirling fashion.   We frequently encounter integral transforms of a different type which may allow interpolation by completely different tools.   A good example is provided by the Bessel functions $J_p(x)$ which depend on the variable x, but also on the order p.   Let us first consider the Bessel functions of *integer* order $J_n(x)$.   They are defined by the integral transform

$$J_n(x) = \frac{e^{in\pi/2}}{\pi} \int_0^{+\pi} e^{-ix \cos \varphi} \cos n\varphi \, d\varphi \tag{1.22.1}$$

We see that the Bessel function $J_n(x)$ is an entire function of x which has the form of the Fourier transform (18.4) if $\cos \varphi$ is introduced as a new variable θ.   Consequently the conditions for the applicability of the interpolation in central differences are fulfilled.

Quite different is the situation with respect to the *order* p of the Bessel functions.   If p is *not* an integer, the definition (1) does not hold, but has to be replaced by the following definition:

$$J_p(x) = \sqrt{\pi} \, \frac{(\tfrac{1}{2}x)^p}{(p - \tfrac{1}{2})!} \, Q_p(x) \tag{1.22.2}$$

where

$$Q_p(x) = \frac{2}{\pi} \int_0^{\pi/2} \cos (x \sin \varphi) \cos^{2p} \varphi \, d\varphi \tag{1.22.3}$$

Now the function $\cos (x \sin \varphi/2)$, considered as a function of φ, is an *even* function of φ and it is *periodic*, with respect to the period 2π.   Such a function can be expanded into a Fourier cosine series:

$$\cos \left(x \sin \frac{\varphi}{2}\right) = \sum_{k=0}^{\infty}{}' c_k \cos k\varphi \tag{1.22.4}$$

where

$$c_k = \frac{1}{\pi} \int_{-\pi}^{\pi} \cos\left(x \sin \frac{\varphi}{2}\right) \cos k\varphi d\varphi$$

$$= \frac{(-1)^k}{\pi} \int_0^{2\pi} \cos\left(x \cos \frac{\varphi}{2}\right) \cos k\varphi d\varphi$$

$$= \frac{(-1)^k 2}{\pi} \int_0^{\pi} \cos\left(x \cos \varphi\right) \cos 2k\varphi d\varphi$$

$$= 2J_{2k}(x) \tag{1.22.5}$$

as we can see from the definition (1) of the Bessel functions, for $n = 2k$.

Hence we obtain the series

$$\cos\left(x \sin \varphi\right) = 2 \sum_{k=0}^{\infty}{}' J_{2k}(x) \cos 2k\varphi \tag{1.22.6}$$

If we substitute this series in (3), and integrate term by term, we obtain

$$Q_p(x) = \sum_{k=0}^{\infty}{}' c_k J_{2k}(x) \tag{1.22.7}$$

where we have put

$$c_k = \frac{4}{\pi} \int_0^{\pi/2} \cos^{2p} \varphi \cos 2k\varphi d\varphi \tag{1.22.8}$$

These integrals are available in closed form:

$$c_k = \frac{2}{\sqrt{\pi}} \frac{(p - \frac{1}{2})! p!}{(p - n)! (p + n)!} \tag{1.22.9}$$

and thus, going back to the original $J_p(x)$ according to (2), we obtain the following interpolation of an arbitrary $J_p(x)$ in terms of the Bessel functions of even order:

$$J_p(x) = \frac{2}{p!} \left(\frac{x}{2}\right)^p \left[\frac{J_0(x)}{2} + \frac{p}{p+1} J_2(x) + \frac{p(p-1)}{(p+1)(p+2)} J_4(x) + \ldots\right]$$

$$= \frac{2}{p!} \left(\frac{x}{2}\right)^p \sum_{n=0}^{\infty}{}' \frac{(p!)^2}{(p - n)! (p + n)!} J_{2n}(x) \tag{1.22.10}$$

This formula can be conceived as a generalisation of the recurrence relation

$$J_1(x) = \frac{x}{2}[J_0(x) + J_2(x)] \tag{1.22.11}$$

which is a special case of (10), for $p = 1$. The series on the right terminates for any integer value of p and expresses the function $J_n(x) x^{-n}$ as a certain *weighted mean* of the Bessel functions of even order, up to $J_{2n}(x)$, with coefficients which are *independent of x*.

Problem 56. What is the maximum tabulation interval $\Delta x = \beta$ for the key-values $J_n(\beta m)$ to allow convergence of the Stirling interpolation? What is the same interval for interpolation by simple differences?
[Answer:

$$\text{a) } \beta = \pi, \qquad \text{b) } \beta = \frac{\pi}{3} \qquad (1.22.12)]$$

Problem 57. Answer the same questions if the tabulated function is e^x.
[Answer:

$$\text{a) } \cosh \beta = 3, \qquad \text{b) } e^\beta = 2 \qquad (1.22.13)]$$

Problem 58. The Harvard Tables* give the following values of the Bessel functions of even order at the point $x = 3.5$:

$$
\begin{aligned}
J_0(3.5) &= -0.3801277400 & J_8(3.5) &= 0.0015430467 \\
J_2(3.5) &= 0.4586291841 & J_{10}(3.5) &= 0.0000560095 \\
J_4(3.5) &= 0.2044052930 & J_{12}(3.5) &= 0.0000013581 \\
J_6(3.5) &= 0.0254289545 & J_{14}(3.5) &= 0.0000000236
\end{aligned} \qquad (1.22.14)
$$

Obtain $J_{3.5}(3.5)$ by interpolation, and compare the result with the correct value. The Bessel functions of half-order are expressible in closed form in terms of elementary functions, in particular:

$$J_{3.5}(x) = \sqrt{\frac{2}{\pi x}}\left[\left(1 - \frac{15}{x^2}\right)\cos x - \left(\frac{6}{x} - \frac{15}{x^3}\right)\sin x\right] \qquad (1.22.15)$$

[Answer: 0.293783539
Correct Value: 0.293783454]

Another aspect of the interpolation properties of the Bessel functions reveals itself if we write the formula (4), (5) in the following form:

$$J_p(x) = \frac{2}{\sqrt{\pi}} \int_0^{\pi/2} \frac{(\frac{1}{2}x \cos^2 \varphi)^p}{(p - \frac{1}{2})!} \cos (x \sin \varphi) d\varphi \qquad (1.22.16)$$

We fix x and consider p as the variable. Then we have an integral transform which has clearly the form (12.10) if we introduce $t = \frac{1}{2}x \cos^2 \varphi$ as a new integration variable and consider $g(t)$ beyond the upper limit $\frac{1}{2}x$ as identically zero. This shows that $J_p(x)$, considered as a function of p, belongs to that class of functions which *allows the application of the Gregory-Newton type of interpolation.* The Bessel functions of non-integer order are thus calculable in terms of the Bessel functions of integer order of the same argument, using the method of simple differences.

Problem 59. Obtain $J_{3.5}(3.5)$ by Gregory-Newton interpolation, using the values of $J_3(3.5), J_4(3.5), \ldots, J_{14}(3.5)$. We complete the table (14) by the following tabular values:

$$
\begin{aligned}
J_1(3.5) &= 0.3867701117 & J_9(3.5) &= 0.0003109276 \\
J_5(3.5) &= 0.0804419866 & J_{11}(3.5) &= 0.0000091267 \qquad (1.22.17) \\
J_7(3.5) &= 0.0067430003 & J_{13}(3.5) &= 0.0000001860
\end{aligned}
$$

* The Annals of the Computation Laboratory of Harvard University (Harvard University Press, 1947).

[Answer: 0.2941956626 (observe the very slow convergence, compared with the result in Problem 58)]

Problem 60. Riemann's zeta-function can be defined by the following definite integral, valid for all $z > 0$:

$$\zeta(z + 1) = \frac{1}{z!} \int_0^\infty \frac{x^z e^{-x}}{1 - e^{-x}} \, dx \tag{1.22.18}$$

a) Show that $\zeta(z + 1)$ has a simple pole at $z = 0$.

b) Show that $z\zeta(z + 1)$ allows in the right half plane the Gregory-Newton type of interpolation.

BIBLIOGRAPHY

[1] Jordan, Ch., *Calculus of Finite Differences* (Chelsea, New York, 1950)
[2] Milne, W. E., *Numerical Calculus* (Princeton University Press, 1949)
[3] Milne-Thomson, L. M., *The Calculus of Finite Differences* (Macmillan, London, 1933)
[4] Whittaker, E. T., and G. Robinson, *The Calculus of Observations* (Blackie & Sons, London, 1924)
[5] Whittaker, J. M., *Interpolatory Function Theory* (Cambridge University Press, 1935)

CHAPTER 2

HARMONIC ANALYSIS

Synopsis. The Fourier series was historically the first example of an expansion into orthogonal functions and retained its supreme importance as the most universal tool of applied mathematics. We study in this chapter some of its conspicuous properties and investigate particularly the "Gibbs oscillations" which arise by terminating the series to a finite number of terms. By the method of the "sigma smoothing" the convergence of the series is increased, due to a reduction of the amplitudes of the Gibbs oscillations. This brings us to a brief investigation of the interesting asymptotic properties of the sigma factors.

2.1. Introduction

In the first chapter we studied the properties of polynomial approximations and came to the conclusion that the powers of x are not well suited to the approximation of equidistant data. A function tabulated or observed at equidistant points does not lend itself easily to polynomial interpolation, even if the points are closely spaced. We have no guarantee that the error oscillations between the points of interpolation will decrease with an increase of the order of the interpolating polynomial. To the contrary, only a very restricted class of functions allows unlimited approximation by powers. If the function does not belong to this special class of functions, the error oscillations will decrease up to a certain point and then increase again.

In marked contrast to the powers are the *trigonometric functions* which we will study in the present chapter. These functions show a remarkable flexibility in their ability to interpolate even under adverse conditions. At the same time they have no "extrapolating" faculty. The validity of the approximation is strictly limited to the *real* range.

The approximations obtainable by trigonometric functions fall into two categories: we may have the function $f(x)$ given in a finite range and our aim may be to find a close approximation—and in the limit representation—with the help of a trigonometric series; or we may have $f(x)$ given in a discrete number of equidistant points and our aim is to construct a well-approximating trigonometric series, in terms of the given discrete data. In the first case the theory of the Fourier series is involved; in the second case, the theory of trigonometric interpolation.

49

The basic theory of harmonic analysis is concerned with the convergence properties of the Fourier series. But in the actual applications of the Fourier series we have to be concerned not only with the convergence of the infinite series but with the *error bounds* of the finite series. It is not enough to know that, taking more and more terms of the series, the error—that is the difference between function and series—tends to zero. We must be able to estimate what the maximum error of the finite expansion is, if we truncate the Fourier series at an arbitrarily given point. We must also have proper estimates in the case of trigonometric interpolation. The present chapter is devoted to problems of this kind.

2.2. The Fourier series for differentiable functions

The elementary theory of the Fourier series proceeds in the following manner. We take the infinite expansion

$$f(x) = \sum_{k=0}^{\infty}{}' a_k \cos kx + \sum_{k=1}^{\infty} b_k \sin kx \qquad (2.2.1)$$

for granted. We assume that this series is valid in the range $[-\pi, +\pi]$, and is uniformly convergent in that range. If we multiply on both sides by $\cos kx$, respectively $\sin kx$ and integrate term by term, we obtain, in view of the orthogonality of the Fourier functions, the well-known expressions

$$a_k = \frac{1}{\pi} \int_{-\pi}^{+\pi} f(x) \cos kx \, dx$$

$$b_k = \frac{1}{\pi} \int_{-\pi}^{+\pi} f(x) \sin kx \, dx \qquad (2.2.2)$$

These coefficients can be constructed if $f(x)$ is merely *integrable*, without demanding differentiability. We do not know yet, however, whether the infinite series (1) thus constructed will truly converge and actually represent $f(x)$. This is in fact not necessarily the case, even if $f(x)$ is everywhere continuous.

Let us, however, assume that $f(x)$ is not only continuous but even sectionally *differentiable* throughout the range which means that $f'(x)$ exists everywhere, although the continuity of $f'(x)$ is not demanded. Furthermore, let us assume the existence of the boundary condition

$$f(\pi) = f(-\pi) \qquad (2.2.3)$$

Then the coefficients (2) (omitting the constant term $\frac{1}{2}a_0$) are expressible with the help of $f'(x)$, by using the method of integrating by parts:

$$a_k = \frac{1}{\pi} \int_{-\pi}^{+\pi} f(\xi) \cos k\xi \, d\xi = \frac{1}{k\pi} \left| f(\xi) \sin k\xi \right|_{-\pi}^{+\pi} - \frac{1}{k\pi} \int_{-\pi}^{+\pi} f'(\xi) \sin k\xi \, d\xi$$

$$\qquad (2.2.4)$$

$$b_k = \frac{1}{\pi} \int_{-\pi}^{+\pi} f(\xi) \sin k\xi \, d\xi = -\frac{1}{k\pi} \left| f(\xi) \cos k\xi \right|_{-\pi}^{+\pi} + \frac{1}{k\pi} \int_{-\pi}^{+\pi} f'(\xi) \cos k\xi \, d\xi$$

The boundary terms vanish—the first automatically, the second in view of the boundary condition (3)—and now, substituting back in the formal series (1) we obtain the infinite sum

$$\frac{1}{\pi}\sum_{k=1}^{\infty}\int_{-\pi}^{+\pi} f'(\xi)d\xi\left[\frac{\cos k\xi}{k}\sin kx - \frac{\sin k\xi}{k}\cos kx\right] \tag{2.2.5}$$

The question is whether or not this infinite sum will converge to $f(x) - \frac{1}{2}a_0$ at all points of the interval.

This calls our attention to the investigation of the following infinite sum which depends on the *two* variables x and ξ but becomes in fact a function of the single variable

$$\xi - x = \theta \tag{2.2.6}$$

alone:

$$\frac{1}{\pi}\sum_{k=1}^{\infty}\frac{\sin kx \cos k\xi - \cos kx \sin k\xi}{k}$$

$$= -\frac{1}{\pi}\sum_{k=1}^{\infty}\frac{\sin k(\xi - x)}{k} = -\frac{1}{\pi}\sum_{k=1}^{\infty}\frac{\sin k\theta}{k} = G_1(\theta) \tag{2.2.7}$$

If this sum converges uniformly, then a term by term integration is permitted and the sum (5) becomes replaceable by the definite integral

$$\int_{-\pi}^{+\pi} f'(\xi)G_1(\xi - x)d\xi \tag{2.2.8}$$

The function $G_1(\xi - x)$ is called the "kernel" of this integral.

Now the simple law of the coefficients of the infinite sum permits us to actually perform the summation and obtain the sum in closed form. The result is as follows (cf. Problem 63):

$$G_1(\theta) = -G_1(-\theta) = -\frac{1}{2} + \frac{\theta}{2\pi} \qquad (0 < \theta < \pi) \tag{2.2.9}$$

The convergence is uniform at all points of θ, excluding only the point of discontinuity $\theta = 0$ where the series gives zero, which is the arithmetic mean of the two limiting ordinates.

We can now proceed to the evaluation of the integral (8). For this and later purposes it is of great convenience to extend the realm of validity of the function $f(x)$ beyond the original range $[-\pi, +\pi]$. We do that by defining $f(x)$ as a *periodic function* of the period 2π:

$$f(x + 2\pi) = f(x) \tag{2.2.10}$$

By this law $f(x)$ is now uniquely determined everywhere. Then the integral (8) can now be put in the following form, introducing $\xi - x = \theta$ as a new integration variable and realising that the integral over a full period can always be normalised to the limits $-\pi, +\pi$:

$$\int_{-\pi}^{+\pi} G_1(\theta)f'(\theta + x)d\theta = \left[G_1(\theta)f(\theta + x) \right]_{-\pi}^{+\pi} - \int_{-\pi}^{\pi} G'_1(\theta)f(\theta + x)d\theta \tag{2.2.11}$$

In the second term $G'_1(\theta)$ can be replaced by the constant $1/2\pi$. In the first term, in view of the discontinuity of $G_1(\theta)$ we have to take the boundary term between $-\pi$ and 0^- and again between 0^+ and π. In view of the periodicity of the boundary term the contribution from the two boundaries at $\pm\pi$ vanishes and what remains becomes

$$[-G_1(0^+) + G_1(0^-)]f(x) = f(x) \tag{2.2.12}$$

Hence

$$\int_{-\pi}^{+\pi} f'(\xi)G_1(\xi - x)d\xi = f(x) - \frac{1}{2\pi}\int_{-\pi}^{+\pi} f(\xi)d\xi$$

$$= f(x) - \tfrac{1}{2}a_0 \tag{2.2.13}$$

We have thus shown that *any continuous and sectionally differentiable function which satisfies the boundary condition* (3), *allows a uniformly convergent Fourier expansion at every point of the range.*

Problem 61.   Let $f(x)$ be defined between 0 and π. How must we define $f(-x)$ if

     a) all cosine terms
     b) all sine terms
     c) all even harmonics
     d) all odd harmonics

shall drop out.

[Answer: a) $f(-x) = -f(x)$
          b) $f(-x) = f(x)$
          c) $f(-x) = -f(\pi - x)$
          d) $f(-x) = f(\pi - x)$]

Problem 62.   What symmetry conditions are demanded of $f(x)$ if we want

     a) a sine series with even harmonics
     b) a sine series with odd harmonics
     c) a cosine series with even harmonics
     d) a cosine series with odd harmonics

[Answer: a) $f(-x) = -f(x)$, $f\left(\dfrac{\pi}{2} + x\right) = -f\left(\dfrac{\pi}{2} - x\right)$

          b) $f(-x) = -f(x)$, $f\left(\dfrac{\pi}{2} + x\right) = f\left(\dfrac{\pi}{2} - x\right)$

c) $f(-x) = f(x)$, $f\left(\dfrac{\pi}{2} + x\right) = f\left(\dfrac{\pi}{2} - x\right)$

d) $f(-x) = f(x)$, $f\left(\dfrac{\pi}{2} + x\right) = -f\left(\dfrac{\pi}{2} - x\right)$]

Problem 63.   Consider the Taylor expansion of $\log (1 - z)$:

$$\log (1 - z) = -z - \frac{z^2}{2} - \frac{z^3}{3} - \cdots \tag{2.2.14}$$

which converges everywhere inside and on the unit circle, excluding the point $z = 1$.   Put $z = e^{it}$ and obtain the infinite sums

$$s_1 = \sum_{k=1}^{\infty} \frac{\cos k\theta}{k} \quad \text{and} \quad s_2 = \sum_{k=1}^{\infty} \frac{\sin k\theta}{k}$$

[Answer:

$$s_1 = -\log 2 - \log \sin \frac{\theta}{2}$$

$$s_2 = \frac{\pi - \theta}{2} \quad (0 < \theta \leq \pi) \tag{2.2.15}]$$

2.3. The remainder of the finite Fourier expansion

To show the uniform convergence of an infinite expansion is not enough. It merely demonstrates that *taking in a sufficient number of terms we can make the difference between $f(x)$ and the nth sum $f_n(x)$ as small as we wish.* It does not answer the more decisive question: *how near are we to $f(x)$ if we stop with a definite $f_n(x)$ where n is not too small but not arbitrarily large either?*   We can answer this question by taking advantage of the favourable circumstance that the kernel $G_1(\xi - x)$ depends on the single variable $\xi - x = \theta$ only.   Let us assume that we can evaluate with sufficient accuracy the infinite sum

$$-\frac{1}{\pi} \sum_{k=n+1}^{\infty} \frac{\sin k\theta}{k} = g_n(\theta) \tag{2.3.1}$$

Then we will immediately possess a definite expression for the remainder of the finite Fourier series

$$f_n(x) = \sum_{k=0}^{n} {}' (a_k \cos kx + b_k \sin kx) \tag{2.3.2}$$

in the following form:

$$\eta_n(x) = f(x) - f_n(x) = \int_{-\pi}^{+\pi} g_n(\theta) f'(\theta + x) d\theta \tag{2.3.3}$$

This integral can now be used for estimation purposes, by replacing the integrand by its absolute value:

$$|\eta_n(x)| \leq |f'(x)|_{\max} \int_{-\pi}^{+\pi} |g_n(\theta)| d\theta \tag{2.3.4}$$

The second factor is quite independent of the function $f(x)$ and a mere numerical constant for every n. Hence we can put

$$\int_{-\pi}^{+\pi} |g_n(\theta)| d\theta = C_n \tag{2.3.5}$$

and obtain the following estimation of the remainder at any point of the range:

$$\eta_n \leq |f'(x)|_{\max} C_n \tag{2.3.6}$$

Our problem is thus reduced to the evaluation of the infinite sum (1). We shall have frequent occasion to find the sum of terms which appear as the product of a periodic function times another function which changes *slowly* as we go from term to term. For example the change of $1/x$ is slow if we go from $1/(n + k)$ to $1/(n + k + 1)$, assuming that n is large. Let us assume that we have to obtain a sum of the following general character:

$$\sum_{k=1}^{\infty} e^{ik\alpha} \varphi(k) \tag{2.3.7}$$

where $\varphi(k)$ changes slowly from k to $k + 1$. Let us integrate around the point $\xi = k$ between $k + \frac{1}{2}$ and $k - \frac{1}{2}$, making use of the fact that $\varphi(k)$ remained practically constant in this interval:

$$\int_{-\frac{1}{2}}^{+\frac{1}{2}} \varphi(k + t) e^{i\alpha(k+t)} dt = e^{i\alpha k} \varphi(k) \frac{\sin \frac{\alpha}{2}}{\frac{\alpha}{2}} \tag{2.3.8}$$

This means that the summation over k is replaceable by the following definite integral:

$$\sum_{k=1}^{\infty} e^{ik\alpha} \varphi(k) = \frac{\alpha}{2 \sin \frac{\alpha}{2}} \int_{\frac{1}{2}}^{\infty} e^{i\alpha t} \varphi(t) dt \tag{2.3.9}$$

Applying this procedure to the series (1) we obtain (in good approximation), for $\theta > 0$:

$$g_n(\theta) = -\frac{1}{\pi} \frac{\theta}{2 \sin \frac{\theta}{2}} \int_{n+\frac{1}{2}}^{\infty} \frac{\sin \theta \xi}{\xi} d\xi$$

$$= -\frac{1}{\pi} \frac{\theta}{2 \sin \frac{\theta}{2}} \int_{(n+\frac{1}{2})\theta}^{\infty} \frac{\sin \xi}{\xi} d\xi$$

$$= -\frac{1}{\pi} \frac{\theta}{\sin \frac{\theta}{2}} \left[\frac{\pi}{2} - \mathrm{Si}(n + \tfrac{1}{2})\theta \right] \tag{2.3.10}$$

where Si (x) is the so-called "sine-integral"

$$\text{Si }(x) = \int_0^x \frac{\sin \xi}{\xi}\, d\xi \tag{2.3.11}$$

(If n is large, we can replace $n + \frac{1}{2}$ by n with a small error.) For estimation purposes it is unnecessary to employ complete accuracy. The expression (10) shows that, except for very small angles, we are almost immediately in the "asymptotic range" of Si (x) where we can put

$$\frac{\pi}{2} - \text{Si }(x) = \frac{\cos x}{x} \tag{2.3.12}$$

Under these circumstances we obtain with sufficient accuracy:

$$2 \int_0^\pi |g_n(\theta)|\, d\theta = \frac{2}{\pi} \frac{\log (n + \frac{1}{2})\pi}{(n + \frac{1}{2})} \tag{2.3.13}$$

and

$$|\eta_n| < 2 \frac{\log (n + \frac{1}{2})\pi}{(n + \frac{1}{2})\pi} \, |f'(x)|_{\max} \tag{2.3.14}$$

Problem 64. If the summation on the left side of (3.9) extends only to $k = n$, the upper limit of the integral becomes $n + \frac{1}{2}$. Derive by this integration method the following trigonometric identities, and check them by the sum formula of a geometrical series:

$$\sum_{k=0}^n{}' \cos kx = \frac{\sin (n + \frac{1}{2})x}{2 \sin \frac{x}{2}} \tag{2.3.15}$$

$$\sum_{k=1}^n \sin kx = \frac{\sin \frac{n+1}{2} x \sin \frac{n}{2} x}{\sin \frac{x}{2}} \tag{2.3.16}$$

$$\sum_{k=1}^n \cos (k - \frac{1}{2})x = \frac{\sin nx}{2 \sin \frac{x}{2}} \tag{2.3.17}$$

$$\sum_{k=1}^n \sin (k - \frac{1}{2})x = \frac{\sin^2 \frac{n}{2} x}{\sin \frac{x}{2}} \tag{2.3.18}$$

Problem 65. Prove that the "mean square error"

$$\eta^2 = \frac{1}{2\pi} \int_{-\pi}^{+\pi} \eta^2{}_n(x)dx \tag{2.3.19}$$

of the finite Fourier expansion is in the following relation to the Fourier coefficients:

$$\eta^2 = \frac{1}{2} \sum_{k=1}^\infty (a_{n+k}{}^2 + b_{n+k}{}^2) \tag{2.3.20}$$

Problem 66. Evaluate the mean square error of the Fourier series (2.7) and prove that, while the maximum of the local error $\eta_n(x)$ remains constantly $\frac{1}{2}$, the mean square error converges to zero with $n^{-1/2}$.

[Answer:

$$\eta^2 = \frac{1}{2\pi^2(n + \frac{1}{2})} \qquad (2.3.21)]$$

2.4. Functions of higher differentiability

It may happen that $f(x)$ belongs to a class of functions of still higher differentiability. Let us assume that $f(x)$ is twice differentiable, although the continuity of $f''(x)$ is not assumed and thus the continuity of $f'(x)$ and *sectional* existence of $f''(x)$ suffices. We now demand the two boundary conditions:

$$\begin{aligned} f(\pi) &= f(-\pi) \\ f'(\pi) &= f'(-\pi) \end{aligned} \qquad (2.4.1)$$

In fact, we will immediately proceed to the general case, in which the existence of the mth derivative is assumed, coupled with the boundary conditions:

$$\begin{aligned} f(\pi) &= f(-\pi) \\ f'(\pi) &= f'(-\pi) \\ &\vdots \\ f^{(m-1)}(\pi) &= f^{(m-1)}(-\pi) \end{aligned} \qquad (2.4.2)$$

(The existence of $f^{(m)}(x)$ *without* these boundary conditions is of no avail since we have to extend $f(x)$ beyond the original range by the periodicity condition (2.10). Without the conditions (2) the mth derivative would fail to exist at the point $x = \pm \pi$.)

We can again follow the reasoning of the previous section, the only difference being that now the integration by parts (2.4) can be repeated, and applied m times. We thus obtain the coefficients a_k, b_k in terms of the mth derivative. It is particularly convenient to consider the combination $a_k + ib_k$ and use the trigonometric functions in complex form:

$$a_k + ib_k = \frac{1}{\pi} i^m \int_{-\pi}^{+\pi} \frac{f^{(m)}(\xi)e^{ik\xi}d\xi}{k^m} \qquad (2.4.3)$$

(We omit $k = 0$ since it is always understood that our $f(x)$ is the modified function $f(x) - \frac{1}{2}a_0$ which has no area.) We can write even the *entire Fourier series* in complex form, namely

$$f(x) = \sum_{k=1}^{\infty} (a_k + ib_k)e^{-ikx} \qquad (2.4.4)$$

with the understanding that we keep only the *real* part of this expression. With this convention we can once more put

$$f(x) = \int_{-\pi}^{+\pi} f^{(m)}(\xi)G_m(\xi - x)d\xi$$

$$= \int_{-\pi}^{+\pi} f^{(m)}(\theta + x)G_m(\theta)d\xi \tag{2.4.5}$$

where

$$G_m(\theta) = \frac{1}{\pi} i^m \sum_{k=1}^{\infty} \frac{e^{ik\theta}}{k^m} \tag{2.4.6}$$

Once more our aim will be to obtain an error bound for the finite expansion (3.2) and for that purpose we can again put

$$\eta_n(x) = \int_{-\pi}^{+\pi} f^{(m)}(\theta + x)g_n{}^m(\theta)d\theta = \int_{-\pi}^{+\pi} f^{(m)}(\xi)g_n{}^m(\xi - x)d\xi \tag{2.4.7}$$

where $g_n{}^m(\theta)$ is now defined by the real part of the infinite sum

$$g_n{}^m(\theta) = \frac{1}{\pi} i^m \sum_{k=n+1}^{\infty} \frac{e^{ik\theta}}{k^m} \tag{2.4.8}$$

The method of changing this sum to an integral is once more available and we obtain

$$g_n{}^m(\theta) = \frac{\theta}{2\pi \sin \frac{\theta}{2}} i^m \int_{n+\frac{1}{2}}^{\infty} \frac{e^{i\theta\xi}}{\xi^m} d\xi$$

$$= \frac{\theta}{2\pi \sin \frac{\theta}{2}} i^m \theta^{m-1} \int_{(n+\frac{1}{2})\theta}^{\infty} \frac{e^{i\xi}}{\xi^m} d\xi \tag{2.4.9}$$

Again we argue that with the exception of a very small range around $\theta = 0$ the asymptotic stage is quickly reached and here we can put

$$\int_x^{\infty} \frac{e^{i\xi}}{\xi^m} d\xi = \frac{ie^{ix}}{x^m} \tag{2.4.10}$$

But then, repeating the argument of the previous section, we get for not too small n:

$$2 \int_0^{\pi} |g_n{}^m(\theta)|d\theta < \frac{2}{(n + \frac{1}{2})^{m-1}} \frac{\log (n + \frac{1}{2})\pi}{(n + \frac{1}{2})\pi} \tag{2.4.11}$$

and

$$|\eta_n{}^m(x)| < \frac{2}{(n + \frac{1}{2})^{m-1}} \frac{\log (n + \frac{1}{2})\pi}{(n + \frac{1}{2})\pi} |f^{(m)}(x)|_{\max} \tag{2.4.12}$$

A simpler method of estimation is based on "Cauchy's inequality"

$$\left[\int_a^b f(x)g(x)dx\right]^2 \le \int_a^b f^2(x)dx \int_a^b g^2(x)dx \tag{2.4.13}$$

which avoids the use of the absolute value. Applying this fundamental inequality to the integral (7) we can make use of the orthogonality of the Fourier functions and obtain the simple expression (which holds *exactly* for all n):

$$\int_{-\pi}^{+\pi} [g_n{}^m(\theta)]^2 d\theta = \frac{1}{\pi} \sum_{k=n+1}^{\infty} \frac{1}{k^{2m}} \qquad (2.4.14)$$

Changing this sum to an integral we obtain the close approximation

$$\sum_{k=n+1}^{\infty} \frac{1}{k^{2m}} = \frac{1}{(2m-1)(n+\frac{1}{2})^{2m-1}} \qquad (2.4.15)$$

and thus we can deduce the following estimation for the local error $\eta_n(x)$ at any point of the range:

$$|\eta_n(x)| < \frac{N_m}{\sqrt{(2m-1)\pi}(n+\frac{1}{2})^{m-\frac{1}{2}}} \qquad (2.4.16)$$

where N_m is the so-called "norm" of the mth derivative of $f(x)$:

$$N_m{}^2 = \int_{-\pi}^{+\pi} f^{(m)2}(x) dx \qquad (2.4.17)$$

Problem 67.   Show that the approximation (15) is "safe" for estimation purposes because the sum on the left side is always *smaller* than the result of the integration given on the right side.

Problem 68.   Prove the following inequalities: for any $f(x)$ which satisfies the boundary conditions (4.2) and whose total area is zero:

$$a_k{}^2 + b_k{}^2 \leq \frac{1}{\pi} \frac{1}{k^{2m}} \int_{-\pi}^{+\pi} f^{(m)2}(x) dx \qquad (2.4.18)$$

$$\frac{1}{2\pi} \int_{-\pi}^{+\pi} f^2(x) dx \leq \frac{\beta_m}{2\pi} \int_{-\pi}^{+\pi} f^{(m)2}(x) dx \qquad (2.4.19)$$

where β_m is a numerical constant, defined by

$$\beta_m = \sum_{k=1}^{\infty} \frac{1}{k^{2m}} = \frac{\pi^{2m} 2^{2m-1}}{(2m)!} B_{2m} \qquad (2.4.20)$$

The B_{2m} are the Bernoulli numbers: $\frac{1}{6}, \frac{1}{30}, \frac{1}{42}, \frac{1}{30}, \ldots$ (starting with $m = 1$).

2.5. An alternative method of estimation

If we consider the expression (4.7), we notice that the remainder of a finite Fourier series appears as a definite integral over the product of two factors. The one is the mth derivative of the given $f(x)$, the other is a function which is independent of θ for each given n. Our policy has been to estimate the error on the basis that we took the integral over the absolute value of $g_n(\theta)$ which is a mere constant, depending on n, say C_n. This C_n multiplied by the maximum value of $|f^{(m)}(x)|$ gave us an upper bound for the remainder.

We may reverse, however, our procedure by *exchanging* the role of the two factors. Under certain circumstances we may fare better by integrating over the absolute value of $f'(x)$, and multiplying by the maximum of $|g_n(\theta)|$. It may happen, for example, that $f'(x)$ becomes infinite at some point of the range, while the integral over $|f'(\theta)|$ remains finite. In such a case it is clear that the second method will be preferable to the first.

What can we say about the maximum of $|g_n(\theta)|$? If for the moment we skip the case of $m = 1$ and proceed to the case of a twice or more differentiable function, considered in Section 4, we arrive at the function (4.8) which has its maximum at $\theta = 0$. By changing the sum to an integral we obtain with sufficient accuracy:

$$|g_n{}^m(\theta)| < \frac{1}{\pi} \int_{n+\frac{1}{2}}^{\infty} \frac{d\xi}{\xi^m} = \frac{(n + \frac{1}{2})^{-m-1}}{\pi(m - 1)} \qquad (2.5.1)$$

which yields the estimation

$$|\eta_n{}^m| < \frac{1}{\pi(m - 1)n^{m-1}} \int_{-\pi}^{+\pi} |f^{(m)}(x)|dx \qquad (2.5.2)$$

However, in the case of a function which is only *once* differentiable, we do not succeed by a similar method because we lose the factor n in the denominator and the error no longer goes to zero. The kernel $G_1(\theta)$—as we have seen in (2.11)—has a point of discontinuity at $\theta = 0$ with which the successive approximations are unable to cope. Hence the maximum of $g_n{}^1(\theta)$ no longer goes to zero with increasing n but remains constantly $\frac{1}{2}$.

And yet even here we succeed if we use the proper precaution. We divide the range of integration into two parts, namely the realm of small θ and the realm of large θ. In particular, let us integrate between $\theta = 0$ and $\theta = 1/\sqrt{n}$ and then between $\theta = 1/\sqrt{n}$ and $\theta = \pi$; on the negative side we do the same. Now in the second realm the function (4) has already attained its asymptotic value and its maximum is available:

$$\frac{1}{\pi} \left| \frac{\pi}{2} - \text{Si} \sqrt{n} \right|_{\max} = \frac{1}{\pi\sqrt{n}} \qquad (2.5.3)$$

In the central section we use the maximum $\frac{1}{2}$. In this way we obtain as an error bound for

$$\eta_n{}^1(x) = \int_{-\pi}^{+\pi} g_n{}^1(\theta)f'(x + \theta)d\theta:$$

$$|\eta_n{}^1(x)| < \frac{1}{2} \int_0^{1/\sqrt{n}} |f'(x + \theta) - f'(x - \theta)|d\theta + \frac{1}{\pi\sqrt{n}} \int_{-\pi}^{+\pi} |f'(\theta)|d\theta \qquad (2.5.4)$$

The first integral is small because the range of integration is small. The second integral is small on account of the $\sqrt{n}$ in the denominator.

The estimation (4) is, of course, more powerful than the previous estimation (3.14), although the numerical factor C_n was smaller in the previous case.

Even a *jump* in the function $f(x)$ is now permitted which would make $f'(x)$ infinite at the point of the jump, but the integral

$$\int_{-\pi}^{+\pi} |f'(\theta)|d\theta \qquad (2.5.5)$$

remains finite. We see, however, that in the *immediate vicinity* of the jump we cannot expect a small error, on account of the first term which remains finite in this vicinity. We also see that under such conditions the estimated error decreases *very slowly* with n.

We fare better in such a case if we first *remove* the jump in the function by adding to $f(x)$ the special function $G_1(x - x_1)$, multiplied by a proper factor α. Since the function $\alpha G_1(x - x_1)$ makes the jump $-\alpha$ at the point $x = x_1$, we can compensate for the jump α of the function $f(x)$ at $x = x_1$ and reduce $f(x)$ to a new function $\varphi(x)$ which is free of any discontinuities. For the new function the more efficient estimation of Section 3 (cf. 3.14) can be employed, while the remainder of the special function $\alpha G_1(x - x_1)$ is explicitly at our disposal and can be considered separately.

2.6. The Gibbs oscillations of the finite Fourier series

If $f(x)$ is a truly periodic and analytical function, it can be differentiated any number of times. But it happens much more frequently that the Fourier series is applied to the representation of a function $f(x)$ which is given only between $-\pi$ and $+\pi$, and which is made artificially periodic by extending it beyond the original range. Then we have to insure the boundary conditions (4.2) by artificial means and usually we do not succeed beyond a certain m. This means that we have constructed a periodic function which is m times differentiable but the mth derivative becomes discontinuous at the point $x = x_1$. Under such conditions we can put this lack of continuity to good advantage for an efficient estimation of the remainder of the finite series and obtain a very definite picture of the manner in which the truncated series $f_n(x)$ approximates the true function $f(x)$.

We will accordingly assume that $f(x)$ possesses all the derivatives up to the order m, but $f^{(m)}(x)$ becomes discontinuous at a certain point $x = x_1$ of the range (if the same occurs in several points, we repeat our procedure for each point separately and obtain the resulting error oscillations by superposition). Now the formula (4.7) shows that it is not $f^{(m)}(x)$ in itself, but the *integral* over $f^{(m)}(x)$ which determines the remainder $\eta_n(x)$ of the truncated series. Hence, instead of stopping with the mth derivative, we could proceed to the $m + 1^{st}$ derivative and consider the jump in the mth derivative as a jump in the integral of the $m + 1^{st}$ derivative. This has the consequence that the major part of the integral which determines $\eta_n(x)$, is reducible to the immediate neighbourhood of the point $x = x_1$. The same will happen in the case of a function whose $m + 1^{st}$ derivative does not become necessarily infinite but merely *very large*, if compared with the values in the rest of the range.

Since our function $f(x)$ became periodic by extending it beyond the original range of definition, we can shift the origin of the period to any point $x = x_1$. Hence we do not lose in generality but gain in simplicity if we place the point of infinity of the $m + 1$st derivative into the point $x = 0$. The integration in the immediate vicinity of the point $\xi = 0$ gives (cf. (4.7)):

$$g_n{}^{m+1}(-x) \int_{-\epsilon}^{+\epsilon} f^{(m+1)}(\xi)d\xi = g_n{}^{m+1}(-x)[f^{(m)}(\epsilon) - f^{(m)}(-\epsilon)] \quad (2.6.1)$$

The second factor is the jump A of the mth derivative at the point $x = 0$:

$$f^{(m)}(0_+) - f^{(m)}(0_-) = A \quad (2.6.2)$$

Moreover, since $f^{(m+1)}(\xi)$ is generally regular and becomes so extreme only in the vicinity of $x = 0$ (the same holds if $f^{(m+1)}(\xi)$ has merely a strong maximum at $x = 0$), we can consider the remaining part of the integral as *small*, compared with the contribution of the neighbourhood of $\xi = 0$. But then the integral (4.7)—replacing m by $m + 1$—becomes reduced to

$$\eta_n(x) = Ag_n{}^{m+1}(-x) \quad (2.6.3)$$

We will normalise the magnitude of the jump to 1, since the multiplication by a constant can be left to the end. Our aim will be to pay closer attention to the integral (4.9) which is a sufficiently close approximation of the sum (4.8). We will try to obtain a satisfactory approximation of this integral in terms of elementary functions. Replacing θ by x, the integral we want to approximate may be written as follows:

$$\int_{n+\frac{1}{2}}^{\infty} \frac{e^{ix\xi}}{\xi^{m+1}} d\xi = e^{i[n+(1/2)]x} \int_0^{\infty} \frac{e^{ix\xi}d\xi}{(n + \frac{1}{2} + \xi)^{m+1}} \quad (2.6.4)$$

Let us now consider the following function of ξ:

$$y(\xi) = \frac{e^{ix\xi}}{(n' + \xi)^m[ix(n' + \xi) - m - 1]} \quad (2.6.5)$$

(n' stands for $n + \frac{1}{2}$). Differentiating logarithmically we obtain

$$y'(\xi) = \frac{e^{ix\xi}}{(n' + \xi)^m} \frac{ix - \dfrac{m}{n' + \xi} - \dfrac{ix}{ix(n' + \xi) - m - 1}}{ix - \dfrac{m + 1}{n' + \xi}} \quad (2.6.6)$$

The last term in the numerator is very nearly $-1/(n' + \xi)$, on account of the largeness of n'. We fail only in the domain of very small x but even there the loss is not too serious if we exclude the case $m = 0$ which we will consider separately. But then the effect of this substitution is that the m of the previous term changes to $m + 1$, with the consequence that now numerator and denominator cancel out and the second factor becomes 1. The resulting expression is now exactly the integrand of (4). We have thus succeeded with the integration and have merely to substitute the limits 0 and ∞, obtaining

$$-\frac{1}{(n + \frac{1}{2})^m} \frac{1}{ix(n + \frac{1}{2}) - m - 1} \tag{2.6.7}$$

which yields for $g_n(x)$ the expression

$$\eta_n(-x) = g_n^{m+1}(x) = \frac{x}{2\pi \sin (x/2)} i^{m+1} \frac{e^{i[n+(1/2)]x}}{(n + \frac{1}{2})^m} \frac{1}{m + 1 - ix(n + \frac{1}{2})} \tag{2.6.8}$$

Only the real part of this expression must be taken, for *positive* x. The transition to negative x occurs according to the following rules:

$$g_n^{m+1}(-x) = g_n^{m+1}(x) \qquad \text{if } m \text{ is } odd$$
$$g_n^{m+1}(-x) = -g_n^{m+1}(x) \qquad \text{if } m \text{ is } even \tag{2.6.9}$$

Let us assume that m is *odd*: $m = 2s + 1$. Then

$$i^{m+1} = (-1)^{s+1} \tag{2.6.10}$$

and the real part of (8) yields the even function

$$\eta_n(x) = \frac{x}{2\pi \sin (x/2)} \frac{(-1)^{s+1}}{(n + \frac{1}{2})^m} \frac{\cos [(n + \frac{1}{2})x + \alpha]}{\sqrt{(m + 1)^2 + (n + \frac{1}{2})^2 x^2}} \tag{2.6.11}$$

where

$$\tan \alpha = \frac{(n + \frac{1}{2})x}{m + 1} \tag{2.6.12}$$

On the other hand, if m is *even*: $m = 2s$, the real part of (8) yields the odd function

$$\eta_n(x) = \frac{(-1)^s}{(n + \frac{1}{2})^m} \frac{x}{2\pi \sin (x/2)} \frac{\sin [(n + \frac{1}{2})x + \alpha]}{\sqrt{(m + 1)^2 + (n + \frac{1}{2})^2 x^2}} \tag{2.6.13}$$

    The formulae (11) and (13) demonstrate in explicit form the remarkable manner in which the truncated series $f_n(x)$ (which terminates with the terms $\sin nx$, $\cos nx$), approximates the true function $f(x)$. The approximation winds itself around the true function in the form of high frequency oscillations (frequently referred to as the "Gibbs oscillations"), which are superimposed on the smooth course of $f(x)$ [by definition $f_n(x) = f(x) - \eta_n(x)$]. These oscillations appear as of the angular frequency $(n + \frac{1}{2})$, with slowly changing phase and amplitude. The phase α starts with the value 0 at $x = 0$ and quickly increases to nearly $\pi/2$, if n is not too small. Accordingly, the nodal points of the sine-oscillations and the maxima-minima of the cosine oscillations are near to the points

$$x_k = \frac{k + \frac{1}{2}}{n + \frac{1}{2}} \pi$$

These points divide the interval between 0 and π into $n + 1$ nearly equal sections.

The amplitude of the Gibbs oscillations decreases slowly as we proceed from the point $x = 0$—where the break in the mth derivative occurs—towards the point $x = \pi$, where the amplitude is smallest. The decrease is slow, however, and the order of the amplitudes is constantly of the magnitude

$$\frac{A}{2\pi(n + \frac{1}{2})^{m+1}} \qquad (2.6.14)$$

where A is the discontinuity of the mth derivative (cf. (2)). Only in the immediate vicinity of the point $x = 0$ are the oscillations of slightly larger amplitude.

We see that the general phenomenon of the Gibbs oscillations is independent of the order m of the derivative in which the discontinuity occurs. Only the *magnitude* of the oscillations is strongly diminished as m increases. But the slow change in amplitude and phase remains of the same character, whatever m is, provided that n is not too small relative to m.

The *phase-shift* between even and odd m—sine vibrations in the first case, cosine vibrations in the second—is also open to a closer analysis. Let us write the function $f(x)$ as the arithmetic mean of the even function $g(x) = f(x) + f(-x)$ and the odd function $h(x) = f(x) - f(-x)$. Now the Fourier functions of an even function are pure cosines, those of an odd function pure sines. Hence the remainder $\eta_n(x)$ shares with the function its even or odd character. Furthermore, the behaviour of an even, respectively odd function in the neighbourhood of $x = 0$ is such that if an even derivative becomes discontinuous at $x = 0$, the discontinuity must belong to $h(x)$. On the other hand, if an odd derivative becomes discontinuous at $x = 0$, that discontinuity must belong to $g(x)$. In the first case $g(x)$ is smooth compared with $h(x)$, in the second $h(x)$ is smooth compared with $g(x)$. Hence in the first case the cosine oscillations of the remainder are negligible compared with the sine oscillations, while in the second case the reverse is true. And thus the discontinuity in an even derivative at $x = 0$ makes the error oscillations to an *odd* function, the discontinuity in an odd derivative to an *even* function.

We can draw a further conclusion from the formulae (11) and (13). If s is even and thus m of the form $4\mu + 1$, the error oscillations will start with a *minimum* at $x = 0$, while if s is odd and thus m of the form $4\mu + 3$, with a *maximum*. Consequently the arrow which goes from $f(0)$ to $f_n(0)$, points *in* the direction of the break if that break occurs in the first, fifth, ninth, ... derivative, and *away* from the break if it occurs in the third, seventh, eleventh, ... derivative. Similarly, if the break occurs in the second, sixth, tenth, ... derivative, the tangent of $f_n(x)$ at $x = 0$ is directed *towards* the break; if it occurs in the fourth, eighth, twelfth, ... derivative, *away* from the break (see Figure).

The case $m = 0$. If the discontinuity occurs in the function itself, we have the case $m = 0$. Here the formula (8) loses its significance in the realm of small x. On the other hand, we have obtained the $g_n^1(x)$ of formula

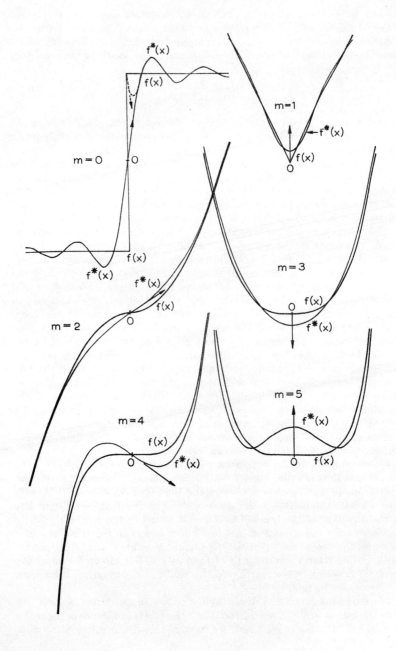

(4.8) (which belongs to our case $m = 0$), at an earlier occasion, in Section 3 (cf. 3.10), with the following result:

$$g_n^1(-x) = -g_n^1(x) = \frac{1}{\pi} \frac{x}{2 \sin (x/2)} \left[\frac{\pi}{2} - \text{Si } (n + \tfrac{1}{2})x \right] \qquad (2.6.15)$$

and thus

$$\eta_n(x) = -\eta_n(-x) = \frac{A}{\pi} \frac{x}{2 \sin (x/2)} \left[\frac{\pi}{2} - \text{Si } (n + \tfrac{1}{2})x \right] \qquad (2.6.16)$$

where A is the discontinuity of the function $f(x)$ at the point $x = 0$:

$$A = f(0_+) - f(0_-) \qquad (2.6.17)$$

In this instance we have an error pattern which is in apparent contradiction to the error pattern observed for general values of m. Since $m = 0$ is an even number, we have to subordinate our case to the sine vibrations of the formula (13). There we have seen that the tangent of $f_n(x)$ at the point of discontinuity is oriented alternately *toward* and *away* from the discontinuity. Since $m = 2$ realises the "toward" case, $m = 0$ should realise the "away" case. And yet, $f_n(x)$ starts its course *in the direction* of the break, rather than away from it.

A closer analysis reveals that even here the contradiction is only apparent. At the very beginning, during the steep ascent of $f_n(x)$, we cannot speak of "Gibbs oscillations" since these oscillations develop only *after* the first nodal point at $nx = 1.9264$ had been reached. If we continue these oscillations backwards to the point $x = 0$, we see that once more the tangent of the sine oscillation points *away* from the jump, in full agreement with the general behaviour of the Gibbs oscillations, established for arbitrary m. The discontinuity in the function merely adds to these oscillations a single steep peak in the neighbourhood of the singularity which is of a non-oscillatory character and is superimposed on the regular pattern of the Gibbs oscillations.

Although the discussions of the present section were devoted to the case of a single break in the mth derivative, actually the Gibbs oscillations of a large class of functions behave quite similarly. A function may not show any discontinuity in its higher derivatives. It is frequently possible, however, to approximate this function quite effectively by local polynomials which fit together smoothly. If we differentiate our approximation, we notice that the higher derivatives become less and less smooth and a certain derivative of the order m becomes composed of mere delta functions. This means that the Gibbs oscillations of this function can be studied by superimposing a relatively small number of oscillation patterns of the type we have exhibited in this section. The resulting pattern will once more show the same features of a fundamentally constant frequency with amplitudes and phases which are nearly constant.

The fact that the Gibbs oscillations are of fairly constant periods, has the following interesting and important consequence. We do not commit any serious change of $f_n(x)$ if we shift the nodal points of its error oscillations into *exactly equidistant* points. Then we no longer have our original truncated Fourier series $f_n(x)$ but another trigonometric series with slightly modified coefficients whose sum, however, does not give anything very different from what we had before. Now this new series has a very definite significance. It is obtainable by the process of *trigonometric interpolation* because we can interpret the new series as a trigonometric expansion of n terms which has zero error in n equidistantly prescribed points, in other words, which fits exactly the functional values of $f(x)$ at the "points of interpolation". This is no longer a problem in infinite series but the algebraic problem of solving n linear equations for n unknowns; it is solvable by mere summation, without any integration. In particular we may prescribe the points of interpolation very near to the nodal points of the Gibbs oscillations if we agree that the *odd* part of the function shall be given in the points

$$x_k = \frac{\pi}{n} k \qquad (k = 1, 2, \ldots, n - 1) \tag{2.6.18}$$

and the *even* part of the function in the points

$$x_k = \frac{\pi}{2n} (2k + 1) \qquad (k = 0, 1, 2, \ldots, n - 1) \tag{2.6.19}$$

The approximating series thus obtained have strictly *equidistant* Gibbs oscillations, with *amplitude* modulations, but without *phase* modulations (cf. Section 18).

Problem 69. Study the Gibbs oscillations by expanding the function

$$f(x) = x(\pi^2 - x^2) \tag{2.6.20}$$

in a sine series of 12 terms. The discontinuity appears here in the *second* derivative at $x = \pm\pi$. Compare the results with the predictions according to the formula (13) (multiplied with the magnitude of the jump).

Problem 70. Study the Gibbs oscillations by expanding the function

$$f(x) = \pi^2 x^2 - \tfrac{1}{2} x^4 \tag{2.6.21}$$

in a cosine series of 12 terms. The discontinuity appears here in the *third* derivative at $x = \pm\pi$.

2.7. The method of the Green's function

Although we have not mentioned it explicitly, the method we have employed in the various remainder estimations is closely related to the method followed in Chapter 1.5, when we were interested in the problem of Lagrangian interpolation and wanted to estimate the closeness of the approximation. It is a method which plays a fundamental role in the solution of differential equations (as we will see later in Chapter 5), and goes under the name of solving a differential equation by means of the "Green's function".

If a function $f(x)$ is m times differentiable, we can consider $f(x)$ as the solution of the differential equation

$$f^{(m)}(x) = \rho(x) \tag{2.7.1}$$

where on the right side we have the given m^{th} derivative of the function. This, however, is not enough for a unique characterisation of $f(x)$ since a differential equation of mth order demands m additional boundary conditions to make the problem unique. In the Lagrangian case we succeeded in characterising the remainder itself by the differential equation (1) and the added boundary conditions—they were in fact inside conditions—followed from the added information that the remainder vanishes at the points of interpolation. In our present problem we will not proceed immediately to the remainder of the Fourier series but stay first with the function $f(x)$ itself. We add as boundary conditions the periodicity conditions (4.2) which are demanded by the nature of the Fourier series. With these added conditions the function $f(x)$ is now uniquely determined, except for an additional constant which is left undetermined. We eliminate this freedom by adding one more condition, namely

$$\int_{-\pi}^{+\pi} f(x)dx = 0 \tag{2.7.2}$$

(This condition is justified since we can always replace $f(x)$ by $f(x) - \tfrac{1}{2}a_0$ which indeed satisfies the condition (2).)

Now under these conditions—as we will prove later—we can solve the given problem in terms of the "Green's function" $G(x, \xi)$:

$$f(x) = \int_{-\pi}^{+\pi} \rho(\xi)G(x, \xi)d\xi \tag{2.7.3}$$

where the auxiliary function $G(x, \xi)$ is quite independent of $f(x)$ and constructed according to certain rules. In our present case it so happens that this function is reducible to a *function of a single variable* because we can show that in fact $G(x, \xi) = G(\xi - x)$.

The operation with the Green's function has the following great advantage. We want to study the degree of approximation obtainable by a given set of functions, such as for example the Fourier functions. Then the resolution (3) has the consequence that we can completely concentrate on the study of the *special function* $G(x, \xi)$, instead of dealing with the general function $f(x)$. If we know how close we can come in approximating $G(x, \xi)$, we also know what can be expected for the general function $f(x)$ because, having the remainder $g_n(x, \xi)$ for the Green's function $G(x, \xi)$, we at once obtain the remainder for $f(x)$, in form of the definite integral

$$\eta_n(x) = \int_{-\pi}^{+\pi} \rho(\xi)g_n(x, \xi)d\xi$$

$$= \int_{+\pi}^{+\pi} f^{(m)}(\xi)g_n(\xi - x)d\xi \tag{2.7.4}$$

This is in fact how we derived earlier Lagrange's remainder formula (1.5.10). But in the Lagrangian case we could take advantage of the fact that $g_n(x, \xi)$, considered as a function of ξ, did not change its sign in the entire interval. Hence it was not necessary to take the absolute value of $g_n(x, \xi)$ and the simple integral

$$\int_a^b g_n(x, \xi)d\xi$$

could be obtained in closed analytical form. This is not the case now because in the case of the Fourier series the remainder of the Green's function has an *oscillatory* character. In order to estimate the value of (4) on the basis of the maximum of $f^{(m)}(\xi)$ we have to replace $g_n(\xi - x)$ by $|g_n(\xi - x)|$ which makes the evaluation of the numerical factor in the remainder formula (4.12) difficult but does not interfere with the fact that an effective estimation of the error is possible. Moreover, the Lagrangian remainder formula (1.3.7) operates specifically with the nth derivative while in our case the order of the derivative m and the order n of the approximating Fourier series are *independent* of each other.

Problem 71. Derive the Green's function $G_2(\theta)$ of (4.6) (for the case $m = 2$) from the $G_1(\theta)$ of Section 2 (cf. 2.9) on the basis that the infinite sum (4.6) which defines $G_2(\theta)$ is the negative integral of the sum (2.7), together with the validity of the condition (2).

[Answer:

$$G_2(\theta) = G_2(-\theta) = -\frac{\pi}{6} + \frac{\theta}{2} - \frac{\theta^2}{4\pi} \tag{2.7.5}]$$

2.8. Non-differentiable functions. Dirac's delta function

We had no difficulty in proving the convergence of the Fourier series for functions which could be differentiated at least *once*. We could also give good error estimates for such functions, in terms of the highest existing derivative. However, the function $f(x)$ may not have a derivative at every point of the range. For example the function $G_1(\theta)$, encountered in Section 2, had a point of discontinuity at the point $\theta = 0$. It we would insist in treating this function as differentiable, we would have to consider the derivative at $\theta = 0$ as infinite and thus the estimation of the error on the basis of the maximum value of $|f'(x)|$ goes out of bound. And yet, this function possesses a Fourier series which converges to $f(x)$ at every point x, even at the point of discontinuity, if we agree that at such a point the value of $f(x)$ shall be defined by the *arithmetic mean* of the two limiting ordinates. How can we estimate the remainder of the Fourier series for an $f(x)$ of this type? The Green's function method is not applicable here since $f(x)$, if not differentiable, cannot be considered as the solution of a differential equation.

The method employed in Section 2 is applicable even without the device of integrating by parts, if we stay with a *finite* expansion. The truncated

Fourier series (3.2), which does not give $f(x)$ but merely defines a certain $f_n(x)$ associated with $f(x)$, can be written in the following form

$$f_n(x) = \int_{-\pi}^{+\pi} f(\xi) K_n(\xi, x) d\xi \qquad (2.8.1)$$

where

$$K_n(\xi, x) = \frac{1}{\pi} \sum_{k=0}^{n}{}' (\cos n\xi \cos nx + \sin n_n\xi \sin nx)$$

$$= \frac{1}{\pi} \sum_{k=0}^{n}{}' \cos k(\xi - x) = K_n(\xi - x) \qquad (2.8.2)$$

According to the trigonometric identity (3.15) we can obtain the function $K_n(\xi - x)$, called the "Dirichlet kernel", in closed form:

$$K_n(\theta) = \frac{\sin (n + \frac{1}{2})\theta}{2\pi \sin \frac{\theta}{2}} \qquad (2.8.3)$$

At any fixed value of x and ξ—and thus also of $\theta = \xi - x$—this function remains *finite* (excluding the point $x = \xi$), no matter how large n may become. However, the function does not approach any limit as n increases but keeps oscillating within the same bounds for ever. It is therefore difficult to apply the function $K_n(\xi - x)$ to an efficient estimation of the error of the Fourier series. We will emphasise specifically that the convergence of the series (2) is *not* demanded for the convergence of the Fourier series. The latter demands only that the integral (1) shall converge. But it is entirely possible that the integration over ξ, with the weight factor $f(\xi)$, smooths out the strong Gibbs oscillations of the kernel $K(\xi, x)$ and yields a definite limit. Hence we should not consider $K_n(\xi, x)$ as a function but rather as an *operator*, in particular as an *integral operator* which has to operate on the function $f(\xi)$ in the sense of a term by term integration. In this sense it is entirely legitimate to write down the infinite sum

$$K(\xi - x) = \frac{1}{\pi} \sum_{k=0}^{\infty}{}' \cos k(\xi - x) \qquad (2.8.4)$$

which has no meaning as a *value*, but has meaning as an *operator*. The equation

$$f(x) = \int_{-\pi}^{+\pi} f(\xi) K(\xi - x) d\xi \qquad (2.8.5)$$

can be considered as a correct operational equation for all functions $f(x)$ which allow a Fourier expansion.

In the literature of modern physics this cautious approach to the problem of a divergent series is often replaced by a more direct approach in which the divergent series (4) assumes a more concrete meaning. Although we realise that the sum (4), as it stands, has no immediate significance, we may interpret it in a different manner. We may find a function $\delta(\xi, x)$ which happens to have the series

$$\frac{1}{\pi} \sum_{k=0}^{\infty} {}' (\cos k\xi \cos kx + \sin k\xi \sin kx) \tag{2.8.6}$$

as its Fourier expansion. Then we could replace the infinite sum (4) by this function and interpret $K(\xi - x)$ not merely as an *operator* but as an actual *function*.

Now the general law (2.2) of the Fourier coefficients tells us that this hypothetical function $\delta(\xi, x)$ must satisfy the following conditions:

$$\begin{aligned}
\int_{-\pi}^{+\pi} \delta(\xi, x) \cos kx\, dx &= \cos k\xi \\
\int_{-\pi}^{+\pi} \delta(\xi, x) \sin kx\, dx &= \sin k\xi
\end{aligned} \tag{2.8.7}$$

Let us consider a function $H(\xi, x)$ of the following properties. It is an even function of the single variable $\xi - x = \theta$:

$$H(\xi, x) = H(x, \xi) = H(\xi - x) = H(\theta) = H(-\theta) \tag{2.8.8}$$

Moreover, this function is everywhere *zero*, with the only exception of the infinitesimal neighbourhood $\pm \epsilon$ of the point $\theta = 0$. Then the expansion coefficients a_k, b_k of this function become:

$$\begin{aligned}
a_k &= \frac{1}{\pi} \int_{-\epsilon}^{+\epsilon} \cos k(\xi + \theta) h(\theta)\, d\theta = \frac{1}{\pi} \cos k\bar{\xi} \int_{-\epsilon}^{+\epsilon} H(\theta)\, d\theta \\
b_k &= \frac{1}{\pi} \int_{-\epsilon}^{+\epsilon} \sin k(\xi + \theta) h(\theta)\, d\theta = \frac{1}{\pi} \sin k\bar{\xi} \int_{-\epsilon}^{+\epsilon} H(\theta)\, d\theta
\end{aligned} \tag{2.8.9}$$

where $\bar{\xi}$ is some intermediate value between $\xi - \epsilon$ and $\xi + \epsilon$. If we demand that

$$\int_{-\epsilon}^{+\epsilon} H(\theta)\, d\theta = 1 \tag{2.8.10}$$

and now let ϵ go towards zero, the point $\bar{\xi}$ becomes in the limit equal to ξ and we have actually obtained the desired expansion coefficients (7). The function thus constructed is Dirac's celebrated "delta function" $\delta(x, \xi) = \delta(\xi, x) = \delta(\xi - x)$. It is comparable to an infinitely sharp *needle* which pinpoints one definite value $f(x)$ of the function, if used under the integral sign as an operator:

$$\int_{-\pi}^{+\pi} f(\xi)\delta(\xi, x)\, d\xi = f(x) \tag{2.8.11}$$

Here the previous equation (5), which was an elegant operational method of writing the Fourier series of a function $f(x)$ (provided that the series converges), now appears in consequence of the definition of the delta function. But we come back to the previous equation (5) if we replace the delta function by its Fourier expansion (6).

Neither the "delta function", nor its Fourier expansion (6) are legitimate concepts if we divest them from their significance as *operators*. The delta function is not a legitimate function because we cannot define a function by a limit process which does not possess a limit. Nor is the infinite series (6) a legitimate Fourier series because an infinite sum which does not converge to a limit is not a legitimate series. Yet this is entirely immaterial if these constructions are used *under the integral sign*, since it suffices that the limits of the performed *operations* shall exist.

2.9. Smoothing of the Gibbs oscillations by Fejér's method

We have mentioned before that the Gibbs oscillations of the Dirichlet kernel (5.8) interfere with an efficient estimation of the remainder $\eta_n(x)$ of the finite Fourier series. Dirichlet succeeded in proving the convergence of the Fourier series if certain restricting conditions called the "Dirichlet conditions", are demanded of $f(x)$. But a much more sweeping result was obtained by Fejér who succeeded in extending the validity of the Fourier series to a much larger class of functions than those which satisfy the Dirichlet conditions. This generalisation became possible by a modification of the summation procedure by which the Fourier series is obtained.

The straightforward method by which the coefficients (2.2) of the Fourier series are derived may lead us to believe that this is the *only* way by which a trigonometric series can be constructed. And yet this is by no means so. What we have proved is only the following: *assuming that we possess a never ending sequence of terms with definite coefficients a_k, b_k whose sum shall converge to $f(x)$, then* these coefficients can be nothing but the Fourier coefficients. This, however, does not interfere with the possibility that for a certain *finite* n we may find much more suitable expansion coefficients since here we are interested in making the error small for *that particular n'* and not in constructing an infinite series with rigid coefficients which in the limit must give us $f(x)$. We may gain greatly in the efficiency of our approximating series if we constantly *modify* the expansion coefficients as n increases to larger and larger values, instead of operating with a *fixed* set of coefficients. And in fact this gives us the possibility by which a much *larger class* of functions becomes expandable than if we operate with fixed coefficients.

Fejér's method of increasing the convergence of a Fourier series consists in the following device. Instead of merely terminating the series after n terms (the terms with a_k and b_k always act together, hence we will unite them as *one* term) and being satisfied with their sum $f_n(x)$, we will construct a *new sequence* by taking the *arithmetic means* of the original sequence:

$$S_1 = f_0, \quad S_2 = \frac{f_0 + f_1}{2}, \ldots, \quad S_n = \frac{f_0 + f_1 + \cdots f_{n-1}}{n} \qquad (2.9.1)$$

This new $S_n(x)$ (the construction of which does not demand the knowledge of the coefficients a_k, b_k beyond $k = n$), has better convergence properties than the original $f_n(x)$. But this $S_n(x)$ may be preferable to $f_n(x)$, quite apart of the question of convergence. The truncated Fourier series $f_n(x)$ has the property that it *oscillates* around the true course of $f(x)$. These "Gibbs oscillations" sometimes interfere with an efficient operation of the Fourier series. Fejér's arithmetic mean method has an excellent influence on these oscillations, by reducing their amplitude and frequently even eliminating them altogether, making the approach to $f(x)$ entirely *smooth*, without any oscillations.

Problem 72. Apply the arithmetic mean method to the Dirichlet kernel (8.3) and show that it becomes transformed into the new kernel (cf. 3.18):

$$\Phi_n(\theta) = \frac{\sin^2 \dfrac{n}{2} \theta}{2\pi n \sin^2 \dfrac{\theta}{2}} \qquad (2.9.2)$$

called "Fejér's kernel".

Problem 73. Show that Fejér's arithmetic mean method is equivalent to a certain *weighting* of the Fourier coefficients which depends on n:

$$\begin{aligned} a'_k &= \rho_k(n)a_k \\ b'_k &= \rho_k(n)b_k \end{aligned} \qquad (2.9.3)$$

with

$$\rho_k(n) = 1 - \frac{k}{n} \qquad (2.9.4)$$

2.10. The remainder of the arithmetic mean method

The great advantage of Fejér's kernel (9.2) compared with the Dirichlet kernel (8.3) is its *increased focusing power*. In the denominator we now find $\sin^2 \theta/2$ instead of $\sin \theta/2$. Since for small angles the sine is replaceable by the angle itself, we can say that the Gibbs oscillations of the new kernel decrease according to the law θ^{-2} instead of θ^{-1}. Hence the new kernel possesses more pronouncedly the properties of the delta function than the Dirichlet kernel. Let us see how we can now estimate the remainder of an nth order approximation.

For this purpose we apply the method that we have studied in Section 5. We have to estimate the integral

$$\varphi_n(x) = \int_{-\pi}^{+\pi} \Phi_n(\theta)f(x + \theta)d\theta \qquad (2.10.1)$$

This is now the nth approximation *itself* and not the *remainder* of that approximation. In order to come to the remainder, let us write $f(x + \theta)$ as follows:

$$f(x + \theta) = f(x) + f_1(x, \theta) \qquad (2.10.2)$$

Substituting we obtain

$$\varphi_n(x) = f(x) \int_{-\pi}^{+\pi} \Phi_n(\theta)d\theta + \int_{-\pi}^{+\pi} f_1(x, \theta)\Phi_n(t)dt \qquad (2.10.3)$$

Now the area under the kernel $\Phi_n(\theta)$ is 1 because $\Phi_n(\theta)$ is a weighted sum of cosines (see 9.4) but each one gives the area zero, except the absolute term $1/2\pi$ which has not been changed by weighting (the corresponding k being zero) and still gives the area 1. Hence the first term is $f(x)$ and thus the second term has now to be interpreted as $-\eta_n(x)$. In this second term we divide the range of integration into two parts: very small θ and larger θ. For the realm of larger θ we can again obtain an expression like the last term of (5.4), although $|f'(\theta)|$ is now to be replaced by $|f(\theta) - f(x)|$ and we have to choose the limiting value of θ, which separates the two domains, not proportional to $1/\sqrt{n}$ but proportional to $1/\sqrt[4]{n}$. In order that this term shall go to zero with increasing n it is only necessary that

$$\int_{-\pi}^{+\pi} |f(\theta)|d\theta = \text{finite} \qquad (2.10.4)$$

Fejér's method demands solely the *absolute integrability* of $f(x)$, without any further conditions.

Now we come to the central region and here we cannot use the estimation based on the maximum of $g_n(\theta)$—which in our earlier case became $\frac{1}{2}$—because $\Phi_n(\theta)$ grows out of bound at $\theta = 0$, as n increases to infinity. But in this region we can *interchange* the role of the two factors and take the maximum value of $|f_1(x, \theta)|$ multiplied by the integral over the absolute value of $\Phi_n(\theta)$ which is certainly less than 1 (Fejér's kernel is *everywhere positive* and needs no change on account of the "absolute value" demand). And thus in the inner domain we have a contribution which is less than the maximum of the absolute value of $f_1(x, \theta)$.

Here we cannot argue that this contribution will be small on account of the small range of integration. But we can argue that we are very near to the point $\theta = 0$ and have to examine the maximum of the quantity

$$|f_1(x, \theta)| = |f(x + \theta) - f(x)| \qquad (2.10.5)$$

Now the continuity of $f(x)$ is *not* demanded for the fulfilment of the condition (9). But if $f(x)$ is not continuous at the point x, then the quantity (10) will not be small and the arithmetic mean method will not converge to $f(x)$. If, however, we are at a point where $f(x)$ is *continuous*, then by the very definition of continuity (without demanding differentiability), the quantity (5) becomes arbitrarily small as the domain of θ shrinks to zero. And thus we have proved that the arithmetic mean method converges to the proper $f(x)$ at any point in which $f(x)$ is continuous, the only restricting condition on the class of admissible $f(x)$ being the absolute integrability (4) of the function.

Problem 74. Carry through the same argument for the case that the continuity of $f(x)$ holds separately to the right of x and to the left of x but $f(x_+)$ and $f(x_-)$

have two different values. Show that in this case the series of Fejér converges to

$$\tfrac{1}{2}[f(x_+) + f(x_-)] \tag{2.10.6}$$

that is the *arithmetic mean of the two limiting ordinates*.

2.11. Differentiation of the Fourier series

Let us assume that the given function $f(x)$ is m times differentiable at all points of the range (including the boundaries). The derivative $f'(x)$ of the original function is a new function in its own right which can also be expanded into a Fourier series. The same series is obtainable by term-by-term differentiation of the original series for $f(x)$.

In Section 6 we have studied the error oscillations of an m times differentiable function. But $f'(x)$ is only $m - 1$ times differentiable and thus we lose the factor $n + \tfrac{1}{2}$ in the denominator of (6.8). The error oscillations of $f'(x)$ have thus increased by the factor $n + \tfrac{1}{2}$. The same will happen whenever we differentiate again, until after $m - 1$ steps we have a function left which is only *once* differentiable. Then we know from the result of Section 2 that $f_n(x)$ will still converge uniformly to $f(x)$ at every point of the range but this is the limit to which we can go without losing the assurance of convergence. And yet it may happen that $f(x)$ can be differentiated many more times—perhaps even an arbitrary number of times—*except* in certain points in which the derivative ceases to exist. It is in the global nature of the Fourier series—shared by all orthogonal expansions—that a single sufficiently strong infinity of the function at any point of the range suffices to destroy the convergence of the series *at all points*. This is the reason that we have to stop with the process of differentiation if even *one* point exists in which the function becomes discontinuous and thus the derivative goes too strongly to infinity.

If we could somehow avoid the magnification of the error oscillations by the factor n in each differentiation, we could obviously greatly increase the usefulness of the Fourier series. Then it would be possible to counteract the global features of the Fourier series and transform it into a locally convergent series. It would not be necessary to demand any differentiability properties of the function $f(x)$. Let us assume that $f(x)$ is merely *integrable*. Then there exists a function $F(x)$ whose derivative is $f(x)$. This $F(x)$ is now differentiable and thus its Fourier series converges uniformly at all points. Now we differentiate this series to obtain $f(x)$. If the error oscillations have not increased in magnitude, the resulting series would give us $f(x)$, possibly with the exception of some singular points in which $f(x)$ ceases to exist. And even less would be sufficient. Let us assume that one integration would not be sufficient to arrive at an absolutely integrable function, but this would happen after m integrations. Then we could obtain the Fourier series of this m times integrated function and return to the original $f(x)$ by m differentiations. The validity of the Fourier series could thus be greatly extended. We will see in the next section how this can actually be accomplished.

2.12. The method of the sigma factors

If we write the Fourier series in the complex form (4.4) and study the remainder of the truncated series, we see that the factor e^{-inx} can be taken in front of the sum. The expression (6.8) obtained for the kernel $g_n{}^m(\xi - x)$ of the integral (4.7) demonstrates that this kernel contains the rapidly oscillating factor $e^{in(\xi-x)}$, multiplied by a second factor which is relatively slowly changing. If we differentiate, we get the factor n on account of the derivative of the first factor, and not on account of the derivative of the second factor. The increase of the error oscillations by the factor n in each step of differentiation is thus caused by the rapidly oscillating character of the Gibbs oscillations. Now we can take advantage of the fortunate circumstance that the functions $\cos nx$ and $\sin nx$ have the exact period $2\pi/n$. Let us write the remainder of the finite Fourier series $f_n(x)$ in the complex form

$$\eta_n(x) = e^{-inx}\gamma(x) \qquad (2.12.1)$$

(with the understanding that only the real part of this expression is to be taken). Instead of the usual differentiation we will now introduce a "curly $\mathscr{D}$ process", defined as follows:

$$\mathscr{D}_n f(x) = \frac{f\left(x + \dfrac{\pi}{n}\right) - f\left(x - \dfrac{\pi}{n}\right)}{\dfrac{2\pi}{n}} \qquad (2.12.2)$$

This is in fact a differencing device, with a Δx which is strictly adjusted to the number of terms with which we operate. While this process introduces an error, it is an error which goes to zero with increasing n and in the limit $n \to \infty$ *coincides* with the ordinary derivative of $f(x)$ if this derivative exists:

$$\lim_{n \to \infty} \mathscr{D}_n f(x) = Df(x) \qquad (2.12.3)$$

The operation $\mathscr{D}_n$ applied to the functions $\cos nx$ and $\sin nx$ has the following effect

$$\begin{aligned} \mathscr{D}_n \cos nx &= 0 \\ \mathscr{D}_n \sin nx &= 0 \end{aligned} \qquad (2.12.4)$$

and we see that the functions $\cos nx$ and $\sin nx$ behave like *constants* with respect to the operation $\mathscr{D}$.

Let us apply this operation to the remainder (1) of the Fourier series. Neglecting quantities of the order n^{-4} we obtain

$$\begin{aligned} \mathscr{D}_n \eta_n(x) &= -e^{-inx} \mathscr{D}_n \gamma(x) \\ &= -e^{-inx}\left[\gamma'(x) + \frac{\pi^2}{6n^2}\gamma'''(x)\right] \end{aligned} \qquad (2.12.5)$$

The Gibbs oscillations have *not* increased in order of magnitude in consequence of this operation, because we have avoided the differentiation of the first factor which would have given the magnification factor n.

If we examine what happens to the terms of the Fourier series in consequence of this operation, we find the following:

$$\mathscr{D}_n \cos kx = - \frac{\sin k \dfrac{\pi}{n}}{\dfrac{\pi}{n}} \sin kx$$

$$\mathscr{D}_n \sin kx = \frac{\sin k \dfrac{\pi}{n}}{\dfrac{\pi}{n}} \cos kx$$

(2.12.6)

We can express the result in the following more striking form. We introduce the following set of factors, called the "sigma factors":

$$\sigma_k = \frac{\sin k \dfrac{\pi}{n}}{k \dfrac{\pi}{n}}$$

(2.12.7)

We apply the ordinary differentiation process to a *modified Fourier series* whose coefficients are multiplied by the sigma factors:

$$a'_k = \sigma_k a_k$$
$$b'_k = \sigma_k b_k$$

(2.12.8)

Notice that this operation leaves the coefficient $\frac{1}{2}a_0$ unchanged (since $\sigma_0 = 1$) while the last terms with a_n, b_n drop out, because $\sigma_n = 0$.

Problem 75. Apply the $\mathscr{D}_n$ process to the Dirichlet kernel (8.3), assuming that n is large.

[Answer:

$$\mathscr{D}_n K_n(\theta) = \frac{\sin n\theta}{4\pi \sin^2 \dfrac{\theta}{2}}$$

(2.12.9)]

2.13. Local smoothing by integration

The application of the $\mathscr{D}_n$ process to the coefficients of a Fourier series can be conceived as the result of two operations. The one is that we differentiate in the ordinary manner, the other is that we multiply the coefficients of the Fourier series by the sigma factors. If we know what the significance of the second process is, we have also found the significance of the operation $\mathscr{D}_n$. Let us now consider the following integration process:

$$\overline{f(x)} = \frac{n}{2\pi} \int_{+\pi/n}^{+\pi/n} f(x + t)dt = \frac{n}{2\pi} \left[F\left(x + \frac{\pi}{n}\right) - F\left(x - \frac{\pi}{n}\right) \right]$$

(2.13.1)

where $F(x)$ is the indefinite integral of $f(x)$. The meaning of this operation is that we replace the value of $f(x)$ by the *arithmetic mean* of all the values in the neighbourhood of $f(x)$, between the limits $\pm \pi/n$.

The operation $\overline{f(x)}$ may be expressed in the following way:

$$\overline{f(x)} = \int_{-\pi}^{+\pi} f(\xi)\delta_n(\xi, x)d\xi \qquad (2.13.2)$$

where the "kernel" $\delta_n(\xi - x)$ is defined as the "square pulse" of the width $2\pi/n$:

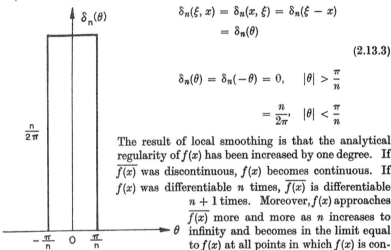

$$\delta_n(\xi, x) = \delta_n(x, \xi) = \delta_n(\xi - x)$$
$$= \delta_n(\theta)$$

$$(2.13.3)$$

$$\delta_n(\theta) = \delta_n(-\theta) = 0, \quad |\theta| > \frac{\pi}{n}$$

$$= \frac{n}{2\pi}, \quad |\theta| < \frac{\pi}{n}$$

The result of local smoothing is that the analytical regularity of $f(x)$ has been increased by one degree. If $\overline{f(x)}$ was discontinuous, $f(x)$ becomes continuous. If $f(x)$ was differentiable n times, $\overline{f(x)}$ is differentiable $n + 1$ times. Moreover, $f(x)$ approaches $\overline{f(x)}$ more and more as n increases to infinity and becomes in the limit equal to $f(x)$ at all points in which $f(x)$ is continuous.

We shall write (2) operationally in the form

$$\overline{f(x)} = I_n f(x) = \int_{-\pi}^{+\pi} f(\xi)\delta_n(\xi, x)d\xi \qquad (2.13.4)$$

A comparison with the equation (8.16) shows that Dirac's "delta function" can be conceived as the limit of the function $\delta_n(\xi, x)$ since the equation (8.16) can now be written (at all points x in which $f(x)$ is continuous) in the form:

$$f(x) = \lim_{n \to \infty} I_n f(x) = \lim_{n \to \infty} \int_{-\pi}^{+\pi} f(\xi)\delta_n(\xi, x)d\xi \qquad (2.13.5)$$

The effect of the operation I_n on the Fourier coefficients a_k, b_k is that they become multiplied by the σ factors, according to (12.8); and vice versa: the operation of multiplying the Fourier coefficients by the σ factors is equivalent to submitting $f(x)$ to the I_n operation.

Problem 76. Show that local smoothing leaves a straight line portion of $f(x)$ unchanged. What is the effect of local smoothing on the parabola $f(x) = x^2$? Express the result in terms of the second derivative.

[Answer:

$$\overline{f(x)} = x^2 + \frac{1}{3}\left(\frac{\pi}{n}\right)^2$$

$$\overline{f(x)} = f(x) + \frac{1}{6}\left(\frac{\pi}{n}\right)^2 f''(x) \tag{2.13.6}$$]

Problem 77. What is the effect of local smoothing on the amplitude, phase and frequency of the oscillation

$$A \cos(\omega t - \varphi) \tag{2.13.7}$$

[Answer: Amplitude changes to

$$\overline{A} = \frac{\sin \omega \dfrac{\pi}{n}}{\omega \dfrac{\pi}{n}} \tag{2.13.8}$$

Frequency and phase remain unchanged.] Show the validity of (6) for small ω.

Problem 78a. Show that at a point of discontinuity $\overline{f(x)}$ approaches in the limit the *arithmetic mean* of the two limiting ordinates.

Problem 78b. Show directly from the definition (12.2) of the $\mathscr{D}$ process the validity of the operational equation

$$\mathscr{D}_n f(x) = DI_n f(x) = I_n Df(x) \tag{2.13.9}$$

(the last equation only if $f(x)$ is differentiable).

2.14. Smoothing of the Gibbs oscillations by the sigma method

We have seen in Section 13 that the operation $\mathscr{D}_n$ could be conceived as the ordinary D operation on a modified function $\overline{f(x)}$ which was obtained by local smoothing. Hence we have the operational equation

$$\mathscr{D}_n f(x) = DI_n f(x) \tag{2.14.1}$$

The excellent qualities of the I_n operator in relation to the Fourier series are based on the strict coordination of the width $2\pi/n$ of the function $\delta_n(\xi - x)$ to the number n of the terms of the truncated Fourier series. We have discussed in Section 12 the fact that by ordinary differentiation the Gibbs oscillation of $f_n(x)$ increase by the factor n, while the $\mathscr{D}_n$ process avoids this increase. Now, since $\mathscr{D}_n$ itself is nothing but the operation DI_n and the operation D increases the Gibbs oscillations by the factor n, the operation I_n must have the effect of *decreasing* the Gibbs oscillations by the factor n. Hence the multiplication (12.8) of the Fourier coefficients by the sigma factors has the beneficial effect of *reducing the Gibbs oscillations* by a considerable factor. The convergence of the Fourier series can thus be *greatly increased*.

Let us assume that the remainder of the truncated series is once more given in the form (12.1):

$$\eta_n(x) = e^{-inx}\gamma_n(x) \tag{2.14.2}$$

Furthermore, let this $\eta_n(x)$ be the derivative of another function

$$\lambda_n(x) = e^{-inx}p_n(x) \tag{2.14.3}$$

This yields the relation

$$p'_n(x) - inp_n(x) = \gamma_n(x) \tag{2.14.4}$$

Now by definition the application of the sigma factors has the following effect on the remainder (see 12.5):

$$\bar{\eta}_n(x) = -e^{inx}\left[p'_n(x) + \frac{\pi^2}{6n^2}p'''_n(x)\right] \tag{2.14.5}$$

On the other hand, the differential equation (4) may be solved without integration for sufficiently large n asymptotically, by expanding into reciprocal powers of n:

$$p_n(x) = \frac{i}{n}\gamma_n(x) + \frac{1}{n^2}\gamma'_n(x) - \frac{i}{n^3}\gamma''_{n}(x) + \ldots \tag{2.14.6}$$

which in view of (5) yields:

$$\bar{\eta}_n(x) = -\frac{i}{n}e^{-inx}\left[\gamma'_n(x) - \frac{i}{n}\gamma''_n(x) + \frac{\pi^2 - 6}{6n^2}\gamma'''_n(x)\right] \tag{2.14.7}$$

Comparison with the original Gibbs oscillations (1) show the following changes: The *phase* of the oscillations has changed by $\pi/2$; the *amplitude* of the oscillations has decreased by the factor n, but coupled with a change of the law of decrease which is no longer $\gamma_n(x)$ but $\gamma'_n(x)$.

The modified remainder $\bar{\eta}_n(x)$ can be conceived as the true Fourier remainder of a modified function $\overline{f(x)} = I_n f(x)$, obtained by the process of local smoothing:

$$\overline{f(x)} = f(x) + \frac{\pi^2}{6n^2}f''(x) \tag{2.14.8}$$

While it is frequently of great advantage that we cut down on the amplitudes of the Gibbs oscillations, we have to sacrifice somewhat on the fidelity of the representation since it is not $f(x)$ but the slightly modified $\overline{f(x)}$ which the truncated Fourier series, weighted by the sigma factors, represents.

Problem 79. Apply the sigma method to the function (2.9). Show that the jump at $x = 0$ is changed to a steep but finite slope of the magnitude $-n/2\pi$. Show that the Gibbs oscillations (3.10) now decrease with $1/n^2$, instead of $1/n$.

Find the position and magnitude of the first two maxima of $\bar{\eta}_n(\theta)$. (The asymptotic procedure (6) is here not applicable, since we are near to the singular point at $\theta = 0$; but cf. (3.10).)

[Answer: with $n\theta = t$:

Position of extremum determined by condition

$$t < \pi: \quad \text{Si}\,(\pi + t) + \text{Si}\,(\pi - t) - \pi = 0$$
$$t > \pi: \quad \text{Si}\,(t + \pi) - \text{Si}\,(t - \pi) = 0$$

Expression of $\eta_n(t)$:

$$t < \pi: \quad \eta_n(t) = -\frac{t}{2\pi} + \frac{(\pi + t)\,\text{Si}\,(\pi + t) - (\pi - t)\,\text{Si}\,(\pi - t)}{2\pi^2}$$

$$t > \pi: \quad \eta_n(t) = -\frac{1}{2} + \frac{(t + \pi)\,\text{Si}\,(t + \pi) - (t - \pi)\,\text{Si}\,(t - \pi)}{2\pi^2}$$

Numerical solution:

$$t_1 = 1.3703, \qquad \eta_1 = 0.02719$$
$$t_2 = 5.1473, \qquad \eta_2 = 0.01187$$

Minimum between at $t = \pi$; $\eta = -\dfrac{1}{2} + \dfrac{\text{Si}\,(2\pi)}{\pi} = -0.04859$]

2.15. Expansion of the delta function

The formal series (8.4) is void of any direct meaning since it diverges at every point. It represents the Fourier series of Dirac's delta function. But let us consider the Fourier series of the function $\delta_n(\xi - x)$; cf. (13.3). Here we obtain the finite sum, weighted by the sigma factors:

$$\rho_n(\theta) = \frac{1}{\pi} \sum_{k=0}^{n}{}' \sigma_k \cos k\theta \tag{2.15.1}$$

which has entirely different properties. If n goes to infinity, $\rho_n(\theta)$ approaches a very definite limit at every point of the interval $[-\pi, +\pi]$, with the only exception of the point $\theta = 0$. At that point $\rho_n(\theta)$ goes strongly to infinity. At all other points, however, $\rho_n(\theta)$ *converges to zero*. Moreover, the area under the curve is constantly 1:

$$\int_{-\pi}^{+\pi} \rho_n(\theta)d\theta = 1 \tag{2.15.2}$$

Hence $\rho_n(\theta)$, as n grows to infinity, satisfies all the conditions of Dirac's delta function and can be considered as the trigonometric expansion of the delta function. We cannot call it the "Fourier series" of the delta function since the coefficients of the expansion are not universal coefficients but the universal coefficients *weighted by the sigma factors*. It is this weighting which makes the series convergent.

The application of local smoothing to Dirichlet's kernel (8.3) yields the new kernel

$$\bar{K}_n(\theta) = \frac{n}{2\pi^2} [\text{Si}\,(n\theta + \pi) - \text{Si}\,(n\theta - \pi)] \qquad (2.15.3)$$

This kernel has the same advantageous properties as Fejér's kernel. The same reasoning we employed in Section 10 for proving that Fejér's method insures the convergence of the Fourier series at all points where $f(x)$ exists, and for all functions which are absolutely integrable, is once more applicable. Hence we obtain the result that *the application of the sigma factors makes the Fourier series of any absolutely integrable function convergent* at all points in which $f(x)$ approaches a definite limit, at least in the sense of $f(x_+)$ and $f(x_-)$. At points where these two limits are different, the series approaches the arithmetic mean of the two limiting ordinates.

The operator I_n can be repeated, of course, which means that now the coefficients a_k, b_k will become multiplied by σ_k^2. At each step the convergence becomes stronger by the factor n. We must not forget, however, that the operation of local smoothing *distorts* the function and we obtain quicker convergence not to the original but to the *modified* function. From the standpoint of going to the limit $n \to \infty$ all these series converge eventually to $f(x)$. But from the standpoint of the *finite* series of n terms we have to compromise between the decrease of the Gibbs oscillations and the modification of the given function due to smoothing. It is an advantage to cut down on the error oscillations, but the price we have to pay is that the basic function to which these oscillations refer is no longer $f(x)$ but $I_nf(x)$, respectively $I_n^kf(x)$, if we multiply by σ^k. The proper optimum will depend on the nature of the given problem.

Problem 80. Show that the function $\delta_n(\theta)$ is obtainable by applying the operation I_n to the function $G_1(\theta)$ (cf. 2.9), taken with a negative sign. Obtain the Gibbs oscillations of the series (1) and the position and magnitude of the maximum amplitude.

[Answer:

$$\eta_n(\theta) = \delta_n(\theta) - \frac{n}{2\pi^2} [\text{Si}\,(n\theta + \pi) - \text{Si}\,(n\theta - \pi)] \qquad (2.15.4)$$

Maximum at $\theta = 0$:

$$\eta_n(0) = \frac{n}{2\pi} - \frac{n}{\pi^2} \text{Si}\,(\pi) \qquad (2.15.5)$$

For θ not too small:

$$\eta_n(\theta) = -\frac{n}{\pi} \frac{\cos n\theta}{n^2\theta^2 - \pi^2} \qquad (2.15.6)]$$

Compare these oscillations with those of the Dirichlet kernel (8.3) and the Fejér kernel (9.2).

2.16. The triangular pulse

In the case of the δ_n-function it is worth while to go one step further and consider the *doubly smoothed* modification of the original delta function.

We now obtain a function which has a *triangular shape*, instead of the square pulse of Section 13. It is defined by

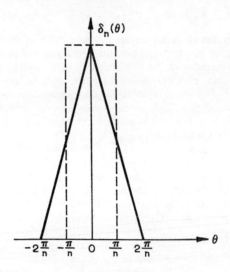

$$\hat{\delta}_n(\theta) = \hat{\delta}_n(-\theta) = \frac{n}{2\pi}\left(1 - \frac{n}{2\pi}\theta\right) \qquad \left(0 \leq \theta \leq 2\frac{\pi}{n}\right)$$

$$= 0 \qquad \left(\theta \geq \frac{2\pi}{n}\right) \tag{2.16.1}$$

comparable to an infinitely sharp *needle*, as n goes to infinity, which pinpoints the special value $f(x)$, if used as an integral operator.

The Fourier series of this new $\hat{\delta}_n(\theta)$ function becomes:

$$\hat{\rho}_n(\theta) = \frac{1}{\pi}\sum_{k=0}^{n}{}' \sigma_k{}^2 \cos k\pi \tag{2.16.2}$$

Problem 81. Obtain the doubly smoothed Dirichlet kernel $\hat{K}_n(\theta)$ by applying the operator I_n to (6). Find again the maximum amplitude of the Gibbs oscillations.

[Answer:

$$\hat{K}_n(\theta) = \frac{n}{4\pi^3}[(n\theta + 2\pi)\operatorname{Si}(n\theta + 2\pi) - 2n\theta\operatorname{Si}(n\theta)$$

$$+ (n\theta - 2\pi)\operatorname{Si}(n\theta - 2\pi)] \tag{2.16.3}$$

$$\eta_n(x)_{\max} = \eta_n(0) = \frac{n}{2\pi}\left(1 - \frac{2}{\pi}\operatorname{Si}(2\pi)\right) = 0.01547n \tag{2.16.4}]$$

2.17. Extension of the class of expandable functions

The smoothing properties of the sigma factors permit us to differentiate a Fourier series, *without* increasing the order of magnitude of the Gibbs oscillations. But this means that a convergent Fourier series can be safely differentiated, without losing its convergence. This process can be repeated any number of times, if at each step we apply a multiplication by the σ_k. Hence a differentiation m times will demand the application of the factors $\sigma_k{}^m$.

As an example let us consider the infinite series s_1 of Problem 63 (cf. 2.15):

$$-\log\left| 2\sin\frac{\theta}{2}\right| = \sum_{k=1}^{\infty} \frac{\cos k\theta}{k} \tag{2.17.1}$$

This series converges everywhere, except at $\theta = 0$ where the function goes to infinity. If we differentiate on both sides formally, we obtain the completely divergent series

$$\frac{1}{2}\cot\frac{\theta}{2} = \sum_{k=1}^{\infty}\sin k\theta \tag{2.17.2}$$

With the sole exception of the point $\theta = \pm\,\pi$, this series diverges everywhere. Nor can we expect a Fourier series for the function $\cot\theta/2$ which is no longer integrable since the area under the curve goes logarithmically to infinity. Hence the Fourier coefficients cannot be evaluated. *The application to the σ_k factors, however, makes the series convergent:*

$$\cot\frac{\theta}{2} = \lim_{n\to\infty} 2\sum_{k=1}^{n}\sigma_k\sin k\theta \tag{2.17.3}$$

This weighted series converges at all points of θ, except at the point of singularity $\theta = 0$. The cotangent, being an odd function, is expanded in a sine series which at $\theta = 0$ gives zero (this can be conceived as the arithmetic mean of the two limiting ordinates $\pm\infty$). In spite of this zero, if we prescribe an arbitrarily small $\theta = \epsilon$, the series (3) manages to rise to an exceedingly large value, if the proper number of terms—excessively large if ϵ is very small—is summed. Then the increase slows down and eventually converges in very small steps to the final value, which is the proper value of $\cot \epsilon/2$.

If we differentiate again and apply the σ operation a second time, we obtain the series

$$-\frac{1}{4\sin^2\dfrac{\theta}{2}} = \lim_{n\to\infty}\sum_{k=1}^{n}\sigma_k{}^2 k\cos k\theta \tag{2.17.4}$$

which is now an even series and which again converges to the proper value at every point of the range, excluding the origin $\theta = 0$.

We see that the sigma factors provide us with a tool of extending the class of functions which allow a harmonic analysis, to a much wider domain, including functions which go very strongly to infinity and are far from being absolutely integrable. These series are *weighted Fourier series* which for every given n behave exactly like an ordinary Fourier series, with the only exception that their coefficients constantly *change* as we increase n to larger and larger values because the *weight factors* by which a rigid set of coefficients is multiplied, keep constantly changing.

Problem 82. Show that the remainder of the series (3) becomes asymptotically

$$\bar{\eta}_n(\theta) = \frac{1}{2\sin^2\frac{\theta}{2}}\left[\frac{\sin n\theta}{n} - \frac{\cos n\theta}{n^2}\cot\frac{\theta}{2}\right] \tag{2.17.5}$$

Demonstrate the formula numerically for $n = 10$, at the point $\theta = \pi/4$ (remembering that $\bar{\eta}_n(\theta)$ is not $f(\theta) - f_n(\theta)$ but $\overline{f(\theta)} - f_n(\theta)$, cf. (14.8). For a table of the sigma factors see Appendix).

[Answer: predicted $\bar{\eta}_{10}\left(\dfrac{\pi}{4}\right) = 0.3414$

        actual $\bar{\eta}_{10}\left(\dfrac{\pi}{4}\right) = 0.3493$]

2.18. Asymptotic relations for the sigma factors

In view of the powerful convergence producing faculties of the σ_k factors we can expect that they should have many interesting mathematical properties. These properties are of an "asymptotic" character, i.e., they hold with increasing accuracy as n increases.

We can make use of the series (15.1) and (16.2) for the $\delta_n(\theta)$ and $\hat{\delta}_n(\theta)$ functions, to derive two asymptotic relations for the σ-factors. In the first case we derive from (15.5), in the second case from (16.4) the asymptotic relations

$$\sum_{k=0}^{n}{}' \sigma_k \to \frac{n}{2}\frac{2}{\pi}\,\mathrm{Si}\,(\pi) = 0.589489872n \tag{2.18.1}$$

$$\sum_{k=0}^{n}{}' \sigma_k^2 \to \frac{n}{2}\frac{2}{\pi}\,\mathrm{Si}\,(2\pi) = 0.451411666n \tag{2.18.2}$$

Further relations are obtainable by substituting in these series for θ some other value. For any fixed value which is not zero, we must get asymptotically zero, since the delta function converges at all points, excluding the origin, to zero. Moreover, in such a relation the σ_k can be replaced by *any* power of σ_k, since the limit value of the series is not influenced by the degree of smoothing. For example the value $\theta = \pi$ gives

$$\sum_{k=0}^{n}{}' \, (-1)^k \sigma_k \;\to 0$$

$$\sum_{k=0}^{n}{}' \, (-1)^k \sigma_k{}^2 \;\to 0 \qquad\qquad (2.18.3)$$

$$\sum_{k=0}^{n}{}' \, (-1)^k \sigma_k{}^m \to 0$$

The values of $\theta = \pi/2, \pi/3, 2\pi/3$ yield the relations

$$
\begin{aligned}
&\tfrac{1}{2} - \sigma_2 + \sigma_4 - \sigma_6 + \ldots && \to 0 \\
&1 + \sigma_1 - \sigma_2 - 2\sigma_3 - \sigma_4 + \sigma_5 + 2\sigma_6 + \sigma_7 - \ldots \to 0 && (2.18.4) \\
&1 - \sigma_1 - \sigma_2 + 2\sigma_3 - \sigma_4 - \sigma_5 + 2\sigma_6 - \ldots && \to 0
\end{aligned}
$$

which hold likewise for all powers of the σ_k. Additional asymptotic relations can be derived from the series (17.3) by substituting for θ the values $\pi/2$, $\pi/3, 2\pi/3, \pi/4$:

$$
\begin{aligned}
&\sigma_1 - \sigma_3 + \sigma_5 - \sigma_7 + \ldots && \to \tfrac{1}{2} \\
&\sigma_1 + \sigma_2 - \sigma_4 - \sigma_5 + \sigma_7 + \sigma_8 - \ldots && \to 1 \\
&\sigma_1 - \sigma_2 + \sigma_4 - \sigma_5 + \sigma_7 - \sigma_8 + \ldots && \to \tfrac{1}{3} \qquad (2.18.5) \\
&\sigma_1 + \sigma_3 - \sigma_5 - \sigma_7 \ldots + \sqrt{2}(\sigma_2 - \sigma_6 + \sigma_{10} - \ldots) \to 1 + \sqrt{2}
\end{aligned}
$$

However, these asymptotic relations can be made much more conclusive if we include the remainder, making use of the previously discussed asymptotic treatment (see Section 14). For example the formula (15.4) was obtained by applying the I_n operation to the function

$$-\frac{\sin n\theta}{\pi\theta} \qquad\qquad (2.18.6)$$

This function originated from the function (see 3.15)

$$-\frac{\sin n\theta}{2\pi} \cot \frac{\theta}{2} \qquad\qquad (2.18.7)$$

We have replaced $\cot \theta/2$ by $\theta/2$ which is permissible for small θ. A more accurate treatment would proceed as follows. We put

$$\frac{1}{2} \sin n\theta \cot \frac{\theta}{2} = \frac{\sin n\theta}{\theta} \left(\frac{\theta}{2} \cot \frac{\theta}{2} \right) \qquad\qquad (2.18.8)$$

and make use of the Taylor series of the second factor, on the basis of the series

$$x \cot x = 1 - \frac{x^2}{2} - \frac{x^4}{45} - \frac{2x^6}{945} - \frac{x^8}{4725} - \ldots \qquad (2.18.9)$$

Generally the following operations are encountered in the asymptotic treatment of the Gibbs oscillations. We may know the remainder of a certain fundamental series which we may put in the complex form (14.2). Now we may want to integrate this series and investigate the new remainder. For this purpose we have to integrate the differential equation (14.4):

$$p'_n - inp_n = \gamma_n \qquad (2.18.10)$$

We do that by the asymptotic expansion*

$$p_n = \frac{i}{n\left(1 - \dfrac{D}{in}\right)} \gamma_n$$

$$= \frac{i}{n}\left(1 + \frac{D}{in} - \frac{D^2}{n^2} - \frac{D^3}{in^3} + \frac{D^4}{n^4} + \ldots\right)\gamma_n \qquad (2.18.11)$$

where D denotes the operation d/dx.

Another basic operation is the smoothing by the sigma factors. This operation has the following effect on the remainder (neglecting higher than third powers of the operator D):

$$\bar{\gamma}_n = -\frac{i}{n}\left(D + \frac{\pi^2}{6n^2} D^3\right)\left(1 + \frac{D}{in} - \frac{D^2}{n^2} - \frac{D^3}{in^3}\right)\gamma_n \qquad (2.18.12)$$

In the case of double smoothing—that is if we multiply the Fourier coefficients by $\sigma_k{}^2$—this operator has to be squared.

As an example let us start with the Gibbs oscillations of the delta function. We will apply the delta function (multiplied by a proper constant) at the point $x = 0$, and the negative delta function at the point $x = \pi$, investigating the following function:

$$f(x) = \frac{\pi}{2}\left[\delta(x, 0) - \delta(x, \pi)\right] \qquad (2.18.13)$$

The truncated Fourier series associated with this function becomes

$$f_n(x) = \cos x + \cos 3x + \ldots \cos (2\nu - 1)x \qquad (2.18.14)$$

with the remainder

$$\eta_n(x) = -\frac{1}{4}\sin nx\left(\cot \frac{x}{2} + \tan \frac{x}{2}\right) \qquad (2.18.15)$$

where $n = 2\nu$ is *even*. The integral of the function (15) yields the "square wave" of the constant value $\pi/4$ for $x > 0$ and $-\pi/4$ for $x < 0$, with a point

---

* A series of this kind converges for sufficiently large n up to a certain point, although it diverges later on. The more descriptive term "semi-convergent" is unfortunately not common in English mathematical literature.

of discontinuity at $x = 0$. We will pay particular attention to the point $x = \pi/2$. Here the truncated Fourier series

$$F_n(x) = \sin x + \frac{\sin 3x}{3} + \ldots \frac{\sin (2\nu - 1)x}{2\nu - 1} \qquad (2.18.16)$$

yields for $\pi/4$ the truncated Leibniz series:

$$1 - \frac{1}{3} + \ldots + \frac{(-1)^{\nu-1}}{2\nu - 1} \qquad (2.18.17)$$

Now in the neighbourhood of $x = \pi/2$ we will put

$$x = \frac{\pi}{2} + \theta \qquad (2.18.18)$$

which yields for the remainder $\eta_n(x)$, if written in complex form:

$$\eta_n(\theta) = \frac{(-1)^{\nu+1}}{4} ie^{-in\theta}\left[\tan\left(\frac{\pi}{4} + \frac{2}{\theta}\right) + \tan\left(\frac{\pi}{4} - \frac{2}{\theta}\right)\right] \qquad (2.18.19)$$

Now the Taylor expansion of the last factor yields

$$\frac{1}{2}\left[\tan\left(\frac{\pi}{4} + \frac{2}{\theta}\right) + \tan\left(\frac{\pi}{4} - \frac{\theta}{2}\right)\right] = 1 + \frac{\theta^2}{2} + \frac{5}{24}\theta^4 + \ldots \qquad (2.18.20)$$

Applying the operator (11) to this expansion we notice first of all that only the *even* powers of D have to be considered (since we focus our attention on the point $\theta = 0$). The result is that the new remainder becomes

$$p_n(0) = \frac{(-1)^\nu}{2n}\left(1 - \frac{1}{n^2} + \frac{5}{n^4}\right) \qquad (2.18.21)$$

This yields the following correction of the slowly convergent Leibniz series:

$$\frac{\pi}{4} = \sum_{k=1}^{\nu} \frac{(-1)^{k-1}}{2k - 1} + \frac{(-1)^\nu}{2n}\left(1 - \frac{1}{n^2} + \frac{5}{n^4}\right) \qquad (2.18.22)$$

The effectiveness of this correction becomes evident if we employ it to the first five terms of the Leibniz series, i.e., $\nu = 5$, $n = 10$:

$$
\begin{array}{l}
0.8349206 \\
-\,0.0495250 \\
\hline
0.7853956 \qquad \left(\frac{\pi}{4} = 0.785398163\right)
\end{array}
$$

The new error is only 3.6 units in the sixth decimal place.

We now come to the application of the sigma factors. This means that the operations (11) and (12) have to be combined, with the following result:

$$\bar{\gamma}_n = \frac{(-1)^{\nu+1}}{n^3}\left[1 - \frac{5}{6n^2}(12 - \pi^2)\right] \qquad (2.18.23)$$

We see that here the sigma smoothing reduced the Gibbs oscillations *quadratically*, instead of linearly, in n. The reason is that before smoothing the point $x = \pi/2$ was a point of maximum amplitude. The shift by 90° changes this maximum to a nodal point, with the result that the term with n^{-2} *drops out* and the error becomes of third order in $1/n$. We thus obtain, up to quantities of the order n^{-7}

$$\sigma_1 - \frac{\sigma_3}{3} + \frac{\sigma_5}{5} - \ldots + (-1)^{\nu-1}\frac{\sigma_{n-1}}{n-1} = \frac{\pi}{4} + \frac{(-1)^{\nu}}{n^3}\left[1 - \frac{5}{6n^2}(12 - \pi^2)\right]$$

(2.18.24)

A second smoothing causes a second phase shift by 90° and the maximum amplitude is once more restored. The reduction by the factor n^2 will cause an error of the order n^{-3}, as we had it in the case of simple smoothing (since here we do not profit by the privileged position of the point $x = \pi/2$). The result of the operation is

$$\bar{\gamma}_n = (-1)^{\nu-1}\frac{1}{2n^3}\left[1 - \frac{5}{3n^2}(18 - \pi^2)\right]$$

(2.18.25)

and thus

$$\sigma_1^2 - \frac{\sigma_3^2}{3} + \ldots + (-1)^{\nu-1}\frac{\sigma_{n-1}^2}{n-1} = \frac{\pi}{4} + \frac{(-1)^{\nu}}{2n^3}\left[1 - \frac{5}{3n^2}(18 - \pi^2)\right]$$ (2.18.26)

Compared with simple smoothing we have not gained more than the factor 2. (As a numerical check, let us apply our formulas to the sigma-weighted Leibniz series, for $\nu = 5$, $n = 10$. The formula gives $(\pi/4) - 0.0009822 = 0.7844160$, against the actual value of 0.7844133, while the calculated value of the doubly smoothed series yields $(\pi/4) - 0.0004322 = 0.7849660$, against the actual value of 0.7849681.)

While in this example we started from a point of maximum amplitude and thus the sigma smoothing gained two powers of n (due to the shift to a nodal point), it may equally happen that we start from a nodal point, in which case the sigma smoothing will not decrease but possibly even increase the local error at that particular point. An example of this kind is encountered in the formulae (31, 32) of Problem 84.

Problem 83. Show the following *exact* relations to be valid for the σ-factors:

$$\sum_{k=1}^{n} k\sigma_k = \frac{n}{\pi}\cot\frac{\pi}{2n}$$

(2.18.27)

$$\sum_{k=1}^{n} (-1)^{k-1}k\sigma_k = 0 \qquad \text{if} \quad n \text{ odd}$$

(2.18.28)

$$= \frac{n}{\pi}\tan\frac{\pi}{2n} \quad \text{if} \quad n \text{ even}$$

Problem 84. Obtain the following improved asymptotic expressions (which hold with an error of the order $n^{-\mu-2}$ if $n^{-\mu}$ is the last power included), and check them numerically for $n = 10$ (making use of the Table of the sigma factors of the Appendix).

$$\sum_{k=0}^{n}{}' \sigma_k = \frac{n}{\pi}\operatorname{Si}(\pi) - \frac{1}{12n}\left(1 + \frac{\pi^2 - 6}{60n^2}\right) \tag{2.18.29}$$

$$\sum{}' \sigma_k{}^2 = \frac{n}{\pi}\operatorname{Si}(2\pi) + \frac{1}{60n^3}\left(1 + \frac{10}{63}\frac{\pi^2 - 6}{n^2}\right) \tag{2.18.30}$$

$$\sum{}' (-1)^k \sigma_k = \frac{(-1)^{n+1}}{4n}\left(1 + \frac{\pi^2 - 6}{12n^2}\right) \tag{2.18.31}$$

$$\sum{}' (-1)^k \sigma_k{}^2 = \frac{(-1)^n}{4n^3}\left(1 + \frac{2}{3}\frac{\pi^2 - 6}{n^2}\right) \tag{2.18.32}$$

$$1 - \frac{1}{2} + \frac{1}{3} - \cdots + \frac{1}{2}\frac{(-1)^{n-1}}{n} = \log 2 + \frac{(-1)^n}{4n^2}\left(1 - \frac{1}{2n^2} + \frac{1}{n^4}\right) \tag{2.18.33}$$

$$\sum_{k=1}^{n} \frac{(-1)^{k+1}\sigma_k}{k} = \log 2 - \frac{\pi^2}{24n^2} - \frac{\pi^4}{960n^4} + \frac{(-1)^n}{4n^2}\left(1 - \frac{18 - \pi^2}{12n^2}\right) \tag{2.18.34}$$

$$\left[\log 2 = 0.6931471806,\quad \pi^2 = 9.869604401,\quad \tan x = x + \frac{x^3}{3} + \frac{2x^5}{15} + \cdots\right]$$

Numerical check:

	Formula	Actual value
(29)	5.886560015	5.886560008
(30)	4.514133437	4.514133438
(31)	− 0.025080617	− 0.025081036
(32)	0.000256449	0.000256612
(33)	0.695634930	0.695634920
(34)	0.691507761	0.691507954

Problem 85. Explain why the third of the asymptotic relations (3) will hold with increasing accuracy as m increases from 1 to n, but ceases to hold if m becomes larger than n. Demonstrate the situation numerically for $n = 6$. [Answer:

$$\Sigma_1 = -0.0420 \qquad \Sigma_5 = 0.0000201$$
$$\Sigma_2 = 0.00125 \qquad \Sigma_6 = 0.000000245$$
$$\Sigma_3 = 0.000525 \qquad \Sigma_7 = \overline{0.0001295}$$
$$\Sigma_4 = -0.000153 \qquad \Sigma_8 = 0.001191\,]$$

2.19. The method of trigonometric interpolation

The exceptional flexibility of the Fourier series in representing a very extensive class of functions makes harmonic analysis to one of the most successful tools of applied mathematics. There is, however, one drawback from the standpoint of practical application: the formulae (2.2) demand the evaluation of a definite integral for every a_k and b_k. If $f(x)$ is not a function of a type which allows such an integration in the sense that the result is expressible in manageable functions, or else if $f(x)$ is not even given in analytical form but may be the result of physical observations, then it is difficult to obtain these coefficients with any degree of accuracy in satisfactory form. It is thus very fortunate that the finite Fourier series of n terms is replaceable by another series whose coefficients are obtained by different tools and which is nevertheless equally effective in the approximation of the same class of functions to which the Fourier series is applicable. For this purpose we need no integration, only *summation*. Moreover, the further fortunate circumstance prevails that it is sufficient to know the function in a number of *discrete points which are equidistantly spaced*. The problem is of a purely algebraic nature and leads to an exceptionally well manageable numerical scheme.

We start with the fundamental trigonometric identity (cf. 3.15)

$$C(x) = \tfrac{1}{2} + \cos x + \cos 2x + \ldots \tfrac{1}{2} \cos nx = \frac{\sin nx}{2} \cot \frac{x}{2} \quad (2.19.1)$$

We choose as points of interpolation the $2n$ equidistant points

$$x_k = \beta + k\,\frac{\pi}{n} \qquad (k = -n, \ldots n - 1) \qquad (2.19.2)$$

where β is any constant between 0 and π/n. Then we see from the relation (1) that the sum

$$U(x) = \frac{1}{n} \sum_{k=-n}^{n-1} f(x_k)\,C(x - x_k) \qquad (2.19.3)$$

has the following properties. It represents a function which at $x = x_k$ assumes the value $f(x_k)$. Moreover, $U(x)$ is a finite trigonometric series of the form

$$U(x) = \tfrac{1}{2}a_0 + a_1 \cos x + \ldots + \tfrac{1}{2}a_n \cos nx$$
$$+ b_1 \sin x + \ldots + \tfrac{1}{2}b_n \sin nx \qquad (2.19.4)$$

with the following values of the coefficients:

$$a_k = \frac{1}{n} \sum_{\alpha=-n}^{n-1} f(x_\alpha) \cos kx_\alpha$$

$$b_k = \frac{1}{n} \sum_{\alpha=-n}^{n-1} f(x_\alpha) \sin kx_\alpha \qquad (2.19.5)$$

If the points of interpolation are to satisfy the principle of left-right symmetry, we have only two choices for the constant β, viz. $\beta = 0$, and $\beta = \pi/2n$. As we have seen in Section 6 (cf. 6.18 and 19), the first choice is advocated for the *odd* part of the function $f(x)$, the second for its *even* part, if our aim is to bring $U(x)$ particularly close to the truncated Fourier series $f_{n-1}(x)$. But this precaution involves the inconvenience that we have to separate the even and the odd parts of the function and prescribe their values in two different sets of points. We do not lose essentially if we abandon this device and choose the *same* points of interpolation for the entire series (usually with the choice $\beta = 0$).* The coefficients (5) of the finite trigonometric series (4) are, of course, not identical with the coefficients a_k, b_k of the truncated Fourier series (cf. 2.2), obtained by integrations But the error oscillations—although strictly equidistant now, approximately equidistant before—are remarkably analogous in both cases and the amplitudes of these oscillations have not increased by the fact that we replace the original truncated Fourier series by the new series (4), obtained by interpolation. The greatly simplified numerical scheme is of inestimable value in the case of complicated functions or empirically observed data.

Problem 86.   Show that the last two coefficients a_n, b_n of the scheme (5) are not independent of each other but related by the condition

$$a_n \sin \beta n - b_n \cos \beta n = 0 \qquad (2.19.6)$$

Problem 87.   Consider the function

$$f(x) = x \qquad (2.19.7)$$

between 0 and π, once defined as an *odd*, and once as an *even* function. (In the first case the function is discontinuous at $x = \pi$ where we define it as $f(\pi) = 0$.) Expand this function by interpolation in a sine and cosine series for $n = 9$ ($\beta = 0$) and compare the resulting Gibbs oscillations with the Gibbs oscillations of the truncated Fourier series with $n = 8$.

2.20. Error bounds for the trigonometric interpolation method

Once more it will be our aim to obtain efficient error estimates for the difference

$$\eta_n(x) = f(x) - f_n(x) \qquad (2.20.1)$$

between the true function and the approximation obtained by trigonometric interpolation. Once more the "method of the Green's function", discussed in Section 7, becomes applicable. We have seen that in the case of an m-times differentiable function we could obtain $f(x)$ in terms of an auxiliary "kernel function" $G(x, \xi)$ (cf. 7.3):

$$f(x) = \int_{-\pi}^{+\pi} f^{(m)}(\xi) G_m(x, \xi) d\xi \qquad (2.20.2)$$

Now let us assume that we have examined the special function $G_m(x, \xi)$, considered as a function of x, and determined the remainder $\eta_n(x)$ for this special function

$$\eta_n(x) = g_n(x, \xi) \qquad (2.20.3)$$

* For the numerical aspects of trigonometric interpolation, cf. A. A., Chapter 4, Sections 11–15.

We have interpolated in the points x_k and thus $\eta_n(x)$ will be automatically zero at all points $x = x_k$. But this fact will not be changed by integrating over ξ. Hence in the end we again have nodal points at the points $x = x_k$ and this means that the function $f_n(x)$ thus determined is exactly the trigonometric series obtained by interpolation, whose deviation from $f(x)$ is thus obtainable with the help of the definite integral

$$\eta_n(x) = \int_{-\pi}^{+\pi} g_n(x, \xi) f^{(m)}(\xi) d\xi \tag{2.20.4}$$

Then again we can make use of "Cauchy's inequality" (4.13) and obtain

$$|\eta_n(x)|^2 \le \int_{-\pi}^{+\pi} g_n^2(x, \xi) d\xi \int_{-\pi}^{+\pi} f^{(m)2}(\xi) d\xi \tag{2.20.5}$$

Now the second factor is once more the square of the "norm" of $f^{(m)}(x)$. In the first factor we encounter conditions which are very similar to those encountered before, when dealing with the Fourier series (cf. 4.14), and in fact the result of the analysis is that *the error bound* (4.16), *found before for the Fourier series, remains valid for the case of trigonometric interpolation.* This result proves once more that the method of trigonometric interpolation is not inferior to the Fourier series of a comparable number of terms. The actual coefficients of the two series may differ considerably but the *closeness of approximation is nearly the same in both cases.*

We can proceed still differently in our problem of comparing the remainder of the trigonometric interpolation with the remainder of the corresponding Fourier series. We will write $f(x)$ in the form

$$f(x) = f_{n-1}(x) + \eta_{n-1}(x) \tag{2.20.6}$$

where $\eta_{n-1}(x)$ is the remainder of the truncated series $f_{n-1}(x)$. Let us now apply the method of trigonometric interpolation to $f(x)$. This can be done by interpolating $f_{n-1}(x)$ and $\eta_{n-1}(x)$ and forming the sum. But the interpolation of $f_{n-1}(x)$ must coincide with $f_{n-1}(x)$ itself, as we can see from the fact that $f_{n-1}(x)$ is already a finite trigonometric series of the form (19.4), and the uniqueness of the coefficients (19.5) of trigonometric interpolation demonstrates that only *one* such series can exist. Hence it suffices to interpolate the remainder $\eta_{n-1}(x)$. Since this remainder is small relative to $f_{n-1}(x)$, we would be inclined to believe that this in itself is enough to demonstrate that the series obtained by trigonometric interpolation cannot differ from either $f_{n-1}(x)$ or $f(x)$ by more than a negligibly small amount.

That this argument is deceptive, is shown by the example of equidistant polynomial interpolation, considered earlier in Chapter 1. If the zeros of interpolation are chosen in a definite non-equidistant fashion—namely as the zeros of the Chebyshev polynomials $T_n(x)$*—we shall obtain a polynomial $p_n(x)$ which approximates the given continuous function $f(x)$ (of bounded variation) to any degree of accuracy. Hence we can put

$$f(x) = p_n(x) + \eta_n(x) \tag{2.20.7}$$

where $\eta_n(x)$ can be made as small as we wish, by going with n to infinity.

Now, if we apply equidistant polynomial interpolation, in $n + 1$ points to $f(x)$, we can again divide our task by applying the procedure to $p_n(x)$ and to $\eta_n(x)$ and then forming the sum. By the same argument as before, the interpolation of $p_n(x)$ once more reproduces $p_n(x)$. If the interpolation of the uniformly small $\eta_n(x)$ were to remain small, we would not get into difficulties. But the peculiar paradox holds that, although $\eta_n(x)$ converges to zero with increasing n at all points of the range, its equidistant interpolation *boosts up* the amplitudes of the oscillations to such an extent that they go to infinity on the periphery of the range.

The paradox comes about by the strong disharmony which exists between the well-distributed zeros and the equidistant zeros. In the case of the trigonometric functions no such disharmony exists because the natural Gibbs oscillations of the truncated Fourier series are automatically nearly equidistant and imitate the behaviour of the imposed equidistancy of the error oscillations of the interpolated series. The natural and the imposed error oscillations are so nearly the same that the interpolation of $\eta_{n-1}(x)$ of the equation (6) can have no strong influence on the error oscillations of $f_{n-1}(x)$. If the remainder $\eta_{n-1}(x)$, caused by the truncation of the Fourier series, is small, it remains small even in its interpolated form.

2.21. Relation between equidistant trigonometric and polynomial interpolations

We have discussed earlier the peculiar fact that a polynomial interpolation of high order was an eminently unsuitable tool for the representation of equidistant data because the error oscillations had the tendency to go completely out of bound around the two ends of the region, even in the case of analytical functions, while in the case of non-analytical functions the method fails completely. In strange contrast to this phenomenon we find that the trigonometric kind of interpolation gives error oscillations which are practically *uniform* throughout the range and which are small even in cases when the function can be differentiated to a very limited degree only. The great superiority of trigonometric versus polynomial interpolation is thus demonstrated.

The more surprising then is the fact that these two important types of interpolation are in fact closely *related* to each other. The Lagrangian type of equidistant interpolation problem can in fact be reformulated as a *trigonometric* type of interpolation problem.

We return to Lagrangian interpolation and construct the fundamental

* Cf. A. A., p. 245.

polynomial $F(x)$. Our data shall be given at the $2n + 1$ points $x_k = 0$, $\pm 1, \pm 2, \ldots, \pm n$. Then

$$F(x) = x(x^2 - 1)(x^2 - 4) \ldots (x^2 - n^2) = \frac{(x + n)!}{(x - n - 1)!} \quad (2.21.1)$$

Now we will make use of a fundamental theorem in the theory of the gamma function [compare the two expansions (1.13.2) (putting $\mu = -x$) and (1.18.9)]:

$$x!(-x)! = \frac{\pi x}{\sin \pi x} \quad (2.21.2)$$

This relation, applied to (1), yields

$$F(x) = (n + x)!(n + 1 - x)! \frac{\sin \pi(n + 1 - x)}{\pi(n + 1 - x)}$$

$$= (-1)^n(n + x)!(n - x)! \frac{\sin \pi x}{\pi} \quad (2.21.3)$$

We introduce the auxiliary function

$$Q_n(x) = Q_n(-x) = \frac{(n + x)!(n - x)!}{n!^2} \quad (2.21.4)$$

and write $F(x)$ in the following form—remembering that a constant factor in $F(x)$ is immaterial:

$$F(x) = Q_n(x) \frac{\sin \pi x}{\pi} \quad (2.21.5)$$

Now we come to the construction of Lagrange's interpolation formula:

$$f_{2n}(x) = F(x) \sum_{k=-n}^{+n} \frac{f(k)}{F'(k)(x - k)} \quad (2.21.6)$$

The function $Q_n(x)$ does not vanish in the given interval. It is the last factor of $F(x)$ which vanishes at the points $x = \pm k$. Hence

$$F'(k) = Q_n(k)(-1)^k \quad (2.21.7)$$

and we can rewrite the interpolation formula (6) as an interpolation for a new function $\varphi(x)$, defined by

$$\varphi(x) = \frac{f(x)}{Q_n(x)} \quad (2.21.8)$$

We obtain

$$\varphi^*(x) = \frac{\sin \pi x}{\pi} \sum_{k=-n}^{n} (-1)^k \frac{\varphi(k)}{x - k} \quad (2.21.9)$$

where $\varphi^*(x)$ denotes the interpolated value of $\varphi(x)$.

We see that *Lagrangian polynomial interpolation of $f(x)$, applied to equidistant data, is equivalent to a trigonometric interpolation of the transformed function $\varphi(x)$*, defined by (8).

We have encountered a similar interpolation formula earlier in Section 1.20 (cf. 1.20.4), when dealing with the Fourier transform. There the limits of summation were infinite, while now they extend only between $-n$ and $+n$. But this only means that we define all $\varphi(k)$ for $|k| > n$ as zero. The formula (9) is closely related to the formula (19.3) of trigonometric interpolation. Let us choose $\beta = 0$. Moreover, let us introduce a new variable t, defined by

$$t = \frac{nx}{\pi} \qquad (2.21.10)$$

$$t_k = \frac{nx_k}{\pi} = k \qquad (2.21.11)$$

Then

$$C(x - x_k) = \frac{\sin \pi(t - k)}{2 \tan \dfrac{\pi}{2n} (t - k)} = (-1)^k \frac{\sin \pi t}{2 \tan \dfrac{\pi}{2n} (t - k)} \qquad (2.21.12)$$

Now, if n increases to infinity, we obtain

$$C(x - x_k) = n(-1)^k \frac{\sin \pi t}{\pi(t - k)} \qquad (2.21.13)$$

and the $U(x)$ of (18.3) becomes

$$U(t) = \frac{1}{\pi} \sin \pi t \sum_{k=-\infty}^{+\infty} (-1)^k \frac{f(k)}{t - k} \qquad (2.21.14)$$

which agrees with (9), except for the limits of summation which do not go beyond $\pm n$, in view of the fact that all the later $f(k)$ vanish.

The relation here established between equidistant polynomial and equidistant trigonometric interpolation permits us to make use of the theory of trigonometric interpolation for the discussion of the error oscillations of the polynomial interpolations of high order. Moreover, the interpolation formula (9) is in fact even *numerically* much simpler than the original Lagrangian formula and may be preferable to it in some cases.

We can now find a new interpretation for the very large error oscillations of high order polynomial approximations. As far as the transformed function $\varphi(x)$ goes, the Gibbs oscillations remain throughout the range of practically constant amplitude. However, when we return to the original function $f(x)$, we have to multiply by the function $Q_n(x)$ defined by (4). This multiplies also the remainder $\eta_n(x)$. Now $Q_n(x)$ can be closely estimated by Stirling's formula:

$$n! = \sqrt{2\pi} \, n^{n+(1/2)} e^{-n} \qquad (2.21.15)$$

which shows that $Q_n(x)$ is very nearly the nth power of a universal function of x/n:

$$Q_n(x) = [Y(\xi)Y(-\xi)]^n \sqrt{1 - \xi^2} \qquad (2.21.16)$$

where

$$Y(\xi) = (1 + \xi)^{1+\xi} \qquad (2.21.17)$$

and

$$\xi = \frac{x}{n} \qquad (2.21.18)$$

The general trend of $Q_n(x)$ is nearly $e^{2n\xi^2}$ which shows the *very strong increase* of $Q_n(x)$ with increasing x, until the maximum 4^n is reached at $x = n$. It is this exponential magnification of the fairly uniform Gibbs oscillations which renders high order polynomial interpolation so inefficient if we leave the central range of interpolation.

The transformed series (9) can be of great help if our aim is to obtain the limit of an infinite Stirling series. In Chapter 1.9 we have encountered an interpolation problem in which the successive terms seemed to converge as more and more data were taken into account, but it seemed questionable that the limit thus obtained would coincide with the desired functional value. In the original form of the Stirling series it is by no means easy to see what happens as more and more terms of the series are taken into account. We fare much better by transforming the series into the form (9) and then make the transition to the limit $n \to \infty$. It is true that this procedure demands a transition from the function $f(x)$ to the new function $\varphi(x)$. But this transformation becomes particularly simple if n is very large and converges to infinity. The relation between $f(x)$ and $\varphi(x)$, as given by (8), requires that we should divide $f(x)$ by $Q_n(x)$ which for any finite x and very large n becomes

$$Q_n(x) = e^{2(x^2/n)} \qquad (2.21.19)$$

We see that for any *finite* point x the functions $f(x)$ and $\varphi(x)$ *coincide* in the limit, as n grows to infinity. This does not absolve us from the obligation to investigate the possible contribution of the points in infinity. But if the nature of the function $f(x)$ is such that we know in advance that the contribution of the points in infinity converges to zero, then it suffices to find the limit of the infinite sum

$$f^*(x) = \frac{\sin \pi x}{\pi} \sum_{k=-\infty}^{+\infty} (-1)^k \frac{f(k)}{x - k} \qquad (2.21.20)$$

In the problem of Chapter 1.9 the given equidistant values had the form (cf. 1.7.3)

$$f(k) = \frac{C}{a^2 + k^2} = \frac{C}{2ia}\left(\frac{1}{k - ia} - \frac{1}{k + ia}\right) \qquad (2.21.21)$$

(with $C = 100$, $a = 2$). Substitution in (20) yields terms of the following character:

$$\frac{1}{x - k}\frac{1}{k - ia} = \frac{1}{x - ia}\left(\frac{1}{x - k} - \frac{1}{ia - k}\right) \qquad (2.21.22)$$

Let us assume that we are able to sum the series

$$p(x) = \sum_{k=-\infty}^{+\infty}\frac{(-1)^k}{x - k} \qquad (2.21.23)$$

Then we will have obtained $f^*(x)$ in the form

$$f^*(x) = \frac{C}{2ia}\left\{\frac{1}{x - ia}[p(x) - p(ia)] - \frac{1}{x + ia}[p(x) - p(-ia)]\right\}\frac{\sin \pi x}{\pi} \qquad (2.21.24)$$

Now the function $f(x) = 1$ certainly allows the Stirling kind of interpolation in an infinite domain and thus, in view of (20), we must have

$$\frac{\sin \pi x}{\pi}\sum_{k=-\infty}^{+\infty}\frac{(-1)^k}{x - k} = 1 \qquad (2.21.25)$$

which gives $p(x)$ in the form

$$p(x) = \frac{\pi}{\sin \pi x} \qquad (2.21.26)$$

Substitution in (24) yields

$$\begin{aligned}
f^*(x) &= f(x) + \frac{C}{2a}\left[\frac{1}{x - ia} + \frac{1}{x + ia}\right]\frac{\sin \pi x}{\sinh \pi a} \\
&= f(x) + \frac{C}{2a}\frac{2x}{x^2 + a^2}\frac{\sin \pi x}{\sinh \pi a} \\
&= f(x)\left[1 + \frac{x}{a}\frac{\sin \pi x}{\sinh \pi a}\right] \qquad (2.21.27)
\end{aligned}$$

This shows that the interpolated value $f^*(x)$ does *not* coincide with $f(x)$ at any point, except at the integer points $x = k$ which provided the key-values of the interpolation procedure. The infinite Stirling expansion of our problem does approach a limit, but it is *not the desired limit*. In our specific problem we have $a = 2$, and we have interpolated at the point $x = \frac{1}{2}$. Since

$$\sinh 2\pi = 267.745$$

we obtain

$$f^*(\tfrac{1}{2}) = 1.0009337 \cdot 23.52941 = 23.55138 \qquad (2.21.28)$$

The difference is small and yet significant. It would easily escape our attention if we were to trust the numerical procedure blindly, without backing it up by the power of a thorough analytical study.

Problem 88. Given the even function

$$f(x) = |x| \qquad (2.21.29)$$

in the range $[-5, +5]$. Apply 11-point Lagrangian interpolation on the basis of the formula (9). Evaluate $\varphi_*(x)$ at the *half-integer* points and obtain the error $\eta_5(x)$ at these points. Then return to the original $f(x)$ by multiplying by $Q_5(x)$ and demonstrate the strong increase of the error oscillations.

2.22. The Fourier series in curve fitting

In the frequent problem of curve fitting of equidistant data we can make excellent use of the uniform error oscillations of the Fourier series, against the exponential increase of the error oscillations, if we try to interpolate by powers. However, we have to overcome the difficulty that the given function in most cases does not satisfy any definite boundary conditions. Furthermore, it is frequently difficult to measure the derivative of the function at the endpoints of the range and we have to rely on the given equidistant ordinates, without further information.

In such cases we can prepare our problem to the application of the Fourier series by the following artifice. We normalise the range of our data to $[0, 1]$, by a proper choice of the independent variable. We then replace $f(x)$ by

$$g(x) = f(x) - f(0) - [f(1) - f(0)]x \qquad (2.22.1)$$

The new $g(x)$ satisfies the boundary conditions

$$g(0) = g(1) = 0 \qquad (2.22.2)$$

Then we reflect $g(x)$ as an *odd* function:

$$g(-x) = -g(x) \qquad (2.22.3)$$

and consider the range $[-1, +1]$. On the boundaries we find that the conditions

$$\begin{aligned} g(-1) &= g(1) = 0 \\ g'(-1) &= g'(1) \end{aligned} \qquad (2.22.4)$$

are automatically satisfied and thus at least function and first derivative can be conceived as continuous. The first break will occur in the *second* derivative.

Under these circumstances we will use the *sine series*

$$g_*{}_n(x) = b_1 \sin \pi x + b_2 \sin 2\pi x + \ldots + b_{n-1} \sin (n-1)\pi x \qquad (2.22.5)$$

for the representation of our data (adding in the end the linear correction to come back to $f(x)$). The coefficients b_k are evaluated according to the formula (19.5) (replacing, however, $\sin kx_\alpha$ by $\sin \pi k x_\alpha$). The expression (6.8) shows that the amplitude of the error oscillations will be of the order of magnitude n^{-3}, except near $x = 0$ and $x = \pi$ where a larger error of the order n^{-2} can be expected.   Hence a Fourier series of 10 to 12 terms will generally give satisfactory accuracy, comparable to the accuracy of the data.*

Problem 89.   A set of 13 equidistant data are given at $x = 0, \frac{1}{12}, \ldots, 1$.   They happen to lie on the curve

$$y = e^{2x} \tag{2.22.6}$$

Fit these data according to the method of Section 22.   Study the Gibbs oscillations of the interpolation obtained.

Problem 90.   Let $f(x)$ be given at n equidistant points between 0 and 1 and let also $f'(0)$ and $f'(1)$ be known.   What method of curve fitting could we use under these circumstances?

[Answer: define

$$g(x) = f(x) - f'(0)x - [f'(1) - f'(0)] \frac{x^2}{2} \tag{2.22.7}$$

and use the cosine series.

Demonstrate that the error oscillations are now of the order n^{-4} respectively near to the ends of the range of the order n^{-3}.]

BIBLIOGRAPHY

[1] Churchill, R. V., *Fourier Series and Boundary Value Problems* (McGraw-Hill, New York, 1941)
[2] Franklin, Ph., *Fourier Methods* (McGraw-Hill, New York, 1949)
[3] Jackson, D., *Fourier Series and Orthogonal Polynomials* (Math. Association of America, Oberlin, 1941)
[4] Sneddon, I. N., *Fourier Transforms* (McGraw-Hill, 1951)

* See also A. A., Chapter 5.11, 12.

CHAPTER 3

MATRIX CALCULUS

Synopsis. The customary theory of matrices is restricted to $n \times n$ square matrices. But the general theory of linear operators can only be based on the general case of $n \times m$ matrices in which the number of equations and number of unknowns do not harmonise, giving rise to either over-determined (too many equations) or under-determined (too few equations) systems. We learn about the two spaces (of n and m dimensions) to which our $n \times m$ matrix belongs and the p-dimensional "eigenspace", in which the matrix is activated. We succeed in extending the customary "principal axis transformation" of symmetric matrices to arbitrary $n \times m$ matrices and arrive at a fundamental "decomposition theorem" of arbitrary matrices which elucidates the behaviour of arbitrarily over-determined or under-determined systems. Every matrix has a unique inverse because every matrix is complete within its own space of activation. Consequently every linear system has in proper interpretation a unique solution. For the existence of this solution it is necessary and sufficient that both right and left vectors shall lie completely within the activated p-dimensional subspaces (one for the right and one for the left vector), which are uniquely associated with the given matrix.

3.1. Introduction

It was around the middle of the last century that Cayley introduced the matrix as an algebraic operator. This concept has become so universal in the meantime that we often forget its great philosophical significance. What Cayley did here parallels the algebraisation of arithmetic processes by the Hindus. While in arithmetic we are interested in getting the answer to a given arithmetic operation, in algebra we are no longer interested in the individual problem and its solution but start to investigate the *properties* of these operations and their effect on the given numbers. In a similar way, before Cayley's revolutionary innovation one was merely interested in the actual numerical solution of a given set of algebraic equations, without paying much attention to the general algebraic properties of the solution. Now came Cayley who said: "Let us write down the scheme of coefficients which appear in a set of linear equations and consider this scheme as one unity":

$$A = \begin{bmatrix} a_{11} & a_{12} & \cdots & a_{1n} \\ a_{21} & a_{22} & \cdots & a_{2n} \\ \vdots & & & \\ a_{n1} & a_{n2} & \cdots & a_{nn} \end{bmatrix} \tag{3.1.1}$$

To call this scheme by the letter A was much more than a matter of notation. It had the significance that we are no longer interested in the numerical values of the coefficients $a_{11} \ldots a_{nn}$. In fact, these numerical values are without any significance in themselves. Their significance becomes established only in the moment when this scheme *operates* on something. The matrix A was thus divested of its arithmetic significance and became an algebraic operator, similar to a complex number $a + ib$, although characterised by a much larger number of components. A large set of linear equations could be written down in the simple form

$$Ay = b \tag{3.1.2}$$

where y and b are no longer simple numbers but a *set* of numbers, called a "vector". That one could operate with sets of numbers in a similar way as with single numbers was the great discovery of Cayley's algebraisation of a matrix and the subsequent development of "matrix calculus".

This development had great repercussions for the field of differential equations. The problems of mathematical physics, and later the constantly expanding industrial research demanded the solution of certain linear differential equations, with given boundary conditions. One could concentrate on these particular equations and develop methods which led to their solution, either in closed form, or in the form of some infinite expansions. But with the advent of the big electronic computers the task of finding the numerical solution of a given boundary value problem is taken over by the machine. We can thus turn to the wider problem of investigating the general *analytical properties* of the *differential operator* itself, instead of trying to find the answer to a given individual problem. If we understand these properties, then we can hope that we may develop methods for the given individual case which will finally lead to the desired numerical answer.

In this search for "properties" the methods of matrix calculus can serve as our guiding light. A linear differential equation does not differ fundamentally from a set of ordinary algebraic equations. The masters of 18th century analysis, Euler and Lagrange, again and again drew exceedingly valuable inspiration from the fact that a differential quotient is not more than a difference coefficient whose Δx can be made as small as we wish. This means that a linear differential equation can be approximated to any degree of accuracy by a set of ordinary linear algebraic equations. But these equations fall in the domain of *matrix calculus*. The "matrix" of these equations is determined by the differential operator itself. And thus the study of linear differential operators and the study of matrices as algebraic operators is in the most intimate relation to one another. The present chapter deals with those aspects of matrix calculus which are of particular importance for the study of linear differential operators. One of the basic things we have to remember in this connection is that the transformation of a differential equation into an algebraic set of equations demands a *limit process* in which the number of equations go to infinity. Hence we can use only those features of matrix calculus which retain their significance if the order of the matrix increases to infinity.

3.2. Rectangular matrices

The scheme (1.1) pictures the matrix of a linear set of equations in which the number of equations is n and the number of unknowns likewise n. Hence we have here an "$n \times n$ matrix". From the standpoint of solving a set of equations it seems natural enough to demand that we shall have just as many equations as unknowns. If the number of equations is *smaller* than the number of unknowns, our data are not sufficient for a unique characterisation of the solution. On the other hand, if the number of equations is *larger* than the number of unknowns, we do not have enough quantities to satisfy all the given data and our equations are generally not solvable. For this reason we consider in the matrix calculus of linear algebraic systems almost exclusively only *square* matrices. However, for the general study of differential operators this restriction is a severe handicap. A differential operator such as y'' for example requires the addition of two "boundary conditions" in order to make the associated *differential equation* well determined. But we may want to study the differential operator y'' *itself*, without any additional conditions. In this case we have to deal with a system of equations in which the number of unknowns exceeds the number of equations by 2. In the realm of *partial differential operators* the discrepancy is even more pronounced. We might have to deal with the operation "divergence" which associates a *scalar* field with a given *vector* field:

$$\operatorname{div} u = \frac{\partial u_1}{\partial x_1} + \frac{\partial u_2}{\partial x_2} + \ldots + \frac{\partial u_n}{\partial x_n} \qquad (3.2.1)$$

The associated set of linear equations is strongly "under-determined", i.e. the number of equations is much *smaller* than the number of unknowns. On the other hand, consider another operator called the "gradient", which associates a *vector* field with a given *scalar* field:

$$\operatorname{grad} \varphi = \frac{\partial \varphi}{\partial x_i} \quad (i = 1, 2, 3) \qquad (3.2.2)$$

The associated set of linear equations is here strongly "over-determined", i.e. the number of equations is much *larger* than the number of unknowns.

Under these circumstances we have to break down our preference for $n \times n$ matrices and extend our consideration to "$n \times m$" matrices of n rows and m columns. The number n of equations and the number m of unknowns are here no longer matched but generally

$$n < m \text{ (under-determined)}$$
$$n = m \text{ (even-determined)} \qquad (3.2.3)$$
$$n > m \text{ (over-determined)}$$

and accordingly we speak of under-determined, even-determined and over-determined linear systems. Our general studies will thus be devoted to the case of *rectangular matrices* of n rows and m columns (called briefly an "$n \times m$ matrix") in which the two numbers n and m are left entirely free.

The operation (1.2) can now be pictured in the following manner, if we choose for the sake of illustration the case $n < m$:

$$(3.2.4)$$

It illustrates the following general situation: "The $n \times m$ matrix A operates on a vector of an m-dimensional space and transforms it into a vector of an n-dimensional space." The vector y on the left side of the equation and the vector b on the right side of the equation *belong generally to two different spaces.* This feature of a general matrix operation is disguised by the special $n \times n$ case in which case both y and b belong to the *same n-dimensional space.*

In the following section we summarise for the sake of completeness the fundamental operational rules with matrices which we assume to be known.*

3.3. The basic rules of matrix calculus

Originally a matrix was conceived as a two-dimensional scheme of $n \times n$ numbers, in contrast to a "vector" which is characterised by a single row of numbers. In fact, however, the operations of matrix calculus become greatly simplified if we consider a matrix as a general scheme of $n \times m$ numbers. Accordingly a vector itself is not more than a one row n-column matrix (called a "row vector"). Since we have the operation "transposition" at our disposal (denoted by the "tilde" $\sim$), it will be our policy to consider a vector basically as a *column* vector and write a row vector x in the form $\tilde{x}$. (Transposition means: exchange of rows and columns and, if the elements of the matrix are complex numbers, simultaneous change of every i to $-i$.)

Basic rule of multiplying matrices: Two general matrices A and B can only be multiplied if A is an $n \times m$ and B an $m \times r$ matrix; the product AB is an $n \times r$ matrix. Symbolically:

$$(m \times n)(n \times r) = (m \times r) \qquad (3.3.1)$$

* Cf. A. A., Chapter 2.

Matrix multiplication is generally *not commutative*.

$$AB \neq BA \tag{3.3.2}$$

but always *associative*:

$$A(BC) = (AB)C \tag{3.3.3}$$

Fundamental transposition rule:

$$\widetilde{AB} = \tilde{B}\tilde{A} \tag{3.3.4}$$

A row vector times a column vector (of the same number of elements) gives a *scalar* (a 1×1 matrix), called the "scalar product" of the two vectors:

$$\tilde{a}b = a^*_1 b_1 + a^*_2 b_2 + \ldots a^*_n b_n \tag{3.3.5}$$

(The asterisk means: "conjugate complex".) The transpose of a scalar coincides with itself, except for a change of i to $-i$. This leads to the following fundamental identity, called the "bilinear identity"

$$\tilde{x}Ay = [\tilde{y}\tilde{A}x]_* \tag{3.3.6}$$

where x is $n \times 1$, A is $n \times m$, and y is $m \times 1$.

Two fundamental matrices of special significance: the "zero matrix" whose elements are all zero, and the "unit matrix", defined by

$$AI = IA = I \tag{3.3.7}$$

whose diagonal elements are all 1, all other elements zero:

$$I = \begin{bmatrix} 1 & & & \\ & 1 & & 0 \\ & & \cdot & \\ & & & \cdot \\ 0 & & & \cdot \\ & & & & 1 \end{bmatrix} \tag{3.3.8}$$

A *symmetric*—or in the complex case *Hermitian*—matrix is characterised by the property

$$\tilde{A} = A \tag{3.3.9}$$

An *orthogonal* matrix U is defined by the property

$$\tilde{U}U = U\tilde{U} = I \tag{3.3.10}$$

If U is not an $n \times n$ but an $n \times p$ matrix ($p < n$), we will call it "semi-orthogonal" if

$$\tilde{U}U = I \tag{3.3.11}$$

but in that case

$$U\tilde{U} \neq I \tag{3.3.12}$$

A *triangular* matrix is defined by the property that all its elements above the main diagonal are zero.

If A is an arbitrary $n \times n$ square matrix, we call the equation

$$Ax = \lambda x \tag{3.3.13}$$

the "*eigenvalue problem*" associated with A. The scalars $\lambda_1, \lambda_2, \ldots, \lambda_n$ for which the equation is solvable, are called the "eigenvalues" (or "characteristic values") of A while the vectors $x_1, x_2, \ldots, x_n$ are called the "eigenvectors" (or "principal axes") of A. The eigenvalues λ_i satisfy the characteristic equation

$$\begin{bmatrix} a_{11} - \lambda, & a_{12}, & \cdots & a_{1n} \\ a_{21} & a_{22} - \lambda, & & a_{2n} \\ \vdots & & & \\ a_{n1} & a_{n2} & \cdots & a_{nn} - \lambda \end{bmatrix} = 0 \tag{3.3.14}$$

This algebraic equation of nth order has always n generally complex roots. If they are all distinct, the eigenvalue problem (13) yields n distinct eigenvectors, whose length can be normalised to 1 by the condition

$$\tilde{x}x = 1 \tag{3.3.15}$$

If some of the eigenvalues coincide, the equation (13) may or may not have n linearly independent solutions. If the number of independent solutions is less than n, the matrix is "defective" in certain eigenvectors and is thus called a "defective matrix".

Any square matrix satisfies its own characteristic equation (the "Hamilton-Cayley identity"):

$$(A - \lambda_1 I)(A - \lambda_2 I) \ldots (A - \lambda_n I) \equiv 0 \tag{3.3.16}$$

Moreover, this is the identity of *lowest order* satisfied by A, if the λ_i are all distinct. If, however, only ρ of the eigenvalues are distinct, the identity of lowest order in the case of a non-defective matrix becomes

$$(A - \lambda_1 I)(A - \lambda_2 I) \ldots (A - \lambda_\rho I) \equiv 0 \tag{3.3.17}$$

Defective matrices, however, demand that some of the root factors shall appear in higher than first power. The difference between the lowest order at which the identity appears and ρ gives the number of eigenvectors in which the matrix is defective.

Problem 91. Let an $n \times n$ matrix M have the property that it commutes with any $n \times n$ matrix. Show that M must be of the form $M = \alpha I$.

Problem 92. Show that if λ is an eigenvalue of the problem (13), it is also an eigenvalue of the "adjoint" problem

$$\tilde{A}y = \lambda y \tag{3.3.18}$$

Problem 93. Let the eigenvalues $\lambda_1, \lambda_2, \ldots, \lambda_n$ of A be all distinct. Show that the matrix

$$A' = (A - \lambda_1 I) \ldots (A - \lambda_\mu I) \tag{3.3.19}$$

has a zero eigenvalue which is μ-fold.

Problem 94.   Show that the eigenvalues of A^m are the mth power of the original eigenvalues, while the eigen*vectors* remain unchanged.

Problem 95.   Show that the (complex) eigenvalues of an orthogonal matrix (10) must lie on the unit circle $|z| = 1$.

Problem 96.   Show that the following properties of a square matrix A remain unchanged by squaring, cubing, ..., of the matrix:

> a) symmetry
> b) orthogonality
> c) triangular quality.

Problem 97.   Show that if two non-defective matrices A and B coincide in eigenvalues and eigenvectors, they coincide altogether: $A - B \equiv 0$.

Problem 98.   Show that two defective matrices A and B which have the same eigenvalues and eigenvectors, need not coincide. (Hint: operate with two triangular matrices whose diagonal elements are all equal.)

Problem 99.   Show that $A^m \equiv 0$ does not imply $A \equiv 0$. Show that if A is an $n \times n$ matrix which does not vanish identically, it can happen that $A^2 \equiv 0$, or $A^3 \equiv 0, \ldots,$ or $A^n \equiv 0$ without any of the lower powers being zero.

Problem 100.   Investigate the eigenvalue problem of a triangular matrix whose diagonal elements are all equal. Show that by a small modification of the diagonal elements all the eigenvalues can be made distinct, and that the eigenvectors thus created are very near to each other in magnitude and direction, collapsing into one as the perturbation goes to zero.

3.4. Principal axis transformation of a symmetric matrix

The principal axis transformation of a symmetric matrix is one of the most fundamental tools of mathematical analysis which became of central importance in the study of linear differential and integral operators. We summarise in this section the formalism of this theory, for the sake of later applications.

While a matrix is primarily an algebraic tool, we gain greatly if the purely algebraic operations are complemented by a geometrical picture. In this picture we consider the scalar equation

$$\tilde{x}Sx = 1 \tag{3.4.1}$$

where S is a symmetric (more generally Hermitian) $n \times n$ matrix:

$$S = \tilde{\overline{S}} \tag{3.4.2}$$

and the components of the vector

$$\tilde{x} = (x_1, x_2, \ldots, x_n) \tag{3.4.3}$$

are conceived as the rectangular coordinates of a point x in an n-dimensional Euclidean space with the distance expression

$$\overline{x\xi}^2 = (x_1 - \xi_1)^2 + \ldots (x_n - \xi_n)^2 = (\tilde{x} - \tilde{\xi})(x - \xi) \tag{3.4.4}$$

In particular the distance from the origin is given by

$$s^2 = \tilde{x}x \tag{3.4.5}$$

The equation (1) can be conceived as the equation of a *second order surface* in an n-dimensional space. The eigenvalue problem

$$Su = \lambda u \tag{3.4.6}$$

characterises those directions in space in which the radius vector and the normal to the surface become parallel. Moreover in consequence of (1) we obtain

$$\lambda = \frac{1}{\tilde{u}u} \tag{3.4.7}$$

which means that these eigenvalues λ_i can be interpreted as the reciprocal square of the distance of those points of the surface in which radius vector and normal are parallel. Hence the λ_i are interpreted in terms of *inherent properties* of the second order surface which proves that they are independent of any special reference system and thus *invariants of an arbitrary orthogonal transformation*, a transformation being "orthogonal" if it leaves the distance expression (5) invariant.

The freedom of choosing our coordinate system is of greatest importance in both physics and mathematics. Many mathematical problems are solved, or at least greatly reduced in complexity, by formulating them in a properly chosen system of coordinates. The transformation law

$$A' = \tilde{U}AU \tag{3.4.8}$$

for an $n \times n$ matrix where U is an arbitrary orthogonal matrix, defined by (3.10), has the consequence that an arbitrary relation between $n \times n$ matrices which involves any combination of the operations *addition*, *multiplication* and *transposition*, remains valid in all frames of reference.

The n solutions of the eigenvalue problem (3.13) can be combined into the single matrix equation

$$SU = U\Lambda \tag{3.4.9}$$

where the eigenvectors defined by (6) are now arranged as successive *columns* of the matrix U, while the diagonal matrix Λ is composed of the eigenvalues $\lambda_1, \lambda_2, \ldots, \lambda_n$:

$$\Lambda = \begin{bmatrix} \lambda_1 & & & \\ & \lambda_2 & & 0 \\ & & \cdot & \\ & & & \cdot \\ 0 & & & & \cdot \\ & & & & & \lambda_n \end{bmatrix} \tag{3.4.10}$$

While in the case of a general matrix A the eigenvalues λ_i are generally complex numbers and we cannot guarantee even the existence of n eigenvectors—they may all collapse into one vector—here we can make much more definite predictions. The eigenvalues λ_i are always *real* and the eigenvectors are always present to the full number n. Moreover, they are in the case of distinct eigenvalues automatically *orthogonal* to each other, while in the case of multiple roots they *can be orthogonalised*—with an arbitrary rotation remaining free in a definite μ-dimensional subspace if μ is the multiplicity of the eigenvalue λ_i. Furthermore, the length of the eigenvectors can be normalised to 1, in which case U becomes an *orthogonal* matrix:

$$\tilde{U}U = U\tilde{U} = I \qquad (3.4.11)$$

But then we can introduce a new reference system in which the eigenvectors —that is the columns of U—are introduced as a new set of coordinate axes (called the "principal axes"). This means the transformation

$$x = Ux' \qquad (3.4.12)$$

Introducing this transformation in the equation (1) we see that the same equation formulated in the new (primed) reference system becomes

$$\tilde{x}'S'x' = 1 \qquad (3.4.13)$$

where

$$S' = \tilde{U}SU \qquad (3.4.14)$$

On the other hand, premultiplication of (10) by $\tilde{U}$ gives the fundamental relation

$$\tilde{U}SU = S' = \Lambda \qquad (3.4.15)$$

This means that in the new reference system (the system of the principal axes), the matrix S is reduced to a *diagonal matrix* and the equation of the second order surface becomes

$$\lambda_1 x'_1{}^2 + \lambda_2 x'_2{}^2 + \ldots + \lambda_n x'_n{}^2 = 1 \qquad (3.4.16)$$

Now we can make use of the fact that the λ_i are *invariants of an orthogonal transformation*. Since the coefficients of an algebraic equation are expressible in terms of the roots λ_i, we see that *the entire characteristic equation* (3.14) *is an invariant of an orthogonal transformation*. This means that we obtain n invariants associated with an orthogonal transformation because the coefficient of every power of λ is an invariant. The most important of these invariants are the coefficient of $(-\lambda)^0$ and the coefficient of $(-\lambda)^{n-1}$. The former is obtainable by putting $\lambda = 0$ and this gives the *determinant* of the coefficients of S, simply called the "determinant of S"

and denoted by $\|S\|$.   The latter is called the "spur" of the matrix and is equal to the *sum of the diagonal terms*:

$$|S| = s_{11} + s_{22} + \ldots + s_{nn} \qquad (3.4.17)$$

But in the reference system of the principal axes the determinant of S' becomes the product of all the λ_i, and thus

$$\|S\| = \lambda_1 \lambda_2 \ldots \lambda_n \qquad (3.4.18)$$

while the "spur" of S' is equal to the *sum* of the λ_i and thus

$$|S| = s_{11} + s_{22} + \ldots + s_{nn} = \lambda_1 + \lambda_2 + \ldots + \lambda_n \qquad (3.4.19)$$

Problem 101.   Writing a linear transformation of the coordinates in the form

$$x = Ux' \qquad (3.4.20)$$

show that the invariance of (5) demands that U satisfy the orthogonality conditions (3.10).

Problem 102.   Show that by considering the principal axis transformation of S, S^2, S^3, $\ldots$, S^n, we can obtain all the n invariants of S by taking the spur of these matrices

$$|S^k| = \lambda_1{}^k + \lambda_2{}^k + \ldots + \lambda_n{}^k \qquad (k = 1, 2, \ldots, n) \qquad (3.4.21)$$

Investigate in particular the case $k = 2$ and show that this invariant is equal to the sum of the squares of the absolute values of all the elements of the matrix S:

$$|s_{11}|^2 + |s_{12}|^2 + \ldots + |s_{nn}|^2 = \lambda_1{}^2 + \lambda_2{}^2 + \ldots + \lambda_n{}^2 \qquad (3.4.22)$$

Problem 103.   Show that the following properties of a matrix are invariants of an arbitrary rotation (orthogonal transformation):

        a) symmetry
        b) anti-symmetry
        c) orthogonality
        d) the matrices 0 and I
        e) the scalar product $\tilde{x}y$ of two vectors.

Problem 104.   Show that for the invariance of the *determinant* and the *spur* the symmetry of the matrix is not demanded: they are invariants of an orthogonal transformation for *any* matrix.

Problem 105.   Show that the eigenvalues of a real anti-symmetric matrix $\tilde{A} = -A$ are purely imaginary and come in *pairs*: $\lambda_i = \pm i\beta_i$. Show that one of the eigenvalues of an anti-symmetric matrix of odd order is always *zero*.

Problem 106.   Show that if all the eigenvalues of a symmetric matrix S collapse into one: $\lambda_i = \alpha$, that matrix must become $S = \alpha I$.

Problem 107.   Find the eigenvalues and principal axes of the following matrix and demonstrate explicitly the transformation theorem (16), together with the validity of the spur equations (21):

$$S = \begin{bmatrix} 1 & 1 & 1 & 1 \\ 1 & 1 & -1 & -1 \\ 1 & -1 & -1 & 1 \\ 1 & -1 & 1 & -1 \end{bmatrix} \tag{3.4.23}$$

[Answer:

$$\lambda = 2 \qquad 2 \qquad -2 \qquad -2$$

$$U = \begin{bmatrix} \dfrac{1}{2} & \dfrac{1}{\sqrt{2}} & \dfrac{1}{2} & 0 \\ -\dfrac{1}{2} & \dfrac{1}{\sqrt{2}} & -\dfrac{1}{2} & 0 \\ \dfrac{1}{2} & 0 & -\dfrac{1}{2} & \dfrac{1}{\sqrt{2}} \\ \dfrac{1}{2} & 0 & -\dfrac{1}{2} & -\dfrac{1}{\sqrt{2}} \end{bmatrix} \tag{3.4.24}]$$

Problem 108.   Find the eigenvalues and principal axes of the following Hermitian matrix and demonstrate once more the validity of the three spur equations (21):

$$S = \begin{bmatrix} 0 & i & -2i \\ -i & 0 & 2i \\ 2i & -2i & 0 \end{bmatrix} \tag{3.4.25}$$

[Answer:

$$\lambda = 0 \qquad 3 \qquad -3$$

$$U = \begin{bmatrix} \dfrac{2}{\sqrt{5}} & \dfrac{1}{2} - \dfrac{i}{6} & \dfrac{1}{2} + \dfrac{i}{6} \\ \dfrac{2}{\sqrt{5}} & -\dfrac{1}{2} - \dfrac{i}{6} & -\dfrac{1}{2} + \dfrac{i}{6} \\ \dfrac{1}{\sqrt{5}} & \dfrac{2}{3}i & -\dfrac{2}{3}i \end{bmatrix} \tag{3.4.26}]$$

Problem 109.   Show that in every principal axis of a Hermitian matrix a complex phase factor of the form $e^{i\theta_k}$ remains arbitrary.

Problem 110.   Show that if a matrix is simultaneously *symmetric and orthogonal*, its eigenvalues can only be ± 1.

Problem 111.   Show that the following class of $n \times n$ matrices are simultaneously symmetric and orthogonal (cf. Section 2.19):

$$b_{\alpha\beta} = \sqrt{\frac{2}{n+1}} \sin \alpha\beta \frac{\pi}{n+1} \tag{3.4.27}$$

Show that for all even n the multiplicity of the eigenvalues ± 1 is even, while for odd n the multiplicity of $+1$ surpasses the multiplicity of -1 by one unit.

Problem 112.   Construct another class of $n \times n$ symmetric and orthogonal matrices by writing down the elements

$$a_{\alpha\beta} = \cos \alpha\beta \, \frac{\pi}{n-1} \qquad (\alpha, \beta = 1, 2, \ldots, n-2)$$

and bordering them by the

upper horizontal $\Big\}$ $\dfrac{1}{2}$, $\dfrac{1}{\sqrt{2}}$, $\dfrac{1}{\sqrt{2}}$, $\dfrac{1}{\sqrt{2}}$, $\cdots$, $\dfrac{1}{2}$
and left vertical

(3.4.28)

lower horizontal $\Big\}$ $\dfrac{1}{2}$, $-\dfrac{1}{\sqrt{2}}$, $\dfrac{1}{\sqrt{2}}$, $-\dfrac{1}{\sqrt{2}}$, $\cdots$, $(-1)^{n-1}\dfrac{1}{2}$
and right vertical

The resulting matrix is multiplied by the scalar $\sqrt{\dfrac{2}{n-1}}$.

Problem 113.   Consider the cases $n = 2$ and 3.   Show that here the sine and cosine matrices coincide.   Obtain the principal axes for these cases.
[Answer:

$$n = 2 \qquad U = \begin{bmatrix} \overset{\lambda = -1}{-\sin \dfrac{\pi}{8}} & \overset{\lambda = 1}{\cos \dfrac{\pi}{8}} \\[2mm] \cos \dfrac{\pi}{8} & \sin \dfrac{\pi}{8} \end{bmatrix} \qquad n = 3 \qquad U = \begin{bmatrix} \overset{\lambda = -1}{-\dfrac{1}{2}} & \overset{1}{\dfrac{1}{\sqrt{2}}} & \overset{1}{\dfrac{1}{2}} \\[2mm] \dfrac{1}{\sqrt{2}} & 0 & \dfrac{1}{\sqrt{2}} \\[2mm] \dfrac{1}{2} & \dfrac{1}{\sqrt{2}} & -\dfrac{1}{2} \end{bmatrix}$$

Problem 114.   Show that an arbitrary matrix which is simultaneously symmetric and orthogonal, can be constructed by taking the product

$$A = U\Lambda\tilde{U} \qquad (3.4.29)$$

where U is an arbitrary orthogonal matrix, while the elements of the diagonal elements of Λ are ± 1, in arbitrary sequence.

3.5. Decomposition of a symmetric matrix

The defining equation (4.9) of the principal axes gives rise to another fundamental relation if we do not post-multiply but *pre*-multiply by $\tilde{U}$, taking into account the orthogonality of the matrix U:

$$S = U\Lambda\tilde{U} \qquad (3.5.1)$$

This shows that *an arbitrary symmetric matrix can be obtained as the product of three factors: the orthogonal matrix U, the diagonal matrix Λ, and the transposed orthogonal matrix $\tilde{U}$.*

A further important fact comes into evidence if it so happens that one or more of the eigenvalues λ_i are *zero*. Let us then separate the zero eigenvalues from the non-zero eigenvalues:

$$
\Lambda = \begin{bmatrix}
\lambda_1 & & & & & & & \\
& \lambda_2 & & & & & & \\
& & \cdot & & & & & \\
& & & \cdot & & & & \\
& & & & \cdot & & & \\
& & & & & \lambda_p & & \\
& & & & & & 0 & \\
& & & & & & & \cdot \\
& & & & & & & & \cdot \\
& & & & & & & & & \cdot \\
& & & & & & & & & & 0
\end{bmatrix} \tag{3.5.2}
$$

We do the same with the eigen*vectors* u_i of the matrix U:

$$
\frac{\lambda_1, \lambda_2, \ldots, \quad \lambda_p, \quad 0, \ldots, 0}{u_1, u_2, \ldots, \quad u_p, \quad u_{p+1}, \ldots, u_n} \tag{3.5.3}
$$

We consider the product (1) and start with post-multiplying U by Λ. This means by the rules of matrix multiplication that the successive columns of U become multiplied in succession by $\lambda_1, \lambda_2, \ldots, \lambda_p, 0, 0, \ldots, 0$. In consequence we have the p columns $\lambda_1 u_1, \lambda_2 u_2, \ldots, \lambda_p u_p$, while the rest of the columns *drop out identically*. Now we come to the post-multiplication by $\tilde{U}$. This means that we should multiply the *rows* of our previous construction by the *columns* of $\tilde{U}$ which, however, is equivalent to the *row by row multiplication* of $U\Lambda$ with U. We observe that *all the vectors $u_{p+1}, u_{p+2}, \ldots, u_n$ are obliterated* and the result can be formulated in terms of the *semi-orthogonal matrix U_p* which is composed of the first p columns of the full matrix U, without any further columns. Hence it is not an $n \times n$ but an $n \times p$ matrix. We likewise omit all the zero eigenvalues of the diagonal matrix Λ and reduce it to the $p \times p$ diagonal matrix

$$
\Lambda = \begin{bmatrix}
\lambda_1 & & & & \\
& \lambda_2 & & & \\
& & \cdot & & \\
& & & \cdot & \\
& & & & \cdot \\
& & & & & \lambda_p
\end{bmatrix} \tag{3.5.4}
$$

Our decomposition theorem now becomes

$$
S = U_p \Lambda_p \tilde{U}_p \tag{3.5.5}
$$

which generates the symmetric $n \times n$ matrix S as a product of the semi-orthogonal $n \times p$ matrix U, the $p \times p$ diagonal matrix Λ_p and the $p \times n$ matrix $\tilde{U}_p$ which is the transpose of the first factor U_p,

Problem 115. Demonstrate the validity of the decomposition theorem (1) for the matrix (4.23) of Problem 107.

Problem 116. Demonstrate the validity of the decomposition theorem (5) for the matrix (4.25) of Problem 108.

3.6. Self-adjoint systems

If we have an arbitrary linear system of equations

$$Ay = b \tag{3.6.1}$$

we obtain the "adjoint system"

$$\tilde{A}x = c \tag{3.6.2}$$

by transposing the matrix of the original system. If the matrices of the two systems coincide, we speak of a "self-adjoint system". In that case we have the condition $A = \tilde{A}$ which means that the matrix of the linear system is a symmetric (in the case of complex elements Hermitian) matrix. But such a matrix has special properties which can immediately be utilised for the solution of a system of linear equations.

We can introduce a new reference system by rotating the original axes into the principal axes of the matrix A. This means the transformation

$$\begin{aligned} y' &= Uy & b' &= Ub \\ y &= \tilde{U}y' & b &= \tilde{U}b' \end{aligned} \tag{3.6.3}$$

$$A'y' = b' \tag{3.6.4}$$

where

$$A' = \tilde{U}AU = \Lambda \tag{3.6.5}$$

Hence in the new reference system the linear system (1) appears in the form

$$\Lambda y' = b' \tag{3.6.6}$$

Since Λ is a mere *diagonal matrix*, our equations are now separated and immediately solvable—provided that they are in fact solvable. This is certainly the case if none of the eigenvalues of Λ is zero. In that case the inverse matrix

$$\Lambda^{-1} = \begin{bmatrix} \lambda_1^{-1} & & & \\ & \lambda_2^{-1} & & 0 \\ & & \cdot & \\ & & & \cdot \\ & 0 & & & \cdot \\ & & & & \lambda_n^{-1} \end{bmatrix} \tag{3.6.7}$$

exists and we obtain the solution y' in the form

$$y' = \Lambda^{-1}b' \tag{3.6.8}$$

This shows that *a self-adjoint linear system whose matrix is free of zero eigenvalues is always solvable and the solution is unique.*

But what happens if some of the eigenvalues λ_i vanish? Since any number multiplied by zero gives zero, the equation

$$\lambda_i y'_i = b'_i \tag{3.6.9}$$

is only solvable for $\lambda_i = 0$ if $b'_i = 0$. Now b'_i can be interpreted as the ith component of the vector b in the reference system of the principal axes:

$$b'_i = \tilde{u}_i b \tag{3.6.10}$$

and thus the condition

$$\tilde{u}_i b = 0 \tag{3.6.11}$$

has the geometrical significance that the vector b is *orthogonal* to the ith principal axis. That principal axis was defined by the eigenvalue equation

$$A u_i = \lambda_i u_i \tag{3.6.12}$$

which, in view of the vanishing of λ_i, is now reduced to

$$A u_i = 0 \tag{3.6.13}$$

If more than one eigenvalue vanishes, then the equation

$$A y = 0 \tag{3.6.14}$$

has more than one linearly independent solution. And since the condition (11) has to hold for *every* vanishing eigenvalue, while on the other hand these are *all* the conditions demanded for the solvability of the linear system (1), we obtain the fundamental result that *the necessary and sufficient condition for the solvability of a self-adjoint linear system is that the right side is orthogonal to every linearly independent solution of the homogeneous equation $Ay = 0$.*

Coupled with these "compatibility conditions" (11) goes a further peculiarity of a zero eigenvalue. The equation

$$0 \cdot y'_i = 0 \tag{3.6.15}$$

is solvable for *any arbitrary y'_i*. The solution of a linear system with a vanishing eigenvalue is *no longer unique*. But the appearance of the free component y'_i in the solution means from the standpoint of the original reference system that the product $y'_i u_i$ can be added to any valid solution of the given linear system. In the case of several principal axes of zero eigenvalue an arbitrary linear combination of these axes can be added and we still have a solution of our linear system. On the other hand, this is *all* the freedom left in the solution. But "an arbitrary linear combination of the zero axes" means, on the other hand, an arbitrary solution of the homogeneous equation (14). And thus we obtain another fundamental result: "*The general solution of a compatible self-adjoint system is obtained by adding to an arbitrary particular solution of the system an arbitrary solution of the homogeneous equation $Ay = 0$.*"

Problem 117. Show that the last result holds for any distributive operator A.

3.7. Arbitrary $n \times m$ systems

We now come to the investigation of an arbitrary $n \times m$ linear system

$$Ay = b \qquad (3.7.1)$$

where the matrix A has n rows and m columns and transforms the column vector y of m components into the column vector b of n components. Such a matrix is obviously associated with *two spaces*, the one of n, the other of m dimensions. We will briefly call them the N-space and the M-space. These two spaces are in a duality relation to each other. If the vector y of the M-space is given, the operator A operates on it and transplants it into the N-space. On the other hand, if our aim is to *solve* the linear system (1), we are given the vector b of the N-space and our task is to find the vector y of the M-space which has generated it through the operator A.

However, in the present section we shall not be concerned with any method of *solving* the system (1) but rather with a general investigation of the basic *properties* of such systems. Our investigation will not be based on the determinant approach that Kronecker and Frobenius employed in their algebraic treatment of linear systems, but on an approach which carries over without difficulty into the field of continuous linear operators.

The central idea which will be basic for all our discussions of the behaviour of linear operators is the following. We will not consider the linear system (1) in isolation but *enlarge it by the adjoint $m \times n$ system*

$$\tilde{A}x = c \qquad (3.7.2)$$

The matrix $\tilde{A}$ has m rows and n columns and accordingly the vectors x and c are in a reciprocity relation to the vectors y and b, x and b being vectors of the N-space, y and c vectors of the M-space.

The addition of the system (2) has no effect on the system (1) since the vectors x and c are entirely independent of the vectors y and b, and vice versa. But the addition of the system (2) to (1) enlarges our viewpoint and has profound consequences for the deeper understanding of the properties of linear systems.

We combine the systems (1) and (2) into the larger scheme

$$Sz = a \qquad (3.7.3)$$

where we now introduce a new $(n + m)$ by $(n + m)$ symmetric (respectively Hermitian) matrix S, defined as follows:

$$S = \begin{bmatrix} O & A \\ \tilde{A} & O \end{bmatrix} \tag{3.7.4}$$

The linear system (3) can now be pictured as follows:

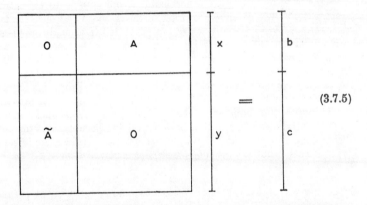

$$\begin{bmatrix} O & A \\ \tilde{A} & O \end{bmatrix} \begin{bmatrix} x \\ y \end{bmatrix} = \begin{bmatrix} b \\ c \end{bmatrix} \tag{3.7.5}$$

The vectors (x, y) combine into the single vector z from the standpoint of the larger system, just as the vectors (b, c) combine into the larger vector a. However, for our present purposes we shall prefer to maintain the individuality of the vectors (x, y) and formulate all our results in vector pairs, although they are derived from the properties of the unified system (5).

Since the unified system has a *symmetric* matrix, we can immediately apply all the results we have found in Sections 4, 5, and 6. First of all, we shall be interested in the principal axis transformation of the matrix S. For this purpose we have to establish the fundamental eigenvalue equation

$$Sw = \lambda w \tag{3.7.6}$$

which in view of the specific character of our matrix (4) appears in the following form, putting $w = (u, v)$:

$$Av = \lambda u$$
$$\tilde{A}u = \lambda v \qquad \qquad (3.7.7)$$

We will call this pair of equations the "shifted eigenvalue problem", since on the right side the vectors u and v are in shifted position, compared with the more familiar eigenvalue problem (3.13), (3.18). It is of interest to observe that the customary eigenvalue problem loses its meaning for $n \times m$ matrices, due to the heterogeneous spaces to which u and v belong, while the shifted eigenvalue problem (7) is always meaningful. We know in advance that it must be meaningful and yield real eigenvalues since it is merely the formulation of the standard eigenvalue problem associated with a symmetric matrix, which is always a meaningful and completely solvable problem. We also know in advance that we shall obtain $n + m$ mutually orthogonal eigenvectors, belonging to $n + m$ independent eigenvalues, although the eigenvalues may not be all distinct (the characteristic equation can have multiple roots).

The orthogonality of two w_i eigenvectors now takes the form

$$\tilde{u}_i u_k + \tilde{v}_i v_k = 0 \qquad (i \neq k) \qquad (3.7.8)$$

but we can immediately add an interesting consequence of the equations (7). Let λ_i be a non-zero eigenvalue. Then, together with the solution (v, u, λ) goes the solution $(v, -u, -\lambda)$ and thus—combining the solutions for λ_i and $-\lambda_k$—we can complement the relation (8) by the equation

$$\tilde{u}_i u_k - \tilde{v}_i v_k = 0 \qquad (i \neq k) \qquad (3.7.9)$$

which yields

$$\tilde{u}_i u_k = 0$$
$$\tilde{v}_i v_k = 0 \qquad (i \neq k) \qquad (3.7.10)$$

thus demonstrating that the vectors u_i and v_i *in themselves* form an orthogonal set of vectors. This holds so far only for all λ_i which are not zero. But we can extend the result to *all* u_i and v_i vectors if we premultiply the first equation (7) by A, respectively the second equation by $\tilde{A}$. This shows that the vectors u_i, and likewise the vectors v_j, can be formulated *independently of each other*, as solutions of the eigenvalue problems

$$A\tilde{A}u = \lambda^2 u \qquad (3.7.11)$$

$$\tilde{A}Av = \lambda^2 v \qquad (3.7.12)$$

Now $A\tilde{A}$ is in itself a symmetric $n \times n$ matrix which operates in the N-space, while $\tilde{A}A$ is a symmetric $m \times m$ matrix which operates in the M-space; (the symmetry of these matrices follows at once by applying the transposition rule (3.4) to a product). Consequently we must obtain n mutually orthogonal u_i vectors as a result of (11) and m mutually orthogonal v_j vectors as a result of (12).   These vectors can serve as an orthogonal set of base vectors which span the entire N-, respectively, M-space.

We will picture these two spaces by their base vectors which are arranged in successive columns.   We thus obtain *two square matrices*, namely the U-matrix, formed out of the n vectors, $u_1, u_2, \ldots, u_n$, and the V-matrix, formed out of the m vectors $v_1, v_2, \ldots, v_m$.   While these two spaces are quite independent of each other, yet the two matrices U and V are related by the coupling which exists between them due to the original eigenvalue problem (7) which may also be formulated in terms of the matrix equations

$$AV = U\Lambda$$
$$\tilde{A}U = V\tilde{\Lambda}$$

$$(3.7.13)$$

This coupling must exist for every non-zero eigenvalue λ_i while for a zero eigenvalue the two equations

$$Av_i = 0 \qquad\qquad\qquad (3.7.14)$$

$$\tilde{A}u_j = 0 \qquad\qquad\qquad (3.7.15)$$

become *independent* of each other.

We will now *separate* the zero eigenvalue from the non-zero eigenvalues. Let us assume that our eigenvalue problem (7) has p independent solutions if we set up the condition that only *positive* eigenvalues λ_i are admitted. They give us p independent solutions $(u_i, v_i; \lambda_i)$, to which we can add the p additional solutions $(u_i, -v_i; -\lambda_i)$, thus providing us with $2p$ solutions of the principal axis problem (6).   On the other hand, the eigenvalue problem (11) must have n independent solutions and, since p was the total number of non-vanishing eigenvalues, the remaining $n - p$ axes must belong to the eigenvalue zero.   Hence the number of independent solutions of (15) must be $n - p$.   By exactly the same reasoning the number of independent solutions of (14) must be $m - p$.   And since these axes are not paired— that is they appear in the form $(u_j, 0)$, respectively $(0, v_i)$—the zero eigenvalue has the multiplicity

$$\mu = m + n - 2p \qquad\qquad (3.7.16)$$

which, together with the $2p$ "paired" axes actually generate the demanded $m + n$ principal axes of the full matrix (4).

Problem 118.   By applying the orthogonality condition (8) to the pair $(u_i, v_i; \lambda_i)$, $(u_i, -v_i; -\lambda_i)$, $(\lambda_i \neq 0)$, demonstrate that the normalisation of the length of u_i to 1 automatically normalises the length of the associated v_i to 1 (or vice versa).

3.8. Solvability of the general $n \times m$ system

The extended matrix S puts us in the position to give a complete answer to the problem of solving arbitrary $n \times m$ linear systems.   This answer was found in Section 6 for arbitrary self-adjoint systems, but now it so happens that the unification of the two problems (7.1) and (7.2) in the form (7.5) actually yields a self-adjoint system and thus the results of our previous investigation become directly applicable. In particular we can state explicitly what are the compatibility conditions to be satisfied by the right side (b, c) of the unified system (7.5) which will make a solution possible. This condition appeared in the form (6.11) and thus demands the generation of the eigenvectors (u_i, v_i) associated with the eigenvalue zero. The necessary and sufficient condition for the solvability of the system (7.5) will thus appear in the general form

$$\tilde{u}_i b_i + \tilde{v}_i c_i = 0 \tag{3.8.1}$$

where (u_i, v_i) is any principal axis associated with the eigenvalue $\lambda = 0$:

$$\begin{aligned} Av &= 0 \\ \tilde{A}u &= 0 \end{aligned} \tag{3.8.2}$$

But now we have seen that these equations fall apart into the two independent sets of solutions

$$u = u_{p+1}, \quad u_{p+2}, \ldots, u_n, \quad v = 0$$

and

$$v = v_{p+1}, \quad v_{p+2}, \ldots, v_m, \quad u = 0 \tag{3.8.3}$$

Consequently the conditions (1) separate into the two sets:

$$\begin{aligned} \tilde{u}_{p+j} b_i &= 0 && (j = 1, 2, \ldots, n - p) \\ \tilde{v}_{p+k} c_i &= 0 && (k = 1, 2, \ldots, m - p) \end{aligned} \tag{3.8.4}$$

and these compatibility conditions can be interpreted as follows: *the necessary and sufficient condition for the solvability of an arbitrary $n \times m$ system is that the right side is orthogonal to all linearly independent solutions of the adjoint homogeneous system.*

The theorem concerning the uniqueness or non-uniqueness ("deficiency") of a solution remains the same as that found before in Section 6: *The general solution of the linear system (7.1) is obtained by adding to an arbitrary particular solution an arbitrary solution of the homogeneous equation $Ay = 0$.* The number $m - p$, which characterises the number of linearly independent solutions of the homogeneous system $Ay = 0$ is called the "degree of deficiency" of the given system (7.1).

Problem 119.   The customary definition of the rank p of a matrix is: "The maximum order of all minors constructed out of the matrix which do not vanish in their totality." In our discussion the number p appeared as "the number of positive eigenvalues for which the shifted eigenvalue problem (7.7) is solvable". Show that these two definitions agree; hence our p coincides with the customary "rank" of a matrix.

Problem 120.   Show that the eigenvalue zero can only be avoided if $n = m$.

Problem 121.   Show that an under-determined system ($n < m$) may or may not demand compatibility conditions, while an over-determined system ($n > m$) always demands at least $n - m$ compatibility conditions.

Problem 122.   Show that the number p must lie between 1 and the smaller of the two numbers n and m:

$$1 \le p \le \min \{n, m\} \qquad (3.8.5)$$

Problem 123.   Prove the following theorem: "The sum of the square of the absolute values of all the elements of an arbitrary (real) $n \times m$ matrix is equal to the sum of the squares of the eigenvalues $\lambda_1, \lambda_2, \ldots, \lambda_p$."

Problem 124.   Given the following 4×5 matrix:

$$A = \begin{bmatrix} 2 & 4 & -1 & 0 & 3 \\ 1 & -2 & 1 & 3 & 2 \\ 4 & 0 & 1 & 6 & 7 \\ 5 & 6 & -1 & 3 & 8 \end{bmatrix} \qquad (3.8.6)$$

Determine the deficiencies (and possibly compatibility conditions) of the linear system $Ay = b$.   Obtain the complete solution of the shifted eigenvalue problem (7.7), to 5 decimal places.

[Hint: Obtain first the solution of the homogeneous equations $Av = 0$ and $\tilde{A}u = 0$.   This gives $3 + 2 = 5$ axes and $p = 2$.   The 4×4 system $A\tilde{A}u = \lambda^2 u$ is reducible to a 2×2 system by making use of the orthogonality of u to the two zero-solutions, thus obtaining λ^2 by solving a quadratic equation.]

[Answer:

$$\text{Deficiency}: \alpha v_3 + \beta v_4 + \gamma v_5 \qquad (3.8.7)$$
$$\text{Compatibility}: \tilde{u}_3 b = 0, \quad \tilde{u}_4 b = 0 \qquad (3.8.8)$$

$\lambda_1 = 15.4347433$	$\lambda_2 = 6.91149043$		$\lambda = 0$	
u_1	u_2	u_3	u_4	
0.28682778	-0.46661531	1	2	
0.15656598	0.52486565	2	1	
0.59997974	0.58311599	-1	0	
0.73023153	-0.40836497	0	-1	
v_1	v_2	v_3	v_4	v_5
0.43935348	-0.01703335	-2	6	14
0.33790963	-0.77644357	3	-3	-1
-0.01687774	0.28690800	8	0	0
0.40559800	0.55678265	0	-4	0
0.72662988	0.06724708	0	0	-8

$$(3.8.9)$$

(Note that the zero axes have not been orthogonalised and normalised.)]

3.9. The fundamental decomposition theorem

In Section 4, when dealing with the properties of symmetric matrices, we derived the fundamental result that by a proper rotation of the frame of reference an arbitrary symmetric matrix can be transformed into a *diagonal* matrix Λ (cf. 4.15). Later, in Section 5, we obtained the result that an arbitrary symmetric matrix could be decomposed into a product of 3 matrices: the semi-diagonal matrix U_p, the diagonal matrix Λ_p, and the transpose $\tilde{U}_p$ of the first matrix. The eigenvalue $\lambda = 0$ was eliminated in this decomposition theorem (5.5). An entirely analogous development is possible for an arbitrary $n \times m$ matrix, on the basis of the eigenvalue problem (7.13). We have seen that both matrices U and V are orthogonal and thus satisfy the orthogonality conditions (3.10). If now we pre-multiply the first equation by $\tilde{U}$ we obtain

$$\tilde{U}AV = \Lambda \tag{3.9.1}$$

The matrix $\tilde{U}$ is $n \times n$, the matrix A $n \times m$, the matrix V $m \times m$. Hence the product is an $n \times m$ matrix, and so is the diagonal matrix Λ, whose elements are all zero, except for the diagonal elements, where we find the positive eigenvalues $\lambda_1, \lambda_2, \ldots, \lambda_p$ and, if the diagonal contains more elements than p, the remaining elements are all zero.

The equation (1) takes the place of the previous equation (4.15). In the special case that A is a symmetric $n \times n$ matrix, the two orthogonal matrices U and V become equal and Λ becomes a diagonal square matrix. In the general case the transformation of A into a diagonal form requires pre-multiplication and post-multiplication by two *different* orthogonal matrices U and V.

We can, however, also *post*-multiply the first equation by V, thus obtaining

$$A = U\Lambda\tilde{V} \tag{3.9.2}$$

The matrix A is now obtained as the product of the $n \times n$ orthogonal matrix U, the $n \times m$ diagonal matrix Λ and the $m \times m$ transpose of the diagonal matrix V.

Here again we want to separate the p positive eigenvalues of Λ from the remaining zero eigenvalues. We define Λ_p as the positive square matrix

$$\Lambda_p = \begin{bmatrix} \lambda_1 & & & & \\ & \lambda_2 & & & \\ & & \cdot & & \\ & & & \cdot & \\ & & & & \cdot \\ & & & & & \lambda_p \end{bmatrix} \tag{3.9.3}$$

The product $U\Lambda$ requires that the successive columns of U be multiplied by $\lambda_1, \lambda_2, \ldots, \lambda_p$, while the remaining part of the matrix *vanishes identically*. Now we shall multiply this product with the columns of $\tilde{V}$, or we may also say: with the rows of V. But then all the columns of V beyond v_p are automatically eliminated and what remains is

$$A = U_p\Lambda_p\tilde{V}_p \tag{3.9.4}$$

where U_p is the semi-orthogonal $n \times p$ matrix which is formed out of the column vectors $u_1, u_2, \ldots, u_p$, while V is a semi-orthogonal $m \times p$ matrix, formed out of the column vectors $v_1, v_2, \ldots, v_p$. The matrix A is thus obtained as a product of an $n \times p$, $p \times p$, and $p \times m$ matrix which actually gives an $n \times m$ matrix.

In order to understand the true significance of the decomposition theorem (4), let us once more pay attention to the two spaces: the N-space and the M-space, with which the matrix A is associated. We represent these spaces with the help of the n eigenvectors $u_1, u_2, \ldots, u_n$, respectively the m eigenvectors $v_1, v_2, \ldots, v_m$. However, we will make a sharp division line beyond the subscript p. The first p vectors u_i, v_i, although belonging to two different spaces, are *paired* with each other. They form the matrices U_p and V_p but we shall prefer to drop the subscript p and call them simply U and V, while the remaining portions of the N-space and M-space, associated with the eigenvalue zero, will be included in the matrices U_0 and V_0:

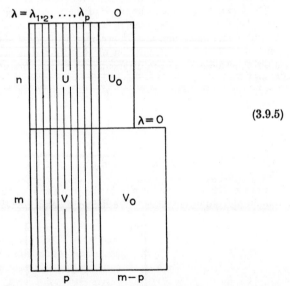

$$(3.9.5)$$

The fundamental matrix decomposition theorem (4) appears now in the form

$$A = U \Lambda \tilde{V} \qquad (3.9.6)$$

and reveals the remarkable fact that *the operator A can be generated without any knowledge of the principal axes associated with the zero eigenvalue*, that is without any knowledge of the solutions of the homogeneous equations $Av = 0$ and $\tilde{A}u = 0$. These solutions gave us vital information concerning the compatibility and deficiency of the linear system $Ay = b$, but exactly these solutions are completely *ignored* by the operator A.

We can understand this peculiar phenomenon and obtain an illuminating view concerning the general character of a linear operator if we form the concept of the "proper space" or "eigenspace" or "operational space" associated with the matrix A as an operator. It is true that the matrix A is associated with an n-dimensional and an m-dimensional vector space. But it so happens that the matrix A is *not activated in all dimensions of these two spaces* but only in a definite *subspace* which is in both cases p-dimensional. Only in the limiting case $n = m = p$ can it happen that the operator A includes the entire N-space and the entire M-space. If $n < m$ (under-determined system) the entire M-space (the space of the right side) may be included but the N-space (the space of the solution) is only partially represented. If $n > m$ (over-determined system) the entire N-space (the space of the solution) may be included but the M-space (the space of the right side) is only partially represented. The reason for the necessity of compatibility conditions and for the deficiency of linear systems is exactly this *partial activation* of the operator A. We will call the principal axes belonging to the positive eigenvalues the "essential axes" of the matrix since they are in themselves sufficient for the construction of A. The remaining axes which belong to the eigenvalue zero (the "zero-field") are the "deficient axes" in which the matrix is not activated, which are in fact *ignored* by the matrix. The matrix A has in a sense a "blind spot" in all these dimensions. We could use the picture that the field spanned by the p axes $u_1, u_2, \ldots, u_p$ and $v_1, v_2, \ldots, v_p$ is "illuminated" by the operator while the remaining fields are left in the dark. In this sense the "eigen-space" of the matrix A as an operator is limited to the p-dimensional subspaces which are spanned by the columns of the matrices U and V while the spaces spanned by the columns of U_0 and V_0 fall outside the operational space of A.

This concept of "activation" is very useful for the understanding of the peculiarities of linear systems. The concept retains its usefulness in the study of continuous operators (differential and integral operators) where the same phenomenon occurs under analogous circumstances.

The separation of the fields U_0, V_0 from the fields U, V has the further advantage that it yields a particularly simple formulation of the compatibility problem and the deficiency problem associated with the solution of the general linear system (7.1). The compatibility conditions can now be written down in the form of the single matrix equation

$$\tilde{U}_0 b = 0 \tag{3.9.7}$$

while the deficiency of the system appears in the form of

$$y = \bar{y} + V_0 \eta \tag{3.9.8}$$

where $\bar{y}$ is any particular solution, and η an arbitrary column vector of $m - p$ components.

Problem 125. Apply the general decomposition theorem (6) to a row vector $\tilde{a}$, considered as a $1 \times m$ matrix. Do the same for the column vector a, considered as an $n \times 1$ matrix.

[Answer:

$$\text{a) } p = 1, \quad \lambda = |a|, \quad u = (1), \quad \tilde{v} = \left(\frac{a_1}{|a|}, \frac{a_2}{|a|}, \dots, \frac{a_m}{|a|}\right)$$

$$\text{b) } p = 1, \quad \lambda = |a|, \quad \tilde{u} = \left(\frac{a_1}{|a|}, \dots, \frac{a_n}{|a|}\right), \quad v = (1) \text{]}$$

Problem 126. Obtain the decomposition of the following 2×3 matrix:

$$A = \begin{bmatrix} 0 & 0 & 0 \\ 0 & 1 & 0 \end{bmatrix} \tag{3.9.9}$$

[Answer: $p = 1, \quad \lambda = 2$

$$A = \begin{bmatrix} 0 \\ 1 \end{bmatrix} [2] [0, 1, 0] \tag{3.9.10}]$$

Problem 127. Construct the 4×5 matrix (8.6) with the help of the two non-zero eigensolutions, belonging to λ_1 and λ_2, of the table (8.9).
[Answer: Carry out numerically the row-by-row operation

$$[\lambda_1 u_1, \lambda_2 u_2] \circ [v_1, v_2] \tag{3.9.11}]$$

3.10. The natural inverse of a matrix

The ordinary inverse of a matrix A is defined by the equation

$$A A^{-1} = A^{-1} A = I \tag{3.10.1}$$

Such an inverse exists only in the limiting case $m = n = p$, that is when the eigen-space of the matrix includes the *complete* vector spaces N and M. In that case the relations

$$\tilde{U} U = \tilde{V} V = I \tag{3.10.2}$$

(which are always valid because U and V are always *semi-orthogonal*) are reversible:

$$U \tilde{U} = V \tilde{V} = I \tag{3.10.3}$$

and U and V become fully orthogonal. Then we can complement the construction of A out of U and V according to (9.6) by the construction of a new matrix B according to the rule

$$B = V \Lambda^{-1} \tilde{U} \tag{3.10.4}$$

This construction is always possible since the matrix

$$\Lambda^{-1} = \begin{bmatrix} \lambda_1^{-1} & & & & \\ & \lambda_2^{-1} & & & \\ & & \cdot & & \\ & & & \cdot & \\ & & & & \cdot \\ & & & & & \lambda_p^{-1} \end{bmatrix} \tag{3.10.5}$$

always exists.   Moreover, the products AB and BA can always be formed:

$$AB = U\tilde{U} \qquad\qquad (3.10.6)$$

$$BA = V\tilde{V} \qquad\qquad (3.10.7)$$

But these products become I only in the case that the relations (3) hold and that is only true if $m = n = p$.

Has the matrix B any significance in the general case in which p is *not* equal to m and n?   Indeed, this is the case and we have good reasons to consider the matrix B as the *natural inverse of A*, even in the general case. Let us namely take in consideration that the operation of the matrix A is restricted to the spaces spanned by the matrices U and V.   The spaces U_0 and V_0 *do not exist* as far as the operator A is concerned.   Now the unit matrix I has the property that, operating on any arbitrary vector u or v, it leaves that vector unchanged.   Since, however, the concept of an "arbitrary vector" is meaningless in relation to the operator A—whose operation is restricted to the eigen-spaces U, V—it is entirely sufficient and adequate to replace the unit matrix I by a less demanding matrix which leaves any vector *belonging to the subspaces U and V* unchanged.

The product AB, being an $n \times n$ matrix, can only operate on a vector of the N-space and if we want this vector to belong to the subspace U, we have to set it up in the form

$$u^{(p)} = U\eta \qquad\qquad (3.10.8)$$

where η is an arbitrary column vector of p elements.   But then, we obtain in view of (6):

$$ABu^{(p)} = U\tilde{U}U\eta = U\eta = u^{(p)} \qquad\qquad (3.10.9)$$

which demonstrates that the product AB has actually the property to *leave any vector belonging to the U-space unchanged.*

On the other hand, the product BA is an $m \times m$ matrix which can only operate on a vector of the M-space.   If again we want to restrict this vector to the subspace V, we have to set it up in the form

$$v^{(p)} = V\eta \qquad\qquad (3.10.10)$$

where once more η is an arbitrary column vector of p elements.   Then we obtain in view of (7):

$$BAv^{(p)} = V\tilde{V}V\eta = V\eta = v^{(p)} \qquad\qquad (3.10.11)$$

and once more we have demonstrated that the product BA has actually the property to *leave any vector belonging to the V-space unchanged.*

The matrix B is thus the natural substitute for the non-existent "strict inverse", defined by (1), and may aptly be called the "natural inverse" of the matrix A.   It is an operator which is uniquely associated with A and whose domain of operation coincides with that of A.   It ignores completely the fields U_0 and V_0.   If B operates on any vector of the subspace U_0, it annihilates that vector. Similarly, if $\tilde{B}$ operates on any vector of the subspace V_0, it likewise annihilates that vector.

Let us now see what happens if we solve the linear equation

$$Ay = b \tag{3.10.12}$$

by

$$y = Bb \tag{3.10.13}$$

Have we found the solution of our equation? Substitution in (12) yields the condition

$$ABb = b \tag{3.10.14}$$

But this condition is actually satisfied—as we have just seen—if b belongs to the subspace U. This, however, was exactly the *compatibility condition* of our linear system. The orthogonality of b to all u-vectors with the eigenvalue zero means that b has *no projection* in the U_0 space and that again means that it belongs completely to the U-space. And thus we see that we *have* found the solution of our system (12), whenever a solution is possible at all. In fact, more can be said. If b does not satisfy the compatibility condition (9.7), then there is still a solution possible *in the sense of least squares*. That is, while the difference $Ay - b$ cannot be made zero, we can at least *minimise the length of the error vector*, that is we can make the scalar

$$r^2 = |Ay - b|^2 \tag{3.10.15}$$

as small as possible. It so happens that the solution (13)—which blots out all projections of b into the U_0 space—automatically coincides with the desired least square solution in the case of incompatible systems.

But what can we say about the possible *deficiency* of the system (12)? If our solution is not unique but allows an infinity of solutions, there is obviously no mathematical trick by which we can overcome this deficiency. And yet the solution (13) seems to give a *definite* answer to our problem. Now it is actually true that we have not eliminated the deficiency of our system. If the homogeneous equation $Ay = 0$ has solutions, any such solution may be added to (13). These, however, are exactly the vectors which constitute the space V_0. And so the deficiency of our system means that our solution is uniquely determined only as far as the space V goes, in which the operator is activated. The solution may have an additional projection into the V_0 space which is ignored by the operator A. This is a piece of information which our linear system (12) is *unable to give* since the space V_0 is outside of its competence. Under these circumstances it is natural to *normalise* our solution by putting it entirely into the well-confined space V in which the operator is activated *and in which the solution is unique*. The projection outside of that space, being outside the competence of the given operator, is equated to zero. This is the significance of the uniqueness of the solution (13) which has the role of a natural normalisation of the solution in the case of incomplete (deficient) systems. The lacking projection of y into V_0 has to be obtained by additional information which our system (12) is unable to give.

Problem 128.   Consider the equation $Ay = b$ where A is the 4 × 5 matrix (8.6) of Problem 124, while b is given as follows:

$$\tilde{b} = (1, 0, 0, 0) \tag{3.10.16}$$

Obtain the *normalised least square solution* (13) of this system, *without* making use of the complete eigenvalue analysis contained in the table (8.9).   Then check the result by constructing the matrix B with the help of the two essential axes belonging to λ_1 and λ_2.

[Hint: Make the right side b orthogonal to u_3 and u_4.   Make the solution y orthogonal to v_3, v_4, v_5, thus reducing the system to a 2 × 2 system which has a unique solution.]

[Answer:

$$\text{Modified } b: \ b' = \left(\frac{3, -2, -1, 4}{10} \right) \tag{3.10.17}$$

$$\tilde{y} = \left(\frac{53, 334, -112, -171, 51}{5690} \right) \tag{3.10.18}]$$

Problem 129.   The least square solution of the problem $Ay = b$ is equivalent to the solution of the system

$$\tilde{A}Ay = \tilde{A}b \tag{3.10.19}$$

Show that for this system the compatibility condition (9.7) is automatically fulfilled.   Show also that for an over-determined system which is free of deficiencies the solution (13), constructed with the help of the B matrix, *coincides with the solution of the system* (19).

3.11. General analysis of linear systems

The study of linear systems with a symmetric matrix is greatly facilitated by the fact that a mere rotation of the reference system can diagonalise the matrix.   The equations are then separated and can be solved at once.   The decomposition theorem (9.6) puts us in the position to extend these advantages to arbitrary $n \times m$ systems.   We cannot expect, of course, that a mere "change of the frame of reference" shall suffice since our problem involves *two* spaces of generally different dimensionality.   We succeed, however, in our endeavour if we apply a proper rotation in the one space and in the other space, although these two rotations are generally *quite different*.   The case of a symmetric $n \times n$ matrix is then distinguished only in that respect that here the vector b and the vector y can be subjected to the *same* orthogonal transformation, while in the general case the two orthogonal transformations *do not coincide*.

We write down our linear equation (7.1), but substituting for A its decomposition (9.6):

$$U\Lambda\tilde{V}y = b \tag{3.11.1}$$

Let us now perform the following orthogonal transformations:

$$b = Ub' \tag{3.11.2}$$

$$y = Vy' \tag{3.11.3}$$

Then, in view of the semi-orthogonality of U and V we obtain, if we pre-multiply by $\tilde{U}$:

$$\Lambda y' = b' \tag{3.11.4}$$

Here we have the full counterpart of the equation (6.6) which we have encountered earlier in the study of $n \times n$ linear systems whose matrix was symmetric. Now we have succeeded in generalising the procedure to arbitrary non-symmetric matrices of the general $n \times m$ type.

But let us notice the peculiar fact that the new equation (4) is a $p \times p$ system while the original system was an $n \times m$ system. How did this reduction come about?

We understand the nature of this reduction if we study more closely the nature of the two orthogonal transformations (2) and (3). Since generally the U and V matrices are not full orthogonal matrices but $n \times p$ respectively $m \times p$ matrices, the transformations (2) and (3) put a definite *bias* on the vectors b and y. We can interpret these two equations as saying that b is inside the U-space, y inside the V-space. The first statement is not necessarily true but if it is not true, then our system is incompatible and allows no solution. The second statement again is not necessarily true since our system may be incomplete in which case the general solution appears in the form (9.8) which shows that the solution y can have an arbitrary projection into V_0. However, we take this deficiency for granted and are satisfied if we find a *particular* solution of our system which can be later augmented by an arbitrary solution of the homogeneous equation. We distinguish this particular solution by the condition that it stays entirely within the V-space. This condition makes our solution unique.

Since the subspaces U and V are both p-dimensional, it is now understandable that our problem was reducible from the original $n \times m$ system to a $p \times p$ system. Moreover, the equations of the new system are separated and are solvable at once:

$$y' = \Lambda^{-1}b' \tag{3.11.5}$$

The matrix Λ^{-1}—encountered before in (10.5)—always exists since the diagonal elements of Λ can never be zero. We obtain one and only one solution. Going back to our original y by rotating back to the original system we obtain

$$y = v\Lambda^{-1}b' \tag{3.11.6}$$

Moreover, the premultiplication of (2) by $\tilde{U}$ gives

$$b' = \tilde{U}b \tag{3.11.7}$$

and thus

$$y = V\Lambda^{-1}\tilde{U}b \tag{3.11.8}$$

We have thus obtained exactly the same solution that we have encountered before in (10.13) when we were studying the properties of the "natural inverse" of a matrix.

We should well remember, however, the circumstances which brought this unique solution in existence:

1. We took it for granted that the compatibility conditions of the system are satisfied. This demands that the right side b shall lie inside the p-dimensional subspace U of the full N-space.

2. We placed the solution in the eigenspace of the matrix A, and that is the p-dimensional subspace V of the full M-space.

Problem 130. Show that for the solution of the adjoint system (7.2) the role of the spaces U and V is exactly *reversed*. Obtain the reduction of this $m \times n$ system to the $p \times p$ system of equation (4).

3.12. Error analysis of linear systems

The spectacular advances in the design of large-scale digital computers brought the actual numerical solution of many previously purely theoretically solved problems into the limelight of interest. While a generation ago systems of 10 to 20 simultaneous linear equations taxed our computing facilities to the utmost, we can now tackle the task of solving linear systems with hundreds of unknowns. Together with this development goes, however, the obligation to understand the peculiarities and idiosyncrasies of large-scale linear systems in order to preserve us from a possible *misuse* of the big machines, caused not by any technical defects of the machines but by the defects of the given mathematical situation.

The problem very frequently encountered can be described as follows. Our goal is to solve the linear system

$$Ay = b \qquad (3.12.1)$$

We have taken care of the proper degree of determination by having just as many equations as unknowns. Furthermore, we are assured that the determinant of the system is not zero and thus the possibility of a zero eigenvalue is eliminated. We then have the ideal case realised:

$$m = n = p \qquad (3.12.2)$$

and according to the rules of algebra our system must have one and only one solution. The problem is entrusted to a big computing outfit which carries through the calculations and comes back with the answer. The engineer looks at the solution and shakes his head. A number which he knows to be positive came out as negative. Something seems to be wrong also with the decimal point since certain components of y go into thousands when he knows that they cannot exceed 20. All this is very provoking and he tells the computer that he must have made a mistake. The computer points out that he has checked the solution, and the equations checked with an accuracy which goes far beyond that of the data. The meeting breaks up in mutual disgust.

What happened here? It is certainly true that the ideal case (2) guarantees a unique and finite answer. It is also true that with our present-day electronic facilities that answer is obtainable with an accuracy which goes far beyond the demands of the engineer or the physicist. Then how could anything go wrong?

The vital point in the mathematical analysis of our problem is that the *data* of our problem are not mere numbers, obtainable with any accuracy we like, but the *results of measurements*, obtainable only with a limited accuracy, let us say an accuracy of 0.1%. On the other hand, the engineer is quite satisfied if he gets the solution with a 10% accuracy, and why should that be difficult with data which are 100 times as good?

The objection is well excusable. The peculiar paradoxes of linear systems have not penetrated yet to the practical engineer whose hands are full with other matters and who argues on the basis of experiences which hold good in many situations but fail in the present instance. We are in the fortunate position that we can completely analyse the problem and trace the failure of that solution to its origins, showing the engineer point by point how the mishap occurred.

We assume the frequent occurrence that the matrix A itself is known with a high degree of accuracy while the *right side b* is the result of measurements. Then the correct equation (1) is actually *not* at our disposal but rather the modified equation

$$Ay_1 = b_1 \qquad (3.12.3)$$

where

$$b_1 = b + \beta \qquad (3.12.4)$$

the given right side, differs from the "true" right side b by the "error vector" β. From the known performance of our measuring instruments we can definitely tell that the length of β cannot be more than a small percentage of the measured vector b—let us say 0.1%.

Now the quantity the computer obtains from the data given by the engineer is the vector

$$y_1 = y + \eta \qquad (3.12.5)$$

since he has solved the equation (3) instead of the correct equation (1). By substituting (5) in (3) we obtain for the error vector η of the solution the following determining equation:

$$A\eta = \beta \qquad (3.12.6)$$

The question now is whether the relative smallness of β will have the relative smallness of η in its wake, and this is exactly the point which has to be answered by "no".

As in the previous section, we can once more carry through our analysis most conveniently in a frame of reference which will separate our equations.

Once more, as before in (11.2) and (11.3), we introduce the orthogonal transformation

$$\beta = U\beta'$$
$$\eta = V\eta' \tag{3.12.7}$$

and reduce our system to the diagonal form

$$\varLambda\eta' = \beta' \tag{3.12.8}$$

Of course, we are now in a better position than before. The U and V matrices are now *full* orthogonal matrices, we have no reduction in the dimensions, all our matrices remain of the size $n \times n$ and thus we need not bother with either compatibility conditions or deficiency conditions—at least this is what we assume.

In actual fact the situation is much less rosy. *Theoretically* it is of the greatest importance that none of the eigenvalues λ_i becomes zero. This makes the system (8) solvable and *uniquely* solvable. But this fact can give us little consolation if we realise that a *very small* λ_i will cause a tremendous *magnification* of the error β'_i in the direction of that principal axis, if we come to the evaluation of the corresponding error caused in the *solution*:

$$\eta'_i = \frac{1}{\lambda_i}\beta'_i \tag{3.12.9}$$

Let us observe that the transformation (7) is an *orthogonal* transformation in both β and η. The length of neither β nor η is influenced by this transformation. Hence

$$|\eta'| = |\eta|$$
$$|\beta'| = |\beta| \tag{3.12.10}$$

Let us assume that the eigenvalue λ_i in the direction of a certain axis happens to be 0.001. Then, according to (9), the small error 0.01 in the data causes immediately the large error 10 in the solution. What does that mean percentage-wise? It means the following. Let us assume that in some other directions the eigenvalues are of the order of magnitude of unity. Then we can say that the order of magnitude of the solution vector y and that of the data vector b is about the same. But *the small error of 1% in one component of the data vector has caused the intolerably large error of 1000% in the solution vector*. That is, the error caused by an inaccuracy of our data which is not more than 1%, has been blown up by solving the linear system to such an extent that it appears in the solution as a vector which is *ten times as big as the entire solution*. It is clear that under such circumstances the "solution" is completely valueless. It is also clear that it was exactly the *correct mathematical inversion of the matrix A* which caused the trouble since it put the tremendous premium of 1000 on that axis and thus magnified the relatively small error in that direction beyond all proportions. Had we used a computing technique which does not invert the matrix but obtains the solution of a linear system in successive approximations, bringing the small eigenvalues *gradually* in appearance, we would have fared much better because we could have stopped at a proper point, before the trouble with the very small eigenvalues developed.*

We see that the critical quantity to which we have to pay attention, is the *ratio of the largest to the smallest eigenvalue*, also referred to as the "condition number" of the matrix:

$$\nu = \frac{\lambda_1}{\lambda_n} \qquad (3.12.11)$$

This popular expression is not without its dangers since it creates the impression that an "ill-conditioned" matrix is merely in a certain mathematical "condition" which could be remedied by the proper know-how. In actual fact we should recognise the general principle that *a lack of information cannot be remedied by any mathematical trickery*. If we ponder on our problem a little longer, we discover that it is actually the *lack of information* that causes the difficulty. In order to understand what a small eigenvalue means, let us first consider what a *zero eigenvalue* means. If in one of the equations of the system (11.4) for example the ith equation, we let λ_i converge to zero, this means that the component y'_i appears in our linear system *with the weight zero*. We can trace back this component to the original vector y, on account of the equation

$$y' = \bar{V}y \qquad (3.12.12)$$

By the rules of matrix multiplication we obtain the component y'_i by multiplying the ith row of $\bar{V}$—and that means the ith *column* of V—by y. And thus we have

$$y'_i = \bar{v}_i y \qquad (3.12.13)$$

Hence we can give a very definite meaning to a vanishing eigenvalue. By forming the scalar product (13), we obtain a definite linear combination of the unknowns $y_1, y_2, \ldots, y_m$, which is *not represented in the given linear system* and which cannot be recovered by any tricks. We have to take it for granted that this information is denied us. We see now that the determination of all the linearly independent solutions of the homogeneous equation $Av = 0$ has the added advantage that it gives very definite information concerning those combinations of the unknowns which are not represented in our system. We have found for example that the 4×5 matrix (8.6) had the three zero-axes v_3, v_4, v_5, tabulated in (8.9). We can now add that if we try to solve the system

$$Ay = b$$

there will be three linear combinations of the 5 unknowns, which are *a priori*

---

* See the author's paper on "Iterative solution of large-scale linear systems" in the *Journal* of SIAM 6, 91 (1958).

unobtainable because they are simple *not represented* in the system. They are:

$$-2y_1 + 3y_2 + 8y_3$$
$$6y_1 - 3y_2 + 4y_3$$
$$14y_1 - y_2 - 7y_5$$

Of course, any linear combinations of these three quantities are likewise unavailable and so we can state the non-existent combinations in infinitely many formulations. If we are interested in the solution of the adjoint system

$$\tilde{A}x = c$$

here the axes u_3 and u_4 come in operation and we see that the two combinations

$$x_1 + 2x_2 - x_3$$
$$2x_1 + x_2 - x_4$$

are *a priori* un-determinable (or any linear aggregate of these two expressions).

Now we also understand what a very *small* eigenvalue means. A certain linear combination of the unknowns, which can be determined in advance, does not drop out completely but is *very weakly represented* in our system. If the data of our system could be trusted with *absolute accuracy*, then the degree of weakness would be quite immaterial. As long as that combination is present *at all*, be it ever so weakly, we can solve our system and it is merely a question of numerical skill to obtain the solution with any degree of accuracy. But the situation is very different if our data are of limited accuracy. Then the very meagre information that our numerical system gives with respect to certain linear combinations of the unknowns is not only unreliable—because the errors of the data do not permit us to make any statement concerning their magnitude—but the indiscriminate handling of these axes ruins our solution even with respect to that information that we could otherwise usefully employ. If we took the values of these weak combinations from some other information—for example by casting a horoscope or by clairvoyance or some other tool of para-psychology—we should probably fare much better because we might be right at least in the order of magnitude, while the mathematically correct solution has no hope of being adequate even in roughest approximation.

This analysis shows how important it is to get a reliable estimate concerning the "condition number" of our system and to reject linear systems whose condition number (11) surpasses a certain danger point, depending on the accuracy of our data. If we admit such systems at all, we should be aware of the fact that they are only theoretically square systems. In reality they are $n \times m$ systems ($n < m$) which are deficient in certain combinations of the unknowns and which are useful only for those combinations of the unknowns which belong to eigenvalues which do not go below a certain limit. Experience shows that large-scale systems especially are often prone to be highly skew-angular ("ill-conditioned") and special care is required in the perusal of their results.

Problem 131. Given the following 3 × 3 matrix:

$$A = \begin{bmatrix} 3 & -1 & 2 \\ -1 & 5 & -3 \\ 0 & 2 & -1 \end{bmatrix} \qquad (3.12.14)$$

what combination of the coordinates is not obtainable by solving the system $Ay = b$, respectively $\tilde{A}x = c$?

[Answer:

$$-y_1 + y_2 + 2y_3; \quad x_1 + 3x_2 - 7x_3 \qquad (3.12.15)]$$

Problem 132. Give the following interpretation of the non-obtainable (or almost non-obtainable) coordinate combinations. Rotate the position vector $(y_1, y_2, \ldots, y_m)$ of the M-space into the (orthonormal) reference system of the principal axes $(v_1, v_2, \ldots, v_m)$; since the axes $v_{p+1}, \ldots, v_m$ are not represented (or almost not represented) in the operator A, the components $y'_{p+1}, \ldots, y'_m$ become multiplied by zero (or exceedingly small numbers) and thus disappear.

3.13. Classification of linear systems

One of the peculiarities of linear systems is that our naive notions concerning enumeration of equations and unknowns fail to hold. On the surface we would think that n equations suffice for the determination of n unknowns. We would also assume that having less equations than unknowns our system will have an infinity of solutions. Both notions can easily be disproved. The following system of three equations with three unknowns

$$x + y + z = 1$$
$$2x + 2y + 2z = 2$$
$$x - y + z = 3$$

is clearly unable to determine the 3 unknowns x, y, z since in fact we have only *two* equations, the second equation being a mere repetition of the first. Moreover, the following system of two equations for three unknowns

$$x + y + z = 1$$
$$2x + 2y + 2z = 3$$

is far from having an infinity of solutions. It has no solution at all since the second equation contradicts the first one. This can obviously happen with *any* number of unknowns, and thus an arbitrarily small number of equations (beyond 1) with an arbitrarily large number of unknowns may be self-contradictory.

The only thing we can be sure of is that a linear system can have no unique solution if the number of equations is less than the number of unknowns. Beyond that, however, we can come to definite conclusions only if in our analysis we pay attention to *three* numbers associated with a matrix:

1. The number of equations: n
2. The number of unknowns: m
3. The rank of the matrix: p.

It is the relation of p to n and m which decides the general character of a given linear system.

The "rank" p can be decided by studying the totality of linearly independent solutions of the homogeneous equation

$$Av = 0 \tag{3.13.1}$$

or

$$\tilde{A}u = 0 \tag{3.13.2}$$

The analysis of Section 9 has shown that these two numbers are not independent of each other. If we have found that the first equation has μ independent solutions, then we know at once the rank of the matrix since

$$\mu = m - p \tag{3.13.3}$$

and thus

$$p = m - \mu \tag{3.13.4}$$

On the other hand, if the number of linearly independent solutions of the second equation is ν, then

$$\nu = n - p \tag{3.13.5}$$

and thus

$$\mu - \nu = m - n \tag{3.13.6}$$

According to the relation of p to n and m we can put the linear systems into various classes. The following two viewpoints in particular are decisive:

1. Are our data sufficient for a unique characterisation of the solution? If so, we will call our system "completely determined", if not, "underdetermined".

2. Are our data independent of each other and thus freely choosable, or are there some linear relations between them so that we could have dropped some of our data without losing essential information? In the first case we will call our system "free" or "unconstrained", in the second case "overdetermined" or "constrained". Over-determination and under-determination can go together since some of our data may be merely linear functions of some basic data (and thus superfluous) and yet the totality of the basic data may not be sufficient for a unique solution.

These two viewpoints give rise to four different classes of linear systems;

1. *Free and complete.* The right side can be chosen freely and the solution is unique. In this case the eigen-space of the operator includes the entire M and N spaces and we have the ideal case

$$m = n = p \qquad (3.13.7)$$

$$(3.13.8)$$

Such a system is sometimes called "well-posed", adopting an expression that J. Hadamard used for the corresponding type of boundary value problems in his "Lectures on the Cauchy-problem". It so happens, however, that so many "ill-posed" problems can be transformed into the "well-posed" category that we prefer to characterise this case as the "well-determined" case.

2. *Constrained and complete.* The right side is subjected to compatibility conditions but the solution is unique. The operator still includes the entire M-space but the N-space extends beyond the confines of the eigen-space U of the operator. Here we have the case

$$p = m < n \qquad (3.13.9)$$

$$(3.13.10)$$

3. *Free and incomplete.* The right side is not subjected to any conditions but the solution is not unique. The operator now includes the entire

N-space but the M-space extends beyond the confines of the eigen-space V of the operator. Here we have the case

$$p = n < m \qquad (3.13.11)$$

The number of equations is smaller than the number of unknowns. Such systems are under-determined.

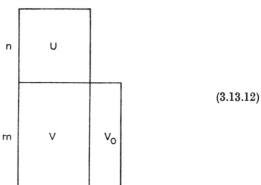

$$(3.13.12)$$

4. *Constrained and incomplete.* The right side is subjected to compatibility conditions and the solution is not unique. The eigen-space of the operator is more restricted than either the M- or the N-space. Here we have the case

$$p < \frac{m}{n} \qquad (3.13.13)$$

$$(3.13.14)$$

The relation of m to n, however, remains undecided and we may have the three sub-cases

$$m < n, \quad m = n, \quad m > n \qquad (3.13.15)$$

Irrespective of this relation, a system of this kind is simultaneously over-determined and under-determined because in some dimensions we have given too much, in some others too little.

Numerical illustrations for these categories are given in the following problems.

Problem 133. Show that the following system is *free and complete* (well-determined):

$$\begin{bmatrix} 1 & 2 & 3 \\ -1 & 4 & 3 \\ 3 & -2 & 2 \end{bmatrix} \quad y = \begin{bmatrix} 6 \\ 0 \\ 13 \end{bmatrix} \qquad (3.13.16)$$

[Answer:

$$\tilde{y} = (1, -2, 3) \qquad n = m = p = 3 \qquad (3.13.17)]$$

Problem 134. Show that the following system is *constrained and complete*:

$$\begin{bmatrix} 1 & 2 & 3 \\ -1 & 4 & 3 \\ 3 & -2 & 2 \\ 0 & 1 & -1 \\ 2 & -1 & -1 \end{bmatrix} \quad y = \begin{bmatrix} 6 \\ 0 \\ 13 \\ -5 \\ 1 \end{bmatrix} \qquad (3.13.18)$$

[Hint: Use successive eliminations.]
[Answer:

2 compat. cond.:
$$7b_1 - 5b_2 - 4b_3 - 2b_4 = 0$$
$$9b_1 - 7b_2 - 4b_3 - 2b_5 = 0$$

Solution: $\tilde{y} = (1, -2, 3) \quad p = m = 3; n = 5]$ $\qquad (3.13.19)$

Problem 135. Show that the following system is *free and incomplete*:

$$\begin{bmatrix} 1 & -1 & 3 & 0 & 2 \\ 2 & 4 & -2 & 1 & -1 \\ 3 & 3 & 2 & -1 & -1 \end{bmatrix} \quad y = \begin{bmatrix} 6 \\ 2 \\ 4 \end{bmatrix} \qquad (3.13.20)$$

[Answer:

$$\tilde{y} = (11, -7, -4, 0, 0) \qquad p = n = 3, \quad m = 5$$
$$+ \alpha(21, -15, -12, -6, 0) \qquad (3.13.21)]$$
$$+ \beta(9, -7, -4, 0, -2)]$$

Problem 136. Show that the following system is *constrained and incomplete*:

$$\begin{bmatrix} 1 & -1 & 3 & 1 \\ 2 & 4 & 5 & 1 \\ -1 & 2 & -1 & 1 \\ 4 & 1 & 9 & 1 \end{bmatrix} \quad y = \begin{bmatrix} 12 \\ 29 \\ 4 \\ 37 \end{bmatrix} \qquad (3.13.22)$$

[Answer:

Compatibility condition: $b_1 + b_2 - b_3 - b_4 = 0$
Solution: $\tilde{y} = (7, 2, 0, 7) + \alpha(-2, 0, 1, -1)$ $\qquad (3.13.23)]$

3.14. Solution of incomplete systems

Since incomplete systems do not allow a unique solution but have an infinity of solutions, we would be inclined to think that such systems are of no significance and need no attention. We cannot take this attitude, however, if we study the properties of differential operators because we encounter very fundamental differential operators in the realm of field operators which from the standpoint of linear equation systems are highly deficient. For example one of the most important linear differential operators in the realm of partial differentiation is the "divergence" of a vector field:

$$\operatorname{div} F \equiv \frac{\partial F_1}{\partial x_1} + \frac{\partial F_2}{\partial x_2} + \ldots \frac{\partial F_n}{\partial x_n} = \rho \qquad (3.14.1)$$

This operator transforms a *vector* of the n-dimensional space into a scalar. Hence from the standpoint of solving this linear differential equation we have the case of determining a vector field from a scalar field which is clearly a highly under-determined problem. And yet, we do encounter the divergence operator as one of the fundamental operators of mathematical physics and it is by no means meaningless to ask: "What consequences can we draw from the fact that the divergence of a vector field is given?"

Now in the previous section we solved such systems by giving an arbitrary particular solution and then adding the general solution of the homogeneous equation. While this procedure is formally correct, it has the disadvantage that the "particular solution" from which we start, can be chosen with a high degree of arbitrariness. This hides the fact that our operator gives a very definite answer to our linear problem in all those dimensions in which the operator is activated, and, on the other hand, fails completely in all those dimensions in which the operator is not activated. But then it seems more adequate to the nature of the operator to give the unique solution in all those dimensions in which this solution exists and ignore those dimensions which are outside the realm of the operator. If we give *some* particular solution, this condition is not fulfilled because in all probability our solution will have some projection in the field V_0 which is not included by the operator. On the other hand, the solution found in Section 10 with the help of the B matrix, considered as the "natural inverse" of A, is unique and satisfies the condition that we make only statements about those co-ordinate combinations which are not outside the realm of the given operator. But to generate this "natural inverse" in every case by going through the task of first solving the shifted eigenvalue problem for all the non-zero axes and then constructing the matrix as a product of three factors, would be a formidable endeavour which we would like to avoid. There may be a simpler method of obtaining the desired particular solution.

This is indeed possible, in terms of a "generating vector". Let us put

$$y = \tilde{A}w \qquad (3.14.2)$$

and shift the task of determining y to the task of determining w. Now we know from the decomposition theorem (9.6) that

$$\tilde{A} = V\Lambda\tilde{U} \tag{3.14.3}$$

But then the vector $\tilde{A}w$—no matter what w may be—is automatically of the form Vq, that is we have a vector which lies completely within the activated field of the operator. The deficiency is thus eliminated and we obtain exactly the solution which we desire to get. The auxiliary vector w may not be unique but the solution y becomes unique.

We can thus *eliminate the deficiency* of any linear system and obtain the "natural solution" of that system by adopting the following method of solution:

$$\begin{aligned} y &= \tilde{A}w \\ A\tilde{A}w &= b \end{aligned} \tag{3.14.4}$$

The matrix $A\tilde{A}$ is an $n \times n$ square matrix. The original $n \times m$ system $(m > n)$ is thus transformed into a "well-posed" $n \times n$ system which can be solved by successive eliminations. If the vector w allows several solutions, this uncertainty is wiped out by the multiplication by $\tilde{A}$. The new solution y is *unique*, no matter how incomplete our original system has been.

For example, in the case of the above-mentioned divergence operator we shall see later that the significance of the substitution (2) is

$$F = -\operatorname{grad} \phi \tag{3.14.5}$$

with the boundary condition

$$\phi(\infty) = 0 \tag{3.14.6}$$

Hence the problem (1) becomes uniquely solvable because now we obtain

$$\operatorname{div} \operatorname{grad} \phi = \Delta\phi = -\rho \tag{3.14.7}$$

which is Poisson's equation and which has (under the condition (6)) a unique solution. This is in fact the traditional method of solving the problem (1). But we now see the deeper significance of the method: we gave a unique solution in all those dimensions of the function space in which the operator is activated and ignored all the other dimensions.

Problem 137. Obtain the solution of the incomplete system (13.20) by the method (2) and show that we obtain a unique solution which is orthogonal to the zero-vectors v_4, v_5.

[Answer:

$$\tilde{w} = \frac{-226, -366, 201}{239}$$

$$\tilde{y} = \frac{-85, -365, 636, -657, -377}{239} \tag{3.14.8}]$$

Problem 138. Do the same for the 4 × 4 system (13.22) and show that the deficiency of w has no influence on the solution y.

Answer:

$$\text{Solution:}\ \tilde{y} = \frac{0, 4, 7, 7}{2} \qquad\qquad (3.14.9)]$$

3.15. Over-determined systems

A similar development is possible for $n \times m$ systems in which the number of equations surpasses the number of unknowns: $n > m$. Such systems came first into prominence when Gauss found an ingenious method of adjusting physical measurements. It often happened that the constants of a certain mathematical law, applied to a physical phenomenon, had to be obtained by actual physical measurements. If more measurements were made than the number of constants warranted, the equation system used for the determination of the unknown parameters became redundant. If the measurements had been free of observational errors, this redundant system would be compatible, and any combination of the minimum number of equations would yield the same results. However, the observational errors have the consequence that each combination yields a different solution and we are in a quandary how to make our choice among the various possibilities. Gauss established the principle (found independently also by Legendre) that it is preferable to keep the redundant system in its totality and determine the "most probable" values of the parameters with the help of the principle that the sum of the squares of the left sides (after reducing the system to zero) should be made as small as possible ("method of least squares"). In the case that our equations are of a *linear* form, we obtain the principle of minimising the necessarily non-negative scalar quantity

$$\eta^2 = (\widetilde{Ay} - b)(Ay - b) = \tilde{y}\tilde{A}Ay - 2\tilde{y}\tilde{A}b + \tilde{b}b \qquad (3.15.1)$$

If the system

$$Ay - b = 0 \qquad\qquad (3.15.2)$$

happens to be *compatible*, then the minimum of (1) is zero and we obtain the correct solution of the system (2). Hence we have not lost anything by replacing the system (2) by the minimisation of (1) which yields

$$\tilde{A}Ay = \tilde{A}b \qquad\qquad (3.15.3)$$

We have gained, however, by the fact that the new system is *always solvable*, no matter how incompatible the original system might have been. The reason is that the decomposition (14.3) of $\tilde{A}$ shows that the new right side

$$\tilde{b} = \tilde{A}b = V\varLambda\tilde{U}b \qquad\qquad (3.15.4)$$

cannot have any component in the field V_0 because the pre-multiplication by V annihilates any such component. The system (3) is thus automatically *compatible.* Moreover, the new system is an $m \times m$ square system, compared with the larger $n \times m$ ($n > m$) system of the original problem. Hence the new problem is *well-posed*, provided that the original problem was only over-determined but not incomplete. As an example let us consider the over-determined system

$$\text{grad } \varphi = F \tag{3.15.5}$$

which transforms a scalar field φ into the vector field F. Let us apply the method (3) to this over-determined system. In Section 14 we have encountered the operator "div" and mentioned that its transpose is the operator "$-\text{grad}$". Accordingly the transpose of the operator "grad" is the operator "$-\text{div}$". Hence the least-square reformulation of the original equation (5) becomes

$$\text{div grad } \varphi = \text{div } F \tag{3.15.6}$$

which yields the scalar equation

$$\Delta\varphi = \text{div } F \tag{3.15.7}$$

and once more we arrive at Poisson's equation. Here again the procedure agrees with the customary method of solving the field equation (5) but we get a deeper insight into the significance of this procedure by seeing that we have applied the least-square reformulation of the original problem.

If we survey the results of the last two sections, we see that we have found the proper remedy against both under-determination and over-determination. In both cases the *transposed operator* $\tilde{A}$ played a vital role. We have eliminated under-determination by transforming the original y into the new unknown w by the transformation $y = \tilde{A}w$ and we have eliminated over-determination (and possibly incompatibility) by the method of multiplying both sides of the given equation by $\tilde{A}$. The unique solution thus obtained coincides with the solution (10.13), generated with the help of the "natural inverse" B.

Problem 139. Two quantities ξ and η are measured in such a way that their sum is measured μ times, their difference ν times. Find the most probable values of ξ and η. Solve the same problem with the help of the matrix B and show the agreement of the two solutions.

[Answer: Let the arithmetic mean of the sum measurements be α, the arithmetic mean of the difference measurements be β. Then

$$\xi = \frac{\alpha + \beta}{2}, \quad \eta = \frac{\alpha - \beta}{2} \tag{3.15.8}]$$

Problem 140. Form the product $\tilde{A}b$ for the system of Problem 128 and show that the vector thus obtained is orthogonal to the zero vectors v_3, v_4 and v_5 (cf. Problem 124).

3.16. The method of orthogonalisation

We can give still another formulation of the problem of removing deficiencies and constraints from our system. The characteristic feature of the solution (14.2) is that the solution is made orthogonal to the field V_0

which is composed of the zero-axes of the M-field. On the other hand the characteristic feature of the least square solution (15.3) is that the right side b is made orthogonal to the field U_0 which is composed of the zero axes of the N-field. If we possess all the zero axes—that is we know all the solutions of the homogeneous equations $Av = 0$, then we can remove the deficiencies and constraints of our system and transform it to a uniquely solvable system by carrying through the demanded orthogonalisation in direct fashion.

1. *Removal of the deficiency of the solution.* Let y_0 be a particular solution of our problem

$$Ay = b \qquad (3.16.1)$$

We want to find the properly normalised solution y which is orthogonal to the zero field V_0. For this purpose we consider the general solution

$$y = y_0 - V_0 q \qquad (3.16.2)$$

and determine the vector q by the orthogonality condition

$$\tilde{V}_0 y = \tilde{V}_0 y_0 - \tilde{V}_0 V_0 q = 0 \qquad (3.16.3)$$

This equation is solvable for q, in the form of a well-posed $(m - p) \times (m - p)$ system:

$$\tilde{V}_0 V_0 q = \tilde{V}_0 y_0 \qquad (3.16.4)$$

and having obtained q, we substitute in (2) and obtain the desired normalised solution.

2. *Removal of the incompatibility of the right side.* We can proceed similarly with the orthogonalisation of the right side b of a constrained system. We must not change anything on the projection of b into the field U but we have to subtract the projection into U_0. Hence we can put

$$\bar{b} = b - U_0 q \qquad (3.16.5)$$

and utilise the condition

$$\tilde{U}_0 \bar{b} = \tilde{U}_0 b - \tilde{U}_0 U_0 q = 0 \qquad (3.16.6)$$

which again yields for the determination of q the well-posed $(n - p) \times (n - p)$ system

$$\tilde{U}_0 U_0 q = \tilde{U}_0 b \qquad (3.16.7)$$

Substituting in (5) we obtain the new right side of the equation

$$Ay = \bar{b} \qquad (3.16.8)$$

which satisfies the necessary compatibility conditions. Now we can *omit* the surplus equations beyond $n = m$ and handle our problem as an $m \times m$ system.

By this method of orthogonalisation we can remain with the original formulation of our linear system and still obtain a solution which is unique and which satisfies the principle of least squares.

Problem 141. Obtain the solution (14.8) of Problem 137 by orthogonalisation. (For the zero-field V_0, cf. (13.21).) Do the same for the problem (13.22), and check it with the solution (14.9).

Problem 142. A certain quantity y has been measured n times:

$$y = \alpha_1$$
$$y = \alpha_2$$
$$\vdots$$
$$y = \alpha_n$$

$$(3.16.9)$$

Find the least square solution of this system (the arithmetic mean of the right sides) by orthogonalisation.

3.17. The use of over-determined systems

In many problems of mathematical analysis, over-determined systems can be used to great advantage. One of the most fundamental theorems in the theory of analytical functions is "Cauchy's integral theorem" which permits us to obtain the value of an analytical function $f(z)$ inside of a given domain of analyticity if $f(z)$ is known everywhere along the boundary:

$$f(z) = \frac{1}{2\pi i} \oint_C \frac{f(\zeta)d\zeta}{\zeta - z}$$

$$(3.17.1)$$

The "given data" are in this case the boundary values of $f(\zeta)$ along the boundary curve C and the relation (1) may be conceived as the solution of the Cauchy-Riemann differential equations, under the boundary condition that $f(z)$ assumes the values $f(\zeta)$ on the boundary. These "given values", however, are by no means freely choosable. It is shown in the theory of analytical functions that giving $f(\zeta)$ along an *arbitrarily small section* of C is sufficient to determine $f(z)$ everywhere inside the domain included by C. Hence the values $f(\zeta)$ along the curve C are by no means independent of each other. They have in fact to satisfy the compatibility conditions

$$\int_C f(\zeta)g(\zeta)d\zeta = 0$$

$$(3.17.2)$$

where $g(z)$ may be chosen as any analytical function of z which is free of singularities inside and on C. There are infinitely many such functions and thus our system is infinitely over-determined. And yet the relation (1) is one of the most useful theorems in the theory of analytical functions.

Another example is provided by the theory of Newtonian potential, which satisfies the Laplace equation

$$\Delta \varphi = 0 \tag{3.17.3}$$

inside of a closed domain. We have then a theorem which is closely analogous to Cauchy's theorem (1):

$$\varphi(\tau) = \frac{1}{4\pi} \oint_C \left[\varphi(S) \frac{\partial G(\tau, S)}{\partial n} - \frac{\partial \varphi(S)}{\partial n} G(\tau, S) \right] dS \tag{3.17.4}$$

where

$$G(\tau, S) = \frac{1}{R_{\tau S}} \tag{3.17.5}$$

$R_{\tau S}$ being the distance between the fixed point τ inside of C and a point S of the boundary surface C.

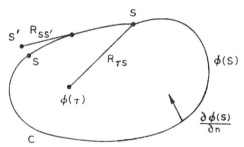

The given data here are the functional values $\varphi(S)$ along the boundary surface S, and the values of the normal derivative $\partial \varphi / \partial n$ along the boundary surface S. This is in fact too much since $\varphi(S)$ alone, or $(\partial \varphi / \partial n)(S)$ alone, would suffice to determine $\varphi(\tau)$ everywhere inside the domain. And thus our problem is once more infinitely over-determined. The given data have to satisfy the compatibility conditions

$$\int_C \left[\varphi(S) \frac{\partial g(S)}{\partial n} - \frac{\partial \varphi(S)}{\partial n} g(S) \right] dS = 0 \tag{3.17.6}$$

where $g(\tau)$ is any potential function which satisfies the Laplace equation (3) everywhere inside and on S, free of singularities. There are infinitely many such functions.

On the surface this abundance of data seems superfluous and handicapped by the constraints to which they are submitted. But the great advantage of the method is that we can operate with such a *simple function* as the *reciprocal distance between two points* as our auxiliary function $G(\tau, S)$. If we want to succeed with the minimum of data, we have first to construct the "Green's function" of our problem—as we will see later—which is for complicated boundaries a very difficult task. On the other hand, of what help is our solution, if in fact our observations give *only* $\varphi(S)$ and $(\partial\varphi/\partial n)(S)$ is not known?

In that case we use the *compatibility conditions themselves* to obtain the surplus data. We can consider the system (17.6) as an infinite set of linear equations for the determination of $\partial\varphi/\partial n$. If we succeed in solving these equations, then we possess now the surplus data $(\partial\varphi/\partial n)(S)$ and the relation (4) solves the problem of obtaining $\varphi(\tau)$ at all inside points.

This solution method is of actual practical advantage since the integration is replaceable by a summation of a large but finite number of terms. We then obtain a finite system of simultaneous algebraic equations of the form

$$Ay = b \tag{3.17.7}$$

Moreover, we can make this system well-conditioned by emphasising the diagonal terms. We do that by choosing for $g(\tau)$ the function

$$g(\tau) = \frac{1}{R_{\tau S'}} \tag{3.17.8}$$

where S' is a point outside of the boundary C but *very near to it*. The sharp increase of this function near to the point S' makes it possible to put the spotlight rather strongly on that particular value $(\partial\varphi/\partial n)(S)$ which belongs to an S directly opposite to S' (see Figure). Although we did not succeed in *separating* $(\partial\varphi/\partial n)(S)$, yet we have a well-conditioned linear system for its evaluation which can be solved with the help of the large electronic computers.

We will demonstrate the value of over-determination by an example within the realm of algebraic equations. Let us assume that we have to solve the system $Ay = b$ which shall be of the order $2n$, i.e. we consider A as a $2n \times 2n$ matrix. We now separate our $2n$ unknowns $(y_1, y_2, \ldots, y_{2n})$ into two groups:

$$\begin{aligned} \xi &= (y_1, y_2, \ldots, y_n) \\ \bar{\eta} &= (y_{n+1}, y_{n+2}, \ldots, y_{2n}) \end{aligned} \tag{3.17.9}$$

All the columns associated with the second group are carried over to the right side, which means that we write our equation in the form

$$X\xi = b - Y\eta \tag{3.17.10}$$

where X is now a $(2n \times n)$ matrix and so is Y. The new system is now *over-determined* because the unknowns η are considered as *given* quantities. Hence the right side has to be orthogonal to every solution of the adjoint homogeneous equation

$$\tilde{X}u = 0 \tag{3.17.11}$$

There are altogether n such solutions, which can be combined into the $2n \times n$ matrix U_0. The compatibility conditions become

$$\tilde{U}_0 Y\eta = \tilde{U}_0 b \tag{3.17.12}$$

which gives us n equations for the determination of η. This requires the inversion of the $n \times n$ matrix $\tilde{U}_0 Y$. Then, having obtained η, we go back to (10) but omitting all equations beyond n, and obtaining ξ by inverting the matrix X_1 where X_1 is an $n \times n$ matrix, composed of the first n rows of X.

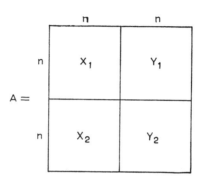

What we needed in this process, is *two* inversions of the order n, instead of *one* inversion of the order $2n$, provided that we possess the solutions of the system (11). This, however, can again be given on the basis of the inversion of X_1. The vector u has $2n$ components which can be split into the two column vectors u_1 and u_2, thus writing (11) in the form

$$\tilde{X}_1 u_1 + \tilde{X}_2 u_2 = 0 \tag{3.17.13}$$

which gives—assuming that we possess $X_1{}^{-1}$, the inverse of X_1:

$$u_1 = -\tilde{X}_1{}^{-1}\tilde{X}_2 u_2 \tag{3.17.14}$$

The full $2n \times n$ matrix U_0 shall also be split into the two $n \times n$ matrices U_1 and U_2, writing U_2 below U_1. We have complete freedom in choosing the matrix U_2, as long as its determinant is not zero. We will identify it with the unit matrix I. Then

$$U_1 = -\tilde{X}_1{}^{-1}\tilde{X}_2 \tag{3.17.15}$$

and

$$U_0 = \begin{bmatrix} U_1 \\ I \end{bmatrix} \tag{3.17.16}$$

If we split the matrix Y and the vector b in a similar way:

$$Y = \begin{bmatrix} Y_1 \\ Y_2 \end{bmatrix} \quad b = \begin{bmatrix} b_1 \\ b_2 \end{bmatrix} \tag{3.17.17}$$

we can write the equation (12) in the form

$$\begin{aligned} \tilde{U}_1 = Q = -X_2 X_1{}^{-1} \\ (QY_1 + Y_2)\eta = Qb_1 + b_2 \end{aligned} \tag{3.17.18}$$

This requires the inversion of the $n \times n$ matrix $QY_1 + Y_2$. Then, having solved this equation, we return to the first half of (10) and obtain

$$\xi = X_1^{-1}(b_1 - Y_1\eta) \tag{3.17.19}$$

We have thus demonstrated that the $2n \times 2n$ system

$$\begin{bmatrix} X_1 & Y_1 \\ X_2 & Y_2 \end{bmatrix} \begin{bmatrix} \xi \\ \eta \end{bmatrix} = \begin{bmatrix} b_1 \\ b_2 \end{bmatrix} \tag{3.17.20}$$

can be solved by *two $n \times n$ inversions* and substitutions.

Problem 143.   Apply this method to the solution of the following 6×6 system:

$$\begin{bmatrix} 1 & 1 & -1 & 1 & 0 & -1 \\ 2 & -1 & 0 & 2 & 1 & 0 \\ -2 & 2 & -1 & 1 & 0 & 2 \\ 1 & 1 & -1 & 2 & 0 & -1 \\ 5 & -1 & -1 & 5 & 3 & -1 \\ 1 & -3 & 2 & -2 & 0 & 0 \end{bmatrix} y = \begin{bmatrix} 2 \\ 3 \\ -1 \\ 2 \\ 9 \\ -1 \end{bmatrix} \tag{3.17.21}$$

[Answer:

$$\tilde{y} = (1, 0, -1, 0, 1, 0) \tag{3.17.22}]$$

3.18. The method of successive orthogonalisation

The matrix decomposition studied in Section 9 showed us that we can construct the matrix A without paying any attention to the solution of the homogeneous equations

$$Av = 0 \tag{3.18.1}$$

or

$$\tilde{A}u = 0 \tag{3.18.2}$$

since the principal axes associated with the zero eigenvalue do not participate in the generation of the matrix. If we possess all the p "essential axes", associated with the non-zero eigenvalues, we possess everything for the solution of the linear system

$$Ay = b \tag{3.18.3}$$

because, together with A, we have obtained also the matrix B which we could conceive as the natural inverse of A. Hence we could give an explicit solution of the system (3).

The drawback of this method is, however, that it requires a very elaborate mechanism to put it into operation. The matrix decomposition theorem (9.6) is a very valuable tool in the study of the general algebraic and analytical *properties* of linear operators but its actual application to the solution of large-scale linear systems and to the solution of linear differential equations demands the complete solution of an eigenvalue problem which is *explicitly* possible only under greatly simplified conditions.

Under these circumstances it is of great interest that there is a remarkable reciprocity between the "essential" principal axes which form the eigenspace in which the matrix operates, and the remaining axes which form the spaces U_0 and V_0 in which the operator is not activated. Although these spaces are apparently ignored by the operator, we can in fact make good use of them and go a long way toward the actual solution of our problem.

The great advantage of the essential axes has been that they bring directly into evidence those p-dimensional subspaces U and V in which the matrix is activated. Hence the original $n \times m$ problem can immediately be reformulated as a $p \times p$ problem and we have at once overcome the handicap of unnecessary surplus equations on the one hand, and deficiency on the other. But in Section 16, we have studied a method which allows the same reduction, purely on the basis of solving the homogeneous equations (1) and (2), without solving the basic eigenvalue problem. We can make this method more efficient by a process known as the "successive orthogonalisation of a set of vectors". This process has the advantage that it proceeds in successive steps, each step being explicitly solvable.

We will apply this orthogonalisation process to both sets of vectors v_i and u_i, found as a solution of the homogeneous systems (1) and (2). Let us observe that in this process the unknown vector y and the right side b of the equation (3) do not come at first into evidence.

We assume that we have found all the linearly independent solutions of the homogeneous system (1). Then any linear combination of these solutions is again a solution. We will use this feature of a homogeneous linear operator to transform the original set of solutions

$$v_{p+1}, v_{p+2}, \ldots, v_m \qquad (3.18.4)$$

into a more useful set. We start with v_{p+1} which we keep unchanged, except that we divide it by the length of the vector:

$$v'_{p+1} = \frac{v_{p+1}}{\sqrt{\tilde{v}_{p+1} v_{p+1}}} \qquad (3.18.5)$$

Now we come to v'_{p+2} which we want to make orthogonal to v'_{p+1}. We put

$$v'_{p+2} = \frac{\alpha v'_{p+1} - v_{p+2}}{\gamma} \qquad (3.18.6)$$

Pre-multiplying by $\tilde{v}'_{p+1}$ we obtain the condition:

$$\alpha = \tilde{v}'_{p+1} v_{p+2} \qquad (3.18.7)$$

Moreover, the condition that the length of v'_{p+2} shall become 1 yields

$$\tilde{v}_{p+2} v_{p+2} = \alpha^2 + \gamma^2 \qquad (3.18.8)$$

which determines γ (except for the sign). The process can be continued. At the kth step we have

$$v'_{p+k} = \frac{1}{\gamma} [\alpha_1 v'_{p+1} + \ldots \alpha_{k-1} v'_{p+k-1} - v_{p+k}] \qquad (3.18.9)$$

The orthogonality conditions give

$$\alpha_i = \tilde{v}'_{p+i} v_{p+k}, \ldots \qquad (i = 1, 2, \ldots, k-1) \tag{3.18.10}$$

The normalisation condition gives

$$\gamma^2 = \alpha_1^2 + \alpha_2^2 + \ldots + \alpha_{k-1}^2 - \tilde{v}_{p+k} v_{p+k} \tag{3.18.11}$$

After $m - p$ steps the entire set (4) is replaced by a new, orthogonal and normalised set of vectors $v'_{p+1}, \ldots, v'_{p+m}$. But we need not stop here. We can continue by choosing p more vectors in any way we like, as long as they are linearly independent of the previous set. They too can be orthogonalised, giving us the p additional

$$v'_1, v'_2, \ldots, v'_p \tag{3.18.12}$$

At the end of the process we possess a set of m orthogonal vectors which can be arranged as the columns of an $m \times m$ orthogonal matrix.

We can proceed in identical fashion with the solutions of the homogeneous equation (2):

$$u_{p+1}, u_{p+2}, \ldots, u_n \tag{3.18.13}$$

They too can be orthogonalised and here again we can continue the process, giving us the p additional vectors

$$u'_1, u'_2, \ldots, u'_p \tag{3.18.14}$$

We obtain an $n \times n$ orthogonal matrix.

Having accomplished this task we are going to establish a relation between these matrices and the critical vectors y and b of our problem (3). We wanted to put the vector y completely into the eigen-space V of the operator. This means that y has to become orthogonal to all the vectors $v'_{p+1}, v'_{p+2}, \ldots, v'_m$. This again means that y has to become a linear combination of the vectors (12). Similarly the right side b has to be orthogonal to the vectors (13) by reason of compatibility and thus must be reducible to a linear combination of the vectors (14). We can express these two statements by constructing the semi-orthogonal $m \times p$ matrix V_p, formed of the column vectors (12) and the semi-orthogonal $n \times p$ matrix U_p, formed of the column vectors (14). Then

$$\begin{aligned} y &= V_p y' \\ b &= U_p b' \end{aligned} \tag{3.18.15}$$

Both y' and b' are free column vectors of p components.

With this transformation our linear system (3) becomes

$$A V_p y' = U_p b' \tag{3.18.16}$$

and, pre-multiplying by $\tilde{U}_p$:

$$\tilde{U}_p A V_p y' = b' \tag{3.18.17}$$

If we put

$$A' = \tilde{U}_p A V_p \qquad (3.18.18)$$

$$A'y' = b' = \tilde{U}_p b \qquad (3.18.19)$$

the new matrix A' is an $n \times n$ *square matrix*.

On the basis of these deductions we come to the following conclusion. Although we have not solved any eigenvalue problem except for the complete solution of the problem associated with the eigenvalue $\lambda = 0$, we have succeeded in reducing our original arbitrarily over-determined and deficient $n \times m$ system to a new well-determined $p \times p$ system which has a unique solution. This is much less than what we attained before in Section 7, where the matrices U_p and V_p were formed in terms of the principal axes of A. There the new matrix A' became a *diagonal matrix* and our equations appeared in separated form. This is not so now because we still have to solve a $p \times p$ linear system. On the other hand, we are relieved of the heavy burden of obtaining the complete solution of the principal axis problem and can use other methods for the solution of a linear system.

This reduction to a free and complete $p \times p$ system holds even if our original $n \times m$ system was incompatible. The transformation

$$b' = \tilde{U}b \qquad (3.18.20)$$

automatically eliminates all the "forbidden" components of b, without altering the "permissible" components. The solution obtained coincides with the least square solution of our incompatible problem. It also coincides with the solution obtained with the help of the natural inverse B of the original matrix A.

Problem 144.   Apply the method of successive orthogonalisation to the zero-solutions of the system (8.6) (cf. Problem 124) and solve Problem 128 by reducing it to a 2×2 system. Show that the solution coincides with the solution (10.17, 18) obtained before.

Problem 145.   Apply the same procedure to the 4×4 overdetermined and incomplete system (13.22) and reduce it to a well-determined 3×3 system.

Problem 146.   Formulate the method of successive orthogonalisation (5–11) if we do not normalise the lengths of the successive vectors, thus avoiding the taking of square roots.

Problem 147.   Demonstrate the following properties of the reduced set (19):

a) The matrix A' has no zero eigenvalues.

b) The eigenvalues of A', belonging to the shifted eigenvalue problem, coincide with the non-zero eigenvalues of the original matrix A.

c) The principal axes of A are obtainable from the principal axes of A' by the following transformation:

$$V = V_p V' \qquad (3.18.21)$$

$$U = U_p U' \qquad (3.18.22)$$

3.19. The bilinear identity

In all our dealings with matrices we have to keep in mind that only such methods and results are applicable to the study of differential equations which remain interpretable even if the number of equations increases to infinity. A differential equation can be conceived as a difference equation with a mesh-size which converges to zero. But with decreasing mesh-size the number of equations is constantly on the increase and going to the limit where ϵ becomes arbitrarily small, the corresponding number of equations becomes arbitrarily large. Fredholm in his fundamental investigation of a certain class of integral equations was able to operate with determinants whose order increased to infinity. While this was a great achievement, its use is limited to a very definite class of problems. A much wider outlook is obtained if we completely avoid the use of determinants and operate with other concepts which carry over to the continuous case in a natural way. The ordinary process of matrix inversion is not amenable to the proper re-interpretation if our linear system has the property that more and more equations have to be taken into consideration, without end. We can base, however, the solution of linear systems on concepts which take no recourse to either determinants or successive eliminations.

The one method which carries over in its totality to the continuous field, is the method of the principal axes (studied in Section 7), together with the decomposition of the operator A into a product of three matrices. But there is also the alternative method of studying the solutions of the homogeneous equation—corresponding to the eigenvalue zero—from which valuable results are obtainable. In fact, this "homogeneous" method can be extended to a point where it can be successfully employed not only for the general exploration of the system—its compatibility and deficiency—but even for its actual *solution*. In this method a certain *identity*, the so-called "bilinear identity", plays a pivotal role. It is of central importance for the entire theory of linear differential operators, no matter whether they belong to the ordinary or the partial type.

The scalar product of two vectors, i.e. $\tilde{v}u$, gives a 1×1 matrix, that is a *pure number*. The transpose of this number coincides with itself, or, if it happens to be a complex number, the transposition changes it to its complex conjugate. Let us apply this principle to the product $\tilde{x}Ay$ where c is an n-component, y an m-component column vector:

$$\tilde{x}Ay - (\tilde{y}\tilde{A}x)^* \equiv 0 \qquad (3.19.1)$$

(the asterisk refers to "conjugate complex"). We call this relation an "identity" because it holds (for a given matrix A) for *any choice* of the two vectors x and y.

The power of this identity lies in its great generality. First of all, it may serve even to identify the transposed matrix $\tilde{A}$. In the realm of finite matrices it is easy enough to transpose the rows and columns of a matrix and thus define the elements of the transposed matrix $\tilde{A}$:

$$\tilde{a}_{ik} = a^*_{ki} \qquad (3.19.2)$$

In the realm of *continuous operators*, where the matrices grow beyond all size, the original definition of $\tilde{A}$ loses its meaning. But the identity (1) maintains its meaning and can serve for the purpose of *defining* $\tilde{A}$.

The next fundamental application of (1) is the derivation of the compatibility conditions of the system $Ay = b$. We will extend this system—as we have done before in Section 7—by the adjoint system, considering the complete system

$$Ay = b \qquad (3.19.3)$$

$$\tilde{A}x = c \qquad (3.19.4)$$

We can now ask: "Can we prescribe the right sides b and c freely?" The application of the bilinear identity yields the relation

$$\tilde{x}b - (\tilde{y}c)^* \equiv 0 \qquad (3.19.5)$$

which holds, whatever the vectors b and c may be. We have the right to specify our vectors in any way we like. Let us choose $c = 0$. Then we have no longer an identity but an *equation* which holds for a special case, namely:

$$\tilde{x}b = 0 \qquad (3.19.6)$$

$$\tilde{A}x = 0 \qquad (3.19.7)$$

The result means that *the right side of the system* (3) *must be orthogonal to any solution of the transposed* (adjoint) *homogeneous equation.* The same result can be derived for the vector c by making b equal to zero.

We have seen in the general theory of linear system that these are the *only* compatibility conditions that the right side b has to satisfy. In the special case that the adjoint homogeneous system (7) has *no* non-vanishing solution, the system (3) becomes unconstrained (the vector b can be chosen freely). These results, obtained before by different tools, follow at once from the bilinear identity (1), which is in fact the *only* identity that can be established between the matrix A and its transpose $\tilde{A}$.

We will now go one step further and consider a linear system which is either well-determined or over-determined but not under-determined:

$$p = m \leq n \qquad (3.19.8)$$

This situation can always be achieved because in the case of an incomplete system we can add the $m - p$ further equations

$$\tilde{V}_0 y = 0 \qquad (3.19.9)$$

(orthogonality to the zero-field) and obtain a new system which is complete (cf. Problem 137). In the case of over-determination we assume that the compatibility conditions are in fact fulfilled. Now we are interested in finding the value of one particular component (coordinate) of y, for example

y_i.   For this purpose we add to our previous equation (3) one more equation, considering the complete system

$$Ay = b$$
$$y_i = \alpha_i$$
                                                                            (3.19.10)

This means that we consider the value of y_i as *one of our data*. This, of course, cannot be done freely, otherwise our problem would not have a unique solution.   But the system (10) is a legitimate over-determined system and it is now the *compatibility condition* which will provide the solution by giving us a linear relation between the components of the right side, i.e. a linear relation between the vector y and α_i which is in fact y_i.

This method is of great interest because it brings into evidence a general feature of solving linear systems which we will encounter again and again in the study of linear differential equations.   There it is called the "method of the Green's function".   It consists in constructing an *auxiliary function* which has nothing to do with the data but is in fact entirely determined by the operator itself.   Moreover, this auxiliary function is obtained by solving a certain *homogeneous equation*.

According to the general theory the compatibility of the system (10) demands in the usual way that we solve the *adjoint homogeneous equation*

$$\tilde{A}u = 0$$
                                                                            (3.19.11)

and then we must have

$$\tilde{u}b = 0$$
                                                                            (3.19.12)

for the solvability of our system.   But in our case the matrix A has been extended by an additional row.   This row has all zeros, except the single element 1 in the ith place.   Considering this row as a row-vector, the geometrical significance of such a vector is that it points in the direction of the ith coordinate axis.   Hence we will call it the ith "base vector" and denote it with $\tilde{e}_i$, in harmony with our general custom of considering every vector as a *column vector* :

$$\tilde{e}_i = (0, 0, \ldots, \overset{(i)}{1}, \ldots, \overset{(m)}{0})$$
                                                                            (3.19.13)

Our system (10) can thus be rewritten in the form

$$Ay = b$$
$$\tilde{e}_i y = \alpha_i$$
                                                                            (3.19.14)

Now we have to find the vector u of the equation (11).   It is a column vector of $n + 1$ elements since the matrix of our extended system is an $(n + 1) \times m$ matrix.   We will separate the last element of this vector and write it in the form

$$\begin{bmatrix} g \\ \gamma \end{bmatrix}$$
                                                                            (3.19.15)

where g is a column vector of n elements. Hence the determining equation (11) for the extended vector u becomes

$$\tilde{A}g + e_i\gamma = 0 \tag{3.19.16}$$

while the compatibility condition (12), applied to our system (14), becomes

$$\tilde{g}b + \gamma^*\alpha_i = 0 \tag{3.19.17}$$

Now the equation (16) represents a homogeneous linear system in the unknowns g, γ which leaves one factor free. Assuming that γ is not zero we can divide by it, which is equivalent to saying that γ can be chosen as 1, or equally as -1. Hence only the two cases $\gamma = 0$ and -1 are of essential interest. But if $\gamma = 0$, we merely get the compatibility conditions of the original system and those have been investigated in advance and found satisfied. Hence we can assume without loss of generality that $\gamma = -1$ and then the system (16) can be written in the inhomogeneous form

$$\tilde{A}g = e_i \tag{3.19.18}$$

and

$$\alpha_i = y_i = \tilde{g}b \tag{3.19.19}$$

We see that we could obtain the coordinate y_i on the basis of constructing an auxiliary vector g which satisfied a definite inhomogeneous equation. But the inhomogeneity came about by transferring a term from the left to the right side. In principle we have only made use of the compatibility condition of a linear system and that again was merely an application of the bilinear identity. We thus see how the bilinear identity becomes of leading significance for the solution problem of linear equations and it is this method which carries over in a most natural and systematic manner to the realm of linear differential operators.

We still have to answer the question whether the system (18) is always solvable. This is indeed so because the adjoint homogeneous equation $Av = 0$ has *no solution* (apart from $v = 0$) according to our basic assumptions.

Another interesting point is that if our original system was an $n \times m$ over-determined system $(m > n)$, the adjoint system (18) is accordingly an $m \times n$ *under-determined* system which allows an infinity of solutions. And yet we know from the well-determined character of our problem that y_i must have a definite value and thus the freedom in the solution g of the equation (18) can have no influence on y_i. This is indeed the case if we realise that according to the general theory the general solution of (18) can be written in the form

$$g = g_1 + g_0 \tag{3.19.20}$$

where g_1 is any particular solution and g_0 the solution of the homogeneous equation

$$\tilde{A}g_0 = 0 \tag{3.19.21}$$

But then the new value of y_i, obtained on the basis of a g which is different from g_1 becomes

$$\bar{y}_i = y_i + \tilde{g}_0 b \qquad (3.19.22)$$

However, the *second term vanishes* since b satisfies the compatibility conditions of the original system. And thus the insensitivity of y_i relative to the freedom of choosing *any* solution of (18) is demonstrated.

In Section 17 we have encountered Cauchy's integral theorem (17.1) which was the prototype of a fundamentally important over-determined system. Here the "auxiliary vector g" is taken over by the auxiliary *function*

$$G(\zeta, z) = \frac{1}{\zeta - z} \qquad (3.19.23)$$

where z is the fixed point at which $f(z)$ shall be obtained. But the conditions demanded of $G(\zeta, z)$ are much less strict than to yield the particular function (23). In fact we could add to this special function any function $g(z)$ which remains analytical within the domain bounded by C. But the contribution generated by this additional function is

$$\frac{1}{2\pi i} \oint f(\zeta) g(\zeta) d\zeta \qquad (3.19.24)$$

and this quantity is *zero* according to (17.2), due to the nature of the admissible boundary values $f(\zeta)$. Quite similar is the situation concerning the boundary value problem (17.4) where again the function $G(\tau, S)$ need not be chosen according to (17.5) but we could add any solution of the Laplacian equation (17.3) which remains everywhere regular in the given domain. It is exactly this great freedom in solving the under-determined equation (21) which renders the over-determined systems so valuable from the standpoint of obtaining explicit solutions. If the system is well-determined, the equation (18) becomes likewise well-determined and we have to obtain a unique, highly specified vector g. This is in the realm of partial differential equations frequently a difficult task.

We return to our original matrix problem (14). Since the solution of the system (16) changes with i—which assumes in succession the values $1, 2, \ldots, m$, if our aim is to obtain the entire y vector—we should indicate this dependence by the subscript i. Instead of one single equation (18) we now obtain m equations which can be solved in succession

$$\begin{aligned} \tilde{A}g_1 &= e_1 \\ \tilde{A}g_2 &= e_2 \\ &\vdots \\ \tilde{A}g_m &= e_m \end{aligned} \qquad (3.19.25)$$

These m vectors of n components can be arranged as columns of an $n \times m$

matrix G. Then the equations (19) (for $i = 1, 2, \ldots, m$) can be replaced by the matrix equation

$$y = \tilde{G}b \qquad (3.19.26)$$

Moreover, the base vectors $e_1, e_2, \ldots, e_m$, arranged as columns of a matrix, yield the $m \times m$ *unit matrix* I :

$$[e_1, e_2, \ldots, e_m] = I \qquad (3.19.27)$$

The m defining equations (25) for the vectors $g_1, g_2, \ldots, g_m$ can be united in the single matrix equation

$$\tilde{A}G = I \qquad (3.19.28)$$

Taking the transpose of this equation we can write it in the form

$$\tilde{G}A = I \qquad (3.19.29)$$

which shows that $\tilde{G}$ can be conceived as the *inverse* of A :

$$\tilde{G} = A^{-1} \qquad (3.19.30)$$

and the solution (26) may be written in the form

$$y = A^{-1}b \qquad (3.19.31)$$

But the *sequence of the factors* in (29) is absolutely essential and cannot be changed to

$$A\tilde{G} = I \qquad (3.19.32)$$

In fact, if we try to obtain a solution of the equation (32), we see that it is impossible because the equation (18) would now appear in the form

$$A\tilde{g} = e_i \qquad (3.19.33)$$

The solvability of this equation for all i would demand that the adjoint homogeneous equation

$$\tilde{A}u = 0 \qquad (3.19.34)$$

has no non-vanishing solution. But this is *not* the case if our system is over-determined $(n > m)$.

Problem 147.   Assume that we have found a matrix C such that for all x and y

$$\tilde{x}Ay - (\tilde{y}Bx)^* = 0 \qquad (3.19.35)$$

Show that in this case we have of necessity

$$B = \tilde{A} \qquad (3.19.36)$$

Problem 148.   Show with the help of the natural inverse (10.4) that in the case of a unique but over-determined linear system the "left-inverse" (30) exists but the "right inverse" (32) does not exist.

Problem 149.   Apply the solution method of this section to the solution of the over-determined (but complete) system (13.18) and demonstrate numerically that the freedom in the construction of $\tilde{G}$ has no effect on the solution.

Problem 150. Do the same for the over-determined but incomplete system (13.22) of Problem 136, after removing the deficiency by adding as a fifth equation the condition

$$\tilde{v}_4 y = 0 \tag{3.19.37}$$

Check the solution with the previous solution (14.9).

3.20. Minimum property of the smallest eigenvalue

We consider a complete $n \times m$ system which is either well-determined or over-determined ($p = m, n \geq m$), and we focus our attention on the smallest eigenvalue of the shifted eigenvalue problem (7.7). Since we have agreed that we enumerate the λ_k (which are all positive) in *decreasing* order, the smallest eigenvalue of our problem will be λ_p for which we may also write λ_m since we excluded the possibility $p < m$. This eigenvalue can be characterised for itself, as a solution of a *minimum problem*.

We ask for the smallest possible value of the ratio

$$r^2 = \frac{|Ay|^2}{|y|^2} \tag{3.20.1}$$

Equivalent with this problem is the following formulation. Find the minimum of the quantity $(Ay)^2$ if the length of the vector y is normalised to 1:

$$(y)^2 = \tilde{y}y = 1 \tag{3.20.2}$$

The solution of this minimum problem is

$$\tilde{A}Ay = \lambda^2 y \tag{3.20.3}$$

for which we can also put

$$\begin{aligned} Av &= \lambda u \\ \tilde{A}u &= \lambda v \end{aligned} \tag{3.20.4}$$

with the added condition that we choose among the possible solutions of this problem the *absolutely smallest* $\lambda = \lambda_m$. In consequence of this minimum problem we have for *any arbitrary choice of y*:

$$\frac{(Ay)^2}{(y)^2} \geq \lambda_m{}^2 \tag{3.20.5}$$

or

$$\frac{(y)^2}{(Ay)^2} \lessgtr \frac{1}{\lambda_m{}^2} \tag{3.20.6}$$

This minimum (or maximum) property of the mth eigenvalue is frequently of great value in estimating the error of an algebraic system which approximates the solution of a continuous problem (cf. Chapter 4.9).

We can parallel this extremum property of λ_m by a corresponding extremum property of the *largest* eigenvalue λ_1. We can define λ_1 as the solution of the problem of making the ratio (1) to a *maximum*. Once more we obtain the equations (4) with the added condition that we choose among all possible solutions the one belonging to the *absolutely largest* $\lambda = \lambda_1$. In consequence of this maximum property we obtain for an arbitrary choice of y:

$$\frac{(Ay)^2}{y^2} \leq \lambda_1{}^2 \tag{3.20.7}$$

And thus we can establish an *upper and lower bound* for the ratio (1):

$$\lambda_1{}^2 \geq \frac{(Ay)^2}{y^2} \geq \lambda_m{}^2 \tag{3.20.8}$$

or also

$$\frac{1}{\lambda_m{}^2} \geq \frac{y^2}{(Ay)^2} \geq \frac{1}{\lambda_1{}^2} \tag{3.20.9}$$

These inequalities lead to an interesting consequence concerning the elements of the inverse matrix A^{-1}. Let us assume that A is a given $n \times n$ square matrix with non-vanishing determinant. Then the inverse A^{-1} is likewise an $n \times n$ square matrix. Occasionally the elements of this matrix are very large numbers which makes the numerical determination of the inverse matrix cumbersome and prone to rounding errors. Now the elements of the inverse matrix cannot be arbitrarily large but they cannot be arbitrarily small either. We obtain a definite check on the inverse matrix if the two extreme eigenvalues λ_1 and λ_m are known.

We have seen in Section 19 that the ith row of the inverse matrix A^{-1} can be characterised by the following equation:

$$\tilde{A}g_i = e_i \tag{3.20.10}$$

Let us apply the inequality (9) to this particular vector $y = g_i$, keeping in mind that the replacement of A by $\tilde{A}$ has no effect on the eigenvalues λ_1 and λ_m, in view of the symmetry of the shifted eigenvalue problem (4):

$$\frac{1}{\lambda_1{}^2} \leq \frac{g_i{}^2}{e_i{}^2} \leq \frac{1}{\lambda_m{}^2} \tag{3.20.11}$$

Since by definition the base vector e_i has the length 1 (cf. 19.13), we obtain for the ith row of the inverse matrix the two bounds

$$\frac{1}{\lambda_1{}^2} \leq g_i{}^2 \leq \frac{1}{\lambda_m{}^2} \tag{3.20.12}$$

This means that *the sum of the squares of the elements of any row of A^{-1} is included between the lower bound $\lambda_1{}^{-2}$ and the upper bound $\lambda_m{}^{-2}$.* Exactly the same holds for the sum of the squares of the elements of any *column*, as we can see by replacing A by $\tilde{A}$.

Similar bounds can be established for A itself. By considering A as the inverse of A^{-1} we obtain for the square of any row or any column of A the inequalities

$$\lambda_m{}^2 \leq \sum_{\alpha=1}^{n} a_{i\alpha}{}^2 \leq \lambda_1{}^2 \qquad (3.20.13)$$

$$\lambda_m{}^2 \leq \sum_{\alpha=1}^{n} a_{\alpha i}{}^2 \leq \lambda_1{}^2 \qquad (3.20.14)$$

(In the case of Hermitian matrices $a_{ik}{}^2$ is to be replaced by $|a_{ik}|^2$.)

While up to now we have singled out the special case of $n \times n$ non-singular square matrices, we shall now extend our discussion to the completely general case of an arbitrary $n \times m$ matrix A. Here the "natural inverse" of A (cf. 10.4) can again be characterised by the equation (10) but with the following restriction. The right side of the equation cannot be simply e_i, in view of the compatibility condition which demands that the right side must be orthogonal to every solution of the transposed homogeneous system, if such solutions exist. We will assume that we have constructed all the $m - p$ linearly independent solutions v^p of the equation

$$Av^p = 0 \qquad (\rho = 1, 2, \ldots, m - p) \qquad (3.20.15)$$

Furthermore, we want to assume that we have *orthogonalised* and *normalised* these solutions (cf. Section 18), so that

$$v^\alpha v^\beta = \delta_{\alpha\beta} \qquad (3.20.16)$$

Then the ith row of the natural inverse (10.4) of the matrix can be characterised by the following equation:

$$\tilde{A}g_i = e_i - \sum_{\alpha=1}^{m-p} v_i{}^\alpha v^\alpha \qquad (3.20.17)$$

where $v_i{}^\alpha$ denotes the ith component of the vector v^α. By adding the correction term on the right side we have blotted out the projection of the base vector e_i into the non-activated portion of the U-space, without changing in the least the projection into the activated portion.

The equation (17) has a unique solution if we add the further condition that g_i must be made orthogonal to all the $n - p$ independent solutions of the homogeneous equation

$$\tilde{A}u^\sigma = 0 \qquad (\sigma = 1, 2, \ldots, n - p) \qquad (3.20.18)$$

It is this vector g_i which provides us with the ith row of the inverse B. By letting i assume the values $1, 2, \ldots, m$ we obtain in succession all the rows of B.

Now we will once more apply the inequality (11) to this vector g_i and once more an upper and a lower bound can be obtained for the length of the vector. The difference is only that now the denominator, which appears in (11), is no longer 1 but the square of the right side of (17) for which we obtain

$$\left(e_i - \sum_\alpha v_i{}^\alpha v^\alpha\right)^2 = 1 - 2 \sum (v_i{}^\alpha)^2 + \sum (v_i{}^\alpha)^2$$

$$= 1 - \sum_{\alpha=1}^{m-p} (v_i{}^\alpha)^2 \qquad (3.20.19)$$

And thus the relation (11) now becomes

$$\frac{1}{\lambda_1{}^2} \leq \frac{g_i{}^2}{1 - \sum_\alpha (v_i{}^\alpha)^2} \leq \frac{1}{\lambda_p{}^2} \qquad (3.20.20)$$

Here again we can extend this inequality to the columns and we can likewise return to the original matrix A which differs from A^{-1} only in having λ_i instead of $\lambda_i{}^{-1}$ as eigenvalues which reverses the role of λ_1 and λ_m. We obtain a particularly adequate general scheme if we arrange the (orthogonalised and normalised) zero-solutions, together with the given matrix A, in the following fashion.

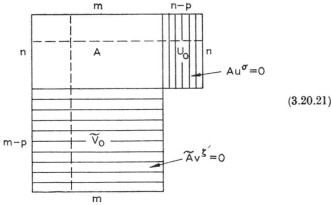

$$(3.20.21)$$

The u^σ vectors are arranged in successive *columns*, the v^ρ vectors in successive *rows*. We now have to form a ratio whose numerator is the square of a certain row of the matrix A while the denominator is the square of the *complementary* row, subtracted from 1:

$$\lambda_p{}^2 \leq \frac{(a^i)^2}{1 - (u_0{}^i)^2} \leq \lambda_1{}^2 \qquad (3.20.22)$$

where

$$(a^i)^2 = \sum_{\alpha=1}^{m} a_{i\alpha}{}^2$$

$$(u_0{}^i)^2 = \sum_{\sigma=1}^{n-p} (u_i{}^\sigma)^2 \qquad (3.20.23)$$

A similar relation is obtainable for the *columns*:

$$\lambda_p{}^2 \le \frac{(\bar{a}^i)^2}{1 - (\bar{v}_0{}^i)^2} \le \lambda_1{}^2 \tag{3.20.24}$$

where

$$(\bar{a}^i)^2 = \sum_{\alpha=1}^{n} a_{\alpha i}{}^2$$

$$(v_0{}^i)^2 = \sum_{\rho=1}^{m-p} (v_i{}^\rho)^2 \tag{3.20.25}$$

In the special case $m = n = p$ the complementary fields disappear and we are right back at the previous formulae (13), (14). If $n = p$, $m > n$, the right complement disappears but not the complement below. If $m = p$, $n > m$, the complement below disappears but not the right complement.

The remarkable feature of these inequalities is their great generality and that they provide simultaneously an upper and a lower bound for the square of every row and every column of an arbitrary $n \times m$ matrix—and likewise its inverse—in terms of the two characteristic numbers λ_1 and λ_p (irrespective of the presence of an arbitrary number of zero eigenvalues, since λ_p is defined as the smallest *positive* eigenvalue for which the equation (4) is solvable).

BIBLIOGRAPHY

[1] Aitken, A. C., *Determinants and Matrices* (Interscience Publishers, New York, 1944)

[2] Ferrar, W. L., *Algebra* (Oxford Press, New York, 1941)

[3] Householder, A. S., *Principles of Numerical Analysis* (McGraw-Hill, New York, 1953)

[4] MacDuffee, C. C., *The Theory of Matrices* (Chelsea, New York, 1946)

CHAPTER 4

THE FUNCTION SPACE

Synopsis. A piecewise continuous function need not be given in an infinity of points but in a sufficiently dense discrete set of points. This leads to the picture of a space of many dimensions in which a function is represented by a *vector*. By this geometrical image the link is established between matrices and linear differential operators, and the previous analysis of linear systems becomes translatable to the analysis of linear differential equations. We learn about the fundamental importance of the "adjoint equation" and likewise about the fundamental role of the given boundary conditions. From now on the given boundary conditions will not be considered as accidental accessories, added to the given differential equation, but the left sides of the boundary conditions become *integral parts* of the operator. From now on, if we write D for a differential operator, we will automatically assume that this operator includes the left sides of the given boundary conditions, no matter how few or how many such conditions may have been prescribed.

4.1. Introduction

The close relation which exists between the solution of differential equations and systems of algebraic equations was recognised by the early masters of calculus. David Bernoulli solved the problem of the completely flexible chain by considering the equilibrium problem of a chain which was composed of a large number of rigid rods of small lengths. Lagrange solved the problem of the vibrating string by considering the motion of discrete masses of finite size, separated by small but finite intervals. It seemed self-evident that this algebraic approach to the problem of the continuum must lead to the right results. In particular, the solution of a problem in partial differential equations seemed obtained if the following conditions prevailed:

1. We replace the continuum of functional values by a dense set of discontinuous values.

2. The partial derivatives are replaced by the corresponding difference coefficients, taken between points which can approach each other as much as we like.

3. We solve the resulting algebraic system and study the behaviour of

the solution as the discrete set of points becomes denser and denser, thus approaching the continuum as much as we wish.

4. We observe that under these conditions the solution of the algebraic system approaches a definite limit.

5. Then this limit is automatically the desired solution of our original problem.

The constantly increasing demands on rigour have invalidated some of the assumptions which seemed self-evident even a century ago. To give an exact existence theorem for the solution of a complicated boundary value problem in partial differential equations can easily tax our mathematical faculties to the utmost. In the realm of ordinary differential equations Cauchy succeeded with the proof that the limit of the substitute algebraic problem actually yields the solution of the original continuous problem. But the method of Cauchy does not carry over into the realm of partial differential operators, and even relatively simple partial differential equations require a thorough investigation if a rigorous proof is required of the kind of boundary conditions which can guarantee a solution. We do not possess any sweeping methods which would be applicable to *all* partial differential equations, even if we restrict ourselves to the realm of *linear* differential operators.

4.2. The viewpoint of pure and applied mathematics

From the standpoint of pure mathematics the existence of the solution of a certain problem in partial differential equations may be more important than the actual construction of the solution. But if our aim is to apply mathematics to the realm of natural phenomena, the shift of emphasis becomes clearly visible. In a given physical situation we feel certain in advance that the solution of a certain boundary value problem must exist since the physical quantity realised in nature is itself the desired solution. And thus we often encounter the expression: "We know for physical reasons that the solution exists." This expression is actually based on wrong premises. While we cannot doubt that a certain physical quantity in fact realises the solution of a certain boundary value problem, we have no guarantee that the *mathematical formulation* of that problem is correct to the last dot. We have neglected so many accessories of the problem, we have simplified so much on the given physical situation that we know in advance that the field equation with which we operate cannot be considered as the final truth. If, instead of getting zero on the right side we get an ϵ which is perhaps of the order 10^{-6}, we are still perfectly satisfied since we know that an accuracy of this order goes far beyond our measuring faculties and also beyond the accuracy of our description. From the standpoint of exact mathematics no error of any order can be tolerated and even the possession of a limit which tolerates an error in the equation which can be made *as small as we wish*, may *not* be enough to demonstrate the existence of the solution of our original problem.

Hence, while on the one hand we have no right to claim that a certain mathematically formulated boundary value problem must have a solution "for physical reasons", we can, on the other hand, dispense with the rigorous existence proofs of pure mathematics, in favour of a more flexible approach which proves the existence of certain boundary value problems under *simplified* conditions.   Pure mathematics would like to extend these conditions to much more extreme conditions and the value of such investigations cannot be doubted.   From the applied standpoint, however, we are satisfied if we succeed with the solution of a *fairly general* class of problems with data which are not too irregular.

The present book is written from the *applied* angle and is thus not concerned with the establishment of existence proofs.   Our aim is not the *solution* of a given differential equation but rather the exploration of the general *properties* of linear differential *operators*.   The solution of a given differential *equation* is of more accidental significance.   But we can hardly doubt that the study of the properties of linear differential operators can be of considerable value if we are confronted with the task of solving a given differential equation because we shall be able to tell in advance what we may and may not expect.   Moreover, certain results of these purely *symptomatic* studies can give us clues which may be even of practical help in the actual construction of the solution.

4.3. The language of geometry

Certain purely analytical relations can gain greatly in lucidity if we express their content in geometrical language.   We have investigated the properties of $n \times m$ matrices and have seen how beneficial it was to associate the two orthogonal vector spaces N and M with the matrix.   Although originally the unknown y is composed of the elements $y_1, y_2, \ldots, y_m$, and represents an aggregate of m numbers, we translate this analytical picture into a geometrical one by thinking of these numbers as the successive rectangular components of a vector.   This vector belongs to a certain *space* of m dimensions while the right side b belongs to another space of n dimensions.   By seeing these spaces in front of us and by populating them with the orthogonal vectors $v_1, v_2, \ldots, v_m$ in the one case, and $u_1, u_2, \ldots, u_n$ in the other, we have gained greatly in our understanding of the basic analytical structure of the matrix A.

The same geometrical ideas will again be of great value in the study of linear differential operators.   Originally we have a certain

$$y = f(x) \tag{4.3.1}$$

and a differential equation by which this function is characterised.   Or we may have a more-dimensional continuum, for example a three-dimensional space, characterised by the coordinates x, y, z, and we may be interested in finding a certain function

$$\Phi = \Phi(x, y, z) \tag{4.3.2}$$

of these coordinates by solving a certain *partial differential equation*, such as the "Laplacian equation"

$$\frac{\partial^2 \phi}{\partial x^2} + \frac{\partial^2 \phi}{\partial y^2} + \frac{\partial^2 \phi}{\partial z^2} = 0 \tag{4.3.3}$$

In the study of such problems great clarification can be obtained by a certain unifying procedure which interprets the originally given problem in geometrical language and emphasises certain features of our problem which hold universally for *all* linear differential operators. The given special problem becomes submerged in a much wider class of problems and we extend certain basic tools of analysis to a much larger class of investigations.

4.4. Metrical spaces of infinitely many dimensions

The characteristic feature of this unification is that the language employed operates with geometrical concepts, taken from our ordinary space conceptions but generalised to spaces of a more abstract character. If we speak of "geometrical concepts", we do not mean that actual geometrical constructions will be performed. We accept Descartes' Analytic Geometry which transforms a given geometrical problem into a problem of algebra, through the use of coordinates. By the method of translating our problem into the language of geometry we make the tools of analytical geometry available to the investigation of the properties of differential operators which originally are far from any direct geometrical significance.

This great tie-up between the analytical geometry of spaces and the study of differential operators came about by the concept of the "function space" which evolved in consequence of Hilbert's fundamental investigation of a certain class of integral equations. While Fredholm, the originator of the theory, formulated the problem in essentially *algebraic* language, Hilbert recognised the close relation of the problem with the analytic geometry of second-order surfaces in a Euclidean space of many (strictly speaking infinitely many) dimensions.

The structure of the space with which we are going to operate, will be of the ordinary *Euclidean* kind. The characteristic feature of this space is that it is homogeneous, by having the same properties at all points and in all directions, and by allowing a similarity transformation without changing anything in the inner relations of figures. The only difference compared with our ordinary space is that the *number of dimensions* is not 3 as in our ordinary space, but an arbitrary number, let us say n. And thus we speak of the "analytic geometry of n-dimensional spaces", which means that a "point" of this space has not 3 but n Cartesian coordinates $x_1, x_2, \ldots, x_n$; and that the "distance" from the origin is not given by the Pythagorean law

$$s^2 = x^2 + y^2 + z^2 \tag{4.4.1}$$

but by the generalised Pythagorean law

$$s^2 = x_1{}^2 + x_2{}^2 + \ldots + x_n{}^2 \tag{4.4.2}$$

While in our geometrical imagination we are somewhat handicapped by not being able to visualise spaces of more than 3 dimensions, for our *analytical* operations it is entirely irrelevant whether we have to extend a sum over 3 or over a thousand terms.   The *large number of dimensions* is a characteristic feature of all these investigations, in fact, we have to keep in mind all the time that strictly speaking the function space has an *infinity of dimensions*.   The essential difference between a discrete algebraic operator and a continuous operator is exactly this that the operation of a continuous operator has to be pictured in a space of *infinitely many dimensions*. However, the transition can be done *gradually*.   We can start with a space of a large number of dimensions and increase that number all the time, letting n grow beyond all bounds.   We then investigate the *limits* to which certain quantities tend.   These limits are the things in which we are really interested, but the very fact that these limits exist means that the difference between the continuous operator and the discrete operator associated with a space of many dimensions can be made as small as we wish.

4.5. The function as a vector

The fundamental point of departure is the method according to which we *tabulate* a certain function $y = f(x)$.   Although x is a *continuous* variable which can assume *any* values within a certain interval, we select a series of *discrete* values at which we tabulate $f(x)$.   Schematically our tabulation looks as follows:

$$\frac{x = \mid x_1, x_2, x_3, \ldots, x_n}{y = \mid y_1, y_2, y_3, \ldots, y_n} \tag{4.5.1}$$

For example in the interval between 0 and 1 we might have chosen 2001 equidistant values of x and tabulated the corresponding functional values of $y = e^x$.   In that case the x_i values are defined by

$$x = 0, 0.0005, 0.001, 0.0015, \ldots, 0.9995, 1 \tag{4.5.2}$$

while the y-values are the 2001 tabulated values of the exponential function, starting from $y = 1$, and ending with $y = 2.718281828$.

We will now associate with this tabulation the following geometrical picture.   We imagine that we have at our disposal a space of 2001 dimensions.   We assign the successive dimensions of this space to the x-values 0, 0.005, 0.001, ... i.e. we set up 2001 mutually orthogonal coordinate axes which we may denote as the axes $X_1, X_2, \ldots, X_n$.   Along these axes we plot the functional values $y_i = f(x_i)$, evaluated at the points $x = x_i$.   These $y_1, y_2, \ldots, y_n$ can be conceived as the coordinates of a certain point Y of an n-dimensional space.   We may likewise connect the point Y with the origin O by a straight line and arrive at the picture of the *vector* $\overrightarrow{OY}$.   The "components" or "projections" of this vector on the successive axes give the successive functional values $y_1, y_2, \ldots, y_n$.

At first sight it seems that the independent variable x has dropped completely out of this picture.   We have plotted the functional values y_i

as the components of the vector OY but where are the values $x_1, x_2, \ldots, x_n$? In fact, these values are present in *latent form*. The role of the independent variable x is that it provides an *ordering principle* for the cataloguing of the functional values. If we want to know for example what the value $f(0.25)$ is, we have to identify this $x = 0.25$ with one of our axes. Suppose we find that $x = 0.25$ belongs to the axis Y_{501}, then we single out that particular axis and see what the projection of the vector $\overrightarrow{OY}$ is on that axis. Our construction is actually isomorphic with every detail of the original tabulation and repeats that tabulation in a new geometrical interpretation.

We can now proceed even further and include in our construction functions which depend on more than one variable. Let a function $f(x, y)$ depend on two variables x and y. We tabulate this function in certain intervals, for example in similar intervals as before, but now, proceeding in equal intervals Δx and Δy, independently. If before we needed 2000 entries to cover the interval $[0, 1]$, we may now need 4 million entries to cover the square $0 \le x < 1, 0 \le y < 1$. But in principle the manner of tabulation has not changed. The independent variables x, y serve merely as an *ordering principle* for the arrangement of the tabular values. We can make a catalogue in which we enumerate all the possible combinations of x, y values in which our function has been tabulated, starting the enumeration with 1 and ending with, let us say, 4 million. Then we imagine a space of 4 million dimensions and again we plot the successive functional values of $u = f(x, y)$ as components of a vector. This one vector is again a perfect substitute for our table of 4 million entries.

We observe that the dimensionality of our original problem is of no immediate concern for the resulting vector picture. The fact that we have replaced a continuum by a discrete set of values abolishes the fundamental difference between functions of one or more variables. No matter how many independent variables we had, as soon as we begin to *tabulate*, we automatically begin to *atomise* the continuum and by this process we can line up any number of dimensions as a *one-dimensional* sequence of values.

Our table may become very bulky but in principle our procedure never changes. We need two things: a *catalogue* which associates a definite cardinal number

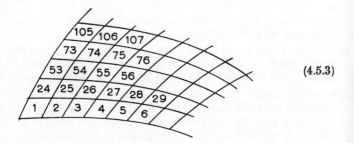

(4.5.3)

with the various "cells" in which our continuum has been broken, and a *table* which associates a definite functional value with these cardinal numbers, from 1 to n, where n may be a tremendously large number. Now we take all these functional values and construct a definite *vector* of the n-dimensional space which is a perfect representation of our function. Another function belonging to the same domain will find its representation in the same n-dimensional space, but will be represented by another vector because the functional values, which are the components of the new vector along the various axes, are different from what they were before.

This concept of a *function as a vector* looks strange and artificial at the first moment and yet it is an eminently useful tool in the study of differential and integral operators. We can understand the inner necessity of this concept if we approach the problem in the same way as Bernoulli and Euler and Lagrange approached the solution of differential equations. Since the derivative is defined as the limit of a difference coefficient, the replacement of a differential equation by a difference equation involves a certain error which, however, can be reduced to as little as we like by making the Δx between the arguments sufficiently small. But it is this replacement of a differential equation by a difference equation which has a profound effect on the nature of our problem. So far as the solution is concerned, we know that we have modified the solution of our problem by a negligibly small amount. But ideologically it makes a very great difference to be confronted by a new problem in which everything is formulated in *algebraic terms*. The unknown is no longer a continuous function of the variables. We have selected a discrete set of points in which we want to obtain the values of the function and thus we have transformed a problem of infinitely many degrees of freedom to a problem of a *finite* number of degrees of freedom. The same occurs with partial differential equations in which the independent variables form a more than one-dimensional manifold. In the problem of a vibrating membrane for example we should find the displacement of an elastic membrane which depends on the three variables x, y, and t. But if we assume that the material particles of the membrane are strictly speaking not distributed continuously over a surface but actually *lumped* in a large number of "mass-points" which exist in isolated spots, then we have the right picture which corresponds to the concepts of the "function space". Because now the displacement of the membrane is no longer a continuous function of x, y, t but a displacement which exists only in a large but finite number of grid-points, namely the points in which the mass-points are concentrated. The new problem is mathematically completely different from the original problem. We are no longer confronted with a partial differential equation but with a large number of *ordinary differential equations*, because we have to describe the elastic vibrations that the n mass-points describe under the influence of the elastic forces which act between them. But now we can go one step still further. We can carry through the idea of atomisation not only with respect to *space* but also with respect to *time*. If we atomise the time variable, these N ordinary differential equations break up into a large number of ordinary algebraic *difference equations* and the concept of the "derivative" disappears. And, yet, if our atomisation is sufficiently microscopic by a sufficiently fine grid-work of points, the difference between the new algebraic problem and the original continuous problem becomes imperceptible.

4.6. The differential operator as a matrix

Let us elucidate the general conditions with the help of a simple but characteristic *example*. We consider the differential equation

$$y''(x) = b(x) \tag{4.6.1}$$

where $b(x)$ is a given function of x. In accordance with the general procedure we are going to "atomise" this equation by breaking it into a large number of ordinary algebraic equations. For this purpose we replace the continuum of x-values by a dense but discrete set of points.

Let us assume that we are interested in the interval $x = [0, 1]$. We start by replacing this continuum of values by the discrete set

$$x_1 = 0, \quad x_2 = \epsilon, \quad x_3 = 2\epsilon, \ldots, x_n = (n-1)\epsilon = 1 \tag{4.6.2}$$

We have chosen for the sake of simplicity an *equidistant* set of points, which is not demanded since generally speaking our $\Delta x = \epsilon$ could *change* from point to point. But a constant Δx is simpler and serves our aims equally well.

Now the function $y(x)$ will also be atomised. We are no longer interested in the infinity of values $y(x)$ but only in the values of $y(x)$ at the selected points x_i:

$$y_1 = y(0), \quad y_2 = y(\epsilon), \quad y_3 = y(2\epsilon), \ldots, y_n = ((n-1)\epsilon) \tag{4.6.3}$$

In transcribing the given differential equation we make use of the definition of a derivative as the limit of a *difference coefficient*. We do not go, however, to the limit $\epsilon = 0$ but let ϵ be a small but *finite* quantity. Then the operation $y''(x)$ has to be interpreted as follows:

$$\frac{y_{k+1} - 2y_k + y_{k-1}}{\epsilon^2} \tag{4.6.4}$$

and now we have everything for the reformulation of our problem as an algebraic problem.

First of all we notice that we cannot write down our equation at the two endpoints $x = 0$ and $x = 1$ because we do not possess the left, respectively right neighbours $y(-\epsilon)$ and $y(1 + \epsilon)$ which go beyond the limitations of the given range. Hence we will write down only the equations at the $n - 2$ points $x_2, x_3, \ldots, x_{n-1}$:

$$\frac{1}{\epsilon^2}(y_1 - 2y_2 + y_3) = b(\epsilon) = b_2$$

$$\frac{1}{\epsilon^2}(y_2 - 2y_3 + y_4) = b(2\epsilon) = b_3$$

$$\vdots \tag{4.6.5}$$

$$\frac{1}{\epsilon^2}(y_{n-2} - 2y_{n-1} + y_n) = b(n\epsilon - 2\epsilon) = b_{n-1}$$

Here we have a simultaneous set of *linear algebraic equations,* exactly of the type studied in the previous chapter:

$$Ay = b \tag{4.6.6}$$

The values $y_1, y_2, y_3, \ldots, y_n$ can be conceived as the components of a *vector* of an n-dimensional space and the same can be said of the "given right side" $b_2, b_3, \ldots, b_{n-1}$ of our equation. We now recognise how natural it is to conceive the values of $y(x)$ at the selected points $x = x_i$ as the components of a *vector* and to do the same with the right side $b(x)$ of the original differential equation. We have re-formulated our original problem as a problem of *algebra.* In this re-formulation the original "function" $y(x)$ disappeared and became transformed into the *vector*

$$y = (y_1, y_2, \ldots, y_n) \tag{4.6.7}$$

The same happened with the given right side $b(x)$ of the differential equation. This $b(x)$ too disappeared in its original entity and re-appeared on the platform as the *vector*

$$b = (b_1, b_2, \ldots, b_n) \tag{4.6.8}$$

Closer inspection reveals that strictly speaking these two vectors do not belong to the same spaces. Our algebraic system (5) is in fact not an $n \times n$ system but an $(n-2) \times n$ system. The number of unknowns *surpasses* the number of unknowns by 2. Here we observe already a characteristic feature of differential operators: They represent in themselves without further data, an *incomplete* system of equations which cannot have a unique solution. In order to remove the deficiency, we have to give some further data and we usually do that by adding some proper *boundary conditions,* that is certain data concerning the behaviour of the solution *at the boundaries.* For example, we could prescribe the values of $y(0)$ and $y(1)$. But we may also give the values $y'(x)$ at the two endpoints which in our algebraic transcription means the two values

$$\frac{1}{\epsilon}(y_2 - y_1) \quad \text{and} \quad \frac{1}{\epsilon}(y_n - y_{n-1}) \tag{4.6.9}$$

There are many other possibilities and we may give two conditions at the point $x = 0$ without any conditions at $x = 1$, or perhaps two conditions at the point $x = 1$ without any conditions at $x = 0$, or two conditions which involve both endpoints simultaneously. The important point is that *the differential equation alone, without boundary conditions, cannot give a unique solution.* This is caused by the fact that a linear differential operator of the order r represents a linear relation between $r + 1$ functional values. Hence, letting the operator operate at every point $x = x_i$, the operator would make use of r additional functional values which go beyond the limits of the given interval. Hence we have to *cross out* r of the equations which makes the algebraic transcription of an rth order linear differential operator to a deficient system of $n - r$ equations between n unknowns. The remaining r equations have to be made up by r additional *boundary conditions*. (In the case of *partial* differential operators the situation is much more complicated and we cannot enumerate so easily the degree of deficiency of a given operator.) In any physical problem the boundary conditions, no less than the differential operators, are dictated by the physical situation. For example, if a differential equation is deducible from the "principle of least action", this principle provides also the *natural boundary conditions* of the problem (cf. Chapter 8.17).

But what happened in our algebraic transcription to the differential operator itself? The function $y(x)$ became a *vector* and the same happened to the function $b(x)$. The given differential equation (1) became resolved in a set of linear algebraic equations (5). This set of equations has a definite *matrix A* and it is this *matrix* which is determined by the given differential operator. Indeed, the coefficients of the linear system (5) came about on account of the fact that the differential operator $y''(x)$ was replaced by the corresponding difference coefficient and the equation written down a sufficient number of times.

It is important to emphasise, however, that the matrix of the final system is determined not only by the differential operator but by the given *boundary conditions*. For example the matrix of the linear system (5) is an $(n-2) \times n$ matrix which is thus 2 equations short of a well-determined system. But let us now add the boundary conditions

$$y(0) = \beta_1$$
$$y(1) = \beta_n \tag{4.6.10}$$

In this case y_1 and y_n *disappear* on the left side since they are no longer unknowns. We now obtain a system of $n-2$ rows and columns and the resulting matrix is a square matrix which can be written out as follows, if we take out the common factor $1/\epsilon^2$:

$$A = \frac{1}{\epsilon^2} \begin{bmatrix} -2 & 1 & & & & \\ 1 & -2 & 1 & & & \\ & 1 & -2 & 1 & & \\ & & & . & & \\ & & & & . & \\ & & & & & . \\ & & & & 1 & -2 \end{bmatrix} \tag{4.6.11}$$

Although we have chosen a particularly simple example, it is clear that by the same method an arbitrarily complicated ordinary differential equation (augmented by the proper boundary conditions) can be transcribed into a linear algebraic system of equations, of the general form (6).

Problem 151. Find the matrix of the previous problem (1) if the boundary conditions are modified as follows: $y(0) = 0$, $y'(0) = 0$. Show that the matrix A now becomes *triangular*.

Problem 152. Transcribe the differential equation

$$xy'(x) - 2y(x) = b(x) \tag{4.6.12}$$

into a matrix equation, assuming a constant $\Delta x = \epsilon$. Let x range between 1 and 2 and let the boundary condition be: $y(1) = 0$.

Problem 153. Find the matrix formulation of Problem 151 if the differential equation (1) is given as a pair of first order equations:

$$\begin{aligned} y'_1(x) - y_2(x) &= 0 \\ y'_2(x) &= b(x) \end{aligned} \tag{4.6.13}$$

[Hint: Combine $y_1(x)$, $y_2(x)$ into one column vector of $2n$ components.]

4.7. The length of a vector

Up to now we have been studying the transcription of a given differential equation into the language of algebra by breaking up the continuum into a dense but discrete set of points. We have not yet discussed what is going to happen as we bring our gridpoints closer and closer together. We are obviously interested in the *limit* to which our construction tends as the number of gridpoints increases to infinity by letting $\Delta x = \epsilon$ decrease to zero. This limit process is a characteristically new feature which has to be added to our ordinary matrix algebra, in order to include the study of linear differential operators within the framework of matrix calculus.

Let us recall that a matrix A operates in conjunction with two vector spaces. The operation $Ay = b$ transforms the vector y into the vector b and thus we have the space M in which y is located and the space N in which b is located. We were able to span these spaces with certain mutually orthogonal *principal axes* which established a natural "frame of reference" in these spaces. It will be our aim to re-interpret these results in relation to those matrices which can be associated with linear differential operators. However, a certain round-aboutness cannot be avoided since we had to *atomise* the continuum in order to arrive at the algebraic matrix picture and now we have to try to bridge the gap by letting the basic grid-parameter ϵ go to zero.

The characteristic feature of a Euclidean metrical space is that the various coordinate axes are on an equal level. We can label our axes by the sequence $1, 2, 3, \ldots, n$, but this labelling is arbitrary and does not correspond to an inherent geometrical property of space. The space is *homogeneous*, it has the same properties in every direction and thus allows arbitrary *rotations* of the basic axes. Hence the components $y_1, y_2, \ldots, y_n$ of a vector are of an accidental character but there is a definite "invariant" in existence, namely the *length of the vector* which is of an absolute and immutable character. This is given by the operation

$$\tilde{y}y = |y_1|^2 + |y_2|^2 + \ldots + |y_n|^2 \tag{4.7.1}$$

It must be our aim to save this valuable feature of a metrical space in relation to the study of continuous operators. If we proceed in the way as we have done before, it is clear that we will not arrive at a useful concept. If we define the components y_i of the vector y simply as the values of $y(x)$ at the points x_i:

$$y_i = y(x_i) \tag{4.7.2}$$

the quantity (1) would be bare of any meaning as $\Delta x = \epsilon$ approaches zero since we would get a steadily increasing quantity which grows to infinity.

We succeed, however, by a slight *modification* of the previous procedure. Let us agree that we include the Δx in the definition of y_i by putting

$$y_i = y(x_i)\sqrt{\Delta x_i} \tag{4.7.3}$$

(the equality of all $\Delta x_i = x_{i+1} - x_i$ is *not* demanded). If we now form the scalar product (1), we actually get something very valuable, namely

$$\tilde{y}y = \sum_{i=1}^{n} |y(x_i)|^2 \Delta x_i \tag{4.7.4}$$

and this quantity approaches a very definite limit as Δx_i decreases to zero, namely the definite integral

$$y^2 = \int_a^b |y(x)|^2 dx \tag{4.7.5}$$

The same quantity retains its significance in the multi-dimensional case, studied in Section 5. We have broken up the continuum into a large number of "cells" which cover the multi-dimensional range of the independent variables $x, y, z, \ldots$. In defining the vector-components Φ_i associated with this manifold we do not take merely the value of $\Phi(x, y, z, \ldots)$ at the midpoint of the cell as our component Φ_i but again we multiply by the square root of ϵ_i where this ϵ_i is now the *volume of the ith cell*:

$$\Phi_i = \Phi(x, y, z, \ldots)\sqrt{\epsilon_i} \tag{4.7.6}$$

This definition has once more the great value that in the limit it leads to a very definite invariant associated with the multi-dimensional function $\Phi(x, y, z, \ldots)$, namely the definite integral

$$\Phi^2 = \int |\Phi(x, y, z, \ldots)|^2 d\tau \tag{4.7.7}$$

where $d\tau$ is the volume element of the region and the integration is extended over the complete range of all the variables.

In fact, this generalisation to the multi-dimensional case is so natural that we often prefer to cover the general case with the same symbolism, denoting by x an arbitrary point of the given multi-dimensional region and by dx the volume element of that region. The formula (7) may then be written in the form

$$\Phi^2 = \int |\Phi(x)|^2 dx \tag{4.7.8}$$

in full analogy to the formula (5), although the symbol x refers now to a much more complicated domain and the integration is extended over a multi-dimensional region.

4.8. The scalar product of two vectors

If *two* vectors f and g are placed in a Euclidean space, their mutual position gives rise to a particularly important invariant, the "scalar product" of these two vectors, expressible in matrix language by the product

$$\tilde{f}g = (\tilde{g}f)^* \tag{4.8.1}$$

In particular, if this product is *zero*, the two vectors are orthogonal to each other. We can expect that the same operation applied to the space of functions will give us a particularly valuable quantity which will be of fundamental significance in the study of differential operators. If we return to Section 7, where we have found the proper definition of the vector components in the space of functions, we obtain—by the same reasoning that gave us the "norm" (length square) of a function—that the "scalar product" of the two functions $f(x)$ and $g(x)$ has the following significance:

$$\tilde{f}g = \int_a^b f^*(x)g(x)dx \tag{4.8.2}$$

The same holds in the multi-dimensional case if we interpret the point "x" in the sense of the formula (7.8) and dx as the volume-element of the domain:

$$\tilde{f}g = \int f^*(x)g(x)dx \tag{4.8.3}$$

We speak of the "orthogonality" of two functions if this integral over the given domain of the one (one or more) variables comes out as *zero*.

4.9. The closeness of the algebraic approximation

The ϵ-method which replaces a given linear differential equation by a finite system of algebraic equations, is not more than an *approximation procedure* whose error we hope to reduce to a negligibly small quantity. Is it possible to estimate how nearly the solution of the algebraic system will approach the solution of the original problem? In order to answer this problem we will consider the *exact* solution $y(x)$ of the given differential equation, but restricted to those specific x_i values which came into existence by the atomisation of the continuum. We form the difference coefficients which in the algebraic treatment take the place of the original derivatives. Writing down all the algebraic equations of the substitute system, we can restore the original differential equation (formulated at the selected discrete points $x = x_i$) if we put on the right side of the substitute system the difference between derivative and difference coefficient. This difference can be estimated on the basis of the repeated application of the truncated Taylor series

$$f(x_\alpha + h) = f(x_a) + h \frac{\partial f}{\partial x_\alpha} + \frac{h^2}{2} \left(\frac{\partial^2 f}{\partial x_\alpha{}^2} \right)_{x_\alpha} \tag{4.9.1}$$

(where $\bar{x}_\alpha$ denotes some intermediate point), which may be written in the form

$$\frac{\partial f}{\partial x_\alpha} = \frac{\Delta f}{\Delta x_\alpha} - \frac{h}{2}\left(\frac{\partial^2 f}{\partial x_\alpha^2}\right)_{\bar{x}_\alpha} \tag{4.9.2}$$

Higher derivatives can be conceived as repeated applications of the $\partial/\partial x_\alpha$ process and thus the same estimation procedure is applicable for the replacement of higher order derivatives by the corresponding algebraic difference quotients.

Under these circumstances we can obtain definite error bounds for the algebraic system

$$Ay = b \tag{4.9.3}$$

which—through the process of atomisation—takes the place of the original differential equation

$$Dy(x) = b(x) \tag{4.9.4}$$

The vector y of the algebraic equation should represent $y(x)$ at the selected points $x = x_i$ but in actual fact y cannot be more than an *approximation* of $y(x_i)$. Hence we will replace y by $\bar{y}$ and write the algebraic system (3) in the form

$$A\bar{y} = b \tag{4.9.5}$$

while the correct $y = y(x_i)$ satisfies the equation

$$Ay = b + \delta \tag{4.9.6}$$

The error vector δ can be estimated on the basis of the given differential operator D and the right side $b(x)$. Assuming continuity of $b(x)$ and excluding any infinities in the coefficients of the operator $Dy(x)$, we can establish an error bound for the value of the component δ_i at the point $x = x_i$:

$$\delta_i < \beta h_i \tag{4.9.7}$$

where h_i is the grid parameter which becomes smaller and smaller as N, the number of grid points in our atomisation process, increases, while β is a constant for the entire domain which remains under a certain finite bound.

The fact that the individual error in each algebraic equation can be made as small as we wish by reducing the grid parameter to an appropriately small value (at all points x_i), does not guarantee that the corresponding error in y will also remain small. We will put

$$y_i = y(x_i) = \bar{y}_i + \eta_i \tag{4.9.8}$$

obtaining for the vector $\eta = (\eta_1, \eta_2, \ldots, \eta_N)$ the equation

$$A\eta = \delta \tag{4.9.9}$$

Now the general method of solving an algebraic system has to be modified to some extent for our present problem, in view of the fact that N, the number of equations, grows to infinity. Without adequate precautions the lengths of the vectors would grow to infinity and we should not be able to obtain any finite limits as N grows larger and larger. We are in fact interested in a *definite point* $x = (x^1, x^2, \ldots, x^s)$ of the continuum, although its *algebraic labelling* x_i changes all the time, in view of the constantly increasing number of points at which the equation is applied.

We avoid the difficulty by a somewhat more flexible formulation of the bilinear identity that we have discussed earlier in Chapter 3.19. In that development the "scalar product" of two vectors x and y was defined on the basis of

$$\tilde{x}y = x_1 y_1 + x_2 y_2 + \ldots + x_N y_N \tag{4.9.10}$$

(omitting the asterisk since we want to stay in the real domain). However, the bilinear identity remains unaffected if we agree that this definition shall be generalised as follows:

$$\tilde{x}y = \rho_1 x_1 y_1 + \rho_2 x_2 y_2 + \ldots + \rho_N x_N y_N \tag{4.9.11}$$

where the weight factors $\rho_1, \rho_2, \ldots, \rho_N$ are freely at our disposal, although we will restrict them in advance by the condition that we will admit only *positive* numbers as weights.

Now in the earlier treatment we made use of the bilinear identity for the purpose of *solving* the algebraic system (3). We constructed a solution of the equation

$$\tilde{A}g_i = e_i \tag{4.9.12}$$

Then the bilinear identity gave us

$$\tilde{e}_i y - \tilde{g}_i b = 0 \tag{4.9.13}$$

and we obtained the special component y_i by defining the vector as the base vector (3.19.13), in which case we got

$$y_i = \tilde{g}_i b \tag{4.9.14}$$

But now, if we agree that the scalar product shall be defined on the basis of (11), we will once more obtain the solution with the help of (14) but the definition of g_i has to occur on the basis of

$$\tilde{A}g_i = \frac{1}{\rho_i} e_i \tag{4.9.15}$$

Now the discrete point x_i was connected with a definite *cell* of our continuum of s dimensions. That cell had the volume

$$\Delta x_i = \tau_i \tag{4.9.16}$$

while the total volume τ of the continuum is the sum total of all the elementary cells:

$$\tau = \sum_{i=1}^{N} \tau_i \tag{4.9.17}$$

With increasing N each one of the individual volumes τ_i shrinks to zero, while the total volume τ remains unchanged.

We will now dispose of our ρ_i in the following manner. We will identify them with the elementary volume τ_i:

$$\rho_i = \tau_i \tag{4.9.18}$$

With this definition the lengths of our vectors do not increase to infinity any more, in spite of the infinity of N. For example the square of the error vector δ on the right side of (9) now becomes

$$\delta^2 = \sum_{i=1}^{N} \tau_i \delta_i^2 < \beta^2 \sum_{i=1}^{N} \tau_i h_i^2 \tag{4.9.19}$$

This can be put in the form

$$\delta^2 < \beta^2 \tau h^2 \tag{4.9.20}$$

if we define h^2 by

$$h^2 = \frac{\sum \tau_i h_i^2}{\sum \tau_i} \tag{4.9.21}$$

This h is a quantity which cannot become larger than the largest of all h_i; hence it goes to zero with ever increasing N.

Let us now solve the system (9) for the ith component, on the basis of (14):

$$\eta_i = \sum_{\alpha=1}^{N} g_{i\alpha} \tau_\alpha \delta_\alpha \tag{4.9.22}$$

Applying to this sum the algebraic form of Cauchy's inequality (2.4.13) we obtain

$$\eta_i^2 \le \left(\sum g_{i\alpha}^2 \tau_\alpha \right)\left(\sum \delta_\beta^2 \tau_\beta \right) \tag{4.9.23}$$

As far as the second sum goes, we have just obtained the bound (20). In order to bound the first sum we follow the procedure of Section 3.20, with the only modification that now the ratio (3.20.1) should be defined in harmony with our extended definition of the scalar product:

$$r^2 = \frac{\sum \tau_\alpha (Ay)_\alpha^2}{\sum \tau_\alpha y_\alpha^2} \tag{4.9.24}$$

However, the problem of minimising this ratio yields once more exactly the same solution as before, namely the eigenvalue problem (3.20.4), to be solved for the smallest eigenvalue λ_m. And thus we obtain, as before in (3.20.6):

$$\frac{\sum \tau_\alpha y_\alpha^2}{\sum \tau_\alpha (Ay)_\alpha^2} \le \frac{1}{\lambda_m^2} \tag{4.9.25}$$

This relation, if applied to the vector $y = g_i$, yields the estimate

$$\sum g_{i\alpha}^2 \tau_\alpha \le \frac{1}{\lambda_m^2} \tag{4.9.26}$$

and substitution in (23) gives the upper bound

$$\eta_l < \frac{\beta \sqrt{\tau} h}{\lambda_m} \tag{4.9.27}$$

We see that the difference between the algebraic and the continuous solution *converges to zero* as N increases to infinity, *provided that λ_m remains finite with increasing N*. If it so happens that λ_m converges to zero as N increases to infinity, the convergence of the algebraic solution to the correct solution can no longer be established. The behaviour of the *smallest eigenvalue of the matrix A with increasing order N of the matrix* is thus of vital importance.*

4.10. The adjoint operator

Throughout our treatment of linear systems in Chapter 3 we have pointed out the fundamental importance of the *transposed matrix $\tilde{A}$* and the associated transposed equation $\tilde{A}u = 0$. Now we deal with the matrix aspects of linear differential operators and the question of the significance of $\tilde{A}$ has to be raised. Since the differential operator itself played the role of the matrix A, we have to expect that the transposed matrix $\tilde{A}$ has to be interpreted as another linear differential operator which is somehow uniquely associated with A.

In order to find this operator, we could proceed in the following fashion. By the method of atomisation we transcribe the given differential equation into a finite system of algebraic equations. Now we abstract from these equations the matrix A itself. We transpose A by exchanging rows and columns. This gives rise to a new linear system and now we watch what happens as ϵ goes to zero. In the limit we obtain a new *differential equation* and the operator of this equation will give the adjoint differential operator (the word "adjoint" taking the place of "transposed").

While this process is rather cumbersome, it actually works and in principle we could obtain in this fashion the adjoint of any given linear differential operator. In practice we can achieve our aim much more simply by a method which we will discuss in the following section. It is, however, of interest to construct the associated operator by actual matrix transposition.

The following general observations should be added. The "transposed" operator is called "adjoint" (instead of transposed). Moreover, it is important to observe that the matrix of the transcribed algebraic system is decisively influenced by the *boundary conditions* of the problem. The same differential operator with different boundary conditions yields a different matrix. For example the problem (6.1) with the boundary conditions (6.10) yields the matrix (6.11) which is *symmetric*. Hence the transposed matrix *coincides* with the original one and our problem is "self-adjoint". But the

---

* It should be pointed out that our conclusion does *not* prove the existence of the solution of the original differential equation (2). What we have proved is only that the algebraic solution converges to the desired solution, *provided that that solution exists*.

change of the boundary conditions to those of Problem 151 transforms the matrix to a *triangular* one and the problem ceases to be self-adjoint. Hence it is important to keep in mind that *the boundary conditions belong to the operator and cannot be separated from it.* Differential equation *plus* boundary conditions determine the matrix A of the problem and correspondingly also the transposed matrix $\tilde{A}$. If we speak of the "adjoint operator", we have in mind a certain differential operator *together* with certain boundary conditions which are corollaries of the originally given boundary conditions.

The originally given boundary conditions may be either of the homogeneous or the inhomogeneous type, that is, certain boundary values may be prescribed as zero ("homogeneous boundary conditions") or as certain given non-vanishing values ("inhomogeneous boundary conditions"). However, the *given value* of a certain quantity always belongs to the *right side* of the equation and not to the left side where we find the matrix A of our problem. Hence from the standpoint of finding the transposed matrix $\tilde{A}$ it is entirely immaterial how the boundary values are prescribed, whether the demand is that a certain combination of function and derivatives on the boundary shall take the value *zero* or some *other* value. Hence in the problem of finding the adjoint operator we can always dispense with inhomogeneous boundary conditions and replace them by the corresponding *homogeneous* conditions. Our problem becomes self-adjoint if the adjoint *differential operator* coincides with the original one, and the adjoint *boundary conditions* coincide with the original ones. For example in the problem (6.11) the adjoint operator coincides with the original operator (the matrix being symmetric) and the adjoint boundary conditions become

$$u(0) = 0$$
$$u(1) = 0 \tag{4.10.1}$$

That the original boundary conditions (6.10) had the non-zero values β_1, β_n on the right side, is operationally immaterial since given numerical values can be no parts of an operator. *What* is prescribed is the decisive question; the accidental numerical values on the right side are immaterial.

Problem 154. Denoting the adjoint operator by $\tilde{D}u(x)$ find the adjoints of the following problems:

1. $Dy(x) = y'(x)$	$x = [0, 1], \quad y(0) = 0$
[Answer: $\tilde{D}u(x) = -u'(x)$	$u(1) = 0$]
2. $Dy(x) = y''(x)$	$x = [0, 1], \quad y(0) = y'(0) = 0$
[Answer: $\tilde{D}u(x) = u''(x)$	$u(1) = u'(1) = 0$]
3. $Dy(x) = xy'(x) - 2y(x)$	$x = [1, 2], \quad y(1) = 0$ (cf. Problem 152)
[Answer: $\tilde{D}u(x) = -xu'(x) - 3u(x)$	$u(2) = 0$]
4. $Dy(x) = y''(x)$	$x = [0, 1]$, no boundary conditions
[Answer: $\tilde{D}u(x) = u''(x)$	$u(0) = u'(0) = 0, \quad u(1) = u'(1) = 0$]

4.11. The bilinear identity

We now turn to the fundamental *bilinear identity* which provides us with a much simpler method of obtaining the adjoint of a given linear differential operator. In the matrix field we encountered this identity in Chapter 3, Section 19. It expresses a fundamental relation between an arbitrary matrix A and its transpose $\tilde{A}$. This relation can be utilised even as a *definition* of $\tilde{A}$, because, if we find a matrix $\tilde{A}$ which satisfies the identity

$$\tilde{u}Av - (\tilde{v}\tilde{A}u)^* \equiv 0 \qquad (4.11.1)$$

we know that $\tilde{A}$ is the transposed matrix. In the realm of finite matrices this procedure is of smaller importance because the exchange of rows and columns is a simple enough operation which needs no substitute. But if we have matrices whose order increases to infinity, the operation loses its simplicity, and its replacement by the bilinear identity may lead to something which can be handled with much greater ease. In fact, we arrive here at a point which becomes of central significance for the general theory of differential operators. By translating the identity (1) into the realm of differential operators we put the basic isomorphism between linear differential operators and matrices to good use, without in every single case having to transcribe the operator into a discrete matrix and then go to the limit where the basic grid-parameter ϵ goes to zero. We can stay from the beginning in the field of differential operators.

There is, however, a peculiarity of differential operators to which special attention has to be paid. In the identity (1) the vectors u and v are completely freely choosable. But if we think of the function space in which the vector u becomes a *function* $u(x)$ and the vector v a *function* $v(x)$, the demands of continuity and differentiability enter. A given differential operator $Dv(x)$ cannot operate on an *arbitrary* function $v(x)$ but only on a function $f(x)$ which is in fact differentiable to the extent demanded by the given operator. For example, if $Du(x)$ is given as $u''(x)$, it is self-evident that the function $u(x)$ must be twice differentiable in order that $u''(x)$ shall have a meaning. Here we observe the first important deviation from the algebraic case, caused by the fact that the "arbitrary vectors" u and v have to be re-interpreted as "arbitrary functions within a certain class of continuous and differentiable functions". The "identity" will then refer to the existence of a certain relation which holds for any pair of functions *chosen from that class.*

Another restriction is caused by the conditions on the boundary. We have seen that "the matrix A" involves more than the differential operator $Dv(x)$. It involves also certain *boundary conditions* demanded of $v(x)$. In a similar way the transposed matrix $\tilde{A}$ will involve more than the adjoint operator $\check{D}u(x)$. It will also involve certain *boundary conditions* demanded of $u(x)$. Hence $\tilde{A}$ includes the *adjoint operator together with the adjoint boundary conditions.* But then it will be of definite advantage to split our task into two parts by finding first the adjoint *differential operator* and then the adjoint *boundary conditions*. Accordingly we shall find *two* formulations of the bilinear identity (1) useful, when transcribing it into the realm of differential operators, the one referring to functions $u(x)$, $v(x)$ which satisfy the natural continuity and differentiability conditions demanded by the operators $Dv(x)$ and $\tilde{D}u(x)$, *without* specific boundary conditions and another referring to functions $u(x)$, $v(x)$ which satisfy the demanded differentiability conditions within the domain *and* the boundary conditions on the boundary of the domain.

4.12. The extended Green's identity

We have seen in Section (8) how the scalar product of two vectors has to be formulated in relation to the demands of the function space. Accordingly we will have to rewrite the bilinear identity (11.1) in the following form

$$\int [u^*(x)Dv(x) - \{v^*(x)\tilde{D}u(x)\}^*]dx \equiv 0 \qquad (4.12.1)$$

which means

$$\int [u^*(x)Dv(x) - v(x)\tilde{D}^*u^*(x)]dx \equiv 0 \qquad (4.12.2)$$

and for the case of *real* functions and operators:

$$\int [u(x)Dv(x) - v(x)\tilde{D}u(x)]dx \equiv 0 \qquad (4.12.3)$$

This fundamental identity, which transcribes the bilinear identity of matrix calculus into the realm of function space, is called "Green's identity". It has the following significance. To any given linear differential operator (ordinary or partial) D we can find a uniquely determined adjoint operator $\tilde{D}$ such that the definite integral of the left side, extended over the given domain, gives zero for any pair of functions $u(x)$, $v(x)$ which are sufficiently differentiable and which satisfy the proper boundary conditions.

A closer analysis of the general complex case (2) reveals that even in the presence of complex elements the formulation (3) of Green's identity has its advantages. Let us observe that the functions $u(x)$, $v(x)$ of the identity (2) are *arbitrary* functions (except for some boundary conditions and the general conditions of continuity and limited differentiability). Hence we could replace $u^*(x)$ by $u(x)$ without loss of generality. If we do so we notice that the entire difference between (2) and (3) is that the $\tilde{D}$ of equation (3) is replaced by $\tilde{D}^*$. Let us now agree to call the operator $\tilde{D}$, as defined by (3), the "algebraic adjoint" of D. Then, if we want to obtain the "Hermitian adjoint" of D (and "self-adjoint" always refers to the identity of the given operator with its *Hermitian* adjoint), all we have to do is to change in $\tilde{D}$ every i to $-i$. For this reason we will henceforth *drop* the formulation (2) of Green's identity and operate consistently with the formulation (3), with the understanding that the Hermitian adjoint of D shall be denoted by $\tilde{D}^*$, while $\tilde{D}$ refers to the algebraic adjoint.

Before we come to closer grips with Green's identity, we will first formulate it in somewhat more general terms. We now assume a pair of functions

$u(x)$, $v(x)$ which satisfy the demanded differentiability conditions but are *not* subjected to any specific boundary conditions:

$$\int [u(x)Dv(x) - v(x)\tilde{D}u(x)]dx = \text{boundary term} \qquad (4.12.4)$$

The result of the integration is no longer zero but something that depends solely on the values of $u(x)$, $v(x)$—and some of their derivatives—taken on the *boundary* of the region. This is the meaning of the expression "boundary term" on the right side of (4). The fundamental identity (4) is called the "extended Green's identity".

In order to see the significance of this fundamental theorem let us first restrict ourselves to the case of a *single* independent variable x. The given operator $Dv(x)$ is now an *ordinary* differential operator, involving the derivatives of $v(x)$ with respect to x. Let us assume that we succeed in showing the validity of the following bilinear relation:

$$u(x)Dv(x) - v(x)\tilde{D}u(x) = \frac{d}{dx} F(u, v) \qquad (4.12.5)$$

where on the right side $F(u, v)$ is an abbreviation for some bilinear function of $u(x)$ and $v(x)$, and their derivatives. If we are able to prove (5), we shall at once have (4) because, integrating with respect to x between the limits a and b we obtain

$$\int_a^b [u(x)dv(x) - v(x)\tilde{D}u(x)]dx = \left[F(u, v) \right]_a^b \qquad (4.12.6)$$

and this equation is exactly of the form (4). Let us then concentrate on the proof of (5).

The operator $Dv(x)$ is generally of the form

$$Dv(x) = \sum_{k=0}^r p_k(x)v^{(k)}(x) \qquad (4.12.7)$$

and it suffices to consider a typical term. The following relation is familiar from the method of integrating by parts:

$$f(x)g^{(k)}(x) - (-1)^k g(x)f^{(k)}(x)$$
$$= \frac{d}{dx} [f(x)g^{(k-1)}(x) - f'(x)g^{(k-2)}(x) + \ldots + (-1)^{k-1}f^{(k-1)}(x)g(x)] \qquad (4.12.8)$$

If we identify $g(x)$ with $v(x)$ and $f(x)$ with $p_k(x)u(x)$, we obtain

$$u(x)p_k(x)v^{(k)}(x) = (-1)^k v(x)[p_k(x)u(x)]^{(k)} + \frac{d}{dx} F(u, v) \qquad (4.12.9)$$

and we see that we have obtained the adjoint operator associated with the term $p_k(x)v^{(k)}(x)$:

$$Dv(x) = p_k(x)v^{(k)}(x), \quad \tilde{D}u(x) = (-1)^k \frac{d^k}{dx^k} [p_k(x)u(x)] \qquad (4.12.10)$$

If we repeat the same procedure with every term, the entire operator $\tilde{D}u(x)$ will be constructed.

We have thus obtained a simple and powerful mechanism by which to any given $Dv(x)$ the corresponding $\tilde{D}u(x)$ can be obtained. The process requires no integrations but only differentiations and combinations of terms. We can even write down explicitly the adjoint of the operator (7):

$$Dv(x) = \sum_{k=0}^{r} p_k(x)v^{(k)}(x)$$

$$(4.12.11)$$

$$\tilde{D}u(x) = \sum_{k=0}^{r} [(-1)^k p_k(x)u(x)]^{(k)}$$

The adjoint *boundary conditions*, however, have not yet been obtained.

Problem 155.   Consider the most general linear differential operator of second order:

$$Dv(x) = A(x)v''(x) + B(x)v'(x) + C(x)v(x) \qquad (4.12.12)$$

Find the adjoint operator.

[Answer:

$$\tilde{D}u(x) = A(x)u''(x) + (2A' - B)u'(x) + (A'' - B' + C)u(x) \quad (4.12.13)]$$

Problem 156.   Find the most general linear differential operator of the second order which is *self-adjoint*.

[Answer:

$$Dv(x) = [A(x)v'(x)]' + C(x)v(x) \qquad (4.12.14)]$$

Problem 157.   Find the adjoint of the system (6.13):

$$Dv(x) = \begin{matrix} v'_1(x) - v_2(x) \\ v'_2(x) \end{matrix} \qquad (4.12.15)$$

[Answer:

$$\tilde{D}u(x) = \begin{matrix} -u'_1(x) \\ -u'_2(x) - u_1(x) \end{matrix} \qquad (4.12.16)]$$

4.13.   The adjoint boundary conditions

We now come to the investigation of the *adjoint boundary conditions* which are of necessity associated with the adjoint operator. Although in our mind differential operator and boundary conditions are separated, in actual fact the boundary conditions are inseparable ingredients of the adjoint operator without which it loses its significance. We have merely divided our task by operating first with the *extended Green's identity* (12.4). This identity avoids the question of boundary conditions by subjecting $u(x)$ and $v(x)$ to some differentiability conditions only, without boundary conditions. But our final goal is to arrive at the theorem (12.3), the Green's identity *without* boundary term, which represents the true transcription of the bilinear identity and thus the true definition of the adjoint operator. The extended Green's theorem (12.4) will change over into the homogeneous form (12.3) if we subject the functions $u(x)$, $v(x)$ to the proper boundary conditions. As far as the function $v(x)$ goes, certain boundary conditions will probably be prescribed since a differential equation without boundary conditions represents an incomplete system which can have no unique solution. For our present purposes we take no notion of the prescribed *values* (which belong to the right side of the equation and have no operational significance). *Any inhomogeneous boundary condition is replaced by the corresponding homogeneous condition.* For example, if the value of $v(x)$ is prescribed as 1 at $x = a$ and as -1 at $x = b$, we will temporarily replace these conditions by the fictitious conditions

$$v(a) = v(b) = 0 \qquad (4.13.1)$$

We shall see later that by knowing how to handle homogeneous boundary conditions we also know how to handle inhomogeneous boundary conditions.

We now examine the boundary term of the extended Green's identity (12.6). Certain terms will vanish on account of the prescribed boundary conditions for $v(x)$ and their derivatives. Some other terms will not drop out. We will now apply the following general principle: *we impose on $u(x)$ and its derivatives the minimum number of conditions which are necessary and sufficient for the vanishing of the boundary term.*

By this principle the adjoint boundary conditions are uniquely determined, irrespective of how complete or incomplete the original set of boundary conditions have been. The transpose of a matrix always exists, no matter how incomplete or over-determined the original matrix may be. From the fact that the transpose of an $n \times m$ matrix is an $m \times n$ matrix we can conclude that the degree of determination of D and $\check{D}$ will be in a *reciprocal* relation to each other: *the more over-determined the original operator D is, the more under-determined is $\check{D}$ and vice versa.* It might happen for example that the original problem has so many boundary conditions prescribed that the boundary term of the extended Green's identity (4) *vanishes* without any further conditions. In this case the adjoint operator $\check{D}u(x)$ is not subjected to any boundary conditions. Again, it might happen that the original problem is completely free of any boundary conditions. In this case the adjoint problem will acquire an overload of boundary conditions, in order to make the boundary term of (12.4) vanish without the help of the function $v(x)$.

As an example let us consider Newton's equation of motion: "mass times acceleration equals moving force". We normalise the mass to 1 and consider motion in one dimension only:

$$v''(x) = \beta(x) \qquad (4.13.2)$$

(x has the significance of the time t). The boundary conditions are that at $x = 0$ the displacement $v(x)$ and the velocity $v'(x)$ are zero:

$$v(0) = 0$$
$$v'(0) = 0 \qquad (4.13.3)$$

The range of interest is $x = [0, 1]$.

We go through the regular procedure of Section 12, obtaining

$$uv'' - vu'' = \frac{d}{dx}(uv' - vu') \qquad (4.13.4)$$

$$\int_0^1 (uv'' - vu'')dx = \Big[uv' - vu'\Big]_0^1 \qquad (4.13.5)$$

We investigate the boundary term of the right side. The given boundary conditions are such that the contribution at the lower limit $x = 0$ becomes zero while at the upper limit $x = 1$ we have

$$u(1)v'(1) - v(1)u'(1) \qquad (4.13.6)$$

Nothing is said about $v(1)$ and $v'(1)$. Hence the vanishing of the boundary term on the right side of (5) demands

$$\begin{aligned} u(1) &= 0 \\ u'(1) &= 0 \end{aligned} \qquad (4.13.7)$$

We have thus found the adjoint operator of the given problem:

$$\tilde{D}u(x) = u''(x) \qquad (4.13.8)$$

with the boundary conditions

$$\begin{aligned} u(1) &= 0 \\ u'(1) &= 0 \end{aligned} \qquad (4.13.9)$$

We notice that our problem is *not self-adjoint* because, although $\tilde{D}$ and D agree, the boundary conditions for $v(x)$ and $u(x)$ do *not* agree since the conditions for $v(x)$ are prescribed at $x = 0$, the conditions for $u(x)$ at $x = 1$.

Let us now assume that we are interested in constructing a mechanism which will guarantee that at the time moment $x = 1$ the moving mass *returns* to the origin, with zero velocity. In this case we have imposed on our problem the two further boundary conditions

$$\begin{aligned} v(1) &= 0 \\ v'(1) &= 0 \end{aligned} \qquad (4.13.10)$$

Now the boundary term of (5) *vanishes automatically* and we do not get any boundary conditions for $u(x)$. The adjoint problem $\tilde{A}u$ is now characterised by the operator $u''(x)$ *alone*, without any boundary conditions.

Problem 158. The results of Problem 154 were obtained by direct matrix transposition. Obtain the same results now on the basis of Green's identity.

Problem 159. Consider the following differential operator:

$$\begin{aligned} Dv(x) &= B(x)v'(x) + C(x)v(x) \\ x &= [a, b] \qquad\qquad B(x) \neq 0 \end{aligned} \qquad (4.13.11)$$

Obtain the adjoint operator $\check{D}u(x)$ and the adjoint boundary conditions under the following circumstances:

a) $v(a) = 0$

b) $v(b) = 0$

c) $v(b) = v(a)$

[Answer:

$$\check{D}u(x) = -Bu' + (C - B')u \qquad (4.13.12)$$

a) $u(b) = 0$

b) $u(a) = 0$

c) $u(b)B(b) = u(a)B(a)$]

Problem 160.   Consider the self-adjoint second-order operator (12.14), $A(x) \neq 0$, range $x = [a, b]$, with the boundary conditions

$$\begin{aligned} v'(a) &= \alpha_1 v(a) + \beta_1 v(b) \\ v'(b) &= \alpha_2 v(a) + \beta_2 v(b) \end{aligned} \qquad (4.13.13)$$

Find the adjoint boundary conditions. When will the system become self-adjoint?

[Answer:

$$\gamma = \frac{A(a)}{A(b)} \qquad \begin{aligned} u'(a) &= \alpha_1 u(a) - \frac{\alpha_2}{\gamma} u(b) \\ u'(b) &= -\beta_1 \gamma u(a) + \beta_2 u(b) \end{aligned} \qquad (4.13.14)$$

Condition of self-adjointness:

$$A(a)\beta_1 + A(b)\alpha_2 = 0 \qquad (4.13.15)]$$

4.14. Incomplete systems

In our general discussions of linear systems we encountered three characteristic numbers which were of decisive importance for the general behaviour of the system: the number of equations n, the number of unknowns m, and the order of the matrix p.   But now we are faced with problems which from the standpoint of algebra represent an infinity of equations for an infinity of unknowns.   The numbers n and m, and likewise the number p, lose their direct significance.   We have seen, however, in Section 7 of Chapter 3 that the solution of the homogeneous equations

$$Av = 0 \qquad (4.14.1)$$

$$\tilde{A}u = 0 \qquad (4.14.2)$$

give us a good substitute for the direct definition of n, m, and p.   The number of independent solutions of the system (1) is always $m - p$, that of the system (2) $n - p$.   These solutions tell us some fundamental facts about the given system, even before we proceed to the task of actually finding the solution.   The system (1) decides the unique or not unique character of the solution while the system (2) yields the compatibility conditions of the system.

The role of the system (1) was: "Add to a particular solution an arbitrary solution of the system (1), in order to obtain the general solution of the given system."

The role of the system (2) was: "The given system is solvable if and only if the right side is orthogonal to every independent solution of (2)."

These results are immediately applicable to the problem of solving linear differential equations or systems of such equations. Before we attempt a solution, there are two questions which we will want to decide in advance:

1. Will the solution be unique? 2. Are the given data such that a solution is possible?

Let us first discuss the question of the uniqueness of the solution. We have seen that an rth order differential equation alone, without additional boundary conditions, represents an $(m - r) \times m$ system and is thus r equations short of a square matrix. But even the addition of r boundary conditions need not necessarily guarantee that the solution will be unique. For example the system

$$v''(x) = \beta(x) \tag{4.14.3}$$

$$v'(a) = 0 \qquad v'(b) = 0 \tag{4.14.4}$$

represents a second order system with two boundary conditions and thus we would assume that the problem is well-determined. And yet this is not so because the homogeneous problem

$$v''(x) = 0 \tag{4.14.5}$$

with the boundary conditions (4) has the solution

$$v(x) = \text{const.} \tag{4.14.6}$$

Hence we could have added one more condition to the system, in order to make it uniquely determined, e.g.

$$v(a) = 0 \tag{4.14.7}$$

But if we have no particular reason for adding a condition of this kind, it is more natural to remove the deficiency in the fashion we have done before in Chapter 3, Section 14 by requiring that *the solution shall be orthogonal to every solution of the homogeneous equation.* In our example it means that we remove the deficiency of our system by adding the condition

$$\int_a^b v(x)dx = 0 \tag{4.14.8}$$

We might think that deficiency is a purely mathematical phenomenon, caused by incomplete information. But this is by no means so. We encounter incomplete systems in well-defined physical situations. Let us consider for example the problem of a *loaded elastic bar.* The mathematical description of this problem leads to a differential equation of *fourth order* which may be formulated in the form of two simultaneous differential equations of *second* order:

$$I(x)v''_1(x) - v_2(x) = 0$$
$$v''_2(x) = \beta(x)$$

$$(4.14.9)$$

($v_1(x)$ is the deflection of the bar, $I(x)$ is the inertial moment of the generally variable cross section, $\beta(x)$ the load density.)   We assume that the bar extends from $x = 0$ to $x = l$.   We will also assume that the bar is not supported at the two endpoints but at points between and we include the forces of support as part of the load distribution, considering them as negative loads.

Now the boundary conditions of a bar which is free at the two endpoints are

$$v_2(0) = 0 \qquad v_2(l) = 0$$
$$v'_2(0) = 0 \qquad v'_2(l) = 0$$

$$(4.14.10)$$

These four conditions seem to suffice for a well-determined solution since the deflection of a bar satisfies a differential equation of fourth order.   But in actual fact the homogeneous system

$$I(x)v''_1(x) - v_2(x) = 0$$
$$v''_2(x) = 0$$

$$(4.14.11)$$

under the boundary conditions (10) demands only the vanishing of $v_2(x)$ while for $v_1(x)$ we obtain two independent solutions

$$v_1(x) = 1$$
$$v_1(x) = x$$

$$(4.14.12)$$

The physical significance of these two solutions is that the bar may be *translated* as a whole and also *rotated* rigidly as a whole.   These two degrees of freedom are in the nature of the problem and not artificially imposed from outside.   We can eliminate this uncertainty by adding the two orthogonality conditions

$$\int_a^b v_1(x)dx = 0 \tag{4.14.13}$$

$$\int_a^b xv_1(x)dx = 0 \tag{4.14.14}$$

These conditions remove the arbitrariness of the frame of reference in which the vertical displacements of the bar are measured, by putting the origin of the Z-axis in the centre of mass of the bar and orienting the Z-axis perpendicular to the "neutral plane" of the bar.

Other conditions can also be chosen for the removal of the deficiency of the system, *provided that they eliminate the solutions of the homogeneous system*.

Problem 161. Show that the added conditions a) or b) are permissible, the conditions c) not permissible for the elimination of the deficiency of the system (9), (10):

$$\text{a)} \quad v_1(0) = v'_1(0) = 0 \tag{4.14.15}$$

$$\text{b)} \quad v_1(0) = v_1(l) = 0 \tag{4.14.16}$$

$$\text{c)} \quad v'_1(0) = v'_1(l) = 0 \tag{4.14.17}$$

Problem 162. Find the adjoint system of the problem (9), (10).

[Answer:

$$\check{D}u = \begin{array}{c} [I(x)u_1(x)]'' \\ u''_2(x) - u_1(x) \end{array} \tag{4.14.18}$$

Boundary conditions:

$$\begin{array}{c} u_1(0) = u_1(l) = 0 \\ u'_1(0) = u'_1(l) = 0 \end{array} \tag{4.14.19}]$$

4.15. Over-determined systems

A counterpart of the homogeneous equation $Av = 0$ is the homogeneous equation $\tilde{A}u = 0$. If this equation has non-vanishing solutions, this is an indication that the given data have to satisfy some definite *compatibility conditions*, and that again means that we have given too many data which cannot be chosen independently of each other. Consider for example the problem of the "free bar", discussed in the previous section. Here the adjoint homogeneous system $\check{D}u(x) = 0$ (cf. 14.18) under the boundary conditions (14.19) yields

$$\begin{array}{c} u_1(x) = 0 \\ u''_2(x) = 0 \end{array} \tag{4.15.1}$$

This system has two independent solutions:

$$\begin{array}{c} u_2(x) = 1 \\ u_2(x) = x \end{array} \tag{4.15.2}$$

The orthogonality of the right side of (14.9) to these solutions demands the following two compatibility conditions:

$$\int_0^l \beta(x)dx = 0 \tag{4.15.3}$$

$$\int_0^l x\beta(x)dx = 0 \tag{4.15.4}$$

These conditions have a very definite physical significance. They express the fact that *the loaded elastic bar can be in equilibrium if and only if the sum of all the loads and the moments of all the loads is zero.*

In Section 13 we considered the motion of a mass 1 under the influence of a force $\beta(x)$. We assumed that the motion started with zero displacement and zero velocity (cf. 13.3). But later we assumed that we succeeded with a mechanism which brought the mass back to the origin with zero velocity at the time moment $x = 1$. Hence we could add the two surplus conditions (13.10) with the result that the adjoint homogeneous system became

$$u''(x) = 0 \qquad (4.15.5)$$

without any additional conditions. Accordingly the adjoint homogeneous system has the two independent solutions $u(x) = 1$ and x, and we obtain the two compatibility conditions

$$\int_0^1 \beta(x)dx = 0 \qquad (4.15.6)$$

$$\int_0^1 x\beta(x)dx = 0$$

They have the following significance. The forces employed by our mechanism satisfy of necessity the two conditions that *the time integral of the force and the first moment of this integral vanishes.*

Problem 163. Find the compatibility conditions of the following system

$$v''(x) = \beta(x)$$
$$v(a) = v(b) \qquad (4.15.7)$$
$$v'(a) = v'(b)$$

[Answer:

$$\int_a^b \beta(x)dx = 0 \qquad (4.15.8)]$$

Problem 164. In Problem 161 the addition of the conditions (14.17) were considered not permissible for the removal of the deficiency of the system. What new compatibility condition is generated by the addition of these boundary conditions? (Assume $I(x) = $ const.)

[Answer:

$$\int_0^l \beta(x)x^2dx = 0 \ ]$$

Problem 165. Given the following system:

$$v^{(n)}(x) = \beta(x)$$
$$v(a) = v'(a) = \ldots v^{n-1}(a) = 0 \qquad (4.15.9)$$
$$v(b) = v'(b) = \ldots v^{n-1}(b) = 0$$

Find the compatibility conditions of this system.

[Answer:

$$\int_a^b \beta(x)x^kdx = 0 \qquad (k = 0, 1, 2, \ldots \ n - 1) \qquad (4.15.10)]$$

4.16. Compatibility under inhomogeneous boundary conditions

In our previous discussions we have assumed that the boundary conditions of our problem were given in *homogeneous* form: some linear combinations of function and derivatives were given as *zero* on the boundary. This may generally not be so. We may prescribe certain linear aggregates of function and derivative on the boundary as some *given value* which is *not zero*. These values belong now to the given "right side" of the equation. The construction of the homogeneous equation $Dv = 0$ and likewise the construction of the *adjoint* homogeneous equation $\tilde{D}u = 0$ is not influenced by the inhomogeneity of the given boundary conditions. What is influenced, however, are the *compatibility conditions* between the data. The given inhomogeneous boundary values belong to the data and participate in the compatibility conditions.

The orthogonality of the given right side to every independent solution of the adjoint homogeneous equation is a direct consequence of "Green's identity" (12.3):

$$\int u(x)Dv(x)dx = 0 \tag{4.16.1}$$

if

$$\tilde{D}u(x) = 0 \tag{4.16.2}$$

But let us now assume that the prescribed boundary conditions for $v(x)$ are of the *inhomogeneous type*. In that case we have to change over to the *extended Green's identity* (12.6). The boundary term on the right side will now be *different from zero* but it will be expressible in terms of the given boundary values and the boundary values of the auxiliary function $u(x)$ which we found by solving the adjoint homogeneous equation:

$$\int_a^b u(x)Dv(x)dx - \left[F(u, v)\right]_a^b = 0 \tag{4.16.3}$$

As an example let us consider once more the problem of the free elastic bar, introduced before in Section 14 (cf. 14.9–10). Let us change the boundary conditions (14.10) as follows:

$$\begin{aligned} v_2(0) &= 0 & v_2(l) &= 0 \\ v'_2(0) &= \rho_1 & v'_2(l) &= \rho_2 \end{aligned} \tag{4.16.4}$$

and let us see what change will occur in the two compatibility conditions (15.3–4). For this purpose we have to investigate the boundary term of the extended Green's identity which in our case becomes

$$\left[u_1Iv'_1 - v_1(Iu_1)' + u_2v'_2 - v_2u'_2\right]_0^l \tag{4.16.5}$$

The first two terms drop out on account of the adjoint boundary conditions. The last two terms dropped out earlier on account of the given homogeneous boundary conditions (14.10) while at present we get

$$\rho_2u_2(l) - \rho_1u_2(0) \tag{4.16.6}$$

and thus the compatibility conditions (15.3) and (15.4) must now be extended as follows:

$$\int_0^l \beta(x)dx + \rho_1 - \rho_2 = 0 \tag{4.16.7}$$

$$\int_0^l x\beta(x)dx - \rho_2 l = 0 \tag{4.16.8}$$

These conditions again have a simply physical significance. We could assume that our bar extends by a very small quantity ϵ beyond the limits 0 and l. Then we could conceive the two conditions (7) and (8) as the previous homogeneous conditions

$$\int_{-\epsilon}^{l+\epsilon} \beta(x)dx = 0 \tag{4.16.9}$$

$$\int_{-\epsilon}^{l+\epsilon} x\beta(x)dx = 0 \tag{4.16.10}$$

which express the mechanical principles that the equilibrium of the bar demands that the sum of all loads and the sum of the moments of all loads must be zero. The loads which we have added to the previous loads are

$$\int_{-\epsilon}^0 \beta(x)dx = \rho_1 \quad \text{and} \quad \int_l^{l+\epsilon} \beta(x)dx = -\rho_2 \tag{4.16.11}$$

We call these loads "point-loads" since they are practically concentrated in one point. In fact the load ρ_2 at $x = l$ and the load $-\rho_1$ at $l = 0$ are exactly the *supporting loads* we have to apply at the two endpoints of the bar ρ in order to keep the load distribution $\beta(x)$ in equilibrium.

As a second example let us modify the boundary conditions (4) as follows:

$$\begin{aligned} v_2(0) &= 0 & v_2(l) &= \mu_2 \\ v'_2(0) &= 0 & v'_2(l) &= \rho_2 \end{aligned} \tag{4.16.12}$$

Here the boundary term (5) becomes

$$\rho_2 u_2(l) - \mu_2 u'_2(l) \tag{4.16.13}$$

and the new compatibility conditions (15.3) and (15.4) become:

$$\int_0^l \beta(x)dx - \rho_2 = 0$$

$$\int_0^l x\beta(x)dx - \rho_2 l + \mu_2 = 0 \tag{4.16.14}$$

Once more we can conceive these equations as the equilibrium conditions

(9) of all the forces, extending the integration from 0 to $l + \epsilon$, and defining the added load distribution by the following two conditions:

$$\int_0^\epsilon \beta(l + \xi)d\xi = -\rho_2$$

$$\int_0^\epsilon \xi\beta(l + \xi)d\xi = \mu_2$$

(4.16.15)

What we have here is a *single force* at the end of the bar *and a force couple* acting at the end of the bar, to balance out the sum of the forces and the moments of the forces distributed along the bar. This means in physical interpretation that the bar is free at the left end $x = 0$ but *clamped* at the right end $x = l$ since the clamping can provide that single force *and* force couple which is needed to establish equilibrium. Support without clamping can provide only a point force which is not enough for equilibrium except if a second supporting force is applied at some other point, for instance at $x = 0$, and that was the case in our previous example.

What we have seen here is quite typical for the behaviour of inhomogeneous boundary conditions. Such conditions can always be interpreted as *extreme distributions* of the right side of the differential equation, taking recourse to point loads and possibly force couples of first and higher order which have the same physical effect as the given inhomogeneous boundary conditions.

Problem 166.   Extend the compatibility conditions (15.6) to the case that the boundary conditions (13.3) and (13.10) of the problem (13.2) are changed as follows:

$$v(0) = \gamma_0 \qquad v(1) = \eta_0$$
$$v'(0) = \gamma_1 \qquad v'(1) = \eta_1$$

(4.16.16)

[Answer:

$$\int_0^1 \beta(x)dx = \eta_1 - \gamma_1$$

$$\int_0^1 x\beta(x)dx = \gamma_0 - \eta_0 + \eta_1 \,]$$

Problem 167.   Consider the following system:

$$v^{(n)}(x) = \beta(x)$$

(4.16.17)

with the boundary conditions

$$v(a) = \alpha_0 \qquad\qquad v(b) = \gamma_0$$
$$v'(a) = \alpha_1$$
$$\vdots$$
$$v^{(n-1)}(a) = \alpha_{n-1}$$

(4.16.18)

Obtain the compatibility condition of this system and show that it is identical with *the Taylor series with the remainder term*.

Problem 168. Consider once more the system of the previous problem, but changing the boundary condition at b to

$$v^{(k)}(b) = \gamma_k \tag{4.16.19}$$

where k is some integer between 1 and $n - 1$:

[Answer:

$$v^{(k)}(b) = v^{(k)}(a) + v^{(k+1)}(a)(b - a) + v^{(k+2)}(a) \frac{(b - a)^2}{2!} + \cdots$$

$$+ v^{(n-1)}(a) \frac{(b - a)^{n-1-k}}{(n - 1 - k)!} + \int_a^b \beta(x) \frac{(b - x)^{n-1-k}}{(n - 1 - k)!} \, dx \tag{4.16.20}]$$

4.17. Green's identity in the realm of partial differential operators

The fundamental Green's identity (12.2) represents the transcription of the bilinear identity which holds in the realm of matrices. Since the isomorphism between matrices and linear differential operators is not restricted to problems of a single variable but holds equally in multi-dimensional domains, we must be able to obtain the Green's identity if our problem involves *partial* instead of ordinary differentiation.

Here again we will insert an intermediate step by not requiring immediately that the functions $u(x)$, $v(x)$—where x now stands for a point of a multi-dimensional domain—shall satisfy the proper boundary conditions. We will again introduce the *extended Green's identity* (12.4) where the right side is not zero but a certain "boundary term" which involves the values of $u(x)$ and $v(x)$ and some of their partial derivatives on the boundary. We have seen that in the realm of ordinary differentiation the integral relation (12.4) could be replaced by the simpler relation (12.5) which involved no integration. It was this relation which gave rise to a boundary term after integrating on both sides with respect to x. But it was also this relation by which the adjoint operator $\check{D}u(x)$ could be obtained (omitting for the first the question of the adjoint boundary conditions).

Now the procedure in the realm of partial operators will be quite similar to the previous one. The only difference is that on the right side of (12.5) we shall not have a single term of the form d/dx but a *sum of terms* of the form $\partial/\partial x_\alpha$. The quantity $F(u, v)$ which was bilinear in u and v, will now become a *vector* which has the components

$$F_1(u, v), F_2(u, v), \ldots, F_\mu(u, v) \tag{4.17.1}$$

if μ is the number of independent variables. We will thus write down the fundamental relation which defines the adjoint operator $\check{D}u(x)$ as follows:

$$u(x)Dv(x) - v(x)\check{D}u(x) = \sum_{\alpha=1}^{\mu} \frac{\partial}{\partial x_\alpha} F_\alpha(u, v) \tag{4.17.2}$$

Let us assume that we have succeeded with the task of constructing the adjoint operator $\breve{D}u(x)$ on this basis. Then we can immediately multiply by the volume-element dx on both sides and integrate over the given domain. On the right side we apply the Gaussian integral transformation:

$$\int \sum_{\alpha=1}^{\mu} \frac{\partial}{\partial x_\alpha} F_\alpha(u, v)dx = \int \sum_{\alpha=1}^{\mu} F_\alpha(u, v)\nu_\alpha dS \qquad (4.17.3)$$

where ν_α is the outside normal (of the length 1) of the boundary surface S. We thus obtain once more the extended Green's theorem in the form

$$\int [u(x)Dv(x) - v(x)\breve{D}u(x)]dx = \int \sum_{\alpha=1}^{\mu} F_\alpha(u, v)\nu_\alpha dS \qquad (4.17.4)$$

The "boundary term" appears now as an integral extended over the boundary surface. From here we continue exactly as we have done before: *we impose the minimum number of conditions on $u(x)$ which are necessary and sufficient to make the boundary integral on the right side of (4) vanish.* This provides us with the adjoint boundary conditions. We have thus obtained the differential operator $\breve{D}u(x)$ and the proper boundary conditions which together form the adjoint operator.

Let us then examine the equation (2). An arbitrary linear differential operator $Dv(x)$ is composed of terms which contain $v(x)$ and its derivatives linearly. Let us pick out a typical term which we may write in the form

$$A(x)\frac{\partial w}{\partial x_i} \qquad (4.17.5)$$

where $A(x)$ is a given function of the x_i while w stands for some partial derivative of v with respect to any number of x_j. We now multiply by $u(x)$ and use the method of "integrating by parts":

$$u(x)A(x)\frac{\partial w}{\partial x_i} = \frac{\partial}{\partial x_i}[u(x)A(x)w] - w\frac{\partial}{\partial x_i}[A(x)u(x)] \qquad (4.17.6)$$

What we have achieved now is that w is no longer differentiated. We might say that we have "liberated" w from the process of differentiation. We can obviously repeat this process any number of times. At every step we reduce the order of differentiation by one, until eventually we must end with a term which contains v itself, without any derivatives. The factor of v is then the contribution of that term to the adjoint operator $\breve{D}u(x)$.

Let us consider for example a term given as follows:

$$A(x, y, z)\frac{\partial^4 v}{\partial x \partial^2 y \partial z} \qquad (4.17.7)$$

Then the "process of liberation" proceeds as follows:

$$uA \frac{\partial^4 v}{\partial x \partial^2 y \partial z} = \frac{\partial}{\partial x}\left(Au \frac{\partial^3 v}{\partial^2 y \partial z}\right) - \frac{\partial^3 v}{\partial^2 y \partial z} \frac{\partial Au}{\partial x}$$

$$-\frac{\partial^3 v}{\partial^2 y \partial z} \frac{\partial Au}{\partial x} = -\frac{\partial}{\partial y}\left(\frac{\partial^2 v}{\partial y \partial z} \frac{\partial Au}{\partial x}\right) + \frac{\partial^2 v}{\partial y \partial z} \frac{\partial^2 Au}{\partial y \partial x}$$

$$\frac{\partial^2 v}{\partial y \partial z} \frac{\partial^2 Au}{\partial y \partial x} = \frac{\partial}{\partial y}\left(\frac{\partial v}{\partial z} \frac{\partial^2 Au}{\partial y \partial x}\right) - \frac{\partial v}{\partial z} \frac{\partial^3 Au}{\partial y^2 \partial x}$$

$$-\frac{\partial v}{\partial z} \frac{\partial^3 Au}{\partial y^2 \partial x} = -\frac{\partial}{\partial z}\left(v \frac{\partial^3 Au}{\partial y^2 \partial x}\right) + v \frac{\partial^4 Au}{\partial z \partial y^2 \partial x}$$

$$(4.17.8)$$

The result of the process is that v is finally liberated and the roles of u and v are exchanged: originally u was multiplied by a derivative of v, now v is multiplied by a derivative of u. Hence the adjoint operator has been obtained as follows:

$$Dv(x) = A \frac{\partial^4 v}{\partial x \partial^2 y \partial z}$$

$$\check{D}u(x) = \frac{\partial^4 Au}{\partial x \partial^2 y \partial z}$$

$$(4.17.9)$$

while the vector $F_\alpha(u, v)$ of the general relation (2) has in our case the following significance:

$$F_1(u, v) = \frac{\partial^3 v}{\partial^2 y \partial z} Au$$

$$F_2(u, v) = -\frac{\partial^2 v}{\partial y \partial z} \frac{\partial Au}{\partial x} + \frac{\partial v}{\partial z} \frac{\partial^2 Au}{\partial y \partial x}$$

$$F_3(u, v) = -v \frac{\partial^3 Au}{\partial y^2 \partial x}$$

$$(4.17.10)$$

We have followed a certain sequence in this process of reducing the degree of differentiation, viz. the sequence x, y, y, z. We might have followed another sequence, in which case the vector $F_\alpha(u, v)$ would have come out differently but the adjoint operator $\check{D}u(x)$ would still be the same. It is the peculiarity of partial differentiation that the same boundary term may appear in various forms, although the complete boundary integral (3) is the same. The vector F_α has no absolute significance, only the complete integral which appears on the right side of the extended Green's identity (4).

Problem 169. Obtain the adjoint operator of the term

$$A(x, y) \frac{\partial^2 v}{\partial x \, \partial y} \qquad (4.17.11)$$

once in the sequence x, y and once in the sequence y, x. Show that the equality of the resulting boundary integral can be established as a consequence of the following identity:

$$\int \left(\frac{\partial w}{\partial x_i} \nu_k - \frac{\partial w}{\partial x_k} \nu_i \right) dS \equiv 0 \tag{4.17.12}$$

Prove this identity by changing it to a volume integral with the help of Gauss's theorem (3).

4.18. The fundamental field operations of vector analysis

In the vector analysis of three dimensional domains

$$x = (x_1, x_2, x_3) \tag{4.18.1}$$

the following operations appear as fundamental in the formulation of the field problems of classical physics:

1. The vector

$$\text{grad } \Phi = \left(\frac{\partial \Phi}{\partial x_1}, \frac{\partial \Phi}{\partial x_2}, \frac{\partial \Phi}{\partial x_3} \right) \tag{4.18.2}$$

2. The scalar

$$\text{div } V = \frac{\partial V_1}{\partial x_1} + \frac{\partial V_2}{\partial x_2} + \frac{\partial V_3}{\partial x_3} \tag{4.18.3}$$

3. The vector

$$\text{curl } V = \left(\frac{\partial V_3}{\partial x_2} - \frac{\partial V_2}{\partial x_3}, \frac{\partial V_1}{\partial x_3} - \frac{\partial V_3}{\partial x_1}, \frac{\partial V_2}{\partial x_1} - \frac{\partial V_1}{\partial x_2} \right) \tag{4.18.4}$$

Amongst the combinations of these operations of particular importance is the "Laplacian operator"

$$\varDelta = \frac{\partial^2}{\partial x_1{}^2} + \frac{\partial^2}{\partial x_2{}^2} + \frac{\partial^2}{\partial x_3{}^2} \tag{4.18.5}$$

which appears in the following constructions:

$$\text{div grad } \Phi = \varDelta \Phi \tag{4.18.6}$$

$$\text{curl curl } V = \text{grad div } V - \varDelta V \tag{4.18.7}$$

According to the general rules we can construct the *adjoints* of these operators:

1. Adjoint of the gradient:

$$\tilde{D}u = -\text{div } U \tag{4.18.8}$$

Boundary term:

$$\int \Phi \sum_{\alpha=1}^{3} U_\alpha \nu_\alpha dS \tag{4.18.9}$$

2. Adjoint of the divergence:

$$\check{D}u = -\operatorname{grad} u \qquad (4.18.10)$$

Boundary term:

$$\int u \sum_{\alpha=1}^{3} V_\alpha \nu_\alpha dS \qquad (4.18.11)$$

3. Adjoint of the curl:

$$\check{D}u = \operatorname{curl} U \qquad (4.18.12)$$

Boundary term:

$$\int \sum_{\alpha=1}^{3} [VU]_\alpha \nu_\alpha dS \qquad (4.18.13)$$

Many important conclusions can be drawn from these formulae concerning the *deficiency* and *compatibility* of some basic field equations. The following two identities are here particularly helpful:

$$\operatorname{div} \operatorname{curl} U \equiv 0 \qquad (4.18.14)$$

$$\operatorname{curl} \operatorname{grad} \Phi \equiv 0 \qquad (4.18.15)$$

Consider for example the problem of obtaining the scalar Φ from the vector field

$$\operatorname{grad} \Phi = F \qquad (4.18.16)$$

What can we say concerning the *deficiency* of this equation? The only solution of the homogeneous equation

$$\operatorname{grad} \Phi = 0 \qquad (4.18.17)$$

is

$$\Phi = \operatorname{const} \qquad (4.18.18)$$

Hence the function Φ will be obtainable from (16), except for an additive constant.

What can we say concerning the *compatibility* of the equation (16)? For this purpose we have to solve the adjoint homogeneous problem. The boundary term (9) yields the boundary condition

$$U_\alpha \nu_\alpha = 0 \qquad (4.18.19)$$

since we gave no boundary condition for Φ itself. Hence we have to solve the field equation

$$\operatorname{div} U = 0 \qquad (4.18.20)$$

with the added condition (19). Now the equation (20) is solvable by putting

$$U = \operatorname{curl} B \qquad (4.18.21)$$

where B is a freely choosable vector field, except for the boundary condition (19) which demands that the normal component of curl B vanishes at all points of the boundary surface S:

$$(\text{curl } B)_\nu = 0 \quad (\text{on } S) \tag{4.18.22}$$

The compatibility of our problem demands that *the right side of* (16) *is orthogonal to every solution of the adjoint homogeneous system.* This means

$$\int F \text{ curl } B d\tau = 0$$

But now the formulae (12), (13) give

$$\int F \text{ curl } B d\tau = \int B \text{ curl } F d\tau + \int [BF]_\nu dS \tag{4.18.23}$$

and, since the vector B is freely choosable inside the domain τ, we obtain the condition

$$\text{curl } F = 0 \tag{4.18.24}$$

as a necessary condition for the compatibility of the system (16). Then we can show that in consequence of this condition the last term on the right side of (23) vanishes too. Hence the condition (24) is *necessary and sufficient* for the solvability of (16).

It is of interest to pursue this problem one step further by making our problem still more over-determined. We will demand that on the boundary surface S the function Φ *vanishes.* In that case we get *no boundary condition* for the adjoint problem since now the boundary term (9) vanishes on account of the given boundary condition.

Here now we obtain once more the condition (24) but this is *not enough.* The vector B can now be freely chosen on the surface S and the vanishing of the last term of (23) demands

$$F = \alpha\nu \tag{4.18.25}$$

that is the *vector F must be perpendicular to S* at every point of the surface. This is indeed an immediate consequence of the fact that the gradient of a potential surface $\Phi = \text{const.}$ is orthogonal to the surface but it is of interest to obtain this condition systematically on the basis of the general compatibility theory of linear systems.

Problem 170. Find the adjoint operator (together with the adjoint boundary conditions) of the following problem:

$$\Delta\varphi = \beta \tag{4.18.26}$$

with the boundary conditions on S:

$$\text{a) } \varphi = 0$$

$$\text{b) } \frac{\partial\varphi}{\partial\nu} = 0 \tag{4.18.27}$$

$$\text{c) } \frac{\partial\varphi}{\partial\nu} - \gamma\varphi = 0$$

Show that in all these cases the problem is *self-adjoint*. Historically this problem is particularly interesting since "Green's identity" was in fact established for this particular problem (George Green, 1793–1841).

Problem 171. Investigate the deficiency and compatibility of the following problem:

$$\text{curl } V = B$$
$$\text{div } V = 0 \tag{4.18.28}$$

Boundary condition: $V = V_0$ on S.
[Answer:

V uniquely determined. Compatibility conditions:

$$\text{div } B = 0$$
$$\int V_0 \nu dS = 0 \tag{4.18.29}]$$

Problem 172. Given the following system

$$\text{curl } V = 0$$
$$\text{div } V = \rho \tag{4.18.30}$$

What boundary conditions are demanded in order that V shall become the gradient of a scalar?
[Answer: Only V_ν can be prescribed on S, with the condition

$$\int \rho d\tau = \int V_\nu dS \tag{4.18.31}]$$

4.19. Solution of incomplete systems

We have seen in the general treatment of linear systems that certain linear aggregates of the unknowns may appear in the given system with zero weight which means that we cannot give a complete solution of the problem, because our problem does not have enough information for a full determination of the unknowns. Such incomplete systems allow nevertheless a unique solution if we agree that we make our solution orthogonal to all the non-activated dimensions. This means that we give the solution in all those dimensions in which a solution is possible and add nothing concerning the remaining dimensions. This unique solution is always obtainable with the help of a "generating function", namely by putting

$$v = \tilde{A}w \tag{4.19.1}$$

We will consider a few characteristic examples for this procedure from the realm of partial differential operators.

Let us possess the equation

$$\text{div } V = \rho \tag{4.19.2}$$

without any further information. Here the adjoint operator is $-\text{grad } \Phi$ with the boundary condition

$$\Phi = 0 \qquad \text{on } S \tag{4.19.3}$$

If we put

$$V = -\operatorname{grad} \Phi$$

we obtain the potential equation

$$\Delta\Phi = -\rho \qquad (4.19.4)$$

with the added boundary condition (3) and this equation has indeed a unique solution.

As a second example let us consider the system

$$\operatorname{curl} V = B$$
$$\operatorname{div} V = \rho \qquad (4.19.5)$$

with the compatibility condition

$$\operatorname{div} B = 0 \qquad (4.19.6)$$

but without further conditions.   Here the adjoint system becomes

$$\tilde{D}u = \operatorname{curl} U - \operatorname{grad} u \qquad (4.19.7)$$

with the boundary conditions

$$U = \alpha\nu \qquad \text{on } S \qquad (4.19.8)$$

(that is U orthogonal to S) and

$$u = 0 \qquad \text{on } S \qquad (4.19.9)$$

According to the general rule we have to put

$$V = \operatorname{curl} U - \operatorname{grad} u \qquad (4.19.10)$$

Substitution in (5) gives first of all the potential equation

$$\Delta u = -\rho \qquad (4.19.11)$$

with the boundary condition (9) and this problem has a unique solution.

Let us furthermore put

$$U = W + \operatorname{grad} \Phi \qquad (4.19.12)$$

where we can determine the scalar field Φ by the conditions

$$\Delta\Phi = \operatorname{div} U$$
$$\frac{\partial\Phi}{\partial\nu} = U_\nu = \alpha \qquad (4.19.13)$$

This leaves a vector W which is free of divergence and which vanishes on S. Substitution in the first equation of (5) yields

$$-\Delta W = B \qquad (4.19.14)$$

We have for every component of W the potential equation with the boundary condition that each component must vanish on S. This problem has a unique solution. Hence we have again demonstrated that the method of the generating function leads to a unique solution.

Finally we consider a system of equations called the "Maxwellian equations" of electromagnetism. In particular we consider only the *first group* of the Maxwellian equations, which represents 4 equations for 6 quantities, the electric and the magnetic field strengths E and H:

$$-\frac{1}{c}\frac{\partial E}{\partial t} + \operatorname{curl} H = J$$

$$\operatorname{div} E = \rho \tag{4.19.15}$$

(J = current density, ρ = charge density). The right sides have to satisfy the compatibility condition

$$\operatorname{div} J + \frac{\partial \rho}{c\partial t} = 0 \tag{4.19.16}$$

Our system is obviously under-determined since we have only 4 equations for 6 unknowns and the compatibility condition (16) reduces these 4 relations to essentially 3 relations.

Let us now obtain the adjoint system and solve our problem in terms of a generating function. The multiplication of the left side of (15) by an undetermined vector U and a scalar u yields:

$$-\frac{U}{c}\frac{\partial E}{\partial t} + U \operatorname{curl} H + u \operatorname{div} E$$

$$= E\frac{\partial U}{c\partial t} + H \operatorname{curl} U - E \operatorname{grad} u + \sum_{\alpha=1}^{4}\frac{\partial F_\alpha}{\partial x_\alpha} \tag{4.19.17}$$

Hence we obtain the adjoint operator $\check{D}u$ of our system in the following form:

$$\check{D}u = \begin{matrix} \dfrac{1}{c}\dfrac{\partial U}{\partial t} - \operatorname{grad} u \\[2ex] \operatorname{curl} U \end{matrix} \tag{4.19.18}$$

At this point, however, we have to take into consideration a fundamental result of the Theory of Relativity. That theory has shown that the true fourth variable of the physical universe is not the time t but

$$x_4 = ict \tag{4.19.19}$$

The proper field quantities of the electromagnetic field are not E and H

but iE and H and the proper way of writing the Maxwellian equations is as follows:

$$-\frac{\partial iE}{\partial x_4} + \text{curl } H = J$$

$$\text{div } iE = i\rho$$

(4.19.20)

$$\text{div } J + \frac{\partial i\rho}{\partial x_4} = 0 \qquad (4.19.21)$$

Accordingly the undetermined multipliers of our equations will be U and iu, and we obtain the (algebraically) adjoint system in the following form:

$$\tilde{D}u = \begin{matrix} \dfrac{\partial U}{\partial x_4} - \text{grad } iu \\[2mm] \\ \text{curl } U \end{matrix} \qquad (4.19.22)$$

The solution of the system (20) in terms of a generating function becomes accordingly:

$$iE = \frac{\partial U}{\partial x_4} - \text{grad } iu$$

$$H = \text{curl } U$$

(4.19.23)

or, replacing U by the more familiar notation A and u by Φ:

$$E = -\frac{1}{c}\frac{\partial A}{\partial t} - \text{grad } \Phi$$

$$H = \text{curl } A$$

(4.19.24)

Here we have the customary representation of E and H in terms of the "vector-potential" A and the "scalar potential" Φ. Customarily this representation is the consequence of the second set of Maxwellian equations:

$$-\frac{\partial H}{\partial x_4} + \text{curl } iE = 0$$

$$\text{div } H = 0$$

(4.19.25)

It is of interest to observe that the same representation is obtainable by a natural normalisation of the solution of the first set of Maxwellian equations. This means that the second set of the Maxwellian equations can be interpreted in the following terms: *The electromagnetic field strength iE, H has no components in those dimensions of the function space which are not activated by the first set of Maxwellian equations.*

Problem 173.   Making use of the Hermitian definition of the adjoint operator according to (12.2) show that the following operator in the four variables x, y, z, t is self-adjoint:

$$Dv(x) = \frac{\partial V}{ic\partial t} + \operatorname{curl} V \qquad (4.19.26)$$

Problem 174.   Given the differential equation

$$v''(x) = \beta(x) \qquad x = [a, b] \qquad (4.19.27)$$

without boundary conditions.   Obtain a unique solution with the help of a generating function and show that the same solution is obtainable by making $v(x)$ orthogonal to the homogeneous solutions

$$v(x) = 1, \quad v(x) = x \qquad (4.19.28)$$

[Answer:

$$v(x) = w''(x) \qquad (4.19.29)$$

with the boundary conditions

$$w(a) = w'(a) = w(b) = w'(b) = 0 \qquad (4.19.30)]$$

Problem 175.   Given the differential equation

$$\Delta\Phi = \beta(x, y, z) \qquad (4.19.31)$$

without boundary conditions.   Obtain a unique solution with the help of a generating function.

[Answer:

$$\Phi = \Delta\psi \qquad (4.19.32)$$

with the boundary conditions

$$\psi = 0$$
$$\frac{\partial\psi}{\partial\nu} = 0 \qquad \text{on } S \qquad (4.19.33)]$$

BIBLIOGRAPHY

[1] Halmos, P. R., *Finite Dimensional Vector Spaces* (Princeton University Press, 1942)

[2] Synge, J. L., *The Hypercircle in Mathematical Physics* (Cambridge University Press, 1957)

CHAPTER 5

THE GREEN'S FUNCTION

Synopsis. With this chapter we arrive at the central issue of our theory. The "Green's Function" represents in fact the *inverse* of the given differential operator. In analogy to the theory of matrices, we can establish—in proper interpretation—a unique inverse to any given sufficiently regular differential operator. The Green's function can thus be defined under very general conditions. In a large class of problems the Green's function appears in the form of a "kernel function" which depends on the position of *two* points of the given domain. This function can be defined as the solution of a certain differential equation which has Dirac's "delta function" on the right side. The "reciprocity theorem" makes it possible to define the Green's function either in terms of the adjoint, or the given differential operator. Over-determined or under-determined systems lead to the concept of the "constrained" Green's function which is constrained to that subspace in which the operator is activated. (In Chapter 8 we will encounter strange cases, in which the inverse operator exists, without being, however, reducible to a kernel function of the type of the Green's function.)

5.1. Introduction

In the domain of simultaneous linear algebraic equations we possess methods by which in a finite number of steps an explicit solution of the given system can be obtained in all cases when such a solution exists. In the domain of linear differential operators we are not in a similarly fortunate position. We have various methods by which we can approximate the solution of a given problem in the realm of ordinary or partial differential equations. But the actual numerical solution of such a specific problem— although perhaps of great importance for the solution of a certain problem of physics or industry—may tell us very little about the interesting *analytical properties* of the given problem. From the analytical standpoint we are not interested in the numerical answer of an accidentally encountered problem but in the general *properties* of the solution. The analytical tools by which a solution is obtained may be of little practical significance but of great importance if our aim is to arrive at a deeper understanding of the nature of linear differential operators and the theoretical conclusions we can draw concerning their behaviour under various circumstances. From the beginning of the nineteenth century the focus of interest shifted from the solution of certain special problems by means of more or less ingenious artifices to a much broader outlook which led to a universal method of solving linear differential equations with the help of an auxiliary function called the "Green's function". This concept was destined to play a central role in the later development. It is this construction which we shall discuss in the present chapter together with a number of typical examples which illustrate the principal properties of this important tool of analysis.

5.2. The role of the adjoint equation

We have seen in the preliminary investigation of a linear system that the solution of the adjoint homogeneous equation played a major role in the problem of deciding whether a given linear system is solvable or not. If the adjoint homogeneous equation had non-vanishing solutions, the orthogonality of the right side to these solutions was the necessary and sufficient condition for the solvability of our problem.

This investigation has not yet touched on the task of actually *finding* the solution; we have merely arrived at a point where we could decide whether a solution exists or not. In actual fact, however, we are already quite near to the solution problem and we need only a small modification of the previous procedure to adapt it to the more pretentious task of actually *obtaining* the solution. We can once more use the artifice of *artificial over-determination* in order to reduce the solution problem to the previous compatibility problem. Let us namely add to our data the additional equation

$$v(x_1) = \rho \tag{5.2.1}$$

The point x_1 is a special point of our domain in which we should like to obtain the value of $v(x)$. Of course, if our problem is well-determined, we have no right to choose ρ freely. The fact that our problem has a unique solution means that ρ is uniquely determined by the data of our problem. But the extended system

$$\begin{aligned} Dv(x) &= \beta(x) \\ v(x_1) &= \rho \end{aligned} \tag{5.2.2}$$

(where we assume that the first equation includes the given boundary conditions) is a perfectly legitimate linear system whose compatibility we can investigate. For this purpose we have to form the adjoint homogeneous equation

$$\tilde{D}u(x) = 0 \tag{5.2.3}$$

and express the compatibility of the system (2) by making the right side orthogonal to the solution $u(x)$ of the system (3) (we know in advance that the adjoint system (3) must have a solution since otherwise the right side of (2) would be freely choosable and that would mean that ρ is not determined). But then the orthogonality of the right side of (2) to the solution $u(x)$ will give a relation of the following form:

$$\int \beta(x)u(x)dx + \rho u_0 = 0 \tag{5.2.4}$$

and this means that we can obtain ρ in terms of the given $\beta(x)$ by the relation

$$\rho = -\frac{1}{u_0}\int \beta(x)u(x)dx \tag{5.2.5}$$

This is the basic idea which leads to the construction of the Green's function and we see the close relation of the construction to the solution of the *adjoint homogeneous equation*.

5.3. The role of Green's identity

Instead of making use of the adjoint homogeneous equation we can return to the fundamental Green's identity (4.12.3) and utilise this identity for our purposes:

$$\int [u(x)Dv(x) - v(x)\check{D}u(x)]dx = 0 \tag{5.3.1}$$

Since $Dv(x)$ is given as $\beta(x)$, we obtain at once the orthogonality of $\beta(x)$ to $u(x)$ if $u(x)$ satisfies the homogeneous equation

$$\check{D}u(x) = 0 \tag{5.3.2}$$

But in our case not only $\beta(x)$ is given but also $v(x)$ at the point $x = x_1$. Consequently the vanishing of $\check{D}u(x)$ is not demanded everywhere. It suffices if $\check{D}u(x)$ is zero everywhere *except at the point* $x = x_1$, since $v(x)$ *at* x_1 *is given as* ρ. Hence the artificial over-determination of our equation in the sense of (2) leads to a *relaxation* of the conditions for the adjoint equation inasmuch as the homogeneous adjoint equation has to be satisfied everywhere *except* at the point $x = x_1$.

5.4. The delta function $\delta(x, \xi)$

In view of the continuous nature of our operator $Dv(x)$ a certain precaution is required. The integration over x brings in the volume-element dx of our domain and we have to pay attention to the difference between "load" and "load density" that we have encountered earlier in our study of the elastic bar. We would like to pinpoint the particular value $v(x_1)$, singling out the definite point $x = x_1$. This, however, does not harmonise with the continuity of our operator. A point-load has to be understood as the result of a *limit-process*. We actually distribute our load over a small but finite domain and then shrink the dimensions of this domain, until in the limit the entire load is applied at the point $x = x_1$.

We shall thus proceed as follows. We take advantage of the *continuity* of the solution $v(x)$ which is demanded by the very fact that the differential operator D must be able to operate on $v(x)$. But if $v(x)$ is continuous at

$x = x_1$, then our assumption that $v(x)$ assumes the value ρ at the point $x = x_1$ can be extended to the immediate neighbourhood of $x = x_1$, with an error which can be made as small as we wish. We will thus extend our system (2) in the following sense. The equation

$$Dv(x) = \beta(x) \qquad (5.4.1)$$

is given everywhere, including the given boundary conditions. We add the equation

$$v(x_1) = \rho \qquad (5.4.2)$$

in the sense that now x_1 is not a definite point but a point *with its neighbourhood* which extends over an arbitrarily small ν-dimensional domain of the small but finite volume ϵ that surrounds the point $x = x_1$. Accordingly in Green's identity (3.1) we can exempt $\tilde{D}u(x)$ from being zero not only at the point $x = x_1$ but also in its immediate neighbourhood. This statement we will write down in the following form

$$\tilde{D}u(x) = \delta_\epsilon(x_1, x) \qquad (5.4.3)$$

The function $\delta_\epsilon(x_1, x)$ has the property that it vanishes everywhere outside of the small neighbourhood ϵ of the point $x = x_1$. Inside of the small neighbourhood ϵ we will assume that $\delta_\epsilon(x_1, x)$ *does not change its sign*: it is either positive or zero.

Then Green's identity yields the compatibility condition of our system (1), (2) in the following form:

$$\int u(x)\beta(x)dx - \int \rho(x)\delta_\epsilon(x_1, x)dx = 0 \qquad (5.4.4)$$

We will now take advantage of the linearity of the operator $\tilde{D}u(x)$. A mere multiplication of $u(x)$ by a constant brings this constant in front of $\tilde{D}u(x)$. We can use the freedom of this constant to a proper *normalisation* of the right side of (3). If we interpret the right side of (3) as a load density, we will make the *total load equal to* 1:

$$\int_\epsilon \delta_\epsilon(x_1, x)dx = 1 \qquad (5.4.5)$$

The integration extends only over the neighbourhood ϵ of the point $x = x_1$, since outside of this neighbourhood the integrand vanishes.

Now we will proceed as follows. In (4) we can replace $\rho(x)$ by $v(x)$ since the significance of $\rho(x)$ is in fact $v(x)$. Then we can make use of the mean value theorem of integral calculus, namely that in the second integral we can take out $v(x)$ in front of the integral sign, replacing x by a certain $\bar{x}_1$ where $\bar{x}_1$ is some unknown point inside of the domain ϵ. Then we obtain

$$v(\bar{x}_1) = \int \beta(x)u(x)dx \qquad (5.4.6)$$

We notice that by this procedure we have not obtained $v(x_1)$ exactly. We have obtained $v(x)$ at a point $\bar{x}_1$ which can come to $x = x_1$ as near as we wish and thus also the uncertainty of $v(\bar{x}_1)$ can be made as small as we wish, considering the continuity of $v(x)$ at $x = x_1$. But we cannot get around this limit procedure. We have to start out with the small neighbourhood ϵ and then see what happens as the domain ϵ shrinks more and more to zero.

For a better description of this limit process, we will employ a more adequate symbolism. We will introduce the symbol ξ as an integration variable. Then we need not distinguish between x and x_1 since for the present x will be an arbitrary fixed point of our domain—later to be identified with that particular point at which the value of $v(x)$ shall be obtained. Hence x becomes a "parameter" of our problem while the active variable becomes ξ. Furthermore, we wish to indicate that the function $u(\xi)$ constructed on the basis of (3) depends on the position of the chosen point x. It also depends on the choice of ϵ. Accordingly we will rewrite the equation (3) as follows:

$$\check{D}G_\epsilon(x, \xi) = \delta_\epsilon(x, \xi) \tag{5.4.7}$$

The previous x has changed to ξ, the previous x_1 to x. The function $u(x)$ has changed to a new function $G(x, \xi)$ in which both x and ξ appear on equal footing because, although originally x is a mere parameter and ξ the variable, we can alter the value of the parameter x at will and that means that the auxiliary function $u(\xi)$ will change with x. (The continuity with respect to x is not claimed at this stage.)

Adopting this new symbolism, the equation (6) appears now in the following form

$$v(\bar{x}) = \int \beta(\xi)G_\epsilon(x, \xi)d\xi \tag{5.4.8}$$

The uncertainty caused by the fact that $\bar{x}$ does not coincide with x can now be eliminated by studying the *limit* approached as ϵ goes to zero. This limit may exist, even if $G_\epsilon(x, \xi)$ does not approach any definite limit since the integration has a smoothing effect and it is conceivable that $G_\epsilon(x, \xi)$, considered as a function of ξ, would not approach a definite limit at *any* point of the domain of integration and yet the integral (8) may approach a definite limit at *every* point of the domain. In actual fact, however, in the majority of the boundary value problems encountered in mathematical physics the function $G_\epsilon(x, \xi)$ *itself* approaches a definite limit (possibly with the exception of certain singular points where $G_\epsilon(x, \xi)$ may go to infinity). This means that in the limit process

$$\check{D}G_\epsilon(x, \xi) = \delta_\epsilon(x, \xi) \qquad \epsilon \to 0 \tag{5.4.9}$$

the function $G_\epsilon(x, \xi)$ with decreasing ϵ approaches a definite limit, called the "Green's function":

$$\lim_{\epsilon \to 0} G_\epsilon(x, \xi) = G(x, \xi) \tag{5.4.10}$$

In that case we obtain from (8):

$$v(x) = \int \beta(\xi)G(x, \xi)d\xi \tag{5.4.11}$$

The limit process here involved is often abbreviated to the equation

$$\tilde{D}G(x, \xi) = \delta(x, \xi) \tag{5.4.12}$$

where on the right side we have "Dirac's delta function", introduced by the great physicist Paul Dirac in his wave-mechanical investigations. Here a function of ξ is considered which vanishes everywhere outside the point $\xi = x$ while at the point $\xi = x$ the function goes to infinity in such a manner that the total volume under the function is 1. Dirac's delta function is a mathematical counterpart of a concept that Clerk Maxwell employed in his elastic investigations where he replaced a continuous distribution of loads by a "point-load" applied at the point x. The elastic deflection caused by this point load, observed at the point ξ, was called by Maxwell the "influence function". It corresponds to the same function that is today designated as the "Green's function" of the given elastic problem.

The literature of contemporary physics frequently advocates discarding all scruples concerning the use of the delta function and assigning an actual meaning to it, without regard to the ϵ-process from which it originated. In this case we assign a limit to a process which in fact has no limit since a function cannot become zero everywhere except in one single point, and yet be associated with an integral which is not zero but 1. However, instead of extending the classical limit concept in order to include something which is basically foreign to it, we can equally well consider all formulas in which the delta function occurs as shorthand notation of a more elaborate statement in which the legitimate $\delta_\epsilon(x, \xi)$ appears, with the added condition that ϵ converges to zero. For example, assuming that $f(x)$ is continuous, we have

$$\int f(\xi)\delta_\epsilon(x, \xi)d\xi = f(\bar{x}) \tag{5.4.13}$$

where $\bar{x}$ is a point in the ϵ-neighbourhood of x. Now, if ϵ converges to zero, we get:

$$\lim_{\epsilon \to 0} \int f(\xi)\delta_\epsilon(x, \xi)d\xi = f(x) \tag{5.4.14}$$

We can now write this limit relation in the abbreviated form:

$$f(x) = \int f(\xi)\delta(x, \xi)d\xi \tag{5.4.15}$$

but what we actually mean is a limit process, in which the delta function never appears as an actual entity but only in the legitimate form $\delta_\epsilon(x, \xi)$, ϵ converging to zero. The truth of the statement (14) does *not* demand that the limit of $\delta_\epsilon(x, \xi)$ shall exist. The same is true of all equations in which the symbol $\delta(x, \xi)$ appears.

5.5. The existence of the Green's function

We know from the general matrix treatment of linear systems that a linear system with arbitrarily prescribed right side is solvable if and only if the adjoint homogeneous equation has no solution which is not identically zero. This principle, applied to the equation (4.7), demands that the equation

$$Dv(\xi) = 0 \tag{5.5.1}$$

shall have no non-zero solution. And that again means that the problem

$$Dv(x) = \beta(x) \tag{5.5.2}$$

under the given boundary conditions, should not have more than *one* solution. Hence the given linear system must belong to the "complete" category, that is the V-space must be completely filled out by the given operator. There is no condition involved concerning the U-space. Our problem can be arbitrarily over-determined. Only under-determination has to be avoided.

The reason for this condition follows from the general idea of a Green's function. To find the solution of a given differential equation (with the proper boundary conditions) means that we establish a linear relation between $v(x)$ and the given data. But if our system is incomplete, then such a relation does not exist because the value of $v(x)$ (at the special point x) may be added freely to the given data. Under such circumstances the existence of the Green's function cannot be expected.

A good example is provided by the strongly over-determined system

$$\text{grad } \Phi = F \tag{5.5.3}$$

where F is a given vector field. We have encountered this problem earlier, in Chapter 4.18, and found that the compatibility of the system demands the condition

$$\text{curl } F = 0 \tag{5.5.4}$$

In spite of the strong over-determination, the problem is under-determined in one single direction of the function space inasmuch as the homogeneous equation permits the solution

$$\Phi = \text{const} \tag{5.5.5}$$

Let us now formulate the problem of the Green's function. The adjoint operator now is

$$-\text{div } U \tag{5.5.6}$$

with the boundary condition

$$U_\nu = 0 \qquad \text{on } S \tag{5.5.7}$$

(ν being the normal at the boundary point considered). Now the differential equation

$$-\text{div } U = \delta(x, \xi) \tag{5.5.8}$$

certainly demands very little of the vector field U since only a *scalar* condition is prescribed at every point of the field and we have a *vector* at our disposal to satisfy that condition. And yet the equation (8) has *no solution* because the application of the Gaussian integral transformation to the equation (8) yields, in view of the boundary condition (7):

$$\int \delta(x, \xi) d\xi = 0 \qquad (5.5.9)$$

while the definition of the delta function demands that the same integral shall have the value 1.

Let us, however, complete the differential equation (3) by the added condition

$$\Phi(0) = 0 \qquad (5.5.10)$$

In this case the adjoint operator (6) ceases to be valid at the point $\xi = 0$ and we have to modify the defining equation of the Green's function as follows

$$-\text{div } U = \delta(x, \xi) + \alpha \delta(0, \xi) \qquad (5.5.11)$$

The condition that the integral over the right side must be zero, determines α to -1 and we obtain finally the determining equation for the Green's function of our problem in the following form:

$$-\text{div } G(x, \xi) = \delta(x, \xi) - \delta(0, \xi) \qquad (5.5.12)$$

This strongly-undetermined problem has infinitely many solutions but every one of these solutions can serve as the Green's function of our problem, giving the solution of our original problem (3), (10) in the form

$$\Phi(x) = \int F(\xi) G(x, \xi) d\xi \qquad (5.5.13)$$

(on the right side the product FG refers to the scalar product of the two vector fields).

One particularly simple solution of the differential equation (12) can be obtained as follows: We choose a narrow tube of constant cross-section q

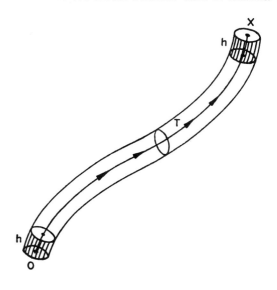

which shall connect the points $\xi = 0$ and $\xi = x$. We define the vector field $G(x, \xi)$ as zero everywhere outside of this tube while inside of the tube we assume that $G(\xi)$ is everywhere perpendicular to the cross-section of the tube and of constant length. We have no difficulty in showing that the divergence of the vector field thus constructed *vanishes* everywhere, except in the neighbourhood of the points $\xi = 0$ and $\xi = x$. But these are exactly the neighbourhoods where we do not want div $G(\xi)$ to vanish since the right side of (12) is not zero in the small neighbourhood ϵ of these points. The infinitesimal volume ϵ in which the delta function $\delta_\epsilon(x, \xi)$ is not zero is given by the product qh. In this volume we can assign to $\delta_\epsilon(x, \xi)$ the constant value

$$\delta_\epsilon(x, \xi) = \frac{1}{qh} \tag{5.5.14}$$

The vector field $G(\xi)$ starts at the lower end of the tube with zero, grows linearly with h and attains the value $1/q$ at the end of the shaded volume. Then it maintains this length throughout the tube T, arrives with this constant value at the upper shaded volume, diminishes again linearly with h and becomes zero at the upper end of the tube.

Let us now form the integral (13). This integral is reduced to the very narrow tube in which $G(\xi)$ is different from zero. If we introduce the line-element ds of the central line of the tube and consider it as an infinitesimal vector $\vec{ds}$ whose length is ds while its direction is tangential to the line, then the product $G(\xi)d\xi$ is replaceable by:

$$G(\xi)d\xi = \frac{1}{q} q\,\vec{ds} = \vec{ds} \tag{5.5.15}$$

because in the present problem $d\xi$—the infinitesimal volume of integration—is qds, while the length of the vector $G(\xi)$ is the constant $1/q$. We see that, as ϵ shrinks to zero, the volume integral (19) becomes more and more the *line-integral*

$$\Phi(x) = \int_0^x F(s)\,\vec{ds} \tag{5.5.16}$$

and this is the well-known elementary solution of our problem, but here obtained quite systematically on the basis of the general theory of the Green's function.

Our problem is interesting from more than one aspect. It demonstrates the correctness of the Green's function method in a case where the Green's function itself evaporates into nothingness. The function $G_\epsilon(x, \xi)$ is a perfectly legitimate function of ξ as long as ϵ is finite but it does not approach any limit as ϵ approaches zero. It assumes in itself the nature of a delta function. But this is in fact immaterial. The decisive question is not whether $G(x, \xi)$ exists as ϵ goes to zero but whether the integral (13) approaches a definite limit as ϵ goes to zero and in our example this is the case, although the limit of $G_\epsilon(x, \xi)$ itself does not exist.

Our example demonstrated that in the limit process, as ϵ goes to zero, peculiar things may happen, in fact the Green's function may lose its significance by not approaching any limit exactly in that region of space over which the integration is extended. Such conditions are often encountered in the integration of the "hyperbolic type" of partial differential equations. The integration is then often extended over a space which is of lower dimensionality than the full space in which the differential equation is stated. We do not hesitate, however, to maintain the concept of the Green's function in such cases by reducing the integration from the beginning to the proper subspace. For example in our present problem we can formulate the solution of our problem in the form

$$\Phi(x) = \int_0^x F(s)G(x, s)ds \tag{5.5.17}$$

and define the Green's function $G(x, s)$ as

$$G(x, s) = T_s \tag{5.5.18}$$

where T_s is the tangent vector to the line s of the length 1.

Even with this allowance we encounter situations in which the existence of the Green's function cannot be saved. Let us consider the partial differential equation in two variables:

$$\frac{\partial v}{\partial t} - \kappa^2 \frac{\partial^2 v}{\partial x^2} = 0 \tag{5.5.19}$$

which is the differential equation of heat conduction in a rod, t being the time, x the distance measured along the rod. We assume that the two ends of the rod, that is the points $x = 0$ and $x = l$, are kept constantly on the temperature $v = 0$. This gives the boundary conditions

$$\begin{aligned} v(0, t) &= 0 \\ v(l, t) &= 0 \end{aligned} \tag{5.5.20}$$

Usually we assume that at the time moment $t = 0$ we observe the temperature distribution $v(x, 0)$ along the rod and now want to calculate $v(x, t)$ at some later time moment. Since, however, there is a one-to-one correspondence between the functions $v(x, 0)$ and $v(x, t)$, it must be possible to salvage $v(x, 0)$ if by accident we have omitted to observe the temperature distribution at $t = 0$ and made instead our observation at the later time moment $t = T$. We now give the "end-condition" (instead of initial condition)

$$v(x, T) = f(x) \tag{5.5.21}$$

and now our problem is to obtain $v(x, t)$ at the *previous* time moments between 0 and T.

In harmony with the general procedure we are going to construct the adjoint problem with its boundary conditions, replacing any given inhomogeneous boundary conditions by the corresponding *homogeneous*

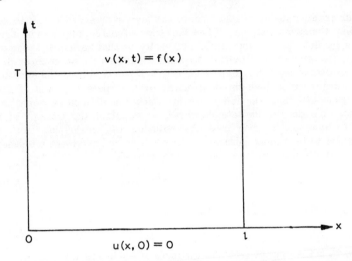

boundary conditions. The adjoint operator becomes if we carry through the routine Green's identity procedure:

$$\tilde{D}u = -\frac{\partial t}{\partial u} - \kappa^2 \frac{\partial u^2}{\partial x^2} \tag{5.5.22}$$

with the boundary conditions

$$u(0, t) = 0$$
$$u(l, t) = 0 \tag{5.5.23}$$

and the initial condition

$$u(x, 0) = 0 \tag{5.5.24}$$

The Green's function of our problem is defined, according to the general method of Section 4, by the differential equation

$$\tilde{D}G(x, t; \xi, \tau) = \delta(x, t; \xi, \tau) \tag{5.5.25}$$

together with the homogeneous boundary conditions (23), (24). This problem has *no solution* and the difficulty is not caused by the extreme nature of the delta function. Even if we replace $\delta(x, t; \xi, t)$ by $\delta_\epsilon(x, t; \xi, \tau)$, we can find *no solution* to our problem. The physical significance of the Green's function defined by (25) and the given boundary conditions would be a heat distribution caused by a unit heat source applied at the point $\xi = x$, $\tau = t$, with the added condition that the temperature is zero all along the rod at a *later* time moment (instead of an earlier time moment). The non-reversibility of heat phenomena prevents the possibility of such a heat distribution.

We have here an example of a boundary value problem which is entirely reasonable and which in the case of properly given data has a solution and in fact a *unique* solution. That solution, however, cannot be given in terms of an auxiliary function which satisfies a given non-homogeneous differential equation. The concept of the Green's function has to be extended to a more general operator in order to include the solution of such problems, as we shall see later in Chapter 8.

Problem 176. Remove the over-determination of the problem (3), (10), by the least square method (application of $\tilde{D}$ on both sides of the equation) and characterise the Green's function of the new problem.

[Answer:

$$\Delta\Phi = \operatorname{div} F$$

$$\frac{\partial\Phi}{\partial\nu} = F_\nu \quad \text{on } S \tag{5.5.26}$$

$$\Delta G(x, \xi) = \delta(x, \xi) - \delta(0, \xi)$$

$$\frac{\partial G}{\partial\nu} = 0 \quad \text{on } S \tag{5.5.27}]$$

Problem 177. Define the Green's function for the Laplacian operator Δv, if the given boundary conditions are

$$v(S) = A(S)$$

$$\frac{\partial v(S)}{\partial\nu} = B(S) \tag{5.5.28}$$

[Answer:

$$\Delta G(x, \xi) = \delta(x, \xi) \tag{5.5.29}$$

$$\textit{no boundary conditions.}]$$

5.6. Inhomogeneous boundary conditions

In the construction of the adjoint operator $\tilde{D}u$—which was needed for the definition of the Green's function $G(x, \xi)$—we have always followed the principle that we have ignored the given right sides and replaced in-homogeneous boundary conditions by the corresponding homogeneous boundary conditions. It so happens that *the solution of an inhomogeneous boundary value problem occurs with the help of the same Green's function $G(x, \xi)$ that holds in the case of homogeneous boundary conditions.* The difference is only that the given right side of the differential equation gives rise to a *volume integral*, extended over the entire domain of the variables, while the given non-zero boundary values give rise to a *surface integral*, extended over the boundary surface S.

Let us return to Green's identity (3.1). The solution (4.11) in terms of the Green's function was an immediate consequence of that identity. But the right side of this identity is zero only if $v(\xi)$ is subjected to the given boundary conditions *in their homogeneous form*, that is if the boundary values are prescribed as *zero*. If the given boundary conditions have non-zero values on the right side, then the "zero" on the right side of Green's identity (3.1) has to be replaced by a certain integral over the boundary surface S. In this integral only the given (inhomogeneous) boundary values will appear, in conjunction with the Green's function $G(x, S)$ and its derivatives on the boundary. Hence the problem of inhomogeneous boundary conditions does not pose a problem which goes beyond the problem of solving the inhomogeneous differential equation with *homogeneous* boundary conditions.

As an example let us consider the solution of the homogeneous differential equation

$$\Delta v = 0 \tag{5.6.1}$$

with the inhomogeneous boundary condition

$$v = v(S) \quad \text{on } S \tag{5.6.2}$$

We go through the regular routine, replacing the condition (2) by the condition $v(S) = 0$ and thus obtaining the adjoint operator

$$\check{D}u = \Delta u \tag{5.6.3}$$

with the boundary condition

$$u = 0 \quad \text{on } S \tag{5.6.4}$$

Now the extended Green's identity—which does not demand any boundary conditions of u and v—becomes in our case:

$$\int (u\Delta v - v\Delta u)d\xi = \int \left(u \frac{\partial v}{\partial \nu} - v \frac{\partial u}{\partial \nu} \right) dS \tag{5.6.5}$$

(ν = outward normal of S). We shall apply this identity to our function $v(\xi)$, defined by the conditions (1) and (2), while $u(\xi)$ will be replaced by the Green's function $G(x, \xi)$, defined by

$$\Delta G(x, \xi) = \delta(x, \xi) \tag{5.6.6}$$

$$G(x, S) = 0 \quad \text{on } S \tag{5.6.7}$$

The integration (5) on the left side yields minus $v(x)$, because $\Delta v = 0$, and Δu is reduced to the delta function which exists only in the ϵ-neighbourhood of the point $\xi = x$, putting in the limit the spotlight on $v(x)$. On the right side the first term drops out due to the boundary condition (7) and the final result becomes

$$v(x) = \int_S v(S) \frac{\partial G(x, S)}{\partial \nu} \, dS \tag{5.6.8}$$

If it so happens that our problem is to solve the inhomogeneous differential equation

$$\Delta v = \beta \tag{5.6.9}$$

with the inhomogeneous boundary condition (2), then we obtain the solution in the form of a *sum of two integrals*, the one extended over the given volume τ, the other over the boundary surface S:

$$v(x) = \int_{\tau} \beta(\xi)G(x,\xi)d\xi + \int_{S} v(S)\frac{\partial G(x,S)}{\partial \nu}\, dS \tag{5.6.10}$$

This form of the solution shows that the inhomogeneous boundary values $v(S)$ can be interpreted as equivalent to a *double layer of surface charges* placed on the boundary surface S.

Problem 178.   Obtain the solution of the boundary value problem (5.28), with

$$\Delta v = 0 \tag{5.6.11}$$

[Answer:

$$v(x) = \int_{S}\left[A(S)\frac{\partial G(x,S)}{\partial \nu} - B(S)G(x,S)\right]dS \tag{5.6.12}]$$

Problem 179.   Obtain the solution of the heat conduction problem (5.19), (5.20), but replacing the end condition (5.21) by the initial condition

$$v(x, 0) = f(x) \tag{5.6.13}$$

[Answer:

$$v(x, t) = \int_{0}^{b} f(\xi)G(x, t; \xi, 0)d\xi \tag{5.6.14}$$

where the Green's function $G(x, t; \xi, \tau)$ satisfies the differential equation (5.25) with the boundary conditions (5.23) and the end-condition

$$G(x, t; \xi, T) = 0 \tag{5.6.15}]$$

Problem 180.   Consider the problem of the elastic bar (4.14.9), with the load distribution $\beta(x) = 0$ and the inhomogeneous boundary conditions

$$\begin{array}{ll} v_1(0) = 0 & v_1(l) = 0 \\ v'_1(0) = \alpha & v'_1(l) = \beta \end{array} \tag{5.6.16}$$

Obtain the deflection of the bar.

[Answer:

$$[I(\xi)G''(x, \xi)]'' = \delta(x, \xi) \tag{5.6.17}$$

boundary conditions

$$\begin{array}{l} G(x, 0) = G'(x, 0) = 0 \\ G(x, l) = G'(x, l) = 0 \end{array} \tag{5.6.18}$$

$$v_1(x) = \alpha I(0)G''(x, 0) - \beta I(l)G''(x, l) \tag{5.6.19}]$$

Problem 181.   Solve the same problem with the boundary conditions

$$\begin{array}{ll} v_1(0) = 0 & v_1(l) = \gamma \\ v'_1(0) = 0 & v'_1(l) = \beta \end{array} \tag{5.6.20}$$

[Answer:

$$v(x) = \gamma I'(l)G''(x, l) + \gamma I(l)G'''(x, l) - \beta I(l)G''(x, l) \tag{5.6.21}]$$

(All differentiations with respect to the second argument.)

5.7. The Green's vector

More than once we have encountered in our discussions the situation that it was not one function but a *set of functions* to which the given differential operator was applied. In the question of obtaining the solution of such a system, we have to specify *which one* of the functions we want to obtain. Consider for example the Laplacian operator

$$\Delta v = \frac{\partial^2 v}{\partial x^2} + \frac{\partial^2 v}{\partial y^2} + \frac{\partial^2 v}{\partial z^2} \tag{5.7.1}$$

The second order differential equation

$$\Delta v = \beta \tag{5.7.2}$$

can be replaced by the following equivalent system of first order equations, denoting v by v_1:

$$\frac{\partial v_1}{\partial x} - v_2 = 0$$

$$\frac{\partial v_1}{\partial y} - v_3 = 0$$

$$\frac{\partial v_1}{\partial z} - v_4 = 0 \tag{5.7.3}$$

$$\frac{\partial v_2}{\partial x} + \frac{\partial v_3}{\partial y} + \frac{\partial v_4}{\partial z} = \beta$$

These systems of four equations for the four unknowns (v_1, v_2, v_3, v_4) is equivalent to the equation (2), as we can see if we substitute for v_2, v_3, v_4 their values into the last equation. But we can equally consider the given system as a simultaneous system of four equations in four unknowns. In the latter case the focus of interest is no longer on v_1 alone. We may equally consider v_2, v_3, v_4 as unknowns. And thus we are confronted with a new situation to which our previous discussions have to be extended. How are we going to construct in this case the "Green's function" which will serve as the auxiliary function for the generation of the solution?

We return once more to the fundamental idea which led to the concept of the "function space" (see Section 4.5). The continuous variables $x^1, x^2, \ldots, x^s$—briefly denoted by the symbol x—were replaced by a set of discrete values in which the function $v(x, y, z)$ was tabulated. Each one of these tabulated values opened up a new dimension in that abstract "function space" in which the function $v(x)$ became represented by a single vector. Now, if we have not merely *one* such function but a *set* of functions

$$(v_1, v_2, \ldots, v_\mu) \tag{5.7.4}$$

we can absorb all these functions in our function space, without giving up the idea of a *single vector* as the representation of a function. We could think, of course, of the μ functions $v_1, v_2, \ldots, v_\mu$ as μ different vectors of the function space. But we can also do something else. We can add to our variable x a *new variable j* which can only assume the values $1, 2, \ldots, \mu$, and is thus automatically a *discrete* variable. Here then it is unnecessary to add a limit process by constantly increasing the density of points in which the function is tabulated. The variable j automatically conforms to the demands of matrix algebra since it is from the very beginning an algebraic quantity. If we replace v_j by the symbol $v(x, j)$ we have only introduced a new notation but this notation suggests a new interpretation. We now extend the previous function space by added dimensions along which we plot the values of $v_1, v_2, \ldots, v_\mu$ at all tabulated points and the entire *set* of vectors is once more represented by *one single vector*. Whether we write the given operator in the form of $Dv_j(x)$, or $Dv(x, j)$, we know at once what we mean: we are going to operate with the dimensions $(1, 2, \ldots, \mu)$ as if they belonged to a surplus variable j which can only assume the μ discrete values $1, 2, \ldots, \mu$.

Let us see what consequences this viewpoint has for the construction of the Green's function. We have first of all to transcribe to our extended problem the solution (4.11) of a differential equation, in terms of the Green's function. For this purpose we shall write down the system (3) in more adequate form. On the right side of this system we have $(0, 0, 0, \beta)$. We can conceive this set of values as accidental and will replace them by the more general set $(\beta_1, \beta_2, \beta_3, \beta_4)$. This means that we will consider a general system of differential equations (ordinary or partial) in the symbolic form

$$Dv(x, j) = \beta(x, j) \qquad (5.7.5)$$

Yet this is not enough. We do not want to exclude from our considerations over-determined systems* of the type

$$\frac{\partial \Phi}{\partial x_1} = \beta_1$$

$$\frac{\partial \Phi}{\partial x_2} = \beta_2$$

$$\vdots \qquad\qquad (5.7.6)$$

$$\frac{\partial \Phi}{\partial x_s} = \beta_s$$

where the unknown is a scalar function Φ while the right side is a vector of s components. In order to cover the general case we have to introduce *two* discontinuous variables k and j, k for the function $v(x)$ and j for the right side $\beta(x)$:

$$Dv(x, k) = \beta(x, j) \qquad (5.7.7)$$

Another thing to remember is that the process of summation is *not* replaced by integration if it comes to the variables k and j, but remains a summation.

* We exclude under-determined systems from our present considerations. For such systems the "constrained Green's vector" comes into operation, as we shall see in Section 22.

The question is now how to obtain the proper interpretation of Green's identity (3.1), in view of our extended system:

$$\int [u(x, j)Dv(x, k) - v(x, k)\tilde{D}u(x, j)]dx = 0 \tag{5.7.8}$$

In order to answer this question, we prefer to write the equation (7) in a more familiar form, replacing the discontinuous variables k and j by *subscripts*:

$$Dv_k(x) = \beta_j(x) \tag{5.7.9}$$

Furthermore, we can conceive the set of functions

$$v_k(x) = (v_1(x), v_2(x), \ldots, v_\mu(x)) \tag{5.7.10}$$

as a "vector" of a μ-dimensional space, associated with *the left side* of the differential equation. Similarly, we can conceive the set of functions

$$\beta_j(x) = (\beta_1(x), \beta_2(x), \ldots, \beta_\nu(x)) \tag{5.7.11}$$

as a "vector" of a ν-dimensional space, associated with the *right side* of the differential equation. But now the danger exists that we have overstressed the use of the word "vector". We have the "function space" in which the entire set of functions $v_1(x), \ldots, v_\mu(x)$ is represented by one single vector. And then we have two additional spaces of μ, respectively ν dimensions. In order to avoid misunderstandings, we will use the word "left-vector" to indicate a μ-dimensional vector of the type (10), and "right-vector" to indicate a ν-dimensional vector of the type (11). Furthermore, we have to remember that the "adjoint operator" $\tilde{D}u(x, j)$ amounts to a *transposition* of rows and columns which has the consequence that a *left vector changes to a right vector and a right vector to a left vector*, that is: in the differential equation which characterises the adjoint operator, we have on the left side a right-vector and on the right side a left-vector. This means that the adjoint system of (9) will have to be written in the following form:

$$\tilde{D}u_j(x) = \gamma_k(x) \tag{5.7.12}$$

This adjoint operator is obtainable with the help of the Green's identity which we can now write down in more definite terms:

$$\int \left[\sum_{j=1}^{\nu} u_j(x)Dv_k(x) - \sum_{k=1}^{\mu} v_k(x)\tilde{D}u_j(x) \right] dx = 0 \tag{5.7.13}$$

Now in the construction of the Green's function we know that this function, as far as the "active variable" ξ is concerned, is nothing but $u_j(\xi)$. The "passive variable" x plays purely the role of a *parameter* in the defining differential equation (4.7). It will be advisable to put the subscript j next to ξ since in fact the subscript j is a portion of the variable ξ. Hence we will use the notation $G(\xi)_j$ in relation to the differential equation which defines the Green's function while the subscript k which is associated with the passive variable x will accordingly appear in conjunction with x. This leads to the notation

$$G(x, k; \xi, j) = G_k(x, \xi)_j \tag{5.7.14}$$

Instead of a "Green's function" we can now speak of a "Green's vector". It is a μ-dimensional left vector with respect to the variable x and a ν-dimensional right vector with respect to the variable ξ. It operates simultaneously in the left space of the vector $v_k(x)$ and the right space of the vector $\beta_j(\xi)$, obtaining the solution of our general system (9) in the following form:

$$v_k(x) = \int \sum_{j=1}^{\nu} G_k(x, \xi)_j \beta_j(\xi) d\xi \tag{5.7.15}$$

What remains is the transcription of the equation (4.10) which defines the Green's function (in our case Green's vector). It now appears in the following form

$$\tilde{D}G_k(x, \xi)_j = \delta_k(x, \xi)_\kappa \tag{5.7.16}$$

On the right side all reference to the subscript j disappears. The delta function of the right side represents a *pure right-vector*, at both points x and ξ. The subscripts k and κ run through *the same set of values* 1, 2, ..., μ. The definition of the delta-function on the right side of (16) is given as follows:

$$\delta_k(x, \xi)_\kappa = \delta(x, \xi)\delta_{k\kappa} \tag{5.7.17}$$

where $\delta_{k\kappa}$ is again "Kronecker's symbol" which is 1, if $k = \kappa$ and zero otherwise, while Dirac's delta function $\delta(x, \xi)$ is once more to be construed by the usual ϵ-process.

In view of the significance of $\delta_{k\kappa}$ it is in fact unnecessary to keep the subscripts k and κ apart. We can write the equation (16) in the simple form

$$\tilde{D}G_k(x, \xi)_j = \delta_k(x, \xi) \tag{5.7.18}$$

with the following interpretation of the right side. We denote with $\delta_k(x, \xi)$ a right side which is composed of zeros, except one single equation, namely the kth equation which has the delta function $\delta(x, \xi)$ as its right side. In order to construct the entire vector $G_k(x, \xi)_j$, $k = 1, 2, \ldots, \mu$, we have to solve a system of μ simultaneous differential equations for ν functions ($\mu \leq \nu$), not once but μ times. We let the $\delta(x, \xi)$ function on the right side glide down gradually from the first to the μth equation and thus obtain in succession the components $G_1(x, \xi)_j$, $G_2(x, \xi)_j$, ..., $G_\mu(x, \xi)_j$ of the complete Green's vector $G_k(x, \xi)_j$ (which is in fact a "vector" in a double sense: μ-dimensional in x and ν-dimensional in ξ).

In frequent applications a system of differential equations originates from one single differential equation of higher order which is transformed into a system of first order equations by the method of surplus variables. In such cases the "given right side" $\beta_j(\xi)$ of the system consists of one single function $\beta(\xi)$ in the jth equation while the right sides of the remaining equations are all zero. In this case the sum on the right side of (7.15) is reduced to one single term and we obtain

$$v_k(x) = \int G_k(x, \xi)_j \beta(\xi)_j d\xi \tag{5.7.19}$$

If we concentrate on the specific function $v_k(x)$, disregarding all the others, the right side of (19) has the form (4.11) of the general theory and we see that the specific component $G_k(x, \xi)_j$ plays the role of the "Green's function" of our problem.

Problem 182. Discuss the solution of the problem (7.6)—with the added condition $\Phi(0) = 0$—from the standpoint of the "Green's vector" and compare it with the solution obtained in section 5.

Problem 183. Write the problem of the elastic bar (4.14.9) in the form of a first order system:

$$\begin{aligned} v'_1(x) - v_2(x) &= 0 \\ I(x)v'_2(x) - v_3(x) &= 0 \\ v'_3(x) - v_4(x) &= 0 \\ v'_4(x) &= \beta(x) \end{aligned} \tag{5.7.20}$$

and solve this system for $v_3(x)$ under the boundary conditions

$$v_1(0) = v_1(l) = v_2(0) = v_2(l) = 0 \tag{5.7.21}$$

[Answer: Adjoint system:

$$\hat{D}u = \begin{bmatrix} -u'_1 \\ -(Iu_2)' - u_1 \\ -u'_3 - u_2 \\ -u'_4 - u_3 \end{bmatrix} \tag{5.7.22}$$

$$u_3(0) = u_3(l) = u_4(0) = u_4(l) = 0 \tag{5.7.23}$$

$$v_3(x) = \int G(x, \xi)\beta(\xi)d\xi \tag{5.7.24}$$

$$G''(x, \xi) - \frac{\alpha(x) + \beta(x)\xi}{I(\xi)} = \delta(x, \xi) \tag{5.7.25}$$

where $\alpha(x)$ and $\beta(x)$ are determined by the boundary conditions prescribed for $G(x, \xi)$ (active variable ξ):

$$G(x, \underline{0}) = G(x, \underline{l}) = G'(x, \underline{0}) = G'(x, \underline{l}) = 0 \tag{5.7.26}$$

Problem 184. Solve the same problem for the function $v_1(x)$ (the elastic deflection).

[Answer:

$$v_1(x) = \int G(x, \xi)\beta(\xi)d\xi \tag{5.7.27}$$

$$[I(\xi)G''(x, \xi)]'' = \delta(x, \xi) \tag{5.7.28}$$

with the boundary conditions (26).]

5.8. Self-adjoint systems

The differential equation of the elastic bar can be given in a variety of different forms. We can give it in the form of the fourth order equation

$$[I(x)v''(x)]'' = \beta(x) \tag{5.8.1}$$

The same differential equation can be formulated as a pair of second order equations (cf. 4.14.9) and we can go still further and replace this second order system by a system of four first order equations, as we have done in (7.20). The adjoint system (7.22) does *not* agree with the original system. But let us now formulate exactly the same system in the following sequence:

$$Dv(x) = \begin{bmatrix} -v'_4(x) = \beta(x) \\ -v'_3(x) - v_4(x) = 0 \\ v'_2(x) - \dfrac{v_3(x)}{I(x)} = 0 \\ v'_1(x) - v_2(x) = 0 \end{bmatrix} \tag{5.8.2}$$

If now we multiply these equations in succession by the undetermined factors $u_1(x)$, $u_2(x)$, $u_3(x)$, $u_4(x)$, form the sum and go through the usual routine of "liberating" the $v_i(x)$, we obtain the adjoint system in the following form:

$$\tilde{D}u = \begin{bmatrix} -u'_4 \\ -u'_3 - u_4 \\ u'_2 - \dfrac{u_3}{I} \\ u'_1 - u_2 \end{bmatrix} \tag{5.8.3}$$

with the boundary term

$$\Big[-u_1v_4 - u_2v_3 + u_3v_2 + u_4v_1 \Big]_0^l \tag{5.8.4}$$

The boundary conditions (7.21) (the bar clamped on both ends) demand for the adjoint system

$$u_1(0) = u_1(l) = u_2(0) = u_2(l) = 0 \tag{5.8.5}$$

and we observe that the new system is *self-adjoint*, inasmuch as the adjoint operator *and* the adjoint boundary conditions coincide with the original operator and its boundary conditions.

The self-adjointness of a certain problem in linear differential equations is a very valuable property which corresponds to the symmetry $A = \tilde{A}$ of the associated matrix problem. Such a symmetry, however, can be destroyed if the equations of the system $Ay = b$ are not written down in the proper order, or even if they are multiplied by wrong factors. Hence it is understandable that the system (7.20) was not self-adjoint, although in proper formulation the problem is in fact self-adjoint. We have to find a method by which we can guarantee in advance that the self-adjoint character of a system will not be destroyed by a false ordering of the equations.

The majority of the differential equations encountered in mathematical physics belong to the self-adjoint variety. The reason is that all the equations of mathematical physics which do not involve any energy losses are deducible from a "principle of least action", that is the principle of making a certain scalar quantity a minimum or maximum. All the linear differential equations which are deducible from minimising or maximising a certain quantity, are automatically self-adjoint and vice versa: *all differential equations which are self-adjoint, are deducible from a minimum-maximum principle.*

In order to study these problems, we will first investigate their algebraic counterpart. Let A be a (real) symmetric matrix

$$A = \tilde{A} \tag{5.8.6}$$

and let us form the scalar quantity

$$s = \tfrac{1}{2}\tilde{y}Ay \tag{5.8.7}$$

We will change the vector y by an arbitrary infinitesimal amount δy, called "variation of y". The corresponding infinitesimal change of s becomes

$$\delta s = \tfrac{1}{2}(\delta\tilde{y}Ay + \tilde{y}A\delta y) \tag{5.8.8}$$

The second term can be transformed in view of the bilinear identity (3.3.6) which in the real case reads

$$\tilde{y}A\delta y = \delta\tilde{y}\tilde{A}y \tag{5.8.9}$$

and thus, in view of (6) we have

$$\delta s = \delta\tilde{y}Ay \tag{5.8.10}$$

If now we modify the scalar s to

$$\bar{s} = \tfrac{1}{2}\tilde{y}Ay - \tilde{y}b \tag{5.8.11}$$

we obtain

$$\delta\bar{s} = \delta\tilde{y}(Ay - b) \tag{5.8.12}$$

and the equation

$$Ay - b = 0 \tag{5.8.13}$$

can be conceived as the consequence of making the variation $\delta\bar{s}$ equal to zero for arbitrary infinitesimal variations of y. The condition

$$\delta\bar{s} = 0 \tag{5.8.14}$$

is equally demanded for maximising or minimising the quantity $\bar{s}$ and will not necessarily make $\bar{s}$ either a maximum or a minimum; we may have a maximum in some directions and a minimum in others and we may even have a "point of inflection" which has an extremum property only if we consider the right and left sides independently. However, the condition (14) can be conceived as a necessary and sufficient condition of a summit in the *local* sense, staying in the infinitesimal neighbourhood of that particular point in which the condition (13) is satisfied. Such a summit in the local sense is called a "stationary value", in order to distinguish it from a true maximum or minimum. *The technique of finding such a stationary value of the scalar quantity $\bar{s}$ is that we put the infinitesimal variation of $\bar{s}$ caused by a free infinitesimal variation of y, equal to zero.*

It is of interest to see what happens to the scalar s in the case of a general (non-symmetric) matrix A. In that case

$$\delta s = \tfrac{1}{2}(\delta\tilde{y}Ay + \delta\tilde{y}\tilde{A}y)$$
$$= \delta\tilde{y}[\tfrac{1}{2}(A + \tilde{A})y]$$

That is, the variational method automatically *symmetrises* the matrix A by replacing it by its symmetric part. An arbitrary square matrix A can be written in the form

$$A = \tfrac{1}{2}(A + \tilde{A}) + \tfrac{1}{2}(A - \tilde{A}) \tag{5.8.15}$$

where the first term is symmetric, the second anti-symmetric

$$A = A_1 + A_2 \tag{5.8.16}$$

$$\begin{aligned} \tilde{A}_1 &= A_1 \\ \tilde{A}_2 &= -A_2 \end{aligned} \tag{5.8.17}$$

Accordingly the quadratic form s becomes

$$s = \tfrac{1}{2}\tilde{y}(A_1 + A_2)y = \tfrac{1}{2}\tilde{y}A_1 y + \tfrac{1}{2}\tilde{y}A_2 y \tag{5.8.18}$$

But the second term *vanishes identically*, due to the bilinear identity

$$\tilde{y}A_2 y = \tilde{y}\tilde{A}_2 y = -\tilde{y}A_2 y = 0 \tag{5.8.19}$$

This explains why the variational method automatically ignores the anti-symmetric part of the matrix A. Exactly the same results remain valid in the complex case, if we replace "symmetric" by "Hermitian" and "anti-symmetric" by "anti-Hermitian". We have to remember, of course, that in the complex case the scalar

$$s = \tfrac{1}{2}\tilde{y}Ay \tag{5.8.20}$$

(which is real for the case $A = \tilde{A}$) comes about by changing in the first factor every i to $-i$ since this change is included in the operation $\tilde{y}$.

These relations can be re-interpreted for the case of differential operators. If Dv is a self-adjoint operator, the differential equation

$$Dv(x) = \beta(x) \tag{5.8.21}$$

can be conceived as the result of a variational problem. For this purpose we have to form a scalar quantity Q (replacing the notation s by the more convenient Q) which under the present circumstances is defined as

$$Q = \tfrac{1}{2}\int v(x)Dv(x)dx \tag{5.8.22}$$

in the real case and

$$Q = \tfrac{1}{2}\int v^*(x)Dv(x)dx \tag{5.8.23}$$

in the complex case. However, the characteristic feature of differential operators is that in the re-interpretation of a matrix relation sometimes an *integration by parts* has to be performed.

Consider for example the problem of the elastic bar. By imitating the matrix procedure we should form the basic scalar $\bar{s}$ in the form

$$\bar{Q} = \int_0^l [\tfrac{1}{2}y(Iy'')'' - y\beta]dx \qquad (5.8.24)$$

But now in the first term an integration by parts is possible by which the order of differentiation can be reduced. The first term is replaceable (apart from a boundary term which is variationally irrelevant) by

$$- \int_0^l \tfrac{1}{2}y'(Iy'')'dx \qquad (5.8.25)$$

and applying the method a second time, by

$$Q = \int_0^l \tfrac{1}{2}I(y'')^2 dx \qquad (5.8.26)$$

The original *fourth order* operator which appeared in (24), could be replaced by a *second order* operator. The *quadratic dependence on $y(x)$ has not changed*. Generally an operator of the order $2n$ can be gradually reduced to an operator of the order n.

The integral (26) represents the "elastic energy" of the bar which in the state of equilibrium becomes a minimum. The additional term in $\bar{Q}$, caused by the load distribution $\beta(x)$:

$$- \int_0^l \beta(x)y(x)dx \qquad (5.8.27)$$

represents the potential energy of the gravitational forces. The minimisation of (24) expresses the mechanical principle that the *state of equilibrium can be characterised as that particular configuration in which the potential energy of the system is a minimum.*

Problem 185. Find the variational integral ("action integral") associated with the Laplacian operator

$$\Delta v = \left[\frac{\partial^2}{\partial x_1^2} + \frac{\partial^2}{\partial x_2^2} + \ldots + \frac{\partial^2}{\partial x_s^2} \right] v \qquad (5.8.28)$$

[Answer:

$$Q = - \frac{1}{2} \int \left[\left(\frac{\partial v}{\partial x_1} \right)^2 + \ldots + \left(\frac{\partial v}{\partial x_s} \right)^2 \right] dx \qquad (5.8.29)]$$

Problem 186. Do the same for the "bi-harmonic operator"

$$\Delta \Delta v \qquad (5.8.30)$$

[Answer:

$$Q = \tfrac{1}{2} \int (\Delta v)^2 dx \qquad (5.8.31)]$$

Problem 187.   Show that the system

$$Dv(x) = \beta(x)$$
$$\check{D}u(x) = \gamma(x) \qquad (5.8.32)$$

is deducible from the principle $\delta Q = 0$ if Q is chosen as follows:

$$Q = \int \tfrac{1}{2}[uDv + v\check{D}u] - (u\beta + v\gamma)]dx \qquad (5.8.33)$$

Problem 188.   Show that the following variational integral yields no differential equation but only boundary conditions:

$$Q = \tfrac{1}{2}\int [uDv - v\check{D}u]dx \qquad (5.8.34)$$

5.9. The calculus of variations

The problem of finding the extremum value of a certain integral is the subject matter of the "calculus of variations". For our purposes it suffices to obtain the "stationary value" of a certain integral, irrespective of whether it leads to a real minimum or maximum. What is necessary is only that the infinitesimal change δQ, caused by an arbitrary infinitesimal change (called "variation") of the function $v(x)$, shall become zero. To achieve this, we need certain techniques which are more systematically treated in the calculus of variations.* The following three procedures are of particular importance:

1. *The method of integrating by parts.* When we were dealing with the problem of finding the adjoint operator $\check{D}u(x)$ on the basis of Green's identity, we employed the technique of "liberation": the derivatives $v^{(k)}(x)$ could be reduced to $v(x)$ itself, by integrating by parts. Finally the resulting factor of $v(x)$ gave us the adjoint operator $\check{D}u(x)$ (cf. Chapter 4.12 for the case of ordinary and 4.17 for the case of partial differential operators). Exactly the same procedure applies to our problem. Whenever a variation of the form $\delta v^{(k)}(x)$ is encountered, we employ the same procedure of integrating by parts, until we have $\delta v(x)$ itself. Then the factor of $\delta v(x)$ put equal to zero is the sufficient and necessary condition of an extremum if augmented by the boundary conditions which follow from the requirement that also the boundary term has to vanish.

2. *The method of the Lagrangian multiplier.* It may happen that our task is to find an extremum value (or stationary value) but with certain *restricting conditions* (called auxiliary conditions or "constraints"), which have to be observed during the process of variation. The method of the Lagrangian multiplier requires that we should add the left sides of the auxiliary conditions (assumed to be reduced to zero), each one multiplied by an undetermined factor λ, to the given variational integrand, and handle the new variational problem as a *free* problem, without auxiliary conditions.

3. *The elimination of algebraic variables.* Let us assume that the variational integrand L, called the "Lagrangian function":

$$q = \int L dx \qquad (5.9.1)$$

---

* Cf., e.g., the author's book [6], quoted in the Bibliography of this chapter.

depends on a certain variable w which is present in L, without any derivatives. Such a variable can be eliminated in advance, by solving for w the equation

$$\frac{\partial L}{\partial w} = 0 \tag{5.9.2}$$

and substituting the w thus obtained into L.

Problem 189. By using integration by parts, obtain the differential equations associated with the following form of the Lagrangian function L:

$$L = L(w; q_1, q'_1; q_2, q'_2, q''_2; x) \tag{5.9.3}$$

[Answer:

$$\frac{\partial L}{\partial w} = 0$$

$$\frac{\partial L}{\partial q_1} - \frac{d}{dx}\frac{\partial L}{\partial q'_1} = 0 \tag{5.9.4}$$

$$\frac{\partial L}{\partial q_2} - \frac{d}{dx}\frac{\partial L}{\partial q'_2} + \frac{d^2}{dx^2}\frac{\partial L}{\partial q''_2} = 0]$$

Problem 190. Derive the differential equation and boundary conditions of the elastic bar which is free at the two ends (i.e. no imposed boundary conditions) by minimizing the integral

$$q = \int_0^l [\tfrac{1}{2}I(x)v''^2(x) - \beta(x)v(x)]dx \tag{5.9.5}$$

[Answer:

$$[I(x)v''(x)]'' - \beta(x) = 0 \tag{5.9.6}$$

$$v''(0) = v'''(0) = v''(l) = v'''(l) = 0] \tag{5.9.7}$$

Problem 191. Do the same for the supported bar; this imposes the boundary conditions

$$v(0) = v(l) = 0 \tag{5.9.8}$$

(and consequently $\delta v(0) = \delta v(l) = 0$).

[Answer: Differential equation (6), together with

$$v(0) = v(l) = v''(0) = v''(l) = 0] \tag{5.9.9}$$

Problem 192. Show that all linear differential equations which are deducible from a variational principle, are automatically self-adjoint in both operator and boundary conditions.

[Hint: Replace $\delta v(x)$ by $u(x)$ and make use of Green's identity.]

5.10. The canonical equations of Hamilton

W. R. Hamilton in 1834 invented an ingenious method by which all ordinary differential equations which are derivable from a variational principle can be put in a normal form, called the "canonical form", which does not involve higher than *first* derivatives. The problem of the elastic bar (9.5) is well suited to the elucidation of the method.

The Lagrangian of our problem is

$$L = \tfrac{1}{2}I(x)v''^2(x) - \beta(x)v(x) \tag{5.10.1}$$

We can make this function *purely algebraic* by introducing the first and the second derivatives of $v(x)$ as new variables. We do that by putting

$$\begin{aligned}
v(x) &= v_1(x) \\
v'_1(x) &= v_2(x) \\
v'_2(x) &= v_3(x)
\end{aligned} \tag{5.10.2}$$

and writing L in the form

$$L = \tfrac{1}{2}I(x)v_3^2(x) - \beta(x)v_1(x) \tag{5.10.3}$$

But in the new formulation we have to add the two auxiliary conditions

$$\begin{aligned}
v'_1(x) - v_2(x) &= 0 \\
v'_2(x) - v_3(x) &= 0
\end{aligned} \tag{5.10.4}$$

which will be treated by the Lagrangian multiplier method. This means that the original L is to be replaced by

$$L' = L + p_1(v'_1 - v_2) + p_2(v'_2 - v_3) \tag{5.10.5}$$

denoting the two Lagrangian multipliers by p_1 and p_2. The new Lagrangian is of the form

$$L' = p_1 v'_1 + p_2 v'_2 - H \tag{5.10.6}$$

where in our problem

$$H = p_1 v_2 + p_2 v_3 - \tfrac{1}{2}Iv_3^2 + \beta v_1 \tag{5.10.7}$$

Since, however, the variable v_3 is *purely algebraic*, we can eliminate it in advance, according to the method (9.2):

$$\frac{\partial L'}{\partial v_3} = -\frac{\partial H}{\partial v_3} = Iv_3 - p_2 = 0 \tag{5.10.8}$$

$$v_3 = \frac{p_2}{I} \tag{5.10.9}$$

$$H = p_1 v_2 + p_2 v_3 + \tfrac{1}{2}\frac{p_2^2}{I} + \beta v_1 \tag{5.10.10}$$

Exactly the same procedure is applicable to differential operators of any order and systems of such operators. By introducing the proper number of surplus variables we shall always end with a new Lagrangian L' of the following form

$$L' = \sum_{i=1}^{n} p_i q'_i - H \tag{5.10.11}$$

where H, the "Hamiltonian function", is an explicitly given function of the variables q_i and p_i, and of the independent variable x:

$$H = H(p_1 \ldots p_n; q_1 \ldots q_n; x) \qquad (5.10.12)$$

Now the process of variation can be applied to the integral

$$Q = \int_a^b \left(\sum_{i=1}^n p_i q'_i - H \right) dx \qquad (5.10.13)$$

and we obtain the celebrated "canonical equations" of Hamilton:

$$q'_i - \frac{\partial H}{\partial p_i} = 0$$

$$-p'_i - \frac{\partial H}{\partial q_i} = 0 \qquad (5.10.14)$$

(The variables were employed in the sequence $p_1 \ldots p_n; q_1 \ldots q_n$.) The boundary term has the form

$$\left[\sum p_i \delta q_i \right]_a^b \qquad (5.10.15)$$

Moreover, in the case of *linear* differential equations the function H cannot contain the variables p_i, q_i in higher than *second* order. The linear part of H gives rise to a "right side" of the Hamiltonian equations (since the resulting terms become constants with respect to the p_i, q_i). The quadratic part gives rise to a matrix of the following structure. Let us combine the variables $(p_1, p_2, \ldots, p_n)$ to the vector p, and the variables $(q_1, q_2, \ldots, q_n)$ to the vector q. Then the Hamiltonian system (14) may be written in the following manner:

$$q' - Pp - Rq = \beta$$

$$-p' - \tilde{R}p - Qq = \gamma \qquad (5.10.16)$$

where the matrices P and Q are symmetric $n \times n$ matrices: $P = \tilde{P}, Q = \tilde{Q}$.

In fact, we can go one step further and put the canonical system in a still more harmonious form. We will unite the two vectors p and q into a single vector of a $2n$-dimensional space (called the "phase-space"). The components of this extended vector shall be denoted in homogeneous notation as follows:

$$(p_1, p_2, \ldots, p_n, \quad q_1, q_2, \ldots, q_n) = (p_1, p_2, \ldots, p_n, \quad p_{n+1}, p_{n+2}, \ldots, p_{2n})$$

Then the matrices P, R, $\tilde{R}$, Q can be combined into *one single* $2n \times 2n$ *symmetric matrix*

$$C = \tilde{C} = \begin{bmatrix} P & R \\ \tilde{R} & Q \end{bmatrix} \qquad (5.10.17)$$

and the entire canonical system may be written in the form of *one unified scheme*:

$$p'_{n+i} - \sum_{\alpha=1}^{2n} C_{i\alpha} p_\alpha = \beta_i \qquad (5.10.18)$$

if we agree that the notation p_{2n+i} shall have the following significance:

$$p_{2n+i} = -p_i \qquad (5.10.19)$$

Hamilton's discovery enables us to put all variational problems into a particularly simple and powerful normal form, namely the "canonical" form. Hence the proper realm of the canonical equations is first of all the field of *self-adjoint* problems. In fact, however, any arbitrary non-self-adjoint problem can be conceived as the result of a variational problem and thus the canonical equations assume *universal significance*. Indeed, an arbitrary linear differential equation $Dv(x) = \beta(x)$ can be conceived as the result of the following variational principle:

$$\delta \int u(x)[Dv(x) - \beta(x)]dx = 0 \qquad (5.10.20)$$

because the variation of $u(x)$ gives at once:

$$Dv(x) - \beta(x) = 0 \qquad (5.10.21)$$

while the variation of $v(x)$ adds the equation

$$\tilde{D}u(x) = 0 \qquad (5.10.22)$$

which can be added without any harm since it has no effect on the solution of the equation (21).

To obtain the resulting canonical scheme we proceed as follows. By the method of surplus variables we introduce the first, second, ..., $n - 1$st derivative as new variables. We will illustrate the operation of the principle by considering the general linear differential equation of third order:

$$A_1(x)v'''(x) + A_2(x)v''(x) + A_3(x)v'(x) + A_4(x)v(x) = \beta(x) \quad (5.10.23)$$

We denote v by p_4 and introduce two new variables p_5 and p_6 by putting

$$\begin{aligned} p'_4 &= p_5 \\ A_1 p'_5 &= p_6 \end{aligned} \qquad (5.10.24)$$

Hence $A_1 v'' = p_6$. The first term of (23) may be written in a slightly modified form:

$$(A_1 v'')' - A'_1 v'' \qquad (5.10.25)$$

and thus the given differential equation (23) may now be formulated as the following first order system:

$$\begin{aligned} p'_4 - p_5 &= 0 \\ p'_5 - \frac{p_6}{A_1} &= 0 \\ p'_6 + \frac{A_2 - A'_1}{A_1} p_6 + A_3 p_5 + A_4 p_4 &= \beta \end{aligned} \qquad (5.10.26)$$

If now we multiply by the undetermined factors p_1, p_2, p_3 and apply in the first term the usual integration by parts technique, we obtain the adjoint system in the form

$$-p'_1 + A_4 p_3 = 0$$
$$-p'_2 - p_1 + A_3 p_3 = 0$$
$$-p'_3 - \frac{1}{A_1} p_2 + \frac{A_2 - A'_1}{A_1} p_3 = 0 \qquad (5.10.27)$$

The two systems (26) and (27) can be combined into the single system

$$p'_{3+i} - \sum_{\alpha=1}^{6} C_{i\alpha} p_\alpha = \beta \delta_{3i} \qquad (5.10.28)$$

where the matrix C_{ik} has the following elements

$$
C = \tilde{C} = \left[
\begin{array}{ccc|ccc}
 & & & 0 & 1 & 0 \\
 & 0 & & 0 & 0 & \dfrac{1}{A_1} \\
 & & & -A_4 & -A_3 & \dfrac{A'_1 - A_2}{A_1} \\
\hline
0 & 0 & -A_4 & & & \\
1 & 0 & -A_3 & & 0 & \\
0 & \dfrac{1}{A_1} & \dfrac{A'_1 - A_2}{A_1} & & &
\end{array}
\right] \qquad (5.10.29)
$$

This is a special case of the general canonical scheme (16), with a matrix (17) in which the $n \times n$ matrices P and Q are missing and thus the operators D and $\tilde{D}$ *fall apart*, without any coupling between them. But *the canonical equations of Hamilton are once more valid.*

The procedure we followed here has one disadvantage. In the case that the given differential operator is self-adjoint, we may destroy the self-adjoint nature of our equation and thus unnecessarily *double* the number of equations, in order to obtain the canonical scheme. An example was given in the discussion of the problem of the *elastic bar* (cf. Section 7), there we used the method of surplus variables and succeeded in reducing the given fourth order equation into a system of first order equations (7.20), which, however, were *not* self-adjoint, although the system (8.2) demonstrated that the same system can also be given in self-adjoint form. This form was deducible if we knew the action integral from which the problem originated by the process of variation. Then the Hamiltonian method gave us the canonical system (8.2). But let us assume that we do *not* know in advance that our problem can be put in self-adjoint form. Our equations are given in the non-self-adjoint form (7.20). Is there a way of transforming this system into the proper canonical form of *four* instead of *eight* equations?

If we can assume that a certain linear differential equation is deducible from a variational principle, the action integral of that principle must have the form (8.22). Hence we can obtain the Lagrangian of the alleged variational problem by putting

$$L = \tfrac{1}{2}vDv - \beta v \qquad (5.10.30)$$

We can now go through with the Hamiltonian scheme and finally, after obtaining the resulting self-adjoint system, compare it with the given system and see whether the two systems are in fact equivalent or not.

In our problem (7.20) we have given our differential equation in the form

$$v'_4 = \beta \qquad (5.10.31)$$

with the three auxiliary conditions

$$
\begin{aligned}
v'_1 - v_2 &= 0 \\
Iv'_2 - v_3 &= 0 \\
v'_3 - v_4 &= 0
\end{aligned}
\qquad (5.10.32)
$$

Hence our Lagrangian L, modified by the Lagrangian multipliers, now becomes:

$$L' = \tfrac{1}{2}v_1v'_4 - \beta v_1 + p_1(v'_1 - v_2) + p_2(Iv'_2 - v_3) + p_3(v'_3 - v_4) \qquad (5.10.33)$$

($v = v_1$ since the fundamental variable of our problem is v_1).

The first term is variationally equivalent to

$$-\tfrac{1}{2}v'_1v_4 \qquad (5.10.34)$$

and then L' appears in the following form:

$$L' = (p_1 - \tfrac{1}{2}v_4)v'_1 + p_2Iv'_2 + p_3v'_3 - p_1v_2 - p_2v_3 - p_3v_4 \qquad (5.10.35)$$

Let us replace p_1 by the notation $\bar{p}_1$ and put

$$\bar{p}_1 - \tfrac{1}{2}v_4 = p_1 \qquad (5.10.36)$$

Let us likewise replace p_2 by $\bar{p}_2$ and put

$$\bar{p}_2I = p_2 \qquad (5.10.37)$$

With these substitutions L' becomes

$$L' = p_1v'_1 + p_2v'_2 + p_3v'_3 - (p_1 + \tfrac{1}{2}v_4)v_2 - \frac{p_2}{I}v_3 - p_3v_4 - \beta v_1 \qquad (5.10.38)$$

But now v_4 is *purely algebraic* and putting the partial derivative with respect to v_4 equal to zero we obtain the condition

$$p_3 + \tfrac{1}{2}v_2 = 0 \qquad (5.10.39)$$

due to which L' is reducible to

$$L' = p_1v'_1 + (p_2 + \tfrac{1}{2}v_3)v'_2 - p_1v_2 - \frac{p_2v_3}{I} - \beta v_1 \qquad (5.10.40)$$

and replacing $p_2 + \frac{1}{2}v_3$ by a new p_2:

$$L' = p_1v'_1 + p_2v'_2 - p_1v_2 - \frac{1}{I}(p_2 - \tfrac{1}{2}v_3)v_3 - \beta v_1 \qquad (5.10.41)$$

Here we have obtained the canonical Lagrangian function, further simplifiable by the fact that v_3 is purely algebraic which leads to the elimination of v_3 by the condition

$$p_2 - v_3 = 0 \qquad (5.10.42)$$

What remains can be written in the form

$$L' = p_1v'_1 + p_2v'_2 - H \qquad (5.10.43)$$

with

$$H = p_1v_2 + \frac{p_2{}^2}{2I} - \beta v_1 \qquad (5.10.44)$$

The canonical equations become, replacing v_1 and v_2 by p_3 and p_4:

$$
\begin{aligned}
p'_3 - p_4 &= 0 \\
p'_4 - \frac{p_2}{I} &= 0 \\
-p'_1 &= \beta \\
-p'_2 - p_1 &= 0
\end{aligned}
\qquad (5.10.45)
$$

The equivalence of this canonical system with the original system (31) and (32) is easily established if we make the following identifications:

$$p_3 = v_1, \quad p_4 = v_2, \quad p_1 = -v_4, \quad p_2 = v_3 \qquad (5.10.46)$$

The matrix C becomes in our case:

$$
C = \begin{bmatrix}
0 & 0 & 0 & 1 \\
0 & \dfrac{1}{I} & 0 & 0 \\
0 & 0 & 0 & 0 \\
1 & 0 & 0 & 0
\end{bmatrix}
\qquad (5.10.47)
$$

The symmetry of this matrix expresses the self-adjoint character of our system.

Problem 193. Assume the boundary conditions

$$q_i(a) = 0, \quad q_i(b) = 0 \qquad (5.10.48)$$

Show the self-adjoint character of the system (14), (15), denoting the adjoint variables by $\bar{p}_i$, $\bar{q}_i$. Do the same for the boundary conditions

$$q_i(a) = 0, \quad p_i(b) = 0 \qquad (5.10.49)$$

Problem 194.   The planetary motion is characterised by a variational principle whose integral is of the following form:

$$Q = \int \frac{m}{2} (r'^2 + r^2\theta'^2) - V(r) \tag{5.10.50}$$

(m = mass of planet, considered as point; r, θ = polar coordinates of its position, $V(r)$ = potential energy of the central force).   Obtain the Hamiltonian form of the equations of motion by considering r, θ as q_1, q_2 and introducing r', θ' as added variables q_3, q_4.

[Answer:

$$L' = p_1 q'_1 + p_2 q'_2 - H$$

$$H = \frac{p_1{}^2}{2m} + \frac{p_2{}^2}{2mq_1{}^2} + V(q_1)$$

$$q'_1 - \frac{p_1}{m} = 0$$

$$q'_2 - \frac{p_2}{mq_1{}^2} = 0 \tag{5.10.51}$$

$$-p'_1 - V(q_1) + \frac{p_2{}^2}{mq_1{}^3} = 0$$

$$-p'_2 = 0 \,]$$

5.11. The Hamiltonisation of partial operators

Exactly the same method is applicable to the realm of partial differential operators, permitting us to reduce differential equations of arbitrary order to first order differential equations which are once more of the self-adjoint type if both differential equation and boundary conditions are the result of a variational principle.   For example the potential equation

$$-\Delta\varphi = \beta \tag{5.11.1}$$

is derivable by minimising the variational integral

$$Q = \int [\tfrac{1}{2}|\text{grad } \varphi|^2 - \beta\varphi]dx \tag{5.11.2}$$

We introduce the partial derivatives of φ as new variables

$$\frac{\partial \varphi}{\partial x_1} - \varphi_1 = 0$$

$$\frac{\partial \varphi}{\partial x_2} - \varphi_2 = 0 \tag{5.11.3}$$

$$\frac{\partial \varphi}{\partial x_3} - \varphi_3 = 0$$

considering these equations as auxiliary conditions of our variational problem.   Then the application of the Lagrangian multiplier method yields the new Lagrangian

$$L' = p_1 \frac{\partial \varphi}{\partial x_1} + p_2 \frac{\partial \varphi}{\partial x_2} + p_3 \frac{\partial \varphi}{\partial x_3} - H \tag{5.11.4}$$

with

$$H = p_1\varphi_1 + p_2\varphi_2 + p_3\varphi_3 - \tfrac{1}{2}(\varphi_1{}^2 + \varphi_2{}^2 + \varphi_3{}^2) + \beta\varphi \quad (5.11.5)$$

Since φ_1, φ_2, φ_3, are purely algebraic variables (their derivatives do not appear in L'), they can be eliminated, obtaining:

$$H = \tfrac{1}{2}(p_2{}^1 + p_2{}^2 + p_3{}^2) + \beta\varphi \quad (5.11.6)$$

The Hamiltonian system for the four variables p_1, p_2, p_3, φ becomes:

$$\frac{\partial\varphi}{\partial x_1} - p_1 = 0$$

$$\frac{\partial\varphi}{\partial x_2} - p_2 = 0$$

$$\frac{\partial\varphi}{\partial x_3} - p_3 = 0 \quad (5.11.7)$$

$$-\left(\frac{\partial p_1}{\partial x_1} + \frac{\partial p_2}{\partial x_2} + \frac{\partial p_3}{\partial x_3}\right) = \beta$$

with the boundary term

$$\int(p_1\nu_1 + p_2\nu_2 + p_3\nu_3)\delta\varphi ds \quad \text{(on } S) \quad (5.11.8)$$

Problem 195.   According to Problem 170 (Chapter 4.18) the differential equation

$$\Delta\varphi = \beta \quad (5.11.9)$$

with the boundary condition c):

$$\frac{\partial\varphi(S)}{\partial\nu} - \gamma(S)\varphi(S) = 0 \quad (5.11.10)$$

belongs to a *self-adjoint operator*.   Hence this system must be deducible from a variational principle $\delta Q = 0$.   Find the action integral Q of this principle.

[Answer:

$$Q = \int[-\tfrac{1}{2}|\text{grad }\varphi|^2 - \beta\varphi]dx + \int_S \tfrac{1}{2}\gamma(S)\varphi^2(S)dS \quad (5.11.11)]$$

Problem 196.   Reduce the biharmonic equation

$$\Delta\,\Delta\varphi = \beta \quad (5.11.12)$$

(cf. Problem 186, Section 8) to a Hamiltonian system of first order equations.
[Answer:

$$L' = \sum_{i=1}^{3} p_i\left(\frac{\partial\varphi}{\partial x_i} + \lambda\frac{\partial q_i}{\partial x_i}\right) - H \quad (5.11.13)$$

$$H = \sum_{i=1}^{3} p_i q_i + \tfrac{1}{2}\lambda^2 + \beta\varphi \quad (5.11.14)$$

$$\frac{\partial \varphi}{\partial x_i} - q_i = 0$$

$$-\frac{\partial \lambda}{\partial x_i} - p_i = 0$$

$$\sum_i \frac{\partial q_i}{\partial x_i} - \lambda = 0 \qquad\qquad (5.11.15)$$

$$-\sum_i \frac{\partial p_i}{\partial x_i} = \beta\,]$$

5.12. The reciprocity theorem

The general treatment of Green's function has shown that the solution of a differential equation by the Green's function method is not restricted to even-determined systems but equally applicable to arbitrarily over-determined systems. The only condition necessary for the existence of a Green's function was that the system shall be *complete*, that is the equation

$$Dv(x) = 0 \qquad\qquad (5.12.1)$$

shall have no solution other than the trivial solution $v(x) = 0$. If the adjoint homogeneous system

$$\tilde{D}u(x) = 0 \qquad\qquad (5.12.2)$$

possessed solutions which did not vanish identically, the Green's function method did not lose its significance. It was merely necessary that the given right side should be orthogonal to every independent solution of the equation (2):

$$\int \beta(\xi)u(\xi)d\xi = 0 \qquad\qquad (5.12.3)$$

We have seen examples where a finite or an infinite number of such "compatibility conditions" had to be satisfied.

However, in a very large number of problems the situation prevails that neither (1) nor (2) has non-vanishing solutions. This cases realises in matrix language the ideal condition

$$n = m = p \qquad\qquad (5.12.4)$$

The number of equations, the number of unknowns and the rank of the matrix all coincide. In that case we have Hadamard's "well-posed" problem: the solution is unique and the given right side can be chosen freely.

Under such conditions the Green's function possesses a special property which leads to important consequences. Let us consider, together with the problem

$$Dv(x) = \beta(x) \qquad\qquad (5.12.5)$$

the adjoint problem

$$\tilde{D}u(x) = \gamma(x) \qquad\qquad (5.12.6)$$

Since the homogeneous equation (2) has (according to our assumption) no non-vanishing solutions, the general condition for the existence of a Green's function is satisfied and we obtain the solution of (6) in the form

$$u(x) = \int \tilde{G}(x, \xi)\gamma(\xi)d\xi \tag{5.12.7}$$

We have chosen the notation $\tilde{G}(x, \xi)$ to indicate that the Green's function of the adjoint problem is meant. The defining equation of this function is (in view of the fact that the adjoint of the adjoint is the original operator):

$$D\tilde{G}(x, \xi) = \delta(x, \xi) \tag{5.12.8}$$

The active variable is again ξ, while x is a mere parameter.

This equation, however, can be conceived as a special case of the equation

$$Dv(\xi) = \beta(\xi) \tag{5.12.9}$$

whose solution is obtainable with the help of the Green's function $G(x, \xi)$, although we have now to replace x by ξ and consequently choose another symbol, say σ, for the integration variable:

$$v(\xi) = \int G(\xi, \sigma)\beta(\sigma)d\sigma \tag{5.12.10}$$

If we now identify $\beta(\sigma)$ with $\delta(x, \sigma)$ in order to apply the general solution method (10) to the solution of the special equation (8), the integration over σ is reduced to the immediate neighbourhood of the point $\sigma = x$ and we obtain in the limit (as ϵ converges to zero), on the right side of (10):

$$G(\xi, x) \tag{5.12.11}$$

while the left side $v(\xi)$ is by definition $\tilde{G}(x, \xi)$. This gives the fundamental result

$$\tilde{G}(x, \xi) = G(\xi, x) \tag{5.12.12}$$

which has the following significance: *The solution of the adjoint problem* (6) *can be given with the help of the same Green's function* $G(x, \xi)$ *which solved the original problem. All we have to do is to exchange the role of "fixed point" and "variable point".*

$$u(x) = \int G(\xi, x)\gamma(\xi)d\xi \tag{5.12.13}$$

In a similar manner we can give the solution of the original problem in terms of the Green's function of the adjoint problem

$$v(x) = \int \tilde{G}(\xi, x)\beta(\xi)d\xi \tag{5.12.14}$$

This result can be expressed in still different interpretation. The defining equation of $\tilde{G}(x, \xi)$ was the equation (8). Since $\tilde{G}(x, \xi)$ is replaceable by $G(\xi, x)$, the running variable ξ becomes now the *first* variable of the function $G(x, \xi)$. We need not change our notation $G(x, \xi)$ if we now agree to consider x as the active variable and ξ as a mere parameter. The equation (8) can then be written in the following form (in view of the fact that the delta function $\delta(x, \xi)$ is *symmetric* in x and ξ):

$$DG(\underline{x}, \xi) = \delta(\underline{x}, \xi) \tag{5.12.15}$$

The underlining of x shall indicate that it is no longer ξ but x which became the active variable of our problem while the point ξ is kept fixed during the solution of the equation.

This remarkable reciprocity between the solution of the original and the adjoint equations leads to a new interpretation of the Green's function in the case of well-posed problems. Originally we had to construct the adjoint differential equation and solve it, with the delta function on the right side. Now we see that we can remain with the original equation and solve it once more with the delta function on the right side, but with the difference that in the first case ξ is the variable and x the fixed point; while in the second case x is the variable and ξ the fixed point. In both cases the same $G(x, \xi)$ is obtained.

Problem 197. Explain why the reciprocity relation (12) cannot be generalised (without modification) to problems which do not satisfy the two conditions that neither (1) nor (2) shall possess non-vanishing solutions.

5.13. Self-adjoint problems. Symmetry of the Green's function

If the given problem is self-adjoint, the operators D and $\tilde{D}$ coincide. In this case the Green's function $\tilde{G}(x, \xi)$ of the adjoint problem must coincide with $G(x, \xi)$ since the solution of the given differential equation is unique and allows the existence of a single Green's function only. The theorem (12.12) now takes the form

$$G(x, \xi) = G(\xi, x) \qquad (5.13.1)$$

This is a fundamental result which has far-reaching consequences. It means that the position of the two points x and ξ can be exchanged without changing the value of the kernel function $G(x, \xi)$. Since all differential equations derivable from a variational principle are automatically self-adjoint and the equations of elastic equilibrium are derivable from the variational principle that the potential energy of the elastic forces must become a minimum, the equations of elastic deflection provide an example of a self-adjoint system. It was in conjunction with this example that the symmetry of the Green's function was first enunciated by J. C. Maxwell who expressed this fundamental theorem in physical language: "The elastic deflection generated at the point P by a point load located at Q is equal to the elastic deflection generated at the point Q by a point load located at the point P." This "reciprocity theorem" of Maxwell is equivalent to the statement (1) since Maxwell's "point load" is a complete physical counterpart of Dirac's delta function $\delta(x, \xi)$.

5.14. Reciprocity of the Green's vector

In the case of systems of differential equations we had to introduce the concept of a "Green's vector" $G_k(x, \xi)_j$ and generally the number of k-components: $k = 1, 2, \ldots, \mu$ and the number of j-components: $j = 1, 2, \ldots, \nu$ did not agree. But if we restrict ourselves to the "well-posed" case of Section 12, we must assume that now the number of equations ν and the number of unknowns μ coincides

$$\mu = \nu \tag{5.14.1}$$

We will once more assume that neither the given homogeneous system, nor the adjoint homogeneous system has non-vanishing solutions. We can now investigate the role of the reciprocity theorem (12.12), if the Green's *function* $G(x, \xi)$ is changed to the Green's *vector* $G_k(x, \xi)_j$. This question can be answered without further discussion since we have seen that the subscripts k and j can be conceived as *extended dimensions* of the variables x and ξ. Accordingly the theorem (12.12) will now take the form

$$G(k, x; \xi, j) = G(j, \xi; x, k) \tag{5.14.2}$$

and this means, if we return to the customary subscript notation:

$$\tilde{G}_k(x, \xi)_j = G_j(\xi, x)_k \tag{5.14.3}$$

Moreover, in transcribing the equation (12.15) to the present case we now obtain the result that the same Green's vector $G_k(x, \xi)_j$ may be characterised in two different ways, once by solving the vectorial system

$$\tilde{D}G_k(x, \xi)_{\underline{j}} = \delta_k(x, \xi) \tag{5.14.4}$$

considering the variable ξ as an integration variable while the point x is a mere parameter (and so is the subscript k which changes only by shifting the delta function on the right side from equation to equation) and once by solving the system

$$DG_{\underline{k}}(x, \xi)_j = \delta(x, \xi)_j \tag{5.14.5}$$

where the active variable is now x—together with the subscript k—while ξ and j are constants during the process of integration.

This means that the same Green's vector $G_k(x, \xi)_j$ can be obtained in two different ways. In the first case we write down the *adjoint* system, putting the delta function in the kth equation and solving the system for $u_j(\xi)$. In the second case we write down the *given* system, putting the delta function in the jth equation and solving the system for $v_k(x)$. The general theory shows that both definitions yields the same function $G_k(x, \xi)_j$. But the two definitions do not coincide in a trivial fashion since in one case ξ, in the other x is the active variable. In a special case it may not even be simple to demonstrate that the two definitions coincide without actually constructing the explicit solution.

If our system is self-adjoint, the second and the first system of equations becomes identical and we obtain the symmetry condition of the Green's vector in the form:

$$G_k(x, \xi)_j = G_j(\xi, x)_k \tag{5.14.6}$$

In proper interpretation the reciprocity theorem (12.12)—and its generalised form (14.3)—can be conceived as a special application of the symmetry theorem (14.6), although the latter theorem is restricted to self-adjoint equations. It so happens that *every equation becomes self-adjoint if we complement it by the adjoint equation.* Let us consider the system of two simultaneous differential equations, arising from uniting (12.5) and (12.6) into a single system

$$Dv(x) = \beta_1(x)$$
$$\tilde{D}u(x) = \beta_2(x) \qquad\qquad (5.14.7)$$

Now we have a *self-adjoint* problem in the variables u, v (the sequence is important: u is the first, v the second variable), with a *pair* of equations. Accordingly we have to solve this system with a Green's *vector* which has altogether 4 components, viz. $G_1(x, \xi)_{1,2}$ and $G_2(x, \xi)_{1,2}$. Considering ξ as the active variable we need the two components $G_1(x, \xi)_{1,2}$ for the solution of the first function—that is $u(x)$—and the two components $G_2(x, \xi)_{1,2}$ for the solution of the second function, that is $v(x)$. The first two components are obtained by putting the delta function in the first equation and solving the system for $u(\xi)$, $v(\xi)$. Since, however, there is *no coupling* between the two equations, we obtain $u(\xi) = 0$, while in the case of the last two components (when the delta function is in the second equation), $v(\xi) = 0$. This yields:

$$G_1(x, \xi)_1 = 0$$
$$G_2(x, \xi)_2 = 0 \qquad\qquad (5.14.8)$$

For the remaining two components we have the symmetry condition

$$G_1(x, \xi)_2 = G_2(\xi, x)_2 \qquad\qquad (5.14.9)$$

But in the original interpretation $G_1(x, \xi)_2$ was denoted by $\tilde{G}(x, \xi)$, while $G_2(x, \xi)_1$ was denoted by $G(x, \xi)$. And thus the symmetry relation (14.9) expresses in fact the *reciprocity theorem*

$$\tilde{G}(x, \xi) = G(\xi, x) \qquad\qquad (5.14.10)$$

obtained earlier (in Section 12) on a different basis. By the same reasoning the generalised reciprocity theorem (3) can be conceived as an application of the symmetry relation (6), if again we complement the given vectorial system by the adjoint vectorial system, which makes the resultant system self-adjoint.

Problem 198.   Define the Green's function $G(x, \xi)$ of Problem 183 (cf. (7.20)), by considering x as the active variable.

[Answer:

$$G''(\underline{x}, \xi) = \delta(\underline{x}, \xi)$$
$$I(\underline{x})v''_1(\underline{x}) = G(\underline{x}, \xi) \qquad\qquad (5.14.11)$$

with the boundary conditions

$$v_1(0) = v_1(l) = v'_1(0) = v'_1(l) = 0 \qquad\qquad (5.14.12)]$$

Problem 199. Consider the same problem in the self-adjoint form (8.2) and carry through the same procedure. Show the validity of the two definitions (7.25–26) and (14.11–12) by considering once ξ and once x as the active variable.

5.15. The superposition principle of linear operators

The fact that $Dv(x)$ is a *linear operator*, has some far reaching consequences, irrespective of the dimensionality of the point x. In particular, the linearity of the operator D finds expression in two fundamental operational equations which are lost in the case that D is non-linear:

$$Dv_1(x) + Dv_2(x) = D[v_1(x) + v_2(x)] \qquad (5.15.1)$$

$$D\alpha v(x) = \alpha Dv(x) \qquad (5.15.2)$$

provided that α is a *constant* throughout the given range. We thus speak of the *superposition principle* of linear operators which finds its expression in the equations (1) and (2). In consequence of these two properties we see that if

$$\beta(x) = \alpha_1\beta_1(x) + \alpha_2\beta_2(x) + \ldots + \alpha_\mu\beta_\mu(x) \qquad (5.15.3)$$

we can solve the equation

$$Dv(x) = \beta(x) \qquad (5.15.4)$$

by solving the special equations

$$Dv_i(x) = \beta_i(x) \qquad (i = 1, 2, \ldots, \mu) \qquad (5.15.5)$$

Then

$$v(x) = \alpha_1 v_1(x) + \alpha_2 v_2(x) + \ldots + \alpha_\mu v_\mu(x) \qquad (5.15.6)$$

We can extend this method to the case that μ goes to infinity. Let us assume that

$$\beta(x) = \lim_{N \to \infty} \sum_{k=1}^{N} \alpha_k \beta_k(x) \qquad (5.15.7)$$

We solve the equations

$$Dv_i(x) = \beta_i(x) \qquad (i = 1, 2, \ldots) \qquad (5.15.8)$$

and form the sum

$$v_N(x) = \sum_{k=1}^{N} \alpha_k v_k(x) \qquad (5.15.9)$$

Then, if $v_N(x)$ approaches a definite limit as N increases to infinity:

$$v(x) = \lim_{N \to \infty} v_N(x) \qquad (5.15.10)$$

this $v(x)$ becomes the solution of the equation (4).

Now an arbitrary sectionally continuous function can be approximated with the help of "pulses" which are of short duration and which follow each other in proper sequence. We can illustrate the principle by considering the one-dimensional case in which x varies between the limits a and b. The "unit-pulse" has the width ϵ and the height $1/\epsilon$. The centre of the pulse can be shifted to any point we like and if we write $\delta_\epsilon[x, a + (\epsilon/2)]$, this will mean that the unit pulse, illustrated between the points O and P of the figure, is shifted to the point $x = a$, extending between the points

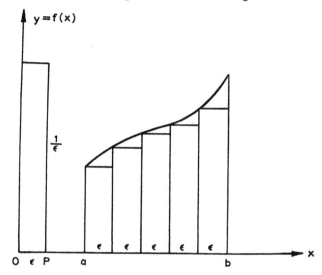

$x = a$ and $x = a + \epsilon$, and being zero everywhere else. The height of this pulse becomes 1 if we multiply by ϵ and we obtain the first panel of our figure by writing

$$\epsilon f(a)\delta_\epsilon\left(x,\, a + \frac{\epsilon}{2}\right)$$

Similarly the second panel of our figure is obtained by writing

$$\epsilon f(a + \epsilon)\delta_\epsilon\left(x,\, a + \frac{3\epsilon}{2}\right)$$

The third panel becomes

$$\epsilon f(a + 2\epsilon)\delta_\epsilon\left(x,\, a + \frac{5\epsilon}{2}\right)$$

and so on. That our continuous function $f(x)$ is crudely represented by these panels should not disturb us since we have ϵ in our hand. By making ϵ smaller and smaller the difference between $f(x)$ and the approximation by pulses can be made as small as we wish and in the limit, as ϵ decreases to zero, we obtain $f(x)$ with the help of a succession of delta functions. We have to *sum*, of course, over all these panels, in order to obtain $f(x)$ but this sum changes to an *integral* as ϵ recedes to zero. And thus we can generate $f(x)$ by a succession of *pulses*, according to the equation

$$f(x) = \int_a^b f(\xi)\delta(x, \xi)d\xi \qquad (5.15.11)$$

The more dimensional case is quite similar and the equation (11) expresses the generation of $f(x)$ by a superposition of pulses if we omit the limits a and b and replace them by the convention that our integral is to be extended over the entire given domain of our problem, $d\xi$ denoting the volume-element of the domain.

Let us now return to our equation (3). We will generate the right side $\beta(x)$ in terms of pulses, according to the equation

$$f(x) = \int \beta(\xi)\delta(x, \xi)d\xi \qquad (5.15.12)$$

The previous α_i corresponds to $\beta(\xi)$, the previous $\beta_i(x)$ to $\delta(x, \xi)$. Accordingly the equation (5) now becomes

$$Dv(x, \xi) = \delta(x, \xi) \qquad (5.15.13)$$

replacing the notation $v_i(x)$ by $v(x, \xi)$. In order to bring into evidence that we have constructed a special auxiliary function which depends not only on x but also on the position of the point ξ (which is a mere constant from the standpoint of solving the differential equation (13)), we will replace the notation $v(x, \xi)$ by $G(x, \xi)$:

$$DG(x, \xi) = \delta(x, \xi) \qquad (5.15.14)$$

The superposition principle (9) now becomes

$$v(x) = \int \beta(\xi)G(x, \xi)d\xi \qquad (5.15.15)$$

Once more we have obtained the standard solution of a linear differential equation, in terms of the Green's function. But now all reference to the adjoint system has disappeared. We have defined $G(x, \xi)$ solely on the basis of the *given* equation, making use of the superposition principle of linear operators. The result is the same as that obtained in Section 12, but now re-interpreted in the light of the superposition principle. It is of interest to observe that the solvability of (14) demands that the adjoint equation $\tilde{D}u(x) = 0$ shall have no non-vanishing solutions while the definition of $G(x, \xi)$ on the basis of the adjoint equation demanded that $Dv(x) = 0$ should have no non-vanishing solutions. Hence the definition of $G(x, \xi)$ as a function of ξ excludes under-determination, the definition of $G(x, \xi)$ as a function of x excludes over-determination.

In the case of a *system* of differential equations (cf. (7.9)):

$$Dv_k(x) = \beta_j(x) \qquad (5.15.16)$$

the right side can once more be obtained as a superposition of pulses. But now we have to write

$$\beta_j(x) = \int \beta_j(\xi)\delta(x, \xi)_j d\xi \qquad (5.15.17)$$

$$DG_k(x, \xi) = \delta(x, \xi)_j \qquad (5.15.18)$$

$$v_k(x) = \int \sum_{j=1} G_k(x, \xi)_j \beta_j(\xi) d\xi \qquad (5.15.19)$$

where again $\delta(x, \xi)_j$ denotes that the delta function $\delta(x, \xi)$ is put in the jth equation (while all the other right sides are zero). Once more the result agrees with the corresponding result (14.5) of our previous discussion, but here again obtained on the basis of the superposition principle.

Problem 200. On the basis of the superposition principle find the solution of the following boundary value problems:

$$v''(x) + \rho^2 v(x) = 0 \qquad x = [0, \pi] \qquad (5.15.20)$$

a) $\begin{aligned} v(0) &= 1 \\ v'(0) &= 1 \end{aligned}$ b) $\begin{aligned} v(0) &= 1 \\ v'(1) &= 0 \end{aligned}$ c) $\begin{aligned} v(0) - v(\pi) &= 0 \\ v'(0) + v'(\pi) &= 1 \end{aligned}$

Explain the peculiar behaviour of c).

[Answer:

$$\text{a)} \quad v(x) = \frac{1}{\rho}[\cos \rho x + \sin \rho x]$$

$$\text{b)} \quad v(x) = \frac{1}{\rho}[\cos \rho x + \tan \rho \pi \sin \rho x]$$

$$\text{c)} \quad v(x) = A \cos \rho x + \frac{\sin \rho x}{2\rho} \quad (\text{if } \rho = 2k)$$

c) holds only if $\rho = 2k =$ even integer, because the adjoint homogeneous equation has a non-zero solution and thus the right side of c) cannot be prescribed freely; the compatibility condition holds only for the special case $\rho = 2k$.]

Problem 201. Given once more the differential equation (20), with the boundary conditions

$$\begin{aligned} v(0) - v(\pi) &= \alpha \\ v'(0) + v'(\pi) &= \beta \end{aligned} \qquad (5.15.21)$$

Find the compatibility condition between α and β and explain the situation found above (why is $\rho = k$ forbidden?).

[Answer: By Green's identity:

$$\alpha\rho(1 + \cos \pi\rho) + \beta \sin \pi\rho = 0 \qquad (5.15.22)]$$

5.16. The Green's function in the realm of ordinary differential equations

While it is simple enough to formulate the differential equation by which the Green's function can be defined, it is by no means so simple to find the explicit *solution* of the defining equation. In the realm of partial differential equations we have only a few examples which allow an explicit construction of the Green's function. In the realm of ordinary differential equations, however, we are in a more fortunate position and we can find numerous examples in which we succeed with the actual construction of the Green's function. This is particularly so if the given differential equation has constant coefficients.

Let us then assume that our x belongs to a one-dimensional manifold which extends from a to b. Then only ordinary derivatives occur. Moreover, the homogeneous equation—with zero on the right side—now possesses only a finite number of solutions, depending on the order of the differential equation. An nth order equation allows n constants of integration and has thus n independent solutions. The same is true of a first order system of n equations. These homogeneous solutions play an important part in the construction of the Green's function, in fact, if these homogeneous solutions are known, the construction of the Green's function is reducible to the solution of a simultaneous set of ordinary algebraic equations.

Our aim is the solution of the differential equation

$$\tilde{D}G(x, \xi) = \delta(x, \xi) \tag{5.16.1}$$

or else the solution of the differential equation

$$DG(x, \xi) = \delta(x, \xi) \tag{5.16.2}$$

provided that the homogeneous equations $Dv(x) = 0$ and $\tilde{D}u(x) = 0$ under the given homogeneous boundary conditions have no non-vanishing solutions. This we want to assume for the time being.

We will now make use of the fact that the delta function on the right side of (1) and (2) is everywhere zero, except in the infinitesimal neighbourhood of the point $x = \xi$. This has the following consequence. Let us focus our attention on the equation (1), considering ξ as the active variable. Then we can divide the range of ξ in two separate parts: $a \leq \xi \leq x$, and $x \leq \xi \leq b$. In both realms the homogeneous equation

$$\tilde{D}u(\xi) = 0 \tag{5.16.3}$$

is valid and thus, if we forget about boundary conditions, we can obtain the solution by the superposition of n particular solutions, each one multiplied by an undetermined constant c_i. This gives apparently n degrees of freedom. In actual fact, however, there is a *dividing line* between the two realms, at the point $\xi = x$ which is a common boundary between the two regions. In consequence of the delta function which exists at the point $\xi = x$, we cannot assume that the same analytical solution will exist on both sides. We have to assume *one* homogeneous solution to the left of the point $\xi = x$, and *another* homogeneous solution to the right of the point $\xi = x$. This means $2n$ free constants of integration, obtaining $u(x)$

as a superposition of n particular solutions to the left, and n particular solutions to the right:

$$u(\xi) = A_1 u^1(\xi) + A_2 u^2(\xi) + \ldots + A_n u^n(\xi) \qquad (a \leq \xi \leq x)$$
$$u(\xi) = B_1 u^1(\xi) + B_2 u^2(\xi) + \ldots + B_n u^n(\xi) \qquad (b \geq \xi \geq x) \qquad (5.16.4)$$

The n prescribed homogeneous boundary conditions will yield n simultaneous homogeneous algebraic equations for the $2n$ constants A_i and B_i. This leaves us with n remaining degrees of freedom which have to be obtained by joining the two solutions (4) at the common boundary $\xi = x$. Let us investigate the nature of this joining.

As we have done before, we will replace the delta function $\delta(x, \xi)$ by the finite pulse $\delta_\epsilon(x, \xi)$, that is a pulse of the width ϵ and the height $1/\epsilon$. We will now solve the differential equation

$$y'(\xi) = \delta(x, \xi) \qquad (5.16.5)$$

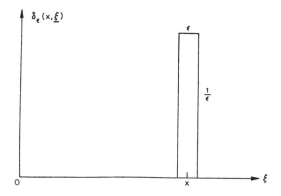

Since the delta function is zero everywhere to the left of x, up to the point $\xi = x - (\epsilon/2)$, $y(\xi)$ must be a *constant* C in this region. But the delta function is zero also to the right of x, beyond the point $\xi = x + (\epsilon/2)$. Hence $y(\xi)$ must again be a constant in this region. But will it be the *same* constant C we had on the left side? No, because the presence of the pulse in the region between $\xi = x - (\epsilon/2)$ and $x + (\epsilon/2)$ changes the course of the function $y(\xi)$. In this region we get

$$y(\xi) = C + \frac{1}{\epsilon}\left(\xi - x + \frac{\epsilon}{2}\right) \qquad (5.16.6)$$

and thus we arrive at the point $\xi = x + (\epsilon/2)$ with the value $C + 1$. If now we let ϵ go to zero, the rate of increase between the points $x \mp (\epsilon/2)$ becomes steeper and steeper, without changing the final value $C + 1$. In the limit $\epsilon = 0$ the solution $y(\xi) = C$ extends up to the point x coming from

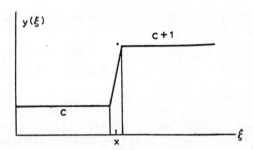

the left, and $y(\xi) = C + 1$ coming from the right, with a *point of discontinuity at the point* $\xi = x$. *The magnitude of the jump is* 1.

Now let us change our differential equation to

$$y'(\xi) + \alpha(\xi)y(\xi) = \delta(x, \xi) \tag{5.16.7}$$

Then $y(\xi)$ is no longer a constant on the two sides of the point x but a function of ξ, obtainable by solving the homogeneous equation

$$y'(\xi) + \alpha(\xi)y(\xi) = 0 \tag{5.16.8}$$

But in the narrow region between $\xi = x \mp (\epsilon/2)$ we can write our equation in the form

$$y'(\xi) = \delta(x, \xi) - \alpha(\xi)y(\xi) \tag{5.16.9}$$

and we find that now the increase of $y(\xi)$ in this narrow region is not exactly 1, in view of the presence of the second term, but the additional term is proportional to ϵ and goes to zero with ϵ going to zero. Hence the previous result that $y(\xi)$ *suffers a jump of the magnitude* 1 *at the point* $\xi = x$ remains once more true. · Nor is any change encountered if we add on the left sides terms which are still smoother than $y(\xi)$, namely $y^{(-1)}(\xi)$, $y^{(-2)}(\xi)$, where these functions are the first, second..., *integrals* of $y(\xi)$. As in (9), all these terms can be transferred to the right side and their contribution to the jump of $y(\xi)$ at $\xi = x$ is of the order ϵ^2, ϵ^3, ..., with the limit zero as ϵ goes to zero.

Now we can go one step further still and assume that the coefficient of $y'(\xi)$ in the equation (7) is not 1 but a certain continuous function $p(\xi)$ which we will assume to remain of the *same sign* throughout the range $[a, b]$. In the neighbourhood of the point $\xi = x$ this function can be replaced by the constant value $p(x)$. By dividing the equation by this constant we change the height of the delta function by the factor $p^{-1}(x)$. Accordingly the jump of $y(\xi)$ at the point $\xi = x$ will no longer be 1 but $1/p(x)$.

Let us now consider an arbitrary ordinary linear differential operator $Du(\xi)$ of nth order. This equation is exactly of the type considered before, if we identify $u^{(n)}(\xi)$ with $y'(\xi)$, that is $y(\xi)$ with $u^{(n-1)}(\xi)$. Translating our previous result to the new situation we arrive at the following result: *The presence of the delta function at the point* $\xi = x$ *has the consequence that* $u(\xi)$,

$u'(\xi)$, $u''(\xi)$, ..., *up to* $u^{(n-2)}(\xi)$ *remain continuous at the point* $\xi = x$, *while* $u^{(n-1)}(\xi)$ *suffers a jump of the magnitude* $1/p(x)$ *if* $p(\xi)$ *is the coefficient of the highest derivative* $u^{(n)}(\xi)$ *of the given differential operator.*

In view of this result we can dispense with the delta function in the case of ordinary differential equations, and characterise the Green's function $G(x, \xi)$ by the homogeneous differential equation

$$\check{D}G(x, \xi) = 0 \qquad (5.16.10)$$

to the left and to the right from the point $\xi = x$ if we complement this equation by the aforementioned continuity and discontinuity conditions.

We can now return to our previous only partially solved problem of obtaining the $2n$ undetermined constants A_i, B_i, of the system (4). The given boundary conditions yield n homogeneous algebraic equations. Now we add n further conditions. The condition that $u(x)$, $u'(x)$, ..., $u^{(n-2)}(x)$ must remain continuous, whether we come from the left or from the right, yields $n - 1$ homogeneous algebraic equations between the A_i and the B_i. The last condition is that $u^{(n-1)}(x)$ is *not* continuous but makes a jump of the magnitude

$$u^{(n-1)}(x_+) - u^{(n-1)}(x_-) = \frac{1}{p(x)} \qquad (5.16.11)$$

$p(\xi)$ being the coefficient of $u^{(n)}(\xi)$ of the adjoint operator $\check{D}u(\xi)$. This last condition is the only *non-homogeneous* equation of our algebraic system of $2n$ unknowns A_i, B_i.

As an example let us construct the Green's function of the differential equation of the vibrating spring

$$v''(\xi) + \rho^2 v(\xi) = \beta(\xi) \qquad \xi = [0, l] \qquad (5.16.12)$$

with the boundary conditions

$$\begin{aligned} v(0) &= v(l) \\ v'(0) &= v'(l) \end{aligned} \qquad (5.16.13)$$

This problem is *self-adjoint*, that is $u(\xi)$ satisfies the same differential equation and the same boundary conditions as $v(\xi)$. The homogeneous equation has the two solutions

$$\sin \rho\xi, \quad \cos \rho\xi \qquad (5.16.14)$$

and thus we set up the system

$$\begin{aligned} u(\xi) &= A_1 \sin \rho\xi + A_2 \cos \rho\xi \qquad \text{(for } 0 \le \xi \le x\text{)} \\ u(\xi) &= B_1 \sin \rho\xi + B_2 \cos \rho\xi \qquad \text{(for } x \le \xi \le l\text{)} \end{aligned} \qquad (5.16.15)$$

The boundary conditions yield the two relations

$$\begin{aligned} A_2 &= B_1 \sin \rho l + B_2 \cos \rho l \\ A_1 &= B_1 \cos \rho l - B_2 \sin \rho l \end{aligned} \qquad (5.16.16)$$

The continuity of $u(\xi)$ at $\xi = x$ yields the further condition

$$(B_1 - A_1) \sin \rho x + (B_1 - A_2) \cos \rho x = 0 \tag{5.16.17}$$

and finally the jump-condition of $u'(\xi)$ at $\xi = x$ demands the relation

$$\rho(B_1 - A_1) \cos \rho x - \rho(B_2 - A_2) \sin \rho x = 1 \tag{5.16.18}$$

These four equations determine the four constants A_1, A_2, B_1, B_2 of our solution as follows:

$$A_1 = \frac{\sin \rho\left(x - \dfrac{l}{2}\right)}{2\rho \sin \rho \dfrac{l}{2}} \qquad A_2 = \frac{\cos \rho\left(x - \dfrac{l}{2}\right)}{2\rho \sin \rho \dfrac{l}{2}}$$

$$\tag{5.16.19}$$

$$B_1 = \frac{\sin \rho\left(x + \dfrac{l}{2}\right)}{2\rho \sin \rho \dfrac{l}{2}} \qquad B_2 = \frac{\cos \rho\left(x - \dfrac{l}{2}\right)}{2\rho \sin \rho \dfrac{l}{2}}$$

and thus

$$G(x, \xi) = \frac{\cos \rho\left(x - \xi - \dfrac{l}{2}\right)}{2\rho \sin \rho \dfrac{l}{2}} \qquad (0 \le \xi \le x)$$

$$\tag{5.16.20}$$

$$= \frac{\cos \rho\left(x - \xi + \dfrac{l}{2}\right)}{2\rho \sin \rho \dfrac{l}{2}} \qquad (l \ge \xi \ge x)$$

We can now test the symmetry condition of the Green's function:

$$G(x, \xi) = G(\xi, x)$$

which must hold in view of the self-adjoint character of our problem. This does not mean that the expressions (20) must remain unchanged for an exchange of x and ξ. An exchange of x and ξ causes the point ξ to come to the *right* of x if it was originally to the *left* and vice versa. What is demanded then is that an exchange of x and ξ changes the left Green's function to the right Green's function, and vice versa:

$$G_l(x, \xi) = G_r(\xi, x) \tag{5.16.22}$$

This, however, is actually the case in our problem since the second expression of (20) may also be written in the form

$$\frac{\cos \rho\left(\xi - x - \dfrac{l}{2}\right)}{2\rho \sin \rho \dfrac{l}{2}} \tag{5.16.23}$$

We also observe that the Green's function ceases to exist if the constant ρ assumes the values

$$\rho = \frac{2k\pi}{l} \tag{5.16.24}$$

because then the denominator becomes zero. But then we have violated the general condition always required for the existence of the Green's function, namely that the homogeneous equation (under the given boundary conditions) must have no solutions which do not vanish identically. If the condition (24) is satisfied, then the homogeneous solutions (14) satisfy the boundary conditions (13) and the homogeneous problem has now non-zero solutions. The modification of our treatment to problems of this kind will be studied in a later section (see Section 22).

Problem 202.   Find the Green's function for the problem

$$v''(\xi) = \beta(\xi) \qquad \xi = [a, b]$$
$$v(a) = 0, \qquad v'(b) = 0 \tag{5.16.25}$$

[Answer:

$$G(x, \xi) = a - \xi \qquad (a \le \xi \le x)$$
$$= a - x \qquad (x \le \xi \le b) \,] \tag{5.16.26}$$

Problem 203.   Do the same for (25) with the boundary conditions

$$v(a) = 0, \qquad v(b) = 0$$

[Answer:

$$G(x, \xi) = \frac{(x - b)(\xi - a)}{b - a} \qquad (\xi \le x)$$
$$= \frac{(x - a)(\xi - b)}{b - a} \qquad (\xi \ge x) \,] \tag{5.16.27}$$

Problem 204.   Find the Green's function for the problem of the vibrating spring excited by an external force:

$$v''(\xi) + \rho^2 v(\xi) = \beta(\xi)$$
$$v(0) = 0, \qquad v'(0) = 0 \tag{5.16.28}$$

Obtain the result by considering once ξ and once x as the active variable.
[Answer:

$$G(x, \xi) = \frac{1}{\rho} \sin \rho(x - \xi) \qquad (\xi \le x)$$
$$= 0 \qquad (\xi \ge x) \,] \tag{5.16.29}$$

Problem 205.   Find the Green's function for the motion of the "ballistic galvanometer"

$$v'(\xi) + \gamma v(\xi) = \beta(\xi)$$
$$v(0) = 0 \tag{5.16.30}$$

[Answer:

$$G(x, \xi) = e^{\gamma(\xi-x)} \qquad (\xi < x)$$
$$= 0 \qquad (\xi > x)]$$

(5.16.31)

Problem 206. If the constant γ in the previous problem becomes very large, the first term of the differential equation becomes practically negligible and we obtain

$$v(x) = \frac{1}{\gamma} \beta(x)$$

Demonstrate this result by solving (30) with the help of the Green's function.

Problem 207. Solve the differential equation of the vibrating spring in a resisting medium:

$$v''(\xi) + 2\alpha v'(\xi) + \rho^2 v(\xi) = \beta(\xi)$$
$$v(0) = v'(0) = 0$$

(5.16.32)

with the help of the Green's function and discuss particularly the case of "critical damping" $\rho = \alpha$.

[Answer:

$$G(x, \xi) = e^{\alpha(\xi-x)} \frac{\sin \sqrt{\rho^2 - \alpha^2}(x - \xi)}{\sqrt{\rho^2 - \alpha^2}} \qquad (\xi \leq x)$$
$$= 0 \qquad (\xi \geq x)]$$

(5.16.33)

Problem 208. Solve with the help of the Green's function the differential equation

$$v'(\xi) + \alpha(\xi)v(\xi) = \beta(\xi)$$
$$v(0) = 0$$

(5.16.34)

and compare the solution with that obtained by the "variation of the constants" method.

[Answer:

$$G(x, \xi) = e^{-\int_\xi^x \alpha(t)dt} \qquad (\xi < x)$$
$$= 0 \qquad (\xi > x)]$$

(5.16.35)

Problem 209. The differential equation of the loaded elastic bar of uniform cross-section [cf. Section 4.14, with $I(x) = 1$] is given by

$$v''''(\xi) = \beta(\xi)$$

(5.16.36)

Obtain the Green's function under the boundary conditions

$$v(0) = v'(0) = 0, \quad v''(l) = v'''(l) = 0$$

(5.16.37)

(bar free at $\xi = l$, clamped at $\xi = 0$).

[Answer:

$$G(x, \xi) = \frac{\xi^2}{2}\left(x - \frac{\xi}{3}\right) \qquad (\xi \leq x)$$
$$= \frac{x^2}{2}\left(\xi - \frac{x}{3}\right) \qquad (\xi \geq x)]$$

(5.16.38)

5.17. The change of boundary conditions

Let us assume that we have obtained the Green's function of an ordinary differential equation problem under certain boundary conditions, and we are now interested in the Green's function of the same differential operator, but under some other boundary conditions. Then it is unnecessary to repeat the entire calculation. In both cases we have to solve the same differential equation (16.1) and consequently the difference of the two Green's functions satisfies the homogeneous equation

$$\tilde{D}u(\xi) = 0 \tag{5.17.1}$$

We thus see that we can come from the one Green's function to the other by adding some solution of the homogeneous equation. The coefficients of this solution have to be adjusted in such a way that the new boundary conditions shall become satisfied.

For example the Green's function of the clamped-free uniform bar came out in the form (16.38). Let us now obtain the Green's function of the clamped-clamped bar:

$$v(0) = v'(0) = v(l) = v'(l) = 0 \tag{5.17.2}$$

For this purpose we will add to the previous $G(x, \xi)$ an arbitrary solution of the homogeneous equation

$$u''''(\xi) = 0 \tag{5.17.3}$$

that is

$$u(\xi) = \alpha + \beta\xi + \gamma\xi^2 + \delta\xi^3 \tag{5.17.4}$$

with the undetermined constants α, β, γ, δ. These four constants have to be determined by the condition that at $\xi = 0$ and $\xi = l$ the given boundary conditions (2) shall be satisfied (our problem is self-adjoint and thus the boundary conditions for $u(\xi)$ are the same as those for $v(\xi)$).

Now at $\xi = 0$ we are to the left of x and thus the *upper* of the two expressions (16.38) has to be chosen:

$$u(\xi) = \frac{\xi^2}{2}\left(x - \frac{\xi}{3}\right) + \alpha + \beta\xi + \gamma\xi^2 + \delta\xi^3 \tag{5.17.5}$$

At this point the previous boundary conditions remain unchanged (namely $u(0) = u'(0) = 0$), and thus α and β have to be chosen as zero. We now come to the point $\xi = l$. Here the *lower* of the two expressions (16.38) comes into operation since now ξ is to the *right* of x:

$$u(\xi) = \frac{x^2}{2}\left(\xi - \frac{x}{3}\right) + \gamma\xi^2 + \delta\xi^3 \tag{5.17.6}$$

The new boundary conditions $u(l) = u'(l) = 0$ determine γ and δ, and we obtain the additional term in the form

$$\frac{x^2\xi^2}{2l^2}\left(x + \xi - 2l - \frac{2x\xi}{3l}\right) \tag{5.17.7}$$

This is a *symmetric function* of x and ξ, as it is indeed demanded by the symmetry of the Green's function $G(x, \xi) = G(\xi, x)$ since the added term is analytic and belongs equally to the left and to the right expression of the Green's function. The resulting Green's function of the elastic uniform bar, clamped at both ends, thus becomes:

$$G(x, \xi) = \frac{\xi^2}{2}\left(x - \frac{\xi}{3}\right) + \frac{x^2\xi^2}{2l^2}\left(x + \xi - 2l - \frac{2x\xi}{3l}\right) \qquad (\xi \leq x)$$

$$= \frac{x^2}{2}\left(\xi - \frac{x}{3}\right) + \frac{x^2\xi^2}{2l^2}\left(x + \xi - 2l - \frac{2x\xi}{3l}\right) \qquad (\xi \geq x)$$

(5.17.8)

Problem 210. Obtain by this method the expression (16.27) from (16.26).

Problem 211. The boundary conditions of a bar simply supported on both ends are

$$v(0) = v''(0) = v(l) = v''(l) = 0 \qquad (5.17.9)$$

Obtain the Green's function of the simply supported bar from (16.38).

[Answer: Added term:

$$x\xi\left(\frac{l}{3} - \frac{x + \xi}{2} + \frac{x^2 + \xi^2}{6l}\right) \qquad (5.17.10)]$$

Problem 212. Obtain from the Green's function (16.29) the Green's function of the same problem but modifying the boundary conditions to

$$v(l) = v'(l) = 0 \qquad (5.17.11)$$

Explain why now the added term is not symmetric in x and ξ.

[Answer: Added term:

$$\frac{1}{\rho}\sin \rho(\xi - x) \qquad (5.17.12)]$$

5.18. The remainder of the Taylor series

Let us expand a function $f(x)$ around the origin $x = 0$ in a power series:

$$f^*(x) = f(0) + f'(0)x + f''(0)\frac{x^2}{2!} + \ldots + f^{(n-1)}(0)\frac{x^{n-1}}{(n-1)!} \qquad (5.18.1)$$

By terminating the series after n terms we cannot expect that $f^*(x)$ shall coincide with $f(x)$. But we can introduce the difference between $f(x)$ and $f^*(x)$ as the "remainder" of the series. Let us call it $v(x)$:

$$v(x) = f(x) - f^*(x) \qquad (5.18.2)$$

This function $v(x)$ can be uniquely characterised by a certain *differential equation*, with the proper boundary conditions. Since $f^*(x)$ is a polynomial of the order $n - 1$, the nth derivative of this function vanishes and thus we obtain

$$v^{(n)}(x) = f^{(n)}(x) \qquad (5.18.3)$$

The unique characterisation of $v(x)$ requires n boundary conditions but these conditions are available by comparing the derivatives of the polynomial (1) with the derivatives of $f(x)$ at the point $x = 0$. By construction the functional value and the values of the first $n - 1$ derivatives coincide at the point $x = 0$ and this gives for $v(x)$ the n boundary conditions

$$v(0) = v'(0) = v''(0) = \ldots v^{(n-1)}(0) = 0 \qquad (5.18.4)$$

We will now find the Green's function of our differential equation (3) and accordingly obtain the solution in the range $[0, x]$ by the integral

$$v(x) = \int_0^x G(x, \xi) f^{(n)}(\xi) d\xi \qquad (5.18.5)$$

Two methods are at our disposal: we can consider $G(x, \xi)$ as a function of ξ and operate with the adjoint equation, or we can consider $G(x, \xi)$ as a function of x and operate with the given equation. We will follow the latter course and solve the equation

$$[G(x, \xi)]^{(n)} = \delta(x, \xi) \qquad (5.18.6)$$

with the boundary conditions (4).

Now between the points $x = 0$ and $x = \xi$ we have to solve the homogeneous equation

$$v^{(n)}(x) = 0 \qquad (5.18.7)$$

This means that $v(x)$ can be an arbitrary polynomial of the order $n - 1$:

$$v(x) = a_0 + a_1 x + a_2 x^2 + \ldots + a_{n-1} x^{n-1} \qquad (5.18.8)$$

But then the boundary conditions (4) make every one of these coefficients equal to zero and thus

$$v(x) = 0 \qquad (0 \leq x \leq \xi) \qquad (5.18.9)$$

Now we come to the range $x \geq \xi$. Here again the homogeneous equation has to be solved and we can write our polynomial in the form

$$v(x) = b_0 + b_1(x - \xi) + b_2(x - \xi)^2 + \ldots + b_{n-1}(x - \xi)^{n-1} \quad (5.18.10)$$

The conditions of continuity demand that $v(\xi), v'(\xi), \ldots, v^{(n-2)}(\xi)$ shall be zero since the function on the left side vanishes identically. This makes $b_0 = b_1 = b_2 = \ldots b_{n-2} = 0$, and what remains is

$$v(x) = b_{n-1}(x - \xi)^{n-1} \qquad (5.18.11)$$

According to the general properties of the Green's function, discussed in Section 16, the $n - 1$st derivative must make the jump 1 at the point $x = \xi$ (since in our problem $p(x) = 1$). And thus we get

$$(n - 1)! b_{n-1} = 1 \qquad (5.18.12)$$

or

$$b_{n-1} = \frac{1}{(n - 1)!} \qquad (5.18.13)$$

which gives

$$v(x) = \frac{(x - \xi)^{n-1}}{(n - 1)!} \qquad (x \geq \xi) \qquad (5.18.14)$$

We thus have obtained the Green's function of our problem:

$$G(x, \xi) = \frac{(x - \xi)^{n-1}}{(n - 1)!} \qquad (\xi \leq x)$$

$$= 0 \qquad (\xi \geq x) \qquad (5.18.15)$$

Returning to our solution (5) we get

$$v(x) = \frac{1}{(n - 1)!} \int_0^x (x - \xi)^{n-1} f^{(n)}(\xi) d\xi \qquad (5.18.16)$$

and this is *the Lagrangian remainder of the truncated Taylor series* which, as we have seen in Chapter 1.3, can be used for an estimation of the error of the truncated Taylor series.

Problem 213. Carry through the same treatment on the basis of the adjoint equation, operating with $G(x, \xi)$.

5.19. The remainder of the Lagrangian interpolation formula

An interesting application of the Green's function method occurs in connection with the Lagrangian interpolation of a given function $f(x)$ by a polynomial of the order $n - 1$, the points of interpolation being given at the arbitrary points (arranged in increasing order),

$$x = x_1, x_2, \ldots, x_n \qquad (5.19.1)$$

of the interval $[a, b]$.

This is a problem we have discussed in the first chapter. We obtained the interpolating polynomial in the form

$$f^*(x) = \sum_{\alpha=1}^n \varphi_\alpha(x) f(x_\alpha) \qquad (5.19.2)$$

where $\varphi_k(x)$ are the Lagrangian "interpolation coefficients"

$$\varphi_\alpha(x) = \frac{F(x)}{F'(x_\alpha)(x - x_\alpha)} \qquad (5.19.3)$$

$F(x)$ being the fundamental root polynomial $(x - x_1)(x - x_2) \ldots (x - x_n)$.

Our task will now be to obtain an estimation of the error of the approximation. For this purpose we deduce a differential equation for the remainder, in full analogy to the procedure of the previous section. If we form the difference

$$v(x) = f(x) - f^*(x) \qquad (5.19.4)$$

we first of all notice that at the points of interpolation (1) $f(x)$ and $f^*(x)$ coincide and thus

$$v(x_i) = 0 \qquad (i = 1, 2, \ldots, n) \tag{5.19.5}$$

These conditions take the place of the previous boundary conditions (18.4). Furthermore, if we differentiate (4) n times and consider that the nth derivative of $f^*(x)$ (being a polynomial of the order $n - 1$) vanishes, we once more obtain the differential equation

$$v^{(n)}(x) = f^{(n)}(x) \tag{5.19.6}$$

We thus have to solve the differential equation (6) with the inside conditions

$$v(x_1) = v(x_2) = \ldots = v(x_n) = 0 \tag{5.19.7}$$

The unusual feature of our problem is that *no boundary conditions are prescribed*. Instead of them the value of the function is prescribed as zero in n *inside points* of the given interval $[a, b]$. These conditions determine our problem just as effectively as if n boundary conditions were given. The homogeneous equation

$$v^{(n)}(x) = 0 \tag{5.19.8}$$

allows as a solution an arbitrary polynomial of the order $n - 1$. But such a polynomial cannot be zero in n points without vanishing identically since a polynomial which is not zero everywhere, cannot have more than $n - 1$ roots.

We now come to the construction of the Green's function $G(x, \xi)$. We will do that by considering $G(x, \xi)$ as a function of ξ. Hence we will construct the adjoint problem. For this purpose we form the extended Green's identity:

$$\int_a^b [uv^{(n)} - (-1)^n vu^{(n)}]d\xi$$
$$= \left| uv^{(n-1)} - u'v^{(n-2)} + \ldots + (-1)^{n-1}u^{(n-1)}v \right|_a^b \tag{5.19.9}$$

Since there are no boundary conditions given for $v(x)$ at $x = a$ and $x = b$, the vanishing of the boundary term demands the conditions

$$u(a) = u'(a) = \ldots = u^{(n-1)}(a) = 0$$
$$u(b) = u'(b) = \ldots = u^{(n-1)}(b) = 0 \tag{5.19.10}$$

These are altogether $2n$ boundary conditions and thus our problem is apparently strongly over-determined. However, we have not yet taken into consideration the n inside conditions (7). The fact that $v(x)$ is given in n points, has the consequence that the adjoint differential equation

$$(-1)^n u^{(n)}(\xi) = 0 \tag{5.19.11}$$

is put *out of action* at these points, just as we have earlier conceived the appearance of the delta function at the point $\xi = x$ as a consequence of the fact that we have added $v(x)$ to the data of our problem.

The full adjoint problem can thus be described as follows. The delta function $\delta(x, \xi)$ appears not only at the point $\xi = x$ but also at the points $\xi = x_1, x_2, \ldots, x_n$. The strength with which the delta function appears at these points has to be left free. The completed adjoint differential equation of our problem will thus become

$$(-1)^n u^{(n)}(\xi) = \alpha_1 \delta(x_1, \xi) + \alpha_2 \delta(x_2, \xi) + \ldots + \alpha_n \delta(x_n, \xi) + \delta(x, \xi) \quad (5.19.12)$$

We have gained the n new constants $\alpha_1, \alpha_2, \ldots, \alpha_n$ which remove the over-determination since we now have these constants at our disposal, together with the n constants of integration associated with a differential equation of the nth order. They suffice to take care of the $2n$ boundary conditions (10).

First we will satisfy the n boundary conditions

$$u(b) = u'(b) = \ldots = u^{(n-1)}(b) = 0 \quad (5.19.13)$$

together with the differential equation

$$(-1)^n u^{(n)}(\xi) = \delta(x, \xi) \quad (5.19.14)$$

This is the problem of the Green's function $G(x, \xi)$, solved in the previous section as a function of x, but now considered as a function of ξ. We can take over the previous solution:

$$u(\xi) = \frac{1}{(n-1)!} (x - \xi)^{n-1} \quad (\xi \leq x) \quad (5.19.15)$$
$$= 0 \quad (\xi \geq x)$$

We will combine these two solutions into one single analytical expression by using the symbol $[t^{n-1}]$ for a function which is equal to t^{n-1} for all *positive* values of t but zero for all *negative* values of t:

$$[t^{n-1}] = t^{n-1} \quad (t \geq 0)$$
$$= 0 \quad (t \leq 0) \quad (5.19.16)$$

With this convention, and making use of the superposition principle of linear operators, the solution of the differential equation (12) under the boundary conditions (13) becomes:

$$u(\xi) = \frac{1}{(n-1)!} \left\{ \sum_{k=1}^{m} \alpha_k [(x_k - \xi)^{n-1}] + [(x - \xi)^{n-1}] \right\} \quad (5.19.17)$$

Now we go to the point $\xi = a$ and satisfy the remaining boundary conditions

$$u(a) = u'(a) = \ldots = u^{(n-1)}(a) \quad (5.19.18)$$

This has the consequence that *the entire polynomial*

$$Q_{n-1}(\xi) = \sum_{k=1}^{n} \alpha_k (x_k - \xi)^{n-1} + (x - \xi)^{n-1} \quad (5.19.19)$$

must vanish identically. We thus see that the Green's function $G(x, \xi)$ will not extend from a to b but only from x_1 to x_n if the point x is inside the interval $[x_1, x_n]$, or from x to x_n if x is to the left of x_1, and from x_1 to x if x is to the right of x_n.

The vanishing of $Q_{n-1}(\xi)$ demands the fulfilment of the following n equations:

$$\alpha_1 + \alpha_2 + \ldots + \alpha_n + 1 = 0$$
$$\alpha_1 x_1 + \alpha_2 x_2 + \ldots + \alpha_n x_n + x = 0 \qquad (5.19.20)$$
$$\vdots$$
$$\alpha_1 x_1^{n-1} + \alpha_2 x_2^{n-1} + \ldots + \alpha_n x_n^{n-1} + x^{n-1} = 0$$

These equations are solvable by the method of undetermined coefficients. We multiply the equations in succession by the undetermined factors $c_0, c_1, c_2, \ldots, c_{n-1}$, and form the sum. Then the last term becomes some polynomial $P_{n-1}(x)$ which can be chosen at will while the factor of α_k becomes the same polynomial, taken at the point $x = x_k$:

$$\sum_{k=1}^{n} \alpha_k P_{n-1}(x_k) + P_{n-1}(x) = 0 \qquad (5.19.21)$$

We will now choose for $P_{n-1}(x)$ the Lagrangian polynomials $\varphi_k(x)$, defined by (3). Then the factor of α_k becomes 1, the factor of all the other α_j zero, and we obtain

$$\alpha_k = -\varphi_k(x) \qquad (5.19.22)$$

Hence we see that the strength with which the delta functions are represented at the points of interpolation x_k, depends on the position of the variable point x. Moreover, this strength is given by exactly the same interpolation coefficients, taken with a negative sign, which appear in the Lagrangian interpolation formula.

We have now constructed our Green's function in explicit form:

$$G(x, \xi) = \frac{1}{(n-1)!} \left\{ [(x - \xi)^{n-1}] - \sum_{k=1}^{n} \varphi_k(x)[(x_k - \xi)^{n-1}] \right\} \qquad (5.19.23)$$

and obtain the remainder $v(x)$ of the Lagrangian interpolation problems in the form of a definite integral

$$v(x) = \int_{x_1}^{x_n} G(x, \xi) f^{(n)}(\xi) d\xi \qquad (5.19.24)$$

(We have assumed that x is an inside point of the interval $[x_1, x_n]$; if x is outside of this interval and to the left of x_1, the lower limit of integration becomes x instead of x_1; if x is outside of $[x_1, x_n]$ and to the right of x_n, the upper limit of integration becomes x instead of x_n.)

It will hardly be possible to actually *evaluate* this integral. We can use it, however, for an *estimation* of the error $\eta(x)$ of the Lagrangian interpolation (this $\eta(x)$ is now our $v(x)$). In particular, we have seen in the first chapter that we can deduce the Lagrangian error formula (1.5.10) if we can show that the function $G(x, \xi)$, taken as a function of ξ, *does not change its sign* throughout the interval of its existence which in our case will be between x_1 and x_n, although the cases x to x_n or x_1 to x can be handled quite analogously.

First of all we know that $G(x, \xi) = u(\xi)$ vanishes at the limiting points $\xi = x_1$ and $\xi = x_n$ with all its derivatives up to the order $n - 2$; we cannot go beyond $n - 2$ because the $(n - 1)$st derivative makes a jump at x_1 and x_n, due to the presence of $\delta(x_1, \xi)$ and $\delta(x_n, \xi)$ in the nth derivative (cf. (12)), and thus starts and ends with a finite value. Now $u(\xi)$, being a continuous function of ξ and starting and ending with the value zero, must have at least *one* maximum or minimum in the given interval $[x_1, x_n]$. But if $u(\xi)$ were to change its sign and thus pass through zero, the number of extremum values would be at least *two*. Accordingly the derivative $u'(\xi)$ must vanish at least once and if we can show that it vanishes in fact *only* once, we have established the non-vanishing of $u(\xi)$ inside the critical interval. Continuing this reasoning we can say that $u^{(k)}(\xi)$ must vanish inside the critical interval *at least* k times and if we can show that the number of zeros is indeed *exactly* k and not more, the non-vanishing of $u(\xi)$ is once more established.

Now let us proceed up to the $(n - 2)$nd derivative and investigate its behaviour. Since the nth derivative of $u(\xi)$ is composed of delta functions, the $(n - 1)$st derivative is composed of *step functions*. Hence it is the $(n - 2)$nd derivative where we first encounter a *continuous* function composed of straight zig-zag lines, drawn between the $n + 1$ points $x_1, x_2, \ldots, x,$ $\ldots, x_n$. The number of intervals is n and since no crossings occur in the

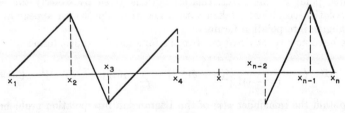

first and in the last interval, the number of zeros cannot exceed $n - 2$. Hence $u^{(n-2)}(\xi)$ cannot vanish *more* than $n - 2$ times while our previous reasoning has shown that it cannot vanish *less* than $n - 2$ times. This establishes the number of zeros as *exactly* $n - 2$ which again has the consequence that an arbitrary kth derivative has exactly k zeros within the given interval while $G(x, \xi)$ itself does not change its sign as ξ varies between x_1 and x_n. The theorem on which the estimation of Lagrange was based is thus established. Furthermore, the formula (23) puts us in the position to construct $G(x, \xi)$ explicitly, on the basis of Lagrange's interpolation formula. The Lagrangian interpolation coefficients $\varphi_k(x)$ ordinarily multiplied by

$f(x_k)$ are now multiplied by the functional values of the special function $[(x - \xi)^{n-1}]/(n - 1)!$ taken at the points $x = x_k$ (considering ξ as a mere parameter).

We see that the Green's function of Lagrangian interpolation can be conceived as *the remainder of the Lagrangian interpolation of the special function* $[(x - \xi)^{n-1}]/(n - 1)!$.

Problem 214.   Show that if $x > x_n$ and ξ is a point between x_n and x:

$$G(x, \xi) = \frac{1}{(n - 1)!} (x - \xi)^{n-1}$$

If $x < x_1$ and ξ is a point between x and x_1:

$$G(x, \xi) = - \frac{1}{(n - 1)!} (x - \xi)^{n-1}$$

Problem 215.   Show from the definition (23) that $G(x, \xi)$ vanishes at all points $x = x_k$.

Problem 216.   Show from the definition (23) that $G(x, \xi)$ vanishes at all values of ξ which are outside the realm of the $n + 1$ points $[x_1, x_2, \ldots, x_n, x]$.

5.20. Lagrangian interpolation with double points

The fundamental polynomial of Lagrangian interpolation:

$$F(x) = (x - x_1)(x - x_2) \ldots (x - x_n) \tag{5.20.1}$$

contains all the root factors once and only once. But in a similar way as in algebra where a root may become a multiple root through the collapsing of several single roots, something similar may happen in the process of interpolation. Let us assume that the functional value $f(x_k)$ is prescribed in two points $x = x_k \pm \epsilon$ which are very close together, due to the smallness of ϵ. Then we can put

$$f(x_k \pm \epsilon) = f(x_k) \pm \epsilon f'(x_k) \tag{5.20.2}$$

Moreover, the fundamental polynomial may be written as follows, by separating the two root factors which belong to the critical points:

$$F(x) = \Phi(x)((x - x_k)^2 - \epsilon^2) \tag{5.20.3}$$

($\Phi(x)$ is composed of all the other root factors.)   The two terms associated with the two critical points become

$$\frac{\Phi(x)(x - x_k + \epsilon)}{\Phi(x_k + \epsilon)2\epsilon} [f(x) + \epsilon f'(x)] + \frac{\Phi(x)(x - x_k - \epsilon)}{\Phi(x_k - \epsilon)(-2\epsilon)} [f(x) - \epsilon f'(x)]$$

$$= \frac{\Phi(x)}{\Phi(x_k)} \frac{x - x_k}{\pm 2\epsilon} f(x_k) \left\{ 1 \pm \epsilon \left[\frac{f'(x_k)}{f(x_k)} + \frac{1}{x - x_k} - \frac{\Phi'(x_k)}{\Phi(x_k)} \right] \right\} \tag{5.20.4}$$

The sign $\pm$ shall mean that we should take the sum of two expressions, the one with the upper, the other with lower sign. The factor of $f(x_k)$ becomes

$$\frac{\Phi(x)}{\Phi(x_k)}\left[1 - (x - x_k)\frac{\Phi'(x_k)}{\Phi(x_k)}\right] \qquad (5.20.5)$$

while the factor of $f'(x_k)$ becomes:

$$\frac{\Phi(x)}{\Phi(x_k)}(x - x_k) \qquad (5.20.6)$$

In the limit, as ϵ goes to zero, we have obtained an interpolation which fits not only $f(x_k)$ *but also* $f'(x_k)$ at the critical point x_k. At the same time the root factor $x - x_k$ of the fundamental polynomial appears now in *second* power. The point $x = x_k$ of the interpolation becomes a double point. If all the points of interpolation are made double points, we fit *at every point* $x = x_i$ the functional value for $f(x_i)$ *and* the derivative $f'(x_i)$ correctly, that is the interpolating polynomial of the order $2n - 1$ coincides with $f(x)$ at all points $x = x_i$, and in addition the *derivative* of the interpolating polynomial coincides with $f'(x)$ at all points $x = x_i$. The fundamental polynomial now becomes a *square*:

$$F(x) = [(x - x_1)(x - x_2) \ldots (x - x_n)]^2 \qquad (5.20.7)$$

which remains positive throughout the range, instead of alternating in sign from point to point.

From the standpoint of a universal formula it will be preferable to operate with a single fundamental polynomial $F(x)$ counting every root factor with its proper multiplicity (one for single points, two for double points). The expression (6) can now be written in the form

$$\frac{2F(x)}{F''(x_k)}\frac{1}{x - x_k} \qquad (5.20.8)$$

and this becomes the factor of $f'(x)$. On the other hand, the expression (5) can now be written in the form

$$\frac{2F(x)}{F''(x_k)(x - x_k)^2}\left[1 - \frac{1}{3}\frac{F'''(x_k)}{F''(x_k)}(x - x_k)\right] \qquad (5.20.9)$$

and thus the Lagrangian interpolation formula in the presence of double points has to be generalised as follows:

Contribution of a single point $x = x_i$:

$$\frac{F(x)}{F'(x_i)}\frac{f(x_i)}{x - x_i} \qquad (5.20.10)$$

Contribution of a double point $x = x_k$:

$$\frac{2F(x)}{F''(x_k)}\left[1 - \frac{1}{3}\frac{F'''(x_k)}{F''(x_k)}(x - x_k)\right]\frac{f(x_k)}{(x - x_k)^2} + \frac{2F(x)}{F''(x_k)}\frac{f'(x_k)}{x - x_k} \qquad (5.20.11)$$

For the estimation of an error bound the Lagrangian formula (1.5.10) holds again, counting every root factor with the proper multiplicity.

Problem 217. By analysing the expression (11) demonstrate that the interpolating polynomial assumes the value $f(x_k)$ at the point $x = x_k$, while its derivative assumes the value $f'(x_k)$ at the point $x = x_k$.

Problem 218. Obtain a polynomial approximation of the order 4 by fitting $f(x)$ at the points $x = \pm 1, 0$, and $f'(x)$ at the points $x = \pm 1$.

[Answer:

$$F(x) = (x^2 - 1)^2 x \tag{5.20.12}$$

$$\begin{aligned} f^*(x) = (x^4 - 2x^2 + 1)f(0) \\ - \tfrac{1}{2}(x^4 - 2x^2)[f(1) + f(-1)] - \tfrac{1}{4}(x^3 - 3x)[f(1) - f(-1)] \\ + \tfrac{1}{4}(x^4 - x^2)[f'(1) - f'(-1)] + \tfrac{1}{4}(x^3 - x)[f'(1) + f'(-1)] \end{aligned} \tag{5.20.13}]$$

Problem 219. Apply this formula to an approximation of $\sin(\pi/2)x$ and $\cos(\pi/2)x$ and estimate the maximum error at any point of the range $[-1, 1]$.

[Answer:

$$\sin\frac{\pi}{2}x = \frac{1}{2}(3x - x^3), \quad |\eta| < \left(\frac{\pi}{2}\right)^5 \frac{1}{120\sqrt{5}}\left(\frac{4}{5}\right)^2 = 0.023 \tag{5.20.14}$$

$$\cos\frac{\pi}{2}x = (1 - x^2)\left(1 - \frac{4 - \pi}{4}x^2\right), \quad |\eta| < 0.023\,]$$

Problem 220. Explain why the error bound for the cosine function can in fact be reduced to

$$|\eta| < \left(\frac{\pi}{2}\right)^5 \frac{1}{4860} = 0.00309$$

[Answer: The point $x = 0$ can be considered as a *double point* if $f(x)$ is even. But then $F(x) = x^2(x^2 - 1)^2$ and the estimated maximum error becomes greatly reduced.]

Problem 221. Obtain the Green's function for the remainder of this approximation.

[Answer:

$$\begin{aligned} G(x, \xi) = \tfrac{1}{24}\left[(x - \xi)^4\right] - (x^3 - 2x^2 + 1)[(-\xi)^4] \\ + \tfrac{1}{4}(2x^4 + x^3 - 4x^2 - 3x)(1 - \xi)^4 \\ - (x^4 + x^3 - x^2 - x)(1 - \xi)^3 \end{aligned} \tag{5.20.15}]$$

Problem 222. Reduce the interpolating polynomial to the order 3 by dropping the point $x = 0$. Show that for any odd function $f(-x) = -f(x)$ the result must agree with that obtained in Problem 218.

[Answer:

$$F(x) = (x^2 - 1)^2 \tag{5.20.16}$$

$$\begin{aligned} f^*(x) = \tfrac{1}{2}[f(1) + f(-1)] + \tfrac{1}{4}(3x - x^3)[f(1) - f(-1)] \\ + \tfrac{1}{4}(x^3 - x)[f'(1) + f'(-1)] + \tfrac{1}{4}(x^2 - 1)[f'(1) - f'(-1)] \end{aligned} \tag{5.20.17}]$$

Problem 223. Apply this interpolation once more to the functions $\sin(\pi/2)x$ and $\cos(\pi/2)x$ and estimate the maximum errors.

[Answer:

$$\sin \frac{\pi}{2} x = \frac{1}{2} (3x - x^3) \qquad |\eta| < 0.023$$

$$\cos \frac{\pi}{2} x = \frac{\pi}{4} (1 - x^2) \qquad |\eta| < \frac{1}{24} \left(\frac{\pi}{2}\right)^4 = 0.254\,]$$

(5.20.18)

Problem 224. Obtain the Green's function for the remainder of this interpolation.

[Answer:

$$G(x, \xi) = \tfrac{1}{6}\{[(x - \xi)]^3 - \tfrac{3}{4}(x^3 + x^2 - x - 1)(1 - \xi)^2$$
$$+ \tfrac{1}{4}(x^3 - 3x - 2)(1 - \xi)^3\}$$

(5.20.19)]

Problem 225. Show that this Green's function is characterised by exactly the same conditions as the Green's function of the clamped bar, considered before in Section 17 except that the new domain extends from -1 to $+1$ while the range of the bar was normalised to $[0, l]$. Replacing x by $x + 1$, ξ by $\xi + 1$, show that the expression (19) is in fact equivalent to (17.8), if we put $l = 2$.

5.21. Construction of the Green's vector

We will now generalise our discussions concerning the explicit construction of the Green's vector to the case of *systems* of differential equations. Since an arbitrary scalar or vectorial kind of problem can be normalised to the Hamiltonian canonical form (cf. Section 10), we will assume that our system is already transformed into the canonical form, that is, we will deal with the system (10.18):

$$p'_{n+i} - \sum_{\alpha=1}^{2n} C_{i\alpha} p_\alpha = \beta_i$$

(5.21.1)

$$(p_{2n+i} = -p_i, \quad C_{ik} = C_{ki})$$

In order to obtain the solution of this system in the form

$$p_k(x) = \int_a^b \sum_{j=1}^{2n} G_k(x, \xi)_j \beta_j(\xi) d\xi$$

(5.21.2)

we need first of all the explicit construction of the Green's vector $G_k(x, \xi)_j$. This means that—considering ξ as the active variable—we should put the delta function in the kth equation

$$p'_{n+i} - \sum_{\alpha=1}^{2n} C_{i\alpha} p_\alpha = \delta_k(x, \xi)$$

(5.21.3)

while all the other equations have zero on the right side.

This again means that the homogeneous equations

$$p'_{n+i} - \sum_{\alpha=1}^{2n} C_{i\alpha} p_\alpha = 0$$

(5.21.4)

are satisfied in both regions $\xi < x$ and $\xi > x$. Assuming that we possess the homogeneous solution with all its $2n$ constants of integration, we can set up the solution for $\xi < x$ with *one* set of constants and the solution for $\xi > x$ with *another* set of constants, exactly as we have done in Section 16. The $2n$ boundary conditions of our problem provide us with $2n$ linear algebraic relations between the $4n$ free constants. Now we come to the joining of the two regions at the point $\xi = x$. In view of the fact that the delta function exists solely in the kth equation, we obtain *continuity in all components* $p_i(x)$, with the only exception of the component p_{n+k} where we get a jump of 1 in going from the left to the right:

$$p_{n+k}(x_+) - p_{n+k}(x_-) = 1 \tag{5.21.5}$$

These continuity conditions yield $2n - 1$ additional linear homogeneous algebraic relations between the constants of integration, *plus one inhomogeneous relation*, in consequence of the jump-condition (5).

As an example we return once more to our problem (16.12) for which we have already constructed the Green's function (cf. 16.20). We will deal with the same problem, but now presented in the canonical form:

$$\begin{matrix} p'_2 \\ -p'_1 \end{matrix} + \begin{bmatrix} 1 & 0 \\ 0 & \rho^2 \end{bmatrix} \begin{bmatrix} p_1 \\ p_2 \end{bmatrix} = \begin{matrix} 0 \\ \beta \end{matrix} \tag{5.21.6}$$

with the boundary conditions

$$\begin{aligned} p_1(0) - p_1(l) &= 0 \\ p_2(0) - p_2(l) &= 0 \end{aligned} \tag{5.21.7}$$

(The previous v is now p_2.) The two fundamental solutions of the homogeneous system are

$$\begin{aligned} p_1 &= \cos \rho\xi & p_1 &= \sin \rho\xi \\ p_2 &= -\frac{1}{\rho} \sin \rho\xi & p_2 &= \frac{1}{\rho} \cos \rho\xi \end{aligned}$$

Accordingly we establish the two separate solutions to the left and to the right from the point $\xi = x$ in the form

$$\left. \begin{aligned} p_1 &= A_1 \cos \rho\xi + A_2 \sin \rho\xi \\ p_2 &= -\frac{A_1}{\rho} \sin \rho\xi + \frac{A_2}{\rho} \cos \rho\xi \end{aligned} \right\} \quad (\xi < x)$$

$$\left. \begin{aligned} p_1 &= B_1 \cos \rho\xi + B_2 \sin \rho\xi \\ p_2 &= -\frac{B_1}{\rho} \sin \rho\xi + \frac{B_2}{\rho} \cos \rho\xi \end{aligned} \right\} \quad (\xi > x) \tag{5.21.8}$$

The boundary conditions (7) yield the following two relations between the four constants A_1, A_2, B_1, B_2:

$$\begin{aligned} A_1 - B_1 \cos \rho l - B_2 \sin \rho l &= 0 \\ A_2 - B_2 \cos \rho l + B_1 \sin \rho l &= 0 \end{aligned} \tag{5.21.9}$$

Let us first obtain the two components $G_1(x, \xi)_{1,2}$. Then p_1 is continuous at $\xi = x$ while $p_2(x)$ makes a jump of the magnitude 1. This yields the two further relations

$$(B_1 - A_1) \cos \rho x + (B_2 - A_2) \sin \rho x = 0$$
$$(B_2 - A_2) \cos \rho x - (B_1 - A_1) \sin \rho x = \rho \qquad (5.21.10)$$

The solution of this algebraic system gives the solution

$$\left. \begin{aligned} p_1(\xi) &= \frac{\rho}{2 \sin \rho \dfrac{l}{2}} \cos \rho\left(x - \xi - \frac{l}{2}\right) \\[2ex] p_2(\xi) &= \frac{\rho}{2 \sin \rho \dfrac{l}{2}} \sin \rho\left(x - \xi - \frac{l}{2}\right) \end{aligned} \right\} \quad (\xi < x)$$

$$(5.21.11)$$

$$\left. \begin{aligned} p_1(\xi) &= \frac{\rho}{2 \sin \rho \dfrac{l}{2}} \cos \rho\left(x - \xi + \frac{l}{2}\right) \\[2ex] p_2(\xi) &= \frac{\rho}{2 \sin \rho \dfrac{l}{2}} \sin \rho\left(x - \xi + \frac{l}{2}\right) \end{aligned} \right\} \quad (\xi > x)$$

Hence we have now obtained the following two components of the full Green's vector:

$$G_1(x, \xi)_1 = \frac{\rho}{2 \sin \rho \dfrac{l}{2}} \cos \rho\left(x - \xi \mp \frac{l}{2}\right)$$

$$(5.21.12)$$

$$G_1(x, \xi)_2 = \frac{1}{2 \sin \rho \dfrac{l}{2}} \sin \rho\left(x - \xi \mp \frac{l}{2}\right)$$

with the convention that the upper sign holds for $\xi < x$, the lower sign for $\xi > x$.

We now come to the construction of the remaining two components $G_2(x, \xi)_{1,2}$, characterised by the condition that now p_2 remains continuous while $-p_1$ makes a jump of 1 at the point $\xi = x$. The equations (9) remain unchanged but the equations (10) have to be modified as follows:

$$(B_1 - A_1) \cos \rho x + (B_2 - A_2) \sin \rho x = -1$$
$$(B_2 - A_2) \cos \rho x - (B_1 - A_1) \cos \rho x = 0 \qquad (5.21.13)$$

The solution yields the required components in the following form:

$$G_2(x, \xi)_1 = - \frac{1}{2 \sin \rho \frac{l}{2}} \sin \rho \left(x - \xi \mp \frac{l}{2} \right)$$

$$G_2(x, \xi)_2 = \frac{1}{2\rho \sin \rho \frac{l}{2}} \cos \rho \left(x - \xi \mp \frac{l}{2} \right)$$

(5.21.14)

Now we have constructed the complete Green's vector (although the solution of the system (6) does not require the two components $G_{1,2}(x, \xi)_1$ since $\beta_1(\xi)$ is zero in our problem). We can now demonstrate the characteristic symmetry properties of the Green's vector. First of all the components $G_1(x, \xi)_1$ and $G_2(x, \xi)_2$ must be *symmetric in themselves*. This means that an exchange of x and ξ must change the left expresssion to the right expression, or vice versa. This is indeed the case. In these components the exchange of x and ξ merely changes the sign of $x - \xi$ but the simultaneous change of the sign of $l/2$ restores the earlier value, the cosine being an *even* function. Then we have the symmetry relation

$$G_1(x, \xi)_2 = G_2(\xi, x)_1 \tag{5.21.15}$$

This relation is also satisfied by our solution because an exchange of x and ξ in the second expression of (12) is equivalent to a minus sign in front of the entire expression, together with a change of $-l/2$ to $+l/2$. But this agrees with the first expression of (14), if we take the lower sign, in view of the fact that left and right has to be exchanged. Hence we have actually tested all the symmetry conditions of the Green's vector.

Problem 226.   The components of the Green's vector of our example show the following peculiarities:

$$G_2(x, \xi)_1 = - G_1(x, \xi)_2$$

$$G_2(x, \xi)_2 = \frac{1}{\rho^2} G_1(x, \xi)_1 \tag{5.21.16}$$

Explain these relations on the basis of the differential equation (21.6), assuming the right sides in the form $\gamma(x)$, $\beta(x)$, instead of 0, $\beta(x)$.

Problem 227.   Consider the canonical system (8.2) for the clamped elastic bar (boundary conditions (7.21)). In (7.24) the Green's function for the solution $v_3(x)$ was defined and the determining equations (7.25–26) deduced, while later the application of the reciprocity theorem gave the determining equations (14.11–12), considering x as the active variable. From the standpoint of the Green's vector the component $G_3(x, \xi)_1$ is demanded, that is, we should evaluate $v_1(\xi)$, putting the delta function in the third equation (which means a jump of 1 at $\xi = x$ in the function $v_2(\xi)$). We can equally operate with $G_1(x, \xi)_3$, that is, evaluate $v_3(\xi)$, putting the delta function in the first equation (which means a jump of -1 at $\xi = x$ in $v_4(\xi)$). In the latter method we have to exchange in the end x and ξ. Having obtained the result—for the sake of simplicity assume a bar of uniform cross-section, i.e., put $I(\xi) = $ const. $ = I$—we can verify the previously obtained properties of the Green's function, deduced on the basis of the defining differential equation but without explicit construction:

1a). Considered as a function of ξ, $G(x, \xi)$ and its first derivative must vanish at the two endpoints $\xi = 0$ and $\xi = l$.

1b). The coefficients of ξ^2 and ξ^3 must remain continuous at the point $\xi = x$.

1c). The function $G(x, \xi)$ must pass through the point $\xi = x$ continuously but the first derivative must make a jump of 1 at the point $\xi = x$.

2a). Considered as a function of x the dependence can only be *linear* in x, with a jump of 1 in the first derivative at the point $x = \xi$.

2b). If we integrate this function twice from the point $x = 0$, we must wind up at the endpoint $x = l$ with the value zero for the integral and its first derivative.

[Answer:

$$IG_3(x, \xi)_1 = \frac{\xi^3}{l^2}\left(1 - \frac{2x}{l}\right) + \frac{\xi^2}{l}\left(3\frac{x}{l} - 2\right) \qquad (\xi \leq x)$$

$$= \frac{\xi^3}{l^2}\left(1 - \frac{2x}{l}\right) + \frac{\xi^2}{l}\left(3\frac{x}{l} - 2\right) + \xi - x \qquad (\xi \geq x)]$$

(5.21.17)

5.22. The constrained Green's function

Up to now we have assumed that we have a "well-posed" system, that is neither the given, nor the adjoint equation could have non-vanishing homogeneous solutions. If we combine both equations to the unified self-adjoint system (as we have done before in Section 14):

$$Dv(x) = \beta(x)$$
$$\tilde{D}u(x) = \gamma(x)$$

(5.22.1)

both conditions are included in the statement that the homogeneous system

$$Dv(x) = 0$$
$$\tilde{D}u(x) = 0$$

(5.22.2)

(under the given boundary conditions) has no non-vanishing solutions.

On the other hand, we have seen in the treatment of general $n \times m$ matrices that the insistence on the "well-posed" case is analytically not justified. If the homogeneous system has non-zero solutions, this fact can be interpreted in a natural way: *The solutions of the homogeneous equation* (2) *trace out those dimensions of the function space in which the given differential operator is not activated.* If we restrict ourselves to the proper subspace, viz. the "eigen-space" of the operator, ignoring the rest of space, we find that *within this subspace the operator behaves exactly like a "well-posed" operator.* Within this space the solution is unique and within this space the right side can be given freely. If we do not leave this space, we do not even notice that there is anything objectionable in the given differential operator and the associated differential equation.

From the standpoint of the full space, however, this restriction to the eigen-space of the operator entails some definite *conditions* or *constraints*. These conditions are two-fold:

1. They restrict the choice of the right side (β, γ) by demanding that the vector (β, γ) *must lie completely within the activated subspace associated with the operator*. This again means that the right side has no components in the direction of the non-activated dimensions, that is in the direction of the homogeneous solutions.

Now we must agree on a notation for the possible homogeneous solutions of the system (2). We may have one or more such solutions. In the case of ordinary differential equations the number of independent solutions cannot be large but in the case of partial differential equations it can be arbitrarily large, or even infinite. We could use subscripts for the designation of these independent solutions but this is not convenient since our operator $Dv(x)$ may involve several components $v_i(x)$ of the unknown function $v(x)$ and the subscript notation has been absorbed for the notation of vector components. On the other hand, we have not used any *upper* indices and thus we will agree that the homogeneous solutions shall be indicated in the following fashion:

$$u^1(x), u^2(x), \ldots \tag{5.22.3}$$

and likewise

$$v^1(x), v^2(x), \ldots \tag{5.22.4}$$

The required orthogonality of the right sides to the homogeneous solutions finds expression in the conditions

$$\int \beta(\xi) u^j(\xi) d\xi = 0 \qquad (j = 1, 2, \ldots, r)$$
$$\int \gamma(\xi) v^k(\xi) d\xi = 0 \qquad (k = 1, 2, \ldots, \rho) \tag{5.22.5}$$

These are the *compatibility conditions* of our system, without which a solution cannot exist.

While we have thus formulated the conditions to which the given right sides have to be submitted for the existence of a solution, we will now add some further conditions which will make our solution *unique*. This is done by demanding that the solution shall likewise be completely within the eigen-space of the operator, not admitting any components in the non-activated dimensions. Hence we will complement the conditions (5) by the added conditions

$$\int u(\xi) u^j(\xi) d\xi = 0 \qquad (j = 1, 2, \ldots, r)$$
$$\int v(\xi) v^k(\xi) d\xi = 0 \qquad (k = 1, 2, \ldots, \rho) \tag{5.22.6}$$

Under these conditions our equation is once more solvable and the solution is unique. Hence we can expect that the solution will again be obtainable with the help of a Green's function $G(x, \xi)$, called the "constrained Green's function":

$$v(\xi) = \int G(x, \xi) \beta(\xi) d\xi$$
$$u(x) = \int \tilde{G}(x, \xi)(\xi) d\xi \tag{5.22.7}$$

The question arises how to define these functions. This definition will occur exactly along the principles we have applied before, if only we remember that our operations are now restricted to the activated subspace of the function space. Accordingly we cannot put simply the delta function on the right side of the equation. We have to put something on the right side which excludes any components in the direction of the homogeneous solutions, although keeping everything unchanged in the activated dimensions. We will thus put once more the delta function on the right side of the defining equation, but *subtracting its projection into the unwanted dimensions.* This means the following type of equation:

$$D\tilde{G}(x, \xi) = \delta(x, \xi) - \sum_{j=1}^{r} \rho_j u^j(\xi) \tag{5.22.8}$$

and likewise

$$\check{D}G(x, \xi) = \delta(x, \xi) - \sum_{k=1}^{\rho} \sigma_k v^k(\xi) \tag{5.22.9}$$

Now we have to find the undetermined constants ρ_j, σ_k. This is simple if we assume that the homogeneous solutions $v^k(\xi)$, respectively $u^j(\xi)$ have been *orthogonalised*, and *normalised*, i.e. we use such linear combinations of the homogeneous solutions that the resulting solutions shall satisfy the orthogonality and normalisation conditions

$$\int v^i(\xi)v^k(\xi)d\xi = \delta_{ik}$$
$$\int u^j(\xi)u^m(\xi)d\xi = \delta_{jm} \tag{5.22.10}$$

Then the demanded orthogonality conditions (5) permit us to determine the constants ρ_j, σ_k explicitly and independently of each other by forming the integrals (5) over the right sides of (8) and (9):

$$\rho_j = \int u^j(\xi)\delta(x, \xi)d\xi$$
$$\sigma_k = \int v^k(\xi)\delta(x, \xi)d\xi \tag{5.22.11}$$

These integrals reduce to something very simple, in view of the extreme nature of the delta function: the integration is restricted to the immediate neighbourhood of the point $\xi = x$, but there $u^i(\xi)$ is replaceable by $u^i(x)$ which comes before the integral sign while the integral over the delta function itself gives 1. And thus

$$\rho_j = u^j(x)$$
$$\sigma_k = v^k(x) \tag{5.22.12}$$

We thus see that the definition of Green's function to constrained systems has to occur as follows:

$$D\tilde{G}(x, \xi) = \delta(x, \xi) - \sum_{j=1}^{r} u^j(x)u^j(\xi)$$

$$\check{D}G(x, \xi) = \delta(x, \xi) - \sum_{k=1}^{\rho} v^k(x)v^k(\xi) \tag{5.22.13}$$

These equations are now solvable (under the proper boundary conditions) but the solution will generally not be unique. *The uniqueness is restored, however, by submitting the solution to the orthogonality conditions* (6).

As an example we return to the problem we have discussed in Section 16. We have seen that the solution went out of order if the constant ρ satisfied the condition (16.24). These were exactly the values which led to the homogeneous solutions

$$\sin \rho x, \quad \cos \rho x \tag{5.22.14}$$

These two solutions *are already orthogonal* and thus we can leave them as they are, except for the normalisation condition which implies in our case the factor $\sqrt{2/l}$:

$$v^1(\xi) = u^1(\xi) = \sqrt{\frac{2}{l}} \sin \rho \xi$$
$$v^2(\xi) = u^2(\xi) = \sqrt{\frac{2}{l}} \cos \rho \xi \tag{5.22.15}$$

Hence for the exceptional values (16.24) the defining equation for the Green's function now becomes

$$v''(\xi) + \rho^2 v(\xi) = \delta(x, \xi) - \frac{2}{l} \sin \rho x \sin \rho \xi$$
$$- \frac{2}{l} \cos \rho x \cos \rho \xi \tag{5.22.16}$$
$$= \delta(x, \xi) - \frac{2}{l} \cos \rho(x - \xi)$$

with the previous boundary conditions (16.13).

Now the solution of this equation is particularly simple if we possess the general solution for arbitrary values of ρ^2 (the "eigenvalues" of the differential equation, with which we will deal later in greater detail). Anticipating a later result, to be proved in Section 29, we describe here the procedure itself, restricting ourselves to self-adjoint systems. We put

$$\check{D}u(x, \xi) + \epsilon u(\xi) = \delta(x, \xi) \tag{5.22.17}$$

assuming that ϵ is small. The difficulty arises only for $\epsilon = 0$; for any finite ϵ the problem is solvable. Now we let ϵ go to zero. Then there is a term which is independent of ϵ and a term which goes with $1/\epsilon$ to infinity. We omit the latter term, while we keep the constant term. *It is this constant term which automatically yields the Green's function of the constrained problem.*

For example in our problem we can put

$$\rho^2 = \left(\frac{2k\pi}{l}\right)^2 + \frac{k\pi}{l}\epsilon$$
$$\rho = \frac{2k\pi}{l} + \epsilon\left(1 - \frac{\epsilon l}{4k\pi}\right) \tag{5.22.18}$$

in which case we know already the solution of the equation (17) since we are in the possession of the Green's function for arbitrary values of ρ (cf. (16.20)); we neglect the negligible powers of ϵ:

$$G(x, \xi) = \frac{\cos\left(\dfrac{2k\pi}{l} + \epsilon\right)\left(x - \xi \mp \dfrac{l}{2}\right)}{2\left(\dfrac{2k\pi}{l} + \epsilon\right)\sin\left[k\pi + \epsilon\dfrac{l}{2}\left(1 - \dfrac{\epsilon l}{4k\pi}\right)\right]}$$

(5.22.19)

$$= \frac{l}{4k\pi}\frac{\cos\dfrac{2k\pi}{l}\left(x - \xi \mp \dfrac{l}{2}\right) - \epsilon\left(x - \xi \mp \dfrac{l}{2}\right)\sin\dfrac{2k\pi}{l}\left(x - \xi \mp \dfrac{l}{2}\right)}{\left(1 + \dfrac{\epsilon l}{2k\pi}\right)(-1)^k\epsilon\dfrac{l}{2}\left(1 - \dfrac{\epsilon l}{4k\pi}\right)}$$

The constant term becomes:

$$G(x, \xi) = \frac{1}{2k\pi}\left[\left(\xi - x \pm \frac{l}{2}\right)\sin\frac{2k\pi}{l}(x - \xi) - \frac{l}{4k\pi}\cos\frac{2k\pi}{l}(x - \xi)\right] \quad (5.22.20)$$

Here then is the Green's function of our problem obtained by a limit process from a slightly modified problem which is unconditioned and hence subjected to the usual treatment, without any modification of the delta function on the right side. This function would go to infinity without the proper precautions. By modifying the right side in the sense of (16) we counteract the effect of the term which goes to infinity and obtain a finite result. The symmetry of the Green's function:

$$G(x, \xi) = G(\xi, x) \quad (5.22.21)$$

remains unaffected because our operator, although it is inactive in two particular dimensions of the function space, behaves within its own activated space exactly like any other self-adjoint operator.

The physical significance of our problem is an exciting force acting on an elastic spring which is periodic and of a period which is exactly k times the characteristic period of the spring. Under such circumstances the amplitudes of the oscillations would constantly increase and could never attain a steady state value except if the exciting force is such that its Fourier analysis has a nodal point at the period of the spring. This is the significance of the orthogonality conditions (5) which in our problem assume the form

$$\int_0^l \beta(\xi)\left.\begin{matrix}\sin \\ \cos\end{matrix}\right\}\frac{2k\pi}{l}\xi\,d\xi = 0 \quad (5.22.22)$$

The special character of a steady state solution under resonance conditions is mathematically expressed by the fact that the boundary conditions (16.13)—which are equivalent to the condition of a steady state—can only be satisfied if the right side is orthogonal to the two homogeneous solutions, that is the characteristic vibrations of the spring.

Problem 228. Apply this method to the determination of the constrained Green's function of the following problem:

$$v''(0) = \beta(x)$$
$$v'(0) = v'(l) = 0 \tag{5.22.23}$$

Show that the solution satisfies the given boundary conditions, the defining differential equation (13), the symmetry condition, and the orthogonality to the homogeneous solution.

[Answer:

$$G(x, \xi) = \frac{l}{6} - \frac{1}{2l}(l - x)^2 - \frac{\xi^2}{2l} \qquad (\xi \le x)$$

$$= \frac{l}{6} - \frac{1}{2l}(l - \xi)^2 - \frac{x^2}{2l} \qquad (\xi \ge x) \tag{5.22.24}$$

Compatibility condition:

$$\int_0^l \beta(\xi)d\xi = 0\,]$$

5.23. Legendre's differential equation

An interesting situation arises in the case of Legendre's differential equation which defines the Legendre polynomials. Here the differential equation is (for the case $n = 0$):

$$\frac{d}{dx}(1 - x^2)v'(x) = \beta(x)$$

$$x = [-1, +1] \tag{5.23.1}$$

The variable x ranges between -1 and $+1$ and the coefficient of the highest derivative: $1 - x^2$ *vanishes* at the two endpoints of the range. This has a peculiar consequence. We do not prescribe any boundary conditions for $v(x)$. Then we would expect that the adjoint problem will be over-determined by having to demand 4 boundary conditions. Yet this is not the case. The boundary term in our case becomes

$$\left| (1 - x^2)(uv' - vu') \right|_{-1}^{+1} \tag{5.23.2}$$

The boundary term vanishes automatically, due to the vanishing of the first factor. Hence we are confronted with the puzzling situation that the adjoint equation remains likewise without boundary conditions which makes our problem self-adjoint since both differential operator and boundary conditions remain the same for the given and the adjoint problem.

In actual fact the lack of boundary conditions is only apparent. The vanishing of the highest coefficient of a differential operator at a certain point makes that point to a *singular point* of the differential equation, where the solution will generally go out of bounds. By demanding *finiteness* (but not vanishing) of the solution we already imposed a restriction on our solution which is equivalent to a *boundary condition*. Since the same occurs at the other end point, we have in fact imposed two boundary conditions on our problem by demanding *finiteness* of the solution at the two points $x = \pm 1$.

Our aim is now to find the Green's function of our problem. Since our differential equation is self-adjoint, we know in advance that the Green's function $G(x, \xi)$ will become *symmetric* in x and ξ. There is, however, the further complication that the homogeneous equation has the solution

$$v(x) = \text{const} \tag{5.23.3}$$

Accordingly we have to make use of the extended definition (22.8) for the Green's function. The normalised homogeneous solution becomes

$$v^1(x) = \frac{1}{\sqrt{2}} \tag{5.23.4}$$

since the integral of $\frac{1}{2}$ between the limits ± 1 becomes 1. The defining equation of the Green's function thus becomes:

$$\frac{d}{d\xi}\left[(1 - \xi^2)\frac{dG}{d\xi}\right] = \delta(x, \xi) - \frac{1}{2} \tag{5.23.5}$$

First of all we obtain the general solution of the homogeneous equation:

$$(1 - \xi^2)v' = A$$

$$v' = \frac{A}{1 - \xi^2}$$

$$v = \frac{1}{2}A \log \frac{1 + \xi}{1 - \xi} + B \tag{5.23.6}$$

We also need the solution of the inhomogeneous equation

$$\frac{d}{d\xi}[(1 - \xi^2)v'] = -\frac{1}{2} \tag{5.23.7}$$

which yields

$$(1 - \xi^2)v' = -\frac{1}{2}\xi$$
$$v = \frac{1}{4}\log(1 + \xi)(1 - \xi) \tag{5.23.8}$$

The complete solution of (5) will be a superposition of the solution (8) of the inhomogeneous equation, plus the solution of the homogeneous equation, adjusted to the boundary conditions and the joining conditions at the point $\xi = x$. We will start on the left side: $\xi < x$. Here we get:

$$v(\xi) = \frac{1}{4}[\log(1 + \xi) + \log(1 - \xi)]$$
$$+ \frac{1}{2}A[\log(1 + \xi) - \log(1 - \xi)] + B_1 \tag{5.23.9}$$

Now at the left endpoint $\xi = -1$ of our range the function $\log(1 - \xi)$

remains regular but the function $\log (1 + \xi)$ goes to infinity. Hence the factor of this function must vanish which gives the condition

$$A_1 = -\tfrac{1}{2} \tag{5.23.10}$$

By the same reasoning, if we set up our solution on the right side: $\xi > x$:

$$v(\xi) = \tfrac{1}{4}[\log (1 + \xi) + \log (1 - \xi)]$$
$$+ \tfrac{1}{2}A_2[\log (1 + \xi) - \log (1 - \xi)] + B_2 \tag{5.23.11}$$

we obtain

$$A_2 = \tfrac{1}{2} \tag{5.23.12}$$

because it is now the function $\log (1 - \xi)$ which goes to infinity and which thus has to be omitted.

So far we have obtained

$$G(x, \xi) = \tfrac{1}{2} \log (1 - \xi) + B_1 \qquad (\xi < x)$$
$$= \tfrac{1}{2} \log (1 + \xi) + B_2 \qquad (\xi > x) \tag{5.23.13}$$

The condition of continuity at the point $\xi = x$ adds the condition

$$B_1 - B_2 = \tfrac{1}{2} \log \frac{1 + x}{1 - x} \tag{5.23.14}$$

The discontinuity of the derivative at the point $x = \xi$ becomes

$$\frac{1}{2(1 + x)} - \frac{1}{2(1 - x)} = \frac{1}{1 - x^2} \tag{5.23.15}$$

But this is exactly what the delta function on the right side demands: the jump of the $n - 1^{\text{st}}$ derivative must be 1 divided by the coefficient of the highest derivative at the point $\xi = x$ (which in our problem is $(1 - x^2)$).

We have now satisfied the differential equation (5) and the boundary conditions (since our solution remains finite at both points $\xi = \pm 1$). And yet we did not get a complete solution because the two constants B_1 and B_2 have to satisfy the single condition (14) only. We can put

$$\frac{B_1 + B_2}{2} = B$$

$$\frac{B_1 - B_2}{2} = \tfrac{1}{4} \log \frac{1 + x}{1 - x} \tag{5.23.16}$$

Then

$$B_1 = B + \tfrac{1}{4} \log \frac{1 + x}{1 - x}$$

$$B_2 = B - \tfrac{1}{4} \log \frac{1 + x}{1 - x} \tag{5.23.17}$$

In fact we know that this uncertainty has to be expected. We still have to satisfy the condition that *the constrained Green's function must become orthogonal to the homogeneous solution* (3):

$$\int_{-1}^{+1} G(x, \xi)d\xi = 0 \qquad (5.23.18)$$

This yields the condition

$$B = \tfrac{1}{4} \log (1 - x^2) + \tfrac{1}{2} - \log 2 \qquad (5.23.19)$$

and substituting back in (13) we obtain finally the uniquely determined Green's function of our problem:

$$\begin{aligned} G(x, \xi) &= \tfrac{1}{2} - \log 2 + \tfrac{1}{2} \log (1 - \xi)(1 + x) \qquad (\xi \leq x) \\ &= \tfrac{1}{2} - \log 2 + \tfrac{1}{2} \log (1 + \xi)(1 - x) \qquad (\xi \geq x) \end{aligned} \qquad (5.23.20)$$

Problem 229. Assuming that $\beta(x)$ is an *even* function: $\beta(x) = \beta(-x)$, the integration is reducible to the range [0, 1]. The same holds if $\beta(x)$ is *odd*: $\beta(x) = -\beta(-x)$. In the first case we get the boundary condition at $x = 0$:

$$v'(0) = 0 \qquad (5.23.21)$$

In the second case:

$$v(0) = 0 \qquad (5.23.22)$$

Carry through the process for the half range with the new boundary conditions, (at $x = 1$ the previous finiteness condition remains) and show that the new result agrees with the result obtained above.

[Answer:

1. $$\begin{aligned} G(x, \xi) &= 1 - 2 \log 2 + \log (1 + x) + \tfrac{1}{2} \log (1 - \xi^2) \qquad (\xi \leq x) \\ &= 1 - 2 \log 2 + \log (1 + \xi) + \tfrac{1}{2} \log (1 - x^2) \qquad (\xi \geq x) \end{aligned} \qquad (5.23.23)$$

Compatibility condition:

$$\int_0^1 \beta(\xi)d\xi = 0 \qquad (5.23.24)$$

2. $$\begin{aligned} G(x, \xi) &= \tfrac{1}{2} \log \frac{1 - \xi}{1 + \xi} \qquad (\xi \leq x) \\ &= \tfrac{1}{2} \log \frac{1 - x}{1 + x} \qquad (\xi \geq x) \end{aligned} \qquad (5.23.25)$$

(Right side free.)]

5.24. Inhomogeneous boundary conditions

Throughout our discussions we have assumed that the given boundary conditions were of the *homogeneous type*, that is that certain data on the boundary were prescribed as zero. Indeed, for the definition of the Green's function the assumption of homogeneous boundary conditions is absolutely essential. This does not mean, however, that the Green's function is applicable solely to problems with homogeneous boundary conditions. In fact, the same Green's function which solved the inhomogeneous equation with homogeneous boundary conditions solves likewise the general case of an inhomogeneous differential equation with inhomogeneous boundary conditions, as we have seen before in Section 6 of this chapter. The only difference is that the inhomogeneous boundary data contribute their own share to the result, not in the form of a volume integral but in the form of a surface integral, extended over the boundary surface. In the case of ordinary differential equations there is no integration at all; each one of the given inhomogeneous boundary values contributes a term to the solution, expressible with the help of the Green's function $G(x, \xi)$ and its derivatives with respect to ξ and substituting for ξ the value $\xi = b$ on the upper limit and $\xi = a$ on the lower limit.

We shall discuss, however, the case of *constrained* systems which under the given but *homogenised* boundary conditions possess non-zero solutions. We have seen the rules of constructing the Green's function under these circumstances (cf. Section 22). The operation with this Green's function remains the same as that for unconstrained systems: once more the integration over the domain is complemented by the proper boundary terms, necessitated by the presence of inhomogeneous data on the boundary. The same data influence, however, the compatibility conditions to which the right side is subjected. The orthogonality to the homogeneous solution involves now the boundary terms, which have to be added to the integration over the given domain. It is now the *sum* of the volume integral, plus the properly constructed surface integral—all obtained with the help of the homogeneous solution as auxiliary function—which has to vanish. This consideration shows that an incompatible system whose right side is not orthogonal to the homogeneous solution, can be made compatible by changing some of the given homogeneous boundary conditions to inhomogeneous boundary conditions.

A good example is provided by the problem of the vibrating spring under resonance conditions (cf. 16.24). In order to satisfy the homogeneous boundary conditions

$$v(l) - v(0) = 0$$
$$v'(l) - v'(0) = 0$$

(5.24.1)

(which expresses the fact that the system constantly repeats its motion under the influence of the periodic exciting force) it is necessary and sufficient that the orthogonality conditions

$$\int_0^l \beta(\xi)e^{2\pi i k\xi/l}d\xi = 0 \qquad \left(l = \frac{2k\pi}{\rho}\right)$$

(5.24.2)

are satisfied. But let us assume that these conditions are *not* satisfied. Then instead of saying that now our given problem is unsolvable, we can go through the regular routine of the solution exactly as before. But finally, when checking the compatibility conditions, we find that we have to make allowances in the *boundary conditions* in order to make our problem solvable. The problem of the vibrating spring, kept in motion by a periodic external force, is a very real physical problem in which we know in advance that the solution exists. But in the case of resonance the return of the system to the original position cannot be expected and that means that the conditions (1) will no longer hold.

Now in the construction of the adjoint system we went through the following moves. We multiplied the given operator $Dv(x)$ by an undetermined factor $u(x)$ and succeeded in "liberating" $v(x)$, which was now multiplied by a new operator $\bar{D}u(x)$. In this process a *boundary term* appeared on the right side. For example in the problem of the vibrating spring:

$$\int_0^l [u(v'' + \rho^2 v) - v(u'' + \rho^2 u)]dx = \left| uv' - vu' \right|_0^l \qquad (5.24.3)$$

Now under the homogeneous boundary conditions (1) the right side dropped out and we obtained the compatibility condition

$$\int_0^l u(\xi)\beta(\xi)d\xi = 0 \qquad (5.24.4)$$

for the case that the homogeneous equation under the given homogeneous boundary conditions possessed non-zero solutions. But if we allow that the given boundary conditions (1) have something on the right side, let us say p_1, p_2, then the compatibility condition (4) has to be modified as follows:

$$\int_0^l u(\xi)\beta(\xi)d\xi = u(l)\, p_2 - u'(l)p_1 \qquad (5.24.5)$$

This means for the case of resonance:

$$\int_0^l \beta(\xi)e^{2\pi i k\xi/l}d\xi = p_2 - \frac{2\pi i k}{l}\, p_1 \qquad (5.24.6)$$

which gives

$$\begin{aligned} p_2 &= \int_0^l \beta(\xi)\cos 2k\pi \frac{\xi}{l}\, d\xi \\ p_1 &= -\frac{l}{2k\pi}\int_0^l \beta(\xi)\sin 2k\pi \frac{\xi}{l}\, d\xi \end{aligned} \qquad (5.24.7)$$

But now the solution with the help of the Green's function (22.20) has also to be modified because the inhomogeneous boundary values p_1 and p_2 will contribute something to the solution. The Green's identity (3) comes once more in operation and we find that we have to add to our previous solution the right side of (3), but with a negative sign:

$$-p_2 G(x, l) + p_1 G'(x, l) = \frac{p_2}{2k\pi}\left[\left(x - \frac{l}{2}\right)\sin\frac{2k\pi}{l}x + \frac{l}{4k\pi}\cos\frac{2k\pi}{l}x\right]$$
$$+ \frac{p_1}{l}\left[\left(x - \frac{l}{2}\right)\cos\frac{2k\pi}{l}x + \frac{l}{4k\pi}\sin\frac{2k\pi}{l}x\right] \qquad (5.24.8)$$

In this manner we can separate the contribution of the resonance terms from the steady state terms (the latter being generated by the non-resonant Fourier components of the exciting force).

Problem 230. Obtain the solution of Problem 228 under the assumption that the boundary conditions are modified as follows:

$$v'(0) = 0, \quad v'(l) = p \tag{5.24.9}$$

[Answer: Compatibility condition:

$$\int_0^l \beta(\xi)d\xi = p \tag{5.24.10}$$

Added term in solution:

$$-G(x, l)p = -p\left(\frac{l}{6} - \frac{x^2}{2l}\right) \tag{5.24.11}]$$

Problem 231. The boundary conditions of an elastic bar, free on both ends, can be given according to (4.14.10) in the form:

$$\begin{aligned}v_2(0) &= v_2(l) = 0\\v'_2(0) &= v'_2(l) = 0\end{aligned} \tag{5.24.12}$$

where

$$I(x)v''_1(x) = v_2(x) \tag{5.24.13}$$

and

$$v''_2(x) = \beta(x) \tag{5.24.14}$$

By modifying these conditions to

$$\begin{aligned}v_2(l) &= v'_2(l) = 0\\v_2(0) &= p_2\\v'_2(0) &= p_1\end{aligned} \tag{5.24.15}$$

obtain the point load and the point torque demanded at the point $x = 0$, to keep the bar—which is free at the other endpoint $x = l$—in equilibrium.

[Answer:

$$\begin{aligned}p_1 &= -\int_0^l \beta(\xi)d\xi\\p_2 &= \int_0^l \xi\beta(\xi)d\xi\ ]\end{aligned} \tag{5.24.16}$$

5.25. The method of over-determination

In Section 22 we have dealt with constrained systems for which the standard definition of the Green's function had to be modified because the given differential equation (under the given boundary conditions) did not allow a unique solution. This happened, whenever the homogeneous equation under homogeneous boundary conditions had solutions which did not vanish identically. As an example we studied the differential equation of the vibrating spring with the boundary conditions

$$v(l) - v(0) = p_1$$
$$v'(l) - v'(0) = p_2 \qquad (5.25.1)$$

In the case of resonance, characterised by the condition (16.24), the homogeneous equation under the homogeneous boundary conditions

$$v(l) - v(0) = 0$$
$$v'(l) - v'(0) = 0 \qquad (5.25.2)$$

allowed two non-vanishing solutions. In such problems we can modify the previous procedure to obtain the solution in simpler manner.

Since the given problem allows the addition of two solutions with two free constants, we do not lose anything if we add *two more boundary conditions*, chosen in such manner that now the problem should become uniquely solvable. For example we could choose the additional conditions

$$v(0) = v'(0) = 0 \qquad (5.25.3)$$

in which case the boundary conditions (1) become

$$v(l) = p_1$$
$$v'(l) = p_2 \qquad (5.25.4)$$

Our problem is now apparently strongly over-determined since we have prescribed 4 instead of 2 boundary conditions. And yet we did not alter our problem, except that we disposed of the two integration constants which were left free in the original formulation.

In the new formulation the previous difficulty with the solution of the adjoint equation

$$\tilde{D}u(\xi) = \delta(x, \xi) \qquad (5.25.5)$$

does not occur any more. We need not modify the right side in order to make the equation solvable. In fact, another strange phenomenon is now encountered. The method of the Green's identity now shows that the adjoint equation becomes

$$u''(\xi) + \rho^2 u(\xi) = \delta(x, \xi) \qquad (5.25.6)$$

without any boundary conditions. Any solution of (6) is acceptable as a Green's function. We may choose for example

$$G_1(x, \xi) = 0 \qquad (\xi \leq x)$$
$$= \frac{1}{\rho} \sin \rho(\xi - x) \qquad (\xi \geq x) \qquad (5.25.7)$$

which satisfies the boundary conditions $u(0) = u'(0) = 0$. With this solution we obtain $v(x)$ in the form of the following integral:

$$v(x) = \frac{1}{\rho} \int_x^l \beta(\xi) \sin \rho(\xi - x)d\xi + p_1 \cos \rho(l - x) - \frac{p_2}{\rho} \sin \rho(l - x) \quad (5.25.8)$$

By putting $x = l$ we see that the boundary conditions at $x = l$ are automatically fulfilled. On the other hand, by putting $x = 0$ we obtain:

$$v(0) = \frac{1}{\rho} \int_0^l \beta(\xi) \sin \rho\xi \, d\xi + p_1 \cos \rho l - \frac{p_2}{\rho} \sin \rho l$$

$$v'(0) = - \int_0^\rho \beta(\xi) \cos \rho\xi \, d\xi + p_1\rho \sin \rho l + p_2 \cos \rho l \quad (5.25.10)$$

but in view of the relation (16.24) between ρ and l we can put

$$v(0) = \frac{l}{2k\pi} \int_0^l \beta(\xi) \sin 2k\pi \frac{\xi}{l} \, d\xi + p_1$$

$$v'(0) = - \int_0^l \beta(\xi) \cos 2k\pi \frac{\xi}{l} \, d\xi + p_2 \quad (5.25.11)$$

By demanding that these two values vanish we arrive at the previous compatibility conditions (24.7), now simply obtained by *applying the solution to the boundary at* $x = 0$:

$$p_1 = - \frac{l}{2k\pi} \int_0^k \beta(\xi) \sin 2k\pi \frac{\xi}{l} \, d\xi$$

$$p_2 = \int_0^l \beta(\xi) \cos 2k\pi \frac{\xi}{l} \, d\xi \quad (5.25.12)$$

The solution (8) does not coincide with the earlier solution (22.20)—complemented by (24.8)—because we followed a different policy in normalising the free homogeneous solutions. But the simpler Green's function (7)—obtained along the usual lines of constructing the Green's function without any modification of the right side—is just as good from the standpoint of solving the originally given problem as the more elaborate function (22.20). If we want to normalise the final solution in a different way, we can still add the homogeneous solution

$$A \cos \rho x + B \sin \rho x \quad (5.25.13)$$

and determine A and B by two further conditions.

But we can go one step further still and transform our $G_1(x, \xi)$ into that unique Green's function (22.20) which is symmetric in x and ξ and which has the property of being orthogonal to the homogeneous solutions in both variables x and ξ. For this purpose we solve the defining equation (22.16) which holds for that function. This means (by the superposition principle of linear operators) that we add to $G_1(x, \xi)$ the solution of the equation

$$u'' + \rho^2 u = -\frac{2}{l} \cos \rho(\xi - x) \qquad (5.25.14)$$

This can be done since the auxiliary function $G_1(x, \xi)$ puts us in the position to solve the inhomogeneous equation (14) (remembering, however, that we have to consider x as the integration variable and this x should preferably be called x_1 in order to distinguish it from the previous x which is a mere constant during the integration process. The ξ on the right side of (14) becomes likewise x_1). In our problem we obtain:

$$u(\xi) = -\frac{2}{\rho l} \int_0^\xi \sin \rho(\xi - x_1) \cos \rho(x_1 - x) dx_1$$

$$= -\frac{2}{\rho l} \int_0^\xi \sin \rho t \cos \rho(\xi - x - t) dt \qquad (5.25.15)$$

$$= -\frac{1}{\rho l} \xi \sin \rho(\xi - x) - \frac{1}{\rho^2 l} \sin \rho \xi \sin \rho x$$

At this stage the constrained Green's function appears in the form

$$G_2(x, \xi) = G_1(x, \xi) + u(\xi) \qquad (5.25.16)$$

But this is not yet the final answer. We still have the uncertainty of the homogeneous solution

$$A \cos \rho \xi + B \sin \rho \xi \qquad (5.25.17)$$

We remove this uncertainty by demanding that the resulting function $G(x, \xi)$ (considered as a function of ξ while x is a mere parameter) be orthogonal to both $\cos \rho \xi$ and $\sin \rho \xi$.

We can simplify our task by writing the solution (15) somewhat differently. Since an arbitrary homogeneous solution can be added to (15), we can omit the last term while the first term may be replaced by

$$\frac{1}{\rho l} (x - \xi) \sin \rho(\xi - x) \qquad (5.25.18)$$

We do that because the natural variable of our problem is not ξ but

$$\xi - x = t \qquad (5.25.19)$$

Hence the function we want to orthogonalise becomes

$$G_2(t) = -\frac{t}{l\rho} \sin \rho t + \frac{1}{\rho} [\sin \rho t] \qquad (5.25.20)$$

where the symbol [ ] shall again indicate (cf. (19.16)) that all negative values inside the bracket are to be replaced by zero. Furthermore, the un-determined combination (17) can equally be replaced by

$$A \cos \rho t + B \sin \rho t \qquad (5.25.21)$$

(with different constants A and B). The sum of (20) plus (21) has to satisfy the condition of orthogonality, integrating with respect to the range $\xi = [0, l]$, that is

$$t = [-x, l - x] \qquad (5.25.22)$$

We thus get the two conditions for A and B:

$$\frac{-1}{\rho l} \int_{-x}^{l-x} t \sin \rho t \cos \rho t \, dt + \frac{1}{\rho} \int_{0}^{l-x} \sin \rho t \cos \rho \, dt + \frac{l}{2} A = 0$$

$$\qquad (5.25.23)$$

$$- \frac{1}{\rho l} \int_{-x}^{l-x} t \sin^2 t \, dt + \frac{1}{\rho} \int_{0}^{l-x} \sin^2 \rho t \, dt + \frac{l}{2} B = 0$$

This gives

$$A = - \frac{1}{2\rho^2 l}$$

$$\qquad (5.25.24)$$

$$B = - \frac{1}{2\rho}$$

and the final result becomes

$$G(x, \xi) = - \frac{1}{\rho l} \left[\left(\xi - x \pm \frac{l}{2} \right) \sin \rho(\xi - x) + \frac{1}{2\rho} \cos \rho(\xi - x) \right] \qquad (5.25.25)$$

where the upper sign holds for $\xi \leq x$, the lower sign for $\xi \geq x$. The symmetry in x, ξ is evident. Moreover, a comparison with the earlier expression (22.20) shows perfect agreement.

Problem 232. Apply the method of over-determination to Problem 228 by adding the boundary condition

$$v(0) = 0$$

Find the Green's function of this problem and construct with its help the constrained Green's function (22.24).

[Answer:

1. $G_1(x, \xi) = [x - \xi]$

2. $G_2(x, \xi) = [x - \xi] - \frac{1}{2l} (l - \xi)^2$

$$\qquad (5.25.26)$$

3. $G(x, \xi) = [x - \xi] + \frac{l}{6} - \frac{x^2}{2l} - \frac{1}{2l} (l - \xi)^2$

$$\qquad = \frac{1}{2} |x - \xi| + \frac{1}{2} (x + \xi) - \frac{l}{3} - \frac{x^2 + \xi^2}{2l} \,]$$

Problem 233.   Consider the problem of the free elastic bar of constant cross-section, putting $I(x) = 1$ (cf. Chapter 4.14 and Problem 231), with the added boundary conditions

$$v(0) = v'(0) = 0 \qquad (5.25.27)$$

Obtain the Green's function of this problem and construct with its help the constrained Green's function $G(x, \xi)$ of the free elastic bar.

[Answer:

$$G_1(x, \xi) = \frac{[x - \xi]^3}{6}$$

$$G_2(x, \xi) = \frac{[x - \xi]^3}{6} + (l - 3x) \frac{(l - \xi)^4}{12l^2}$$

$$- (l - 2x) \frac{(l - \xi)^5}{20l^3} \qquad (5.25.28)$$

$$G(x, \xi) = \frac{|x - \xi|^3}{12} + \frac{1}{20l^2} (x^5 + \xi^5) - \frac{1}{6l} (x^4 + \xi^4)$$

$$+ \frac{1}{12} (x^3 + \xi^3) - \frac{11l^2}{210} (x + \xi) + \frac{l^3}{105}$$

$$- x\xi \left(\frac{x^4 + \xi^4}{10l^3} - \frac{x^3 + \xi^3}{4l^2} + \frac{x + \xi}{4} - \frac{13}{35} l \right)]$$

5.26. Orthogonal expansions

In the realm of ordinary differential equations the explicit construction of the Green's function is a relatively simple task and always solvable if we are in the possession of the homogeneous solutions of the given differential operation (that is the right side is made zero while the boundary conditions are left free). In the realm of partial differential operators the conditions are much less favourable for the explicit construction of the Green's function and we know in fact only a few examples in which the Green's function is known in finite and closed form. Here another approach is frequently more adequate which elucidates the nature of the Green's function from an entirely different angle.

If we recall our basic discussions in introducing the Green's function (cf. Section 2), we observe that our deductions were essentially based on the *absence* of something. By adding the value of $v(x)$ at the point x to our data, we have added a surplus dimension to the U-space and obtained a compatibility condition for our over-determined system by demanding that the right side must be orthogonal to a principal axis which is associated with the eigenvalue zero, because that axis is not included in the eigen-space of the operator. It was this compatibility condition between $v(x)$ and the given data which led to the solution of our system in terms of the Green's function. It is thus understandable that the defining equation of the Green's function is *almost* completely the adjoint *homogeneous* equation excluding only an arbitrarily small neighbourhood of the point $\xi = x$ at which the functional value $v(x)$ was prescribed.

On the other hand, in our discussions of matrices we have seen that a matrix as an operator could be completely characterised in terms of those principal axes which are *positively* represented in the matrix. To find all the principal axes of the eigenvalue problem (3.7.7) is generally a much more elaborate task than solving the linear system (3.6.1) by direct matrix inversion. But if we possess all these principal axes, then we possess also the solution of our linear system, without having recourse to determinants or matrix inversion. In Chapter 3.11 we have seen that by a proper orthogonal transformation of both the unknown vector and the right side an arbitrary $n \times m$ matrix could be diagonalised. In the diagonal form the equations are separated and automatically solvable.

Here we have a method of solving a linear system which omits the zero axes—in which the operator is not activated—and operates solely with those axes which are actively present in the operator. These axes were characterised by the "shifted eigenvalue problem" (3.7.7):

$$\begin{aligned} Av &= \lambda u \\ \tilde{A}u &= \lambda v \end{aligned} \qquad (5.26.1)$$

which, if translated into the field of differential operators, has to be interpreted as the following pair of equations:

$$\begin{aligned} Dv(x) &= \lambda u(x) \\ \tilde{D}u(x) &= \lambda v(x) \end{aligned} \qquad (5.26.2)$$

Here $v(x)$ is subjected to the given homogeneous (or homogenised) boundary conditions, $u(x)$ to the adjoint homogeneous conditions.

The new feature which enters in dissimilitude to the matrix problem is that the function space has an *infinity of dimensions* and accordingly we obtain an infinity of solutions for the system (2). The eigenvalues λ_i of our system remain usually *discrete*, that is our system is solvable only for a definite sequence of discrete eigenvalues*

$$\lambda_i = \lambda_1, \lambda_2, \lambda_3, \ldots \qquad (5.26.3)$$

which we will arrange in increasing order, starting with the smallest eigenvalue λ_1, and continuing with the larger eigenvalues which eventually become arbitrarily large. In harmony with our previous policy we will omit all the negative λ_i and also all the zero eigenvalues, which in the case of partial operators may be present with infinite multiplicity.

Now the corresponding eigenfunctions

$$u_1(x), u_2(x), \ldots, u_i(x), \ldots$$

and

$$v_1(x), v_2(x), \ldots, v_i(x), \ldots \qquad (5.26.4)$$

---

* We consider solely *finite* domains; the confluence of eigenvalues in an infinite space is outside the scope of our discussions.

form an ortho-normal set of functions if we agree that the "length" of all these functions (cf. Chapter 4.7) shall be normalised to 1:

$$\int u_i(x)u_k(x)dx = \delta_{ik}$$
$$\int v_i(x)v_k(x)dx = \delta_{ik} \tag{5.26.5}$$

(The normalisation of the $u_i(x)$ automatically entails the normalisation of the $v_i(x)$ and vice versa; cf. Problem 118.)

These functions represent in function space an orthogonal set of base vectors of the length 1. The fact that their number is infinity does not guarantee automatically that they include the *entire* function space. In fact, if the eigenvalue problem (2) allows solutions for $\lambda = 0$—which makes our system either over-determined or incomplete or both—we know in advance that by throwing away these solutions our function system (4) will *not* cover the entire function space. But these functions will cover the entire *eigen-space* of the operator and this is sufficient for the solution of the problem

$$Dv(x) = \beta(x) \tag{5.26.6}$$

or

$$\tilde{D}u(x) = \gamma(x) \tag{5.26.7}$$

The omission of the zero-axes is a property of the operator itself and not the fault of the (generally incomplete) orthogonal system (4)—which, however, must not omit any of the eigenfunctions which belong to a positive λ_i, including possible multiplicities on account of two or more eigenvalues collapsing into one, in which case the associated eigenfunctions have to be properly ortho-normalised.

The eigenfunctions $u_i(x)$ are present in sufficient number to allow an expansion of $\beta(x)$ into these functions, in the form of an infinite convergent series:

$$\beta(x) = \sum_{i=1}^{\infty} \beta_i u_i(x) \tag{5.26.8}$$

We want to assume that $\beta(x)$ belongs to a class of functions for which the expansion converges (if $\beta(x)$ is everywhere in the given domain finite, sectionally continuous, and of bounded variation, this condition is certainly satisfied. It likewise suffices that $\beta(x)$ shall be sectionally differentiable). We now multiply this expansion by a certain $u_k(x)$ and integrate over the given domain term by term. This is permissible, as we know from the theory of convergent infinite series. Then on the right side every term except the kth drops out, in consequence of the orthogonality conditions (5), while in the kth term we get β_k. And thus

$$\beta_k = \int \beta(x)u_k(x)dx \tag{5.26.9}$$

On the other hand, the unknown function $v(x)$ can likewise be expanded, but here we have to use the functions $v_i(x)$ (which may belong to a completely different functional domain, for example $v(x)$ may be a scalar, $\beta(x)$ a vector, cf. Section 5, and Problem 236):

$$v(x) = \sum_{i=1}^{\infty} b_i v_i(x) \qquad (5.26.10)$$

Here again we obtain for the expansion coefficients

$$b_k = \int v(x) v_k(x) dx \qquad (5.26.11)$$

but this relation is of no help if $v(x)$ is an unknown function. However, the expansion coefficients β_k are available if $\beta(x)$ is a given function, and now the differential equation (6) establishes the following simple relation between the expansion coefficients b_k and β_k:

$$\lambda_k b_k = \beta_k \qquad (5.26.12)$$

from which

$$b_k = \frac{\beta_k}{\lambda_k} \qquad (5.26.13)$$

We will assume that the eigenvalue spectrum (3) starts with a definite finite smallest eigenvalue λ_1; this is not self-evident since the eigenvalue spectrum may have a "*condensation point*" or "*limit point*" at $\lambda = 0$ in which case we have an infinity of eigenvalues which come arbitrarily near to $\lambda = 0$ and a minimum does not exist. Such problems will be our concern in a later chapter. For the present we exclude the possibility of a limit point at $\lambda = 0$. Then the infinite series

$$v(x) = \sum_{i=1}^{\infty} \frac{\beta_i}{\lambda_i} v_i(x) \qquad (5.26.14)$$

is even better convergent than the original series (8)—because we *divide* by the λ_i which increase to infinity—and we can consider the series (14) as the solution of our problem. This solution is unique since we have put the solution into the eigen-space of the operator, in case the homogeneous equation

$$Dv^k(x) = 0 \qquad (5.26.15)$$

allows solutions which do not vanish identically. The uniqueness is established by demanding that $v(x)$ shall be orthogonal to every solution of the homogeneous equation:

$$\int v(x) v^k(x) dx = 0 \qquad (k = 1, 2, \ldots, \rho) \qquad (5.26.16)$$

The corresponding condition for the given right side:

$$\int \beta(x) u^j(x) dx = 0 \qquad (j = 1, 2, \ldots, \sigma) \qquad (5.26.17)$$

where $u^j(x)$ is any independent solution of the adjoint homogeneous equation:

$$\check{D} u^j(x) = 0 \qquad (5.26.18)$$

is demanded by the compatibility of the system: the right side must have no components in those dimensions of the function space which are not included by the operator. We have to test the given function $\beta(x)$ as to the validity of these conditions because, if these conditions are not fulfilled, we know in advance that the given problem is not solvable.

Problem 234. Given the partial differential equation

$$\frac{\partial v}{\partial x} = \beta(x, y) \tag{5.26.19}$$

in a closed two-dimensional domain with the boundary condition

$$v(S) = 0 \tag{5.26.20}$$

on the boundary S. Find the compatibility conditions of this problem.

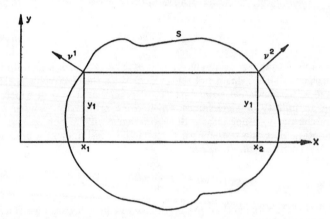

[Answer:

$$\int_{x_1}^{x_2} \beta(x, y_1)dx = 0 \tag{5.26.21}$$

for every fixed y_1.]

Problem 235. Given the same homogeneous problem

$$\frac{\partial v}{\partial x} = 0 \tag{5.26.22}$$

with inhomogeneous boundary conditions

$$v(S) = f(S) \tag{5.26.23}$$

find the compatibility conditions for the boundary data $f(S)$.

[Answer:

$$f(x_1, y_1)\nu_x{}^1 + f(x_2, y_1)\nu_x{}^2 = 0 \tag{5.26.24}$$

for all $y = y_1$.]

Problem 236.   Formulate the eigenvalue problem for the scalar-vector problem (5.3) and obtain the orthogonal expansions associated with it.

[Answer:

$$\Delta\varphi_i + \lambda_i^2\varphi_i = 0 \qquad (\lambda_i > 0) \tag{5.26.25}$$

with the boundary condition

$$\frac{\partial\varphi}{\partial\nu} = 0 \qquad (\text{on } S) \tag{5.26.26}$$

$$F = \sum_{i=1}^{\infty} f_i \operatorname{grad} \varphi_i \tag{5.26.27}$$

where

$$f_i = \int F \cdot \operatorname{grad} \varphi_i \cdot d\tau \tag{5.26.28}$$

($d\tau$ = volume element of 3-dimensional space).

$$\Phi(\tau) = \sum_{i=1}^{\infty} \frac{f_i}{\lambda_i} \varphi_i(\tau) \tag{5.26.29}]$$

5.27. The bilinear expansion

We consider the solution of the differential equation (26.6) which we have obtained in the form

$$v(x) = \sum_{i=1}^{\infty} \frac{\beta_i}{\lambda_i} v_i(x) \tag{5.27.1}$$

where

$$\beta_i = \int \beta(\xi) u_i(\xi) d\xi \tag{5.27.2}$$

If we substitute this value of β_i into the expansion (1), we obtain:

$$v(x) = \sum_{i=1}^{\infty} \frac{1}{\lambda_i} \int \beta(\xi) u_i(\xi) v_i(x) d\xi \tag{5.27.3}$$

In this sum the process of summation and integration is not necessarily interchangeable.   But if we do not go with i to infinity but only up to n, the statement of equation (3) can be formulated as follows:

$$v(x) = \lim_{n\to\infty} S_n(x) \tag{5.27.4}$$

where

$$S_n(x) = \sum_{i=1}^{n} \frac{1}{\lambda_i} \int \beta(\xi) u_i(\xi) v_i(x) d\xi$$

$$= \int \beta(\xi) G_n(x, \xi) d\xi \tag{5.27.5}$$

the auxiliary function $G_n(x, \xi)$ being defined as follows:

$$G_n(x, \xi) = \sum_{i=1}^{n} \frac{v_i(x)u_i(\xi)}{\lambda_i} \tag{5.27.6}$$

From the fact that $S_n(x)$ converges to a definite limit—namely $v(x)$—we cannot conclude that the sequence $G_n(x, \xi)$ must converge to a definite limit. Even if the limit of $G_n(x, \xi)$ (n growing to infinity) does not exist, the integration over ξ, after multiplying by $\beta(\xi)$, will entail convergence. But even assuming that $G_n(x, \xi)$ diverges, this divergence is so weak that a very small modification of the coefficients $1/\lambda_i$ assures convergence. This modification is equivalent to a local smoothing of the same kind that changes $\delta(x, \xi)$ to $\delta_\epsilon(x, \xi)$. Hence we can say that with an arbitrarily small modification of the coefficients the sum (6) converges to a function $G_\epsilon(x, \xi)$ which differs only of the order of magnitude ϵ from a definite function $G(x, \xi)$. With this understanding it is justified to put

$$G(x, \xi) = \sum_{i=1}^{\infty} \frac{v_i(x)u_i(\xi)}{\lambda_i} \tag{5.27.7}$$

and replace the sum (5) in the limit by

$$v(x) = \int G(x, \xi)\beta(\xi)d\xi \tag{5.27.8}$$

But then we are back at the standard solution of a differential equation by the Green's function and we see that our function (7) has to be identified with the Green's function of the given differential operator. The important expansion (7) is called the "bilinear expansion" since it is linear in the functions $v_i(x)$ and likewise in the functions $u_i(\xi)$.

In the literature the bilinear expansion of the Green's function appears usually in the form

$$G(x, \xi) = G(\xi, x) = \sum_{i=1}^{\infty} \frac{v_i(x)v_i(\xi)}{\lambda_i} \tag{5.27.9}$$

and is restricted to *self-adjoint* operators. The eigenvalues λ_i and the associated eigenfunctions $v_i(x)$ are then defined in terms of the traditional eigenvalue problem

$$Dv(x) = \lambda v(x) \tag{5.27.10}$$

once more omitting the solutions for $\lambda = 0$ but keeping all the positive and negative λ_i for which a solution is possible. The transition to the "shifted eigenvalue problems" (26.2) permits us to generalise the usual self-adjoint expansion to a much wider class of operators which includes not only the "well-posed", although not self-adjoint problems but even the case of arbitrarily over-determined or under-determined problems. Hence the functions $v_i(x)$, $u_i(\xi)$ need not belong to the same domain of the function space but may operate in completely different domains.

For example in our Problem 236 the functions $v_i(x)$ are the scalar functions $\varphi_i(x)$, while the functions $u_i(\xi)$ are the vectorial functions grad $\varphi_i(\xi)$. The Green's function of our problem—which is a scalar with respect to the point x and a vector with respect to the point ξ—is obtainable with the help of the following infinite expansion:

$$G(x, \xi) = \sum_{i=1}^{\infty} \frac{\varphi_i(x) \text{ grad } \varphi_i(\xi)}{\lambda_i} \qquad (5.27.11)$$

If we compare this Green's function with the much simpler solution found in Section 5, we are struck by the simplicity of the previous result and the complexity of the new result. The use of the Green's function (11) entails an integration over a three-dimensional domain:

$$\Phi(x) = \sum_{i=1}^{\infty} \frac{\varphi_i(x)}{\lambda_i} \int F(\xi) \text{ grad } \varphi_i(\xi) d\xi \qquad (5.27.12)$$

while previously a simple *line-integral* (5.16) gave the answer. Moreover, the previous Green's function could be given explicitly while the new Green's function can only be given as the limit of an infinite sum whose actual construction would require an exceedingly elaborate scheme of calculations. What is the cause of this discrepancy?

The problem we have studied is strongly over-determined since we have given a vector field for the determination of a scalar field. This means that the $U_M = U + U_0$ space of the right side extends far beyond the confines of the U space in which the vector is activated. The Green's function (11) is that particular Green's function which spans the complete eigen-space of the operator but has no components in any of the dimensions which go beyond the limitations of the U space. On the other hand, there is no objection to the use of a Green's function which spills over into the U_0 space. We can add to our constrained $G(x, \xi)$, defined by (11), an arbitrary sum of the type

$$\sum \rho_i(x) u^i(\xi) \qquad (5.27.13)$$

where the $\rho_i(x)$ can be chosen freely as any functions of x. This additional sum will not contribute anything to the solution $v(x)$ since the right side satisfies the compatibility conditions

$$\int \rho(\xi) u^j(\xi) d\xi = 0 \qquad (5.27.14)$$

(cf. 26.17) and thus automatically *annuls* the contribution from the added sum (13). As we have seen in Section 5, over-determined systems possess the great advantage that we can choose our Green's function much more liberally than we can in a well-determined problem $n = m = p$, where in fact the Green's function is uniquely defined.

Another interesting conclusion can be drawn from the bilinear expansion concerning the reciprocity theorem of the Green's function, encountered earlier in Section 12. Let us assume that we want to solve the *adjoint* equation (26.7). Then our shifted eigenvalue problem (26.2) shows at once that in this case we get exactly the same eigenvalues and eigenfunctions, with the only difference that the role of the functions $u_i(x)$ and $v_i(x)$ is now *exchanged*. Hence the bilinear expansion of the new Green's function $\tilde{G}(x, \xi)$ becomes:

$$\tilde{G}(x, \xi) = \sum_{i=1}^{\infty} \frac{u_i(x)v_i(\xi)}{\lambda_i} \tag{5.27.15}$$

but this is exactly the previous expansion (7), except that the points x and ξ are exchanged; and thus

$$\tilde{G}(x, \xi) = G(\xi, x) \tag{5.27.16}$$

which is in fact the fundamental *reciprocity theorem* of the Green's function.

In the case of *self-adjoint* systems the bilinear expansion (9) becomes in itself symmetric in x and ξ and we obtain directly the *symmetry theorem* of the Green's function for such systems:

$$G(x, \xi) = G(\xi, x) \tag{5.27.17}$$

All these results hold equally for ordinary as for partial differential operators since they express a basic behaviour which is common to all linear operators. But in the case of ordinary differential equations a further result can be obtained. We have mentioned that generally the convergence of the bilinear expansion (7) cannot be guaranteed without the proper modifications. The difficulty arises from the fact that the Green's function of an arbitrary differential operator need not be a very smooth function. If we study the character of the bilinear expansion, we notice that we can conceive it as an ordinary orthogonal expansion into the ortho-normal system $v_i(x)$, if we consider x as a variable and keep the point ξ fixed, or another orthogonal expansion into the eigenfunctions $u_i(\xi)$, if we consider ξ as the variable and x as a fixed point. The expandability of $G(x, \xi)$ into a convergent bilinear series will then depend on whether or not the function $G(x, \xi)$ belongs to that class of functions which allow an orthogonal expansion into a complete system of ortho-normal functions. This "completeness" is at present of a restricted kind since the functions $v_i(x)$ and $u_i(x)$ are generally complete only with respect to a certain *subspace* of the function space. However, this subspace coincides with the space in which the constrained Green's function finds its place. Hence we have no difficulty on account of the completeness of our functions. The difficulty arises from the fact that $G(x, \xi)$ may not be quadratically integrable or may be for other reasons too unsmooth to allow an orthogonal expansion.

In the domain of ordinary differential equations, however, such an unsmoothness is excluded by the fact that the Green's function, considered as a function of x, satisfies the homogeneous differential equation $Dv(x) = 0$ with the only exception of the point $x = \xi$. Hence $G(x, \xi)$ is automatically a sectionally continuous and even differentiable function which remains everywhere finite. The discontinuity at the point $x = \xi$ (in the case of first order operators) is the only point where the smoothness of the function suffers. But this discontinuity is not sufficient to destroy the convergence of an orthogonal expansion, although naturally the convergence cannot be *uniform* at the point of discontinuity. And thus we come to the conclusion that in the case of *ordinary* differential operators we can count on the convergence (and even uniform convergence if the point $x = \xi$ is excluded) of the bilinear expansion, without modifying the coefficients of the expansion by local smoothing.

Problem 237. In Section 23 we have studied Legendre's differential operator which is self-adjoint. Its eigenvalues are

$$\lambda_k = -k(k + 1), \qquad (k = 0, 1, 2, \ldots) \tag{5.27.18}$$

with the normalised eigenfunctions

$$v_k(x) = \sqrt{\frac{2n + 1}{2}} \, P_k(x) \tag{5.27.19}$$

where $P_k(x)$ are the "Legendre polynomials", defined by

$$P_k(x) = \frac{1}{2^k k!} [(x^2 - 1)]^{(k)} \tag{5.27.20}$$

These polynomials are alternately even and odd, e.g.

$$\begin{aligned} P_0(x) &= 1 \\ P_1(x) &= x \\ P_2(x) &= \tfrac{1}{2}(3x^2 - 1) \\ P_3(x) &= \tfrac{1}{2}(5x^3 - 3x) \end{aligned} \tag{5.27.21}$$

They have the common characteristics that they assume at $x = 1$ their maximum value 1.

On the basis of the results of Section 23 obtain the following infinite expansions:

$$2 \log 2 - \log [(1 - \xi)(1 + x)] = 1 + \sum_{n=1}^{\infty} P_n(x) P_n(\xi) \left(\frac{1}{n} + \frac{1}{n + 1} \right) \qquad (\xi \le x) \tag{5.27.22}$$

$$\log 2 - \log (1 + x) = 1 + \sum_{n=1}^{\infty} (-1)^n P_n(x) \left(\frac{1}{n} + \frac{1}{n + 1} \right) \tag{5.27.23}$$

$$\log 2 - \log (1 - x) = 1 + \sum_{n=1}^{\infty} P_n(x) \left(\frac{1}{n} + \frac{1}{n + 1} \right) \tag{5.27.24}$$

$$2 \log 2 - \log (1 - x^2) = 1 + \sum_{n=1}^{\infty} P_n^2(x) \left(\frac{1}{n} + \frac{1}{n + 1} \right) \tag{5.27.25}$$

$$2 \log 2 = 1 + \left(\frac{3.4}{1.2}\right)^2 \frac{1}{4^4}\left(\frac{1}{2} + \frac{1}{3}\right) + \left(\frac{4.5.6}{1.2.3}\right)^2 \frac{1}{4^6}\left(\frac{1}{4} + \frac{1}{5}\right) + \cdots$$

$$= 1 + \sum_{n=1}^{\infty} \left[\frac{(2n)!}{(n!)^2 4^n}\right]^2 \left(\frac{1}{2n} + \frac{1}{2n+1}\right) \tag{5.27.26}$$

$$\log \frac{1+x}{1-x} = \sum_{n=0}^{\infty} P_{2n+1}(x)\left(\frac{1}{2n+1} + \frac{1}{2n+2}\right) \qquad (x \ge 0)$$

$$= \sum_{n=0}^{\infty} P^2{}_{2n+1}(x)\left(\frac{1}{2n+1} + \frac{1}{2n+2}\right) \tag{5.27.27}$$

Problem 238. Legendre's differential operator can be obtained by starting with the first order operator

$$Dv = \sqrt{1-x^2}\,v' \tag{5.27.28}$$

Then the operator on the left side of (23.1) becomes $-\check{D}\,Dv$ which shows that the eigenvalues of the shifted eigenvalue problem (26.2) associated with (28) are equal to

$$\lambda_k = \sqrt{k(k+1)} \tag{5.27.29}$$

Obtain the Green's function of the operator (28) for the range [0, 1], with the boundary condition

$$v(0) = 0 \tag{5.27.30}$$

and apply to it the bilinear expansion (6).

[Answer:

$$G(x, \xi) = \frac{1}{\sqrt{1-\xi^2}} \qquad (0 \le \xi < x) \tag{5.27.31}$$

$$= 0 \qquad (1 \ge \xi > x)$$

$$\frac{2}{1-x^2} = \sum_{n=0}^{\infty} P_{2n+1}(\xi) P'{}_{2n+1}(x)\left(\frac{1}{2n+1} + \frac{1}{2n+2}\right) \qquad (\xi > x) \tag{5.27.32}$$

$$\sum_{n=0}^{\infty} P_{2n+1}(\xi) P'{}_{2n+1}(x)\left(\frac{1}{2n+1} + \frac{1}{2n+2}\right) = 0 \qquad (\xi < x) \tag{5.27.33}]$$

Problem 239. Show that at the point of discontinuity $\xi = 0$ the series (32) yields the arithmetic mean of the two limiting ordinates, and thus:

$$\frac{1}{1-x^2} = \sum_{k=0}^{\infty} P_{2n+1}(x) P'{}_{2n+1}(x)\left(\frac{1}{2n+1} + \frac{1}{2n+2}\right) \qquad (x > 0) \tag{5.27.34}$$

Problem 240. Obtain the Green's function and its bilinear expansion for the following operator:

$$Dv(x) = v'(x) \qquad x = [0, 1] \tag{5.27.35}$$

$$v = 0 \tag{5.27.36}$$

Do the same for the operator $\check{D}D$.

[Answer:

$$v_k(x) = \sqrt{2} \sin \lambda_k x$$
$$u_k(x) = \sqrt{2} \cos \lambda_k x \tag{5.27.37}$$

$$\lambda_k = \frac{\pi}{2}(2k + 1) \tag{5.27.38}$$

1.

$$G(x, \xi) = 1 \quad (\xi < x)$$
$$\qquad\quad = 0 \quad (\xi > 0) \tag{5.27.39}$$

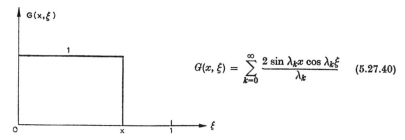

$$G(x, \xi) = \sum_{k=0}^{\infty} \frac{2 \sin \lambda_k x \cos \lambda_k \xi}{\lambda_k} \tag{5.27.40}$$

2.

$$G(x, \xi) = \xi \quad (\xi > x)$$
$$\qquad\quad = 1 \quad (\xi \geq x) \tag{5.27.41}$$

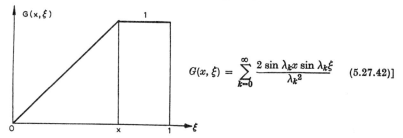

$$G(x, \xi) = \sum_{k=0}^{\infty} \frac{2 \sin \lambda_k x \sin \lambda_k \xi}{\lambda_k{}^2} \tag{5.27.42}]$$

Problem 241. Solve the same problem for the boundary condition

$$v(1) - v(0) = 0 \tag{5.27.43}$$

[Answer:

$$v_k(x) = \sqrt{2} \sin \lambda_k x, \quad \sqrt{2} \cos \lambda_k x$$
$$u_k(x) = \sqrt{2} \cos \lambda_k(x), \quad -\sqrt{2} \sin \lambda_k x \tag{5.27.44}$$

$$\lambda_k = 2k\pi \qquad (k = 1, 2, \ldots) \tag{5.27.45}$$

$$G(x, \xi) = \xi - x + \tfrac{1}{2} \quad (\xi < x)$$
$$\qquad\quad = \xi - x - \tfrac{1}{2} \quad (\xi > x) \tag{5.27.46}$$

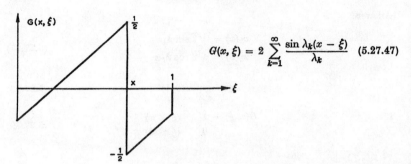

$$G(x, \xi) = 2 \sum_{k=1}^{\infty} \frac{\sin \lambda_k (x - \xi)}{\lambda_k} \quad (5.27.47)$$

2. $$G(x, \xi) = \tfrac{1}{2}(\xi - x)^2 + \tfrac{1}{12} - \tfrac{1}{2}|x - \xi| \quad (5.27.48)$$

$$G(x, \xi) = 2 \sum_{k=1}^{\infty} \frac{\cos \lambda_k (x - \xi)}{\lambda_k^2} \quad (5.27.49)]$$

Problem 242.   Solve the same problem for the boundary condition

$$v(1) + v(0) = 0 \quad (5.27.50)$$

[Answer:

$$v_k(x) = \sqrt{2}\, \sin \lambda_k x, \quad \sqrt{2}\, \cos \lambda_k x$$
$$u_k(x) = \sqrt{2}\, \cos \lambda_k x, \quad -\sqrt{2}\, \sin \lambda_k x \quad (5.27.51)$$
$$\lambda_k = (2k + 1)\pi \quad (k = 0, 1, 2, \ldots)$$

1. $$G(x, \xi) = \tfrac{1}{2} \quad (\xi < x)$$
$$= -\tfrac{1}{2} \quad (\xi > x) \quad (5.27.52)$$

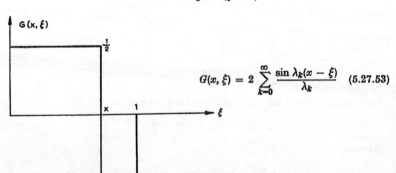

$$G(x, \xi) = 2 \sum_{k=0}^{\infty} \frac{\sin \lambda_k (x - \xi)}{\lambda_k} \quad (5.27.53)$$

2. $$G(x, \xi) = \tfrac{1}{4} - \tfrac{1}{2}|x - \xi| \quad (5.27.54)$$

$$G(x, \xi) = 2 \sum_{k=0}^{\infty} \frac{\cos \lambda_k (x - \xi)}{\lambda_k^2} \quad (5.27.55)]$$

Problem 243. Consider the same problem with the boundary condition

$$v(1) - \alpha v(0) = 0 \qquad (\alpha \neq 1) \tag{5.27.56}$$

where α is an arbitrary real constant (excluding the value $\alpha = 1$ which was treated before). Study particularly the expansions for the point $\xi = x$.

[Answer:

Define an angle λ_0 between $\pm \pi$ by putting

$$\tan \frac{\lambda_0}{2} = \frac{1 - \alpha}{1 + \alpha}, \quad \theta_0 = \frac{\pi}{4} - \frac{\lambda_0}{2} \tag{5.27.57}$$

Then the eigenvalues and associated eigenfunctions become:

$$|\lambda_0|, \quad v_0(x) = \sqrt{2} \cos (\lambda_0 x + \theta_0)$$
$$\lambda_k = \lambda_0 + 2k\pi, \quad v_k(x) = \sqrt{2} \cos [(\lambda_0 + 2k\pi)x + \theta_0] \tag{5.27.58}$$
$$\bar{\lambda}_k = -\lambda_0 + 2k\pi, \quad \bar{v}_k(x) = \sqrt{2} \cos [(-\lambda_0 + 2k\pi)x - \theta_0]$$

1.
$$G(x, \xi) = \frac{\alpha}{\alpha - 1} \qquad (\xi < x)$$
$$= \frac{1}{\alpha - 1} \qquad (\xi > x) \tag{5.27.59}$$

$$G(x, x) = - \frac{\sin(2\lambda_0 x + 2\theta_0)}{\lambda_0} - \sum_{k=1}^{\infty} \left[\frac{\sin(2\lambda_k x + \theta_0)}{\lambda_k} + \frac{\sin (2\bar{\lambda}_k x - \theta_0)}{\bar{\lambda}_k} \right]$$
$$= \frac{1}{2} \frac{\alpha + 1}{\alpha - 1} \tag{5.27.60}$$

2.
$$G(x, \xi) = \frac{1}{(1 - \alpha)^2} + \frac{1}{2} \frac{\alpha + 1}{\alpha - 1} (x + \xi) - \tfrac{1}{2}|x - \xi| \tag{5.27.61}$$

$$G(x, x) = \frac{1 + \cos (2\lambda_0 x + 2\theta_0)}{\lambda_0^2} + \sum_{k=1}^{\infty} \left(\frac{1}{\lambda_k^2} + \frac{1}{\bar{\lambda}_k^2} \right)$$
$$+ \sum_{k=1}^{\infty} \left[\frac{\cos (2\lambda_k x + 2\theta_0)}{\lambda_k^2} + \frac{\cos (2\bar{\lambda}_k x - 2\theta_0)}{\bar{\lambda}_k^2} \right] \tag{5.27.62}$$

5.28. Hermitian problems

In all our previous dealings we have restricted ourselves to the case of *real* operators with *real* boundary conditions. However, in applied problems the more general case of complex elements is of frequent occurrence. For example the fundamental operations of wave-mechanics have the imaginary unit i inherently built into them. These operations are self-adjoint in the Hermitian sense. But even in classical physics we encounter the need for complex elements. In all diffraction problems we solve the time-dependent wave equation by taking out the factor $e^{i\omega t}$, thus reducing the wave equation to the equation

$$\Delta v + k^2 v = 0 \tag{5.28.1}$$

This equation does not reveal any complex elements. However, the *boundary condition* of "outgoing waves" demands in infinity the condition

$$\frac{\partial v}{\partial \nu} + ikv = 0 \tag{5.28.2}$$

This condition would be self-adjoint in the algebraic sense but is *not* self-adjoint in the Hermitian sense since the adjoint boundary condition becomes

$$\frac{\partial u}{\partial \nu} - iku = 0 \tag{5.28.3}$$

If we want to make use of the method of eigenfunctions for the solution of our diffraction problem, we have to complement the given problem by the adjoint problem which demands *incoming* instead of outgoing waves.

The general procedure of obtaining the adjoint operator in the presence of complex elements is as follows. We go through the regular procedure of obtaining the adjoint operator $\tilde{D}$ and the adjoint boundary conditions, paying no attention to the fact that some of the coefficients encountered in this process are complex numbers. Now, after obtaining our $\tilde{D}u$, we consider this expression as a preliminary result and obtain the final $\tilde{D}^*$ by *changing every i to $-i$*. For example in the above diffraction problem the given differential operator is self-adjoint and thus

$$\tilde{D}u = \Delta u + k^2 u \tag{5.28.4}$$

There is no change here since the imaginary unit does not occur anywhere. However, the adjoint boundary condition—obtained in the usual fashion, with the help of the extended Green's identity—becomes

$$\frac{\partial u}{\partial \nu} + iku = 0 \tag{5.28.5}$$

This condition has to be changed to

$$\frac{\partial u}{\partial \nu} - iku = 0 \tag{5.28.6}$$

and we see that our problem loses its self-adjoint character. Without this change of i to $-i$, however, our eigenvalue problem would lose its significance by not yielding real eigenvalues or possibly not yielding any eigensolutions at all. On the other hand, we know in advance from the general analytical theory that the shifted eigenvalue problem with the proper boundary conditions will yield an infinity of *real* eigenvalues and a corresponding set of eigenfunctions which, although complex in themselves, form an *orthonormal set of functions* in the sense that

$$\int \varphi^*_i(\xi)\varphi_k(\xi)d\xi = \int \varphi_i(\xi)\varphi^*_k(\xi)dx = \delta_{ik} \tag{5.28.7}$$

In this section we will study the nature of such problems with the help of an over-simplified model which is nevertheless instructive by demonstrating the basic principles without serious technical complications. We return to the previous problem 243 with the only change that in the boundary condition (27.56) we will now assume that α is a *complex* constant. Since the adjoint boundary condition came out previously in the form

$$u(1) - \frac{1}{\alpha} u(0) = 0 \tag{5.28.8}$$

we now have to change this condition to

$$u(1) - \frac{1}{\alpha^*} u(0) = 0 \tag{5.28.9}$$

Since the differential operator has not changed, we know in advance that our eigensolutions $v_k(x)$ will once more be of the form

$$v_k(x) = A \cos (\lambda_k x + \theta) \tag{5.28.10}$$

and the λ_k must again become positive real numbers. Moreover, the shifted eigenvalue problem yields

$$u_k(x) = -A \sin (\lambda_k x + \theta) \tag{5.28.11}$$

The two boundary conditions (27.56) and (8) give the following two conditions:

$$\cos (\lambda + \theta) - \alpha \cos \theta = 0$$
$$\sin (\lambda + \theta) - \frac{1}{\alpha^*} \sin \theta = 0 \tag{5.28.12}$$

which, if expanded, gives the determinant condition

$$1 - \left(\alpha + \frac{1}{\alpha^*}\right) \cos \lambda + \frac{\alpha}{\alpha^*} = 0 \tag{5.28.13}$$

and thus

$$\cos \lambda = \frac{1 + \dfrac{1}{\alpha^*}}{\alpha + \dfrac{1}{\alpha^*}} = \frac{\alpha + \alpha^*}{1 + \alpha\alpha^*} \tag{5.28.14}$$

We see that in spite of the complex nature of α the eigenvalues λ_k become *real*. In fact, we obtain once more the same system of eigenvalues as before in (27.58):

$$\lambda = \lambda_0 \quad \text{and} \quad \lambda = 2k\pi \pm \lambda_0 \qquad (k = 1, 2, 3, \ldots) \tag{5.28.15}$$

where λ_0 is defined as an angle between 0 and π, satisfying the equation

$$\cos \lambda_0 = \frac{\alpha + \alpha^*}{1 + \alpha\alpha^*} \tag{5.28.16}$$

The eigenfunctions are likewise similarly constructed as those tabulated in

(27.58) but the previous angle θ_0 becomes now complex, being determined by the relation

$$\tan \theta = \frac{\cos \lambda_0 - \alpha}{\sin \lambda_0} \tag{5.28.17}$$

The eigenfunctions $v_k(x)$, $u_k(x)$ now become

$$\begin{aligned} v_k(x) &= A_k \cos (\lambda_k x \pm \theta) \\ u_k(x) &= -A_k \sin (\lambda_k x \pm \theta) \end{aligned} \tag{5.28.18}$$

where the upper sign holds for $\lambda_k = 2k\pi + \lambda_0$ and the lower sign for $\lambda_k = 2k\pi - \lambda_0$. The amplitude factor A_k follows from the condition

$$A^2 \int_0^1 |\cos (\lambda x \pm \theta)|^2 dx = 1 \tag{5.28.19}$$

This condition leads to the following normalisation of the functions $v(x)$ and $u(x)$:

$$\begin{aligned} v(x) &= \sqrt{2}[\cos (\lambda x + \theta_0) \cos \gamma - i \sin (\lambda x + \theta_0) \sin \gamma] \\ u(x) &= -\sqrt{2}[\sin (\lambda x + \theta_0) \cos \gamma + i \cos (\lambda x + \theta_0) \sin \gamma] \end{aligned} \tag{5.28.20}$$

with

$$\lambda = \lambda_0 \pm 2k\pi \tag{5.28.21}$$

$$\theta_0 = \frac{\pi}{4} - \frac{\lambda_0}{2} \tag{5.28.22}$$

The boundary conditions (12) establish the following relation between λ_0, γ and the original complex constant α:

$$\tan \frac{\lambda_0}{2} e^{2\gamma i} = \frac{1 - \alpha}{1 + \alpha} \tag{5.28.23}$$

(For the sake of formal simplicity we have departed from our usual convention of consistently positive eigenvalues. If we want to operate with consistently positive λ_k, we have to change the sign of λ, θ_0, and γ for the second group of eigenvalues which belong to the negative sign of the formula (21).)

The Green's function can again be constructed along the usual lines. However, in view of the complex elements of the operator (which in our problem come into evidence only in the boundary conditions) some characteristic modifications have to be observed. First of all, the Green's function corresponds to the inverse operator and this feature remains unaltered even in the presence of complex elements. Since the proper algebraic adjoint is not the Hermitian adjoint $\check{D}^*$ but $\check{D}$, the definition of the Green's function —considered as a function of ξ—must occur once more in terms of $\check{D}$:

$$\check{D}u(\xi) = \delta(x, \xi) \tag{5.28.24}$$

Furthermore, if the homogeneous equation

$$Dv^k(\xi) = 0 \qquad (5.28.25)$$

possesses non-vanishing (orthogonalised and normalised) solutions, the modification of the right side has to be made as follows:

$$\tilde{D}u(\xi) = \delta(x, \xi) - \sum_{k=1}^{p} v^k(x)v^{k*}(\xi) \qquad (5.28.26)$$

while in the case of the adjoint Green's $\tilde{G}(x, \xi)$ the corresponding equation becomes:

$$D^*v(\xi) = \delta(x, \xi) - \sum_{j=1}^{r} u^j(x)u^{j*}(\xi) \qquad (5.28.27)$$

(Notice that the *asterisk appears consistently in connection with the variable ξ*.)   The bilinear expansion (27.7) becomes now modified as follows:

$$G(x, \xi) = \sum_{i=1}^{\infty} \frac{v_i(x)u^*_i(\xi)}{\lambda_i} \qquad (5.28.28)$$

The corresponding expansion of the adjoint Green's function appears in the form

$$\tilde{G}(x, \xi) = \sum_{i=1}^{\infty} \frac{u_i(x)v^*_i(\xi)}{\lambda_i} \qquad (5.28.29)$$

while the symmetry theorem of a self-adjoint problem becomes

$$G(x, \xi) = G^*(\xi, x) \qquad (5.28.30)$$

The expansion of the right side $\beta(x)$ of the differential equation

$$Dv(x) = \beta(x) \qquad (5.28.31)$$

occurs once more in terms of the eigenfunctions $u_i(x)$:

$$\beta(x) = \sum_{i=1}^{\infty} \beta_i u_i(x) \qquad (5.28.32)$$

but the expansion coefficients are obtained in terms of the integrals

$$\beta_i = \int \beta(x)u^*_i(x)dx \qquad (5.28.33)$$

Similar is the procedure with respect to the expansion of $v(x)$ into the ortho-normal eigenfunctions $v_i(x)$.

We will apply these formulae to our problem (8-9).   Let us first construct the Green's function associated with the given operator $Dv = v'$.   The rule

(24) demonstrates that we obtain once more the result of Problem 243 (cf. 27.59), although the constant α is now complex:

$$G(x, \xi) = \frac{\alpha}{\alpha - 1} \qquad (\xi < x) \tag{5.28.34}$$

$$= \frac{1}{\alpha - 1} \qquad (\xi > x) \tag{5.28.35}$$

We now come to the construction of the self-adjoint operator $\breve{D}Dv(x)$. Here the defining differential equation becomes

$$-u''(\xi) = \delta(x, \xi) \tag{5.28.36}$$

with the boundary conditions

$$u(1) = \alpha^* u(0)$$
$$u'(1) = \frac{1}{\alpha} u'(0) \tag{5.28.37}$$

The solution of the four conditions at $\xi = 0$, $\xi = 1$, and $\xi = x$ yields

$$G(x, \xi) = \frac{1}{(1 - \alpha)(1 - \alpha^*)} - \frac{1}{2} \frac{1 + \alpha^*}{1 - \alpha^*} x - \frac{1}{2} \frac{1 + \alpha}{1 - \alpha} \xi - \frac{1}{2} |x - \xi| \tag{5.28.38}$$

A comparison with the previous expression (27.61) demonstrates that for the case of *real* values of α the previous result is once more obtained. We can demonstrate, furthermore, that our $G(x, \xi)$ satisfies the given boundary conditions in the variable x, while in the variable ξ the same holds if every i is changed to $-i$. We also see that the symmetry condition (31) of a Hermitian Green's function is satisfied.

We now come to the study of the bilinear expansion (28) of the Green's function (35). Making use of the eigenfunctions (20) and separating real and imaginary parts we encounter terms of the following kind:

Real Part:

$$-2 \cos (\lambda x + \theta_0) \sin (\lambda \xi + \theta_0) \cos^2 \gamma + 2 \sin (\lambda x + \theta_0) \cos (\lambda \xi + \theta_0) \sin^2 \gamma$$
$$= \sin \lambda(x - \xi) - \cos 2\gamma \sin [\lambda(x + \xi) + 2\theta_0] \tag{5.28.39}$$

Imaginary part:

$$\sin 2\gamma \cos \lambda(x - \xi) \tag{5.28.40}$$

If now we divide by λ and form the sum, we notice that the result is expressible in terms of two functions $f(t)$ and $g(t)$:

$$f(t) = \sum_{k=0}^{\infty} \frac{\sin (\lambda_0 \pm 2k\pi)t}{\lambda_0 \pm 2k\pi} \tag{5.28.41}$$

$$g(t) = \sum_{k=0}^{\infty} \frac{\cos (\lambda_0 \pm 2k\pi)t}{\lambda_0 \pm 2k\pi} \tag{5.28.42}$$

Then the real part of the sum becomes

$$f(x - \xi) - \cos 2\gamma \cos 2\theta_0 \, f(x + \xi) - \cos 2\gamma \sin 2\theta_0 \, g(x + \xi) \qquad (5.28.43)$$

and the imaginary part

$$\sin 2\gamma \, g(x - \xi) \qquad (5.28.44)$$

In order to identify the two functions $f(t)$ and $g(t)$, we will make use of the fact that both systems $v_k(x)$ and $u_k(x)$ represent a complete ortho-normal function system, suitable for the representation of arbitrary section-ally continuous and differentiable functions. Let us choose the function

$$v(x) = 1 \qquad (5.28.45)$$

and expand it into a series of $v_i(x)$ functions:

$$v(x) = \sum_{i=0}^{\infty} c_i v_i(x) \qquad (5.28.46)$$

where

$$c_i = \int_0^1 v(x) v^*_i(x) dx \qquad (5.28.47)$$

In our problem we obtain

$$2\sqrt{2} \sin \frac{\lambda_0}{2} (\cos \gamma + i \sin \gamma)(A \cos \gamma - iB \sin \gamma) = 1 \qquad (5.28.48)$$

with the abbreviations

$$\begin{aligned}
A &= \sum_{k=0}^{\infty} \frac{\cos \left[(\lambda_0 \pm 2k\pi)x + \theta_0\right]}{\lambda_0 \pm 2k\pi} \\
B &= \sum_{k=0}^{\infty} \frac{\sin \left[(\lambda_0 \pm 2k\pi)x + \theta_0\right]}{\lambda_0 \pm 2k\pi}
\end{aligned} \qquad (5.28.49)$$

Since the imaginary part of the left side of (48) must vanish, we get

$$A = B = \frac{1}{2\sqrt{2} \sin \dfrac{\lambda_0}{2}} \qquad (5.28.50)$$

and taking out the constants $\cos \theta_0$ and $\sin \theta_0$ in the trigonometric sums (49), we finally obtain for the two sums (41) and (42);

$$g(x) = \sum_{k=0}^{\infty} \frac{\cos (\lambda_0 \pm 2k\pi)x}{\lambda_0 \pm 2k\pi} = \tfrac{1}{2} \cot \frac{\lambda_0}{2} \qquad (0 \le x \le 1) \quad (5.28.51)$$

$$f(x) = \sum_{k=0}^{\infty} \frac{\sin (\lambda_0 \pm 2k\pi)x}{\lambda_0 \pm 2k\pi} = \tfrac{1}{2} \qquad (0 < x < 1) \quad (5.28.52)$$

Now we return to our formulae (43), (44), substituting the proper values for $f(x)$ and $g(x)$. Assuming that $\xi < x$ (and $x + \xi < 1$), we obtain for (43):

$$\tfrac{1}{2} - \tfrac{1}{2} \cos 2\gamma \sin \lambda_0 - \tfrac{1}{2} \cos 2\gamma \cos \lambda_0 \cot \lambda_0 = \tfrac{1}{2} - \tfrac{1}{2} \cos 2\gamma \cot \frac{\lambda_0}{2} \quad (5.28.53)$$

and for (44):

$$\tfrac{1}{2} \sin 2\lambda \cot \frac{\lambda_0}{2} \quad (5.28.54)$$

Combining real and imaginary parts into one complex quantity we finally obtain

$$\tfrac{1}{2} - \tfrac{1}{2} e^{-2\gamma \imath} \cot \frac{\lambda_0}{2} \quad (5.28.55)$$

which in view of (23) yields

$$\frac{1}{2} \left(1 - \frac{1 + \alpha}{1 - \alpha} \right) = \frac{\alpha}{\alpha - 1} \quad (5.28.56)$$

in full accordance with the value of $G(x, \xi)$ for $\xi < x$ (cf. 34). If $\xi > x$, the only change is that the first term of (53) changes its sign and we obtain the correct value of $G(x, \xi)$ for $\xi > x$.

Problem 244. In the above proof the restricting condition $x + \xi < 1$ was made, although in fact $x + \xi$ varies between 0 and 2. Complement the proof by obtaining the values of $f(t)$ and $g(t)$ for the interval $1 < t < 2$. (Hint: put $x = 1 + x'$.) Show that at the point of discontinuity $t = 1$ the series yield the arithmetic mean of the two limiting ordinates.
[Answer:

$$f(1 + t) = \frac{1}{2} \frac{\sin \frac{3}{2} \lambda_0}{\sin \frac{\lambda_0}{2}}$$

$$\quad (0 < t < 1) \quad (5.28.57)]$$

$$g(1 + t) = \frac{1}{2} \frac{\cos \frac{3}{2} \lambda_0}{\sin \frac{\lambda_0}{2}}$$

Problem 245. By specifying the values of x to 0 and $\tfrac{1}{2}$ obtain from (51) and (52) generalisations of the Leibniz series

$$\frac{\pi}{4} = 1 - \frac{1}{3} + \frac{1}{5} - \frac{1}{7} + \dots \quad (5.28.58)$$

[Answer:

$$\sum_{k=0}^{\infty}{}' \left(\frac{1}{k + \frac{\lambda_0}{2\pi}} - \frac{1}{k - \frac{\lambda_0}{2\pi}} \right) = \pi \cot \frac{\lambda_0}{2} \quad (5.28.59)$$

$$\sum_{k=0}^{\infty}{}' (-1)^k \left(\frac{1}{k + \frac{\lambda_0}{2\pi}} - \frac{1}{k - \frac{\lambda_0}{2\pi}} \right) = \frac{\pi}{\sin \frac{\lambda_0}{2}} \quad (5.28.60)$$

In particular:

$$1 + \frac{1}{3} - \frac{1}{5} - \frac{1}{7} + \frac{1}{9} + \frac{1}{11} - \ldots = \frac{\sqrt{2}\pi}{4} \qquad (5.28.61)]$$

Problem 246. Consider α as purely imaginary $\alpha = i\omega$ and demonstrate for this case the validity of the bilinear expansion of the second Green's function (38).

Problem 247. Obtain for the interval $x = [0, 1]$ the most general Hermitian operator of first order and find its Green's function $G(x, \xi)$.

[Answer:

$$Dv(x) = ip(x)v'(x) + [p(x)q'(x) + \frac{i}{2}p'(x)]v(x) \qquad (5.28.62)$$

$$p(x) > 0, \quad q(x) \text{ real}$$

Boundary condition:

$$v(1) - \mu v(0) = 0 \qquad (5.28.63)$$

where

$$\mu = \sqrt{\frac{p(0)}{p(1)}} \, e^{i\alpha} \qquad (5.28.64)$$

(α an arbitrary real constant).

$$G(x, \xi) = \frac{e^{i[q(x)-q(\xi)-(\omega/2)]}}{2\sqrt{p(x)p(\xi)} \sin \frac{\omega}{2}} \qquad (\xi < x)$$

$$= \frac{e^{i[q(x)-q(\xi)+(\omega/2)]}}{2\sqrt{p(x)p(\xi)} \sin \frac{\omega}{2}} \qquad (\xi > x) \qquad (5.28.65)$$

where

$$\omega = q(1) - q(0) - \alpha \qquad (5.28.66)]$$

Problem 248. Putting

$$q(x) = 0, \quad p(x) = \frac{1}{\rho'(x)} \qquad (5.28.67)$$

where $\rho(x)$ is a monotonously increasing function, prove that the following set of functions form a *complete Hermitian ortho-normal set in the interval* $[0, 1]$:

$$v_k(x) = \sqrt{\frac{\rho'(x)}{\rho(1) - \rho(0)}} \, e^{2k\pi i\rho(x)/[\rho(1)-\rho(0)]} \qquad (5.28.68)$$

Boundary condition:

$$v(1) = \sqrt{\frac{\rho'(1)}{\rho'(0)}} \, v(0) \qquad (5.28.69)$$

Problem 249. Choose

$$\rho(x) = \log(1 + x) \qquad (5.28.70)$$

and obtain an expansion of the function

$$f(x) = \sqrt{1 + x} \qquad (5.28.71)$$

into the ortho-normal functions $v_k(x)$. Investigate the behaviour of the expansion at the two endpoints $x = 0$ and $x = 1$.

[Answer:

$$1 + x = 2 \sum_{k=0}^{\infty}{}' \frac{\log 2 \cos 2k\pi\theta - 2k\pi \sin 2k\pi\theta}{(\log 2)^2 + 4k^2\pi^2} \qquad (5.28.72)$$

where

$$\theta = \frac{\log (1 + x)}{\log 2} \qquad (0 < x < 1) \qquad (5.28.73)$$

In particular, if we put $x = \sqrt{2} - 1$, $\theta = \frac{1}{2}$, we obtain the interesting series

$$\sqrt{2} = \frac{1}{\log 2} + 2 \log 2 \sum_{k=1}^{\infty} \frac{(-1)^k}{4k^2\pi^2 + (\log 2)^2} \qquad (5.28.74)$$

which explains the numerical closeness of $\sqrt{2} = 1.41421$ and $1/\log 2 = 1.44269$.

At $x = 0 : f(0) = 1$; the series gives $3/2$

At $x = 1 : f(1) = 2$; the series gives $3/2$.]

Problem 250.   By choosing

$$f(x) = (1 + x)^{p-(1/2)} \qquad (p > 0) \qquad (5.28.75)$$

obtain a similar expansion for the function $(1 + x)^p$.

[Answer:

$$\frac{(1 + x)^p}{2^p - 1} = 2 \sum_{k=0}^{\infty}{}' \frac{p \log 2 \cos 2k\pi\theta - 2k\pi \sin 2k\pi\theta}{(p \log 2)^2 + 4k^2\pi^2} \qquad (5.28.76)]$$

Problem 251.   Make the implicit transformation of x into t by

$$\frac{\rho(x)}{\rho(1) - \rho(0)} = t \qquad (5.28.77)$$

and show that the expansion into the functions (68) is equivalent to the *Fourier series* in its complex form.

5.29. The completion of linear operators

We had many occasions to point out that the eigen-space of a linear operator is generally incomplete by including only p dimensions of the m-dimensional V-space and likewise p dimensions of the n-dimensional U-space. We have considered a linear system incomplete only if $p < m$ because the condition $p < n$ had merely the consequence that the right side of the given system had to be subjected to the compatibility conditions of the system but had no influence on the uniqueness of the solution. If $p = m$, the solution of our problem was unique, irrespective of whether p was smaller or equal to n.   From the standpoint of the operator it makes no difference whether the eigen-space of the operator omits certain dimensions in either the one or the other space, or possibly in both spaces.   The operator is incomplete in all these cases.

We will now discuss the remarkable fact that an *arbitrarily small modification of an incomplete operator suffices to make the operator complete in all possible dimensions of both U and V spaces.*

First we restrict ourselves to the *self-adjoint* case. Instead of considering the equation

$$Dv(x) = \beta(x) \tag{5.29.1}$$

we will modify our equation by putting

$$D'v(x) = Dv(x) + \epsilon v(x) = \beta(x) \tag{5.29.2}$$

where ϵ is a small parameter which we have at our disposal. We see at once that the new eigenvalue problem

$$Dv(x) + \epsilon v(x) = \lambda'v(x) \tag{5.29.3}$$

has exactly the same eigenfunctions as our previous problem, while the new eigenvalues λ'_i have changed by the constant amount ϵ:

$$\lambda'_i = \lambda_i + \epsilon \tag{5.29.4}$$

Now we have assumed that the eigenvalue $\lambda = 0$ shall not be a limit-point of the eigenvalue spectrum, that is, the eigenvalue $\lambda = 0$ is a *discrete* value of the eigenvalue spectrum, of arbitrarily high multiplicity. Then the addition of ϵ to the eigenvalues, if ϵ is sufficiently small, definitely eliminates the eigenvalue $\lambda = 0$. But it was precisely the presence of the eigenvalue $\lambda = 0$ which caused the incompleteness of the operator D. The new modified operator $D + \epsilon$ is free of any incompleteness and includes the entire function space. The associated eigenfunction system is now complete and the right side need no longer satisfy the condition to be orthogonal to all the solutions $v^j(x)$ of the homogeneous equation

$$Dv^j(x) = 0 \tag{5.29.5}$$

These solutions belong now to the eigenvalue ϵ and cease to play an exceptional role. This remains so even if we diminish ϵ to smaller and smaller values. *No matter how small ϵ becomes, the modified operator includes the entire function space.*

The previous (constrained) Green's function could be expanded into the eigenfunctions $v_i(x)$, *without* the $v^j(x)$:

$$G(x, \xi) = \sum_{i=1}^{\infty} \frac{v_i(x)v_i(\xi)}{\lambda_i} \tag{5.29.6}$$

The new Green's function becomes:

$$G'(x, \xi) = \sum_{i=1}^{\infty} \frac{v_i(x)v_i(\xi)}{\lambda_i + \epsilon} + \sum_{j=1}^{r} \frac{v^j(x)v^j(\xi)}{\epsilon} \tag{5.29.7}$$

This function is defined in the usual fashion, paying no attention to the modification demanded by the existence of homogeneous solutions:

$$Dv(\xi) + \epsilon v(\xi) = \delta(x, \xi) \tag{5.29.8}$$

We may find it more convenient to solve this equation—in spite of the ϵ-term on the left side—because it has the δ-function *alone* on the right side, without any modifications. Then we can return to the Green's function (6) of the modified problem by the following limit process. We consider ϵ as a small parameter and expand our $G'(x, \xi)$ into powers of ϵ. There is first a term which is inversely proportional to ϵ. Then there is a term which is independent of ϵ, and then there are higher order terms, proportional to ϵ, ϵ^2, ..., which are of no concern. We need not go beyond the first two powers: ϵ^{-1} and ϵ^0. We *omit* the term with ϵ^{-1} and keep only the *constant* term. This gives us automatically the correct Green's function (6), characterised by the fact that it contains no components in the direction of the missing axes. In Section 22 we made use of this method for obtaining the Green's function of the constrained problem.

If we now investigate the solution of our problem (2), it is in fact true that the right side $\beta(x)$ is no longer subjected to any constraints and freely choosable. We can expand it into the complete function system $v_i(x)$, $v^j(x)$:

$$\beta(x) = \sum_{i=1}^{\infty} \beta_i v_i(x) + \sum_{j=1}^{r} \beta^j v^j(x) \tag{5.29.9}$$

with

$$\beta_i = \int \beta(\xi) v_i(\xi) d\xi$$
$$\beta^j = \int \beta(\xi) v^j(\xi) d\xi \tag{5.29.10}$$

It is only when we come to the solution $v(x)$ and the limit process involved in the gradual decrease of ϵ that the difficulties arise:

$$v(x) = \sum_{i=1}^{\infty} \frac{\beta_i v_i(x)}{\lambda_i + \epsilon} + \sum_{j=1}^{r} \frac{\beta^j v^j(x)}{\epsilon} \tag{5.29.11}$$

For every finite ϵ the solution is unique and finite. But this solution *does not approach any limit* as ϵ converges to zero, *except* if all the β^j disappear which means the conditions

$$\int \beta(\xi) v^j(\xi) = 0 \qquad (j = 1, 2, \ldots, r) \tag{5.29.12}$$

We are thus back at our usual compatibility conditions but here approached from a different angle. By a small modification of the operator we have restored to it all the missing axes and extended our operator to the utmost limits of the function space. We have no difficulty any more with the eigenvalue zero which is in fact abolished. But now we watch what happens to our solution as ϵ decreases to zero. We observe that in the limit a *unique solution* is obtained but only if the right side satisfies the demanded compatibility conditions. If any of these conditions is not fulfilled, the solution does not converge to any limit and our original problem (which corresponds to $\epsilon = 0$) has in fact no solution.

We will now extend our considerations to *arbitrary* linear operators, no matter how under-determined or over-determined they may be. We start out with the equation (1) which we complement, however, by the adjoint homogeneous equation. We thus consider the pair of equations

$$Dv(x) = \beta(x)$$
$$\breve{D}u(x) = 0$$
(5.29.13)

By this procedure we have done no harm to our problem since the second equation is completely independent of the first one and can be solved by the trivial solution

$$u(x) = 0$$
(5.29.14)

But now we will establish a weak *coupling* between the two equations by modifying our system as follows:

$$Dv(x) + \epsilon u(x) = \beta(x)$$
$$\breve{D}u(x) + \epsilon v(x) = 0$$
(5.29.15)

This means from the standpoint of the shifted eigenvalue problem

$$Dv(x) + \epsilon u(x) = \lambda u(x)$$
$$\breve{D}u(x) + \epsilon v(x) = \lambda v(x)$$
(5.29.16)

that once more the eigenfunctions have remained unchanged while the eigen*values* have changed by the constant amount ϵ, exactly as in (4).

Once more the previous eigenvalue $\lambda = 0$ has changed to the eigenvalue $\lambda = \epsilon$ and the zero eigenvalue can be avoided by making ϵ sufficiently small. Hence the previously incomplete operator becomes once more complete and *spans the entire U space and the entire V space*. We know from the general theory that now the right side can be given freely and the solution becomes unique, no matter how small ϵ may be chosen. In matrix language we have changed our original $n \times m$ matrix of the rank p to an $(n + m) \times (n + m)$ matrix of the rank $n + m$. The conditions of a "well-determined" and "well-posed" problem are now fulfilled: the solution is unique and the right side can be chosen freely.

The right side $\beta(x)$ can be analysed in terms of the complete orthonormal function system $u_i(x)$, $u^j(x)$:

$$\beta(x) = \sum_{i=1}^{\infty} \beta_i u_i(x) + \sum_{j=1}^{r} \beta^j u^j(x)$$
(5.29.17)

where

$$\beta_i = \int \beta(\xi) u_i(\xi) d\xi$$
$$\beta^j = \int \beta(\xi) u^j(\xi) d\xi$$
(5.29.18)

while the solution $u(x)$, $v(x)$ can be analysed in terms of the complete ortho-normal function systems $u_i(x)$, $u^j(x)$, respectively $v_i(x)$, $v^k(x)$:

$$u(x) = \sum_{i=1}^{\infty} a_i u_i(x) + \sum_{j=1}^{r} a^j u^j(x)$$

$$v(x) = \sum_{i=1}^{\infty} b_i v_i(x) + \sum_{j=1}^{\rho} b^k v^k(x)$$

(5.29.19)

Then the differential equation (15) establishes the following relation between the expansion coefficients a_μ, b_μ on the one hand and β_μ on the other

$$a_i = - \frac{\epsilon \beta_i}{\lambda_i^2 - \epsilon^2}$$

$$b_i = \frac{\beta_i \lambda_i}{\lambda_i^2 - \epsilon^2}$$

(5.29.20)

These formulae hold without exceptions, including the eigenvalue $\lambda = 0$, for which our previous conventions employed the upper indices j and k. Hence we have to complement the formulae (20) by the additional formulae (substituting $\lambda_i = 0$):

$$a^j = \frac{\beta^j}{\epsilon}$$

$$b^k = 0$$

(5.29.21)

The second equation shows that *none of the eigenfunctions $v^k(x)$ appear in the expansion* (19) *which are not represented in the operator $Dv(x)$*. The normalisation we have employed before, namely to put the solution completely into the activated V-space of the operator, is upheld by the perturbed system (15) which keeps the solution constantly in the normalised position, without adding components in the non-activated dimensions.

The new system includes the function $u(x)$ on equal footing with the function $v(x)$. Now the first formula of (2) shows that the solution $u(x)$ is weakly excited in all the activated dimensions of the U-space and converges to zero with ϵ going to zero. This, however, is not the case with respect to the non-activated dimensions $u^j(x)$. Here the first formula of (21) shows that the solution *increases to infinity* with ϵ going to zero, except if the compatibility conditions $\beta^j = 0$ are satisfied. Once more we approach our problem from a well-posed and well-determined standpoint which does not involve any constraints. These constraints have to be added, however, if we want our solution to approach a definite limit with ϵ going to zero.

These results can also be stated in terms of a Green's function which now becomes a "Green's vector" because a pair of equations is involved. Since the second equation of the system (15) has zero on the right side, only the two components $G_1(x, \xi)_1$ and $G_2(x, \xi)_1$ are demanded:

$$u(x) = \int G_1(x, \xi)_1 \beta(\xi) d\xi$$

$$v(x) = \int G_2(x, \xi)_1 \beta(\xi) d\xi$$

(5.29.22)

The formulae (20) and (21) establish the following bilinear expansions for the two components of the Green's function:

$$G_1(x, \xi)_1 = \sum_{i=1}^{\infty} \frac{\epsilon u_i(x) u_i(\xi)}{\epsilon^2 - \lambda_i{}^2} + \frac{1}{\epsilon} \sum_{j=1}^{r} u^j(x) u^j(\xi)$$

$$G_2(x, \xi)_1 = \sum_{i=1}^{\infty} \frac{\lambda_i v_i(x) u_i(\xi)}{\lambda_i{}^2 - \epsilon^2}$$

(5.29.23)

On the other hand, if our aim is to solve the *adjoint* equation

$$\tilde{D}u(x) = \gamma(x)$$

(5.29.24)

with the perturbation (15), we need the other two components of the Green's vector:

$$G_1(x, \xi)_2 = \sum_{i=1}^{\infty} \frac{\lambda_i u_i(x) v_i(\xi)}{\lambda_i{}^2 - \epsilon^2}$$

$$G_2(x, \xi)_2 = \sum_{i=1}^{\infty} \frac{\epsilon v_i(x) v_i(\xi)}{\epsilon^2 - \lambda_i{}^2} + \frac{1}{\epsilon} \sum_{k=1}^{\rho} v^k(x) v^k(\xi)$$

(5.29.25)

The relation

$$G_2(x, \xi)_1 = G_1(\xi, x)_2$$

(5.29.26)

is once more fulfilled. We can, as usual, define the Green's function $G_2(x, \xi)_1$ by considering ξ as the active variable and solving the adjoint equation. $\tilde{D}u(\xi) = \delta(x, \xi)$. But in our case that equation takes the form

$$\tilde{D}u(\xi) + \epsilon v(\xi) = \delta(x, \xi)$$

(5.29.27)

and we obtain a new motivation for the modification of the right side which is needed in the case of a constrained system. The expression (25) for $G_2(x, \xi)_2$ shows that the first term goes to zero while the second term is proportional to $1/\epsilon$ and thus $\epsilon v(\xi)$ will contribute a finite term. If we write (27) in the form

$$\tilde{D}u(\xi) = \delta(x, \xi) - \epsilon v(\xi)$$

(5.29.28)

we obtain, as ϵ goes to zero:

$$\tilde{D}u(\xi) = \delta(x, \xi) - \sum_{k=1}^{\rho} v^k(x) v^k(\xi)$$

(5.29.29)

Hence we are back at the earlier equation (22.13) of Section 22, which defined the differential equation of the constrained Green's function. We see that the correction term which appears on the right side of the equation can actually be conceived as belonging to the *left side*, due to the small modification of the operator by the ϵ-method which changes the constrained operator to a free operator and makes its Green's function amenable to the general definition in terms of the delta function. Hence the special position of a constrained operator disappears and returns only when we demand that the solution shall approach a *definite limit* as ϵ converges to zero.

We will add one more remark in view of a certain situation which we shall encounter later. It can happen that the eigenvalue $\lambda = 0$ has the further property that it is a *limit point* of the eigenvalue spectrum. This means that $\lambda = 0$ is not an isolated eigenvalue of the eigenvalue spectrum but there exists an infinity of λ_i-values which come arbitrarily near to zero. In this case we find an infinity of eigenvalues between 0 and ϵ, no matter how small we may choose ϵ. We are then unable to eliminate the eigenvalue $\lambda = 0$ by the ϵ-method discussed above.

The difficulty can be avoided, however, by choosing ϵ as *purely imaginary*, that is by replacing ϵ by $-i\epsilon$. In this case the solution $v(x)$ *remains real*, while $u(x)$ becomes *purely imaginary*. The Green's functions (23) now become

$$G_1(x, \xi)_1 = i\left[\sum_{i=1}^{\infty} \frac{\epsilon u_i(x)u_i(\xi)}{\lambda_i{}^2 + \epsilon^2} + \frac{1}{\epsilon} \sum_{j=1}^{\infty} u^j(x)u^j(\xi) \right]$$

$$G_2(x, \xi)_1 = G_1(\xi, x)_2 = \sum_{i=1}^{\infty} \frac{\lambda_i v_i(x)u_i(\xi)}{\lambda_i{}^2 + \epsilon^2}$$

(5.29.30)

Although the eigenvalues of the problem (16) have now the complex values $\lambda_k - i\epsilon$, this is in no way damaging, as the expressions (30) demonstrate. The eigenvalues of the modified problem cannot be smaller in absolute value than $|\epsilon|$ and the infinity of eigenvalues which originally crowded around $\lambda = 0$, now crowd around the eigenvalue $-i\epsilon$ but cannot interfere with the existence of the Green's function and its bilinear expansion in the sense of (30). We are thus able to handle problems—as we see later—for which the ordinary Green's function method loses its significance, on account of the limit point of the eigenvalue spectrum at $\lambda = 0$.

Problem 252. The Green's function (28.65) goes out of bound for $\omega = 0$ but at the same time $\lambda = 0$ becomes an eigenvalue. Obtain for this case the proper expression for the constrained Green's function.

[Answer:

$$G(x, \xi) = \frac{i}{2} \frac{e^{i[q(x)-q(\xi)]}}{\sqrt{p(x)p(\xi)}} \qquad (\xi < x)$$

$$= -\frac{i}{2} \frac{e^{i[q(x)-q(\xi)]}}{\sqrt{p(x)p(\xi)}} \qquad (\xi > x)]$$

(5.29.31)

BIBLIOGRAPHY

[1] Cf. {1}, pp. 351–96
[2] Cf. {3}, Chapter 3 (pp. 134–94)
[3] Cf. {7}, Part I, pp. 791–895
[4] Fox, C., *An Introduction to the Calculus of Variations* (Oxford University Press, 1950)
[5] Kellogg, O. D., *Foundations of Potential Theory* (Springer, Berlin, 1929)
[6] Lanczos, C., *The Variational Principles of Mechanics* (University of Toronto Press, 1949)

CHAPTER 6

COMMUNICATION PROBLEMS

Synopsis. Heaviside's "unit step function response" was the first appearance of a Green's function in electrical engineering. The input–output relations of electric networks provide characteristic examples for the application of the method of the Green's function, although frequently the auxiliary functions employed are the first or second *integrals* of the mathematical Green's function. We inquire particularly into the "fidelity problem" of communications devices which can be analysed in terms of the Green's function, paying special attention to the theory of the galvanometer. This leads to a brief discussion of the fidelity problem of acoustical engineering. The steady state versus transient analysis demonstrates the much more stringent requirements which are demanded for the fidelity recording of noise, compared with proper recording of the sustained notes of symphonic instruments.

6.1. Introduction

Even before the Green's function received such a prominent position in the mathematical literature of our days, a parallel development took place in electrical engineering, by the outstanding discoveries of the English engineer O. Heaviside (1850–1925). Although his scientific work did not receive immediate recognition—due to faulty presentation and to some extent also due to personal feuds—his later influence on the theory of electric networks was profound. The input–output relation of electric networks can be conceived as an excellent example of the general theory of the Green's function and Green's vector and the relation of Heaviside's method to the standard Green's function method will be our concern in this chapter. Furthermore, we shall include the general mathematical treatment of the galvanometer problem, as an interesting example of a mathematically well-defined problem in differential equations which has immediate significance in the design of scientific instruments. This has repercussions also in the fidelity problem of acoustical recording techniques.

6.2. The step function and related functions

We have seen in the general theory of the Green's function that the right side of a differential equation could be conceived as a linear superposition of delta functions (in physical interpretation "pulses", cf. Chapter 5.5). In the pulse we recognise a fundamental building block from which even the most complicated functions may be generated. If then we know the solution of the differential equation to the pulse as input, we also know the solution for arbitrary right sides.

This idea of a "fundamental building block" has further implications, particularly in the realm of *ordinary* differential equations where the independent variable x covers a simple one-dimensional manifold (which in electric network theory has the significance of the "time t"). When we wrote $f(x)$ in the form

$$f(x) = \int_a^b f(\xi)\delta(x, \xi)d\xi \tag{6.2.1}$$

we expressed the construction of $f(x)$ as a superposition of unit-pulses in mathematical form. The "pulses" $\delta(x, \xi)$ which appear in this construction are comparable to infinitely sharp and infinitely thin needles. They are far from any analytical properties, in fact they cannot be conceived as legitimate functions in the proper sense of the word. In order to interpret the equation (1) properly, a double limit process has to be employed. The integral is defined as the limit of a sum but the notation $\delta(x, \xi)$ itself hides a second limit process since in fact we should operate with $\delta_\epsilon(x, \xi)$ and let ϵ converge to zero.

While we have succeeded in introducing a universal building block in the form of the delta function $\delta(x, \xi)$, this function cannot be interpreted properly without the inconvenience of a limit process. The fundamental building block used by Heaviside in generating functions, namely the "unit step function", is free of this objection. It is in a very simple relation to the delta function by being its *integral*. The basic character of the construction remains: once more we are in the position of generating $f(x)$ as a linear superposition of a base function which is transported from point to point, multiplied by a suitable constant and then integrated.

This new base function, introduced by Heaviside, is defined as follows:

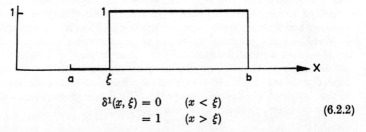

$$\delta^1(x, \xi) = 0 \qquad (x < \xi)$$
$$= 1 \qquad (x > \xi) \tag{6.2.2}$$

It remains zero between a and ξ, and then jumps to the constant value 1 between ξ and b. Exactly as the pulse was a universal function which was rigidly transported to the point ξ—which made $\delta(x, \xi)$ to $\delta(x - \xi)$—the same can be said of the new function $\delta^1(x, \xi)$, which can be written in the form $\delta^1(x - \xi)$ where the universal function $\delta^1(t)$ is defined as follows:

$$\delta^1(t) = 0 \quad (t < 0) \qquad \delta^1(t) = 1, \quad (t > 0) \tag{6.2.3}$$

The derivative of $\delta^1(t)$ is the delta function $\delta(t)$.

Let us integrate by parts in the formula (1):

$$f(x) = -\left[f(\xi)\delta^1(x - \xi) \right]_a^b + \int_a^b f'(\xi)\delta^1(x - \xi)d\xi \tag{6.2.4}$$

The boundary term vanishes at $\xi = b$ (since $x - b$ is negative), while at the lower limit $\delta^1(x - a)$ becomes 1. Thus

$$f(x) = f(a) + \int_a^b f'(\xi)\delta^1(x - \xi)d\xi \tag{6.2.5}$$

The significance of this superposition principle becomes clear if we conceive the integral as the limit of a sum. We then see that $f(x)$ is generated as a superposition of small step functions of the height $f'(x)\Delta x$.

$$f(x) = f(a) + f'(a)\Delta x_1\delta^1(x_1) + f'(x_1)\Delta x_2\delta^1(x_2) + \ldots \tag{6.2.6}$$

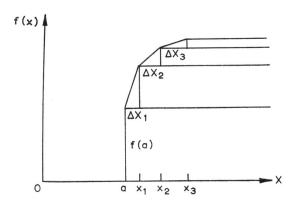

The disadvantage of Heaviside's method is that it presupposes the differentiability of $f(x)$ while before only the continuity of $f(x)$—in fact not more than piecewise continuity—was demanded. The advantage is that $\delta^1(t)$ is a legitimate function which requires no limit process for its definition.

Although we have now generated $f(x)$ by a superposition of step functions, the basic building block is still rather rugged. We need a very large number of these building blocks for a fairly satisfactory approximation of the function $f(x)$, although $f(x)$ itself is not only continuous but even differentiable. We will now repeat the process and integrate a second time, assuming that even the second derivative of $f(x)$ exists:

$$\int_a^b f'(\xi)\delta^1(x - \xi)d\xi = - \left[f'(\xi)\delta^2(x - \xi)\right]_a^b + \int_a^b f''(\xi)\delta^2(x - \xi)d\xi \quad (6.2.7)$$

The new building block is now the integral of the previous step function. This new function $\delta^2(t)$ is already *continuous*, although its tangent is discontinuous at $t = 0$:

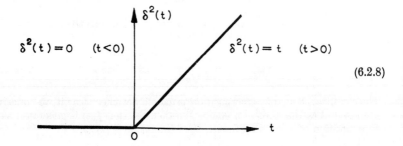

$$\delta^2(t) = 0 \quad (t < 0) \qquad \delta^2(t) = t \quad (t > 0) \qquad (6.2.8)$$

The boundary term of (7) becomes $f'(a)(x - a)$ which yields the following formula:

$$f(x) = f(a) + f'(a)(x - a) + \int_a^b f''(\xi)\delta^2(x - \xi)d\xi \quad (6.2.9)$$

Now our function is put together with the help of *straight line portions* and the ruggedness has greatly decreased. We will now succeed with a much

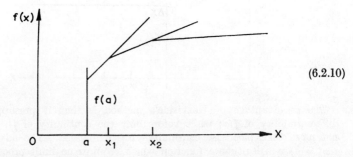

$$(6.2.10)$$

smaller number of building blocks for a satisfactory approximation of $f(x)$.

One further integration leads us to the new building block

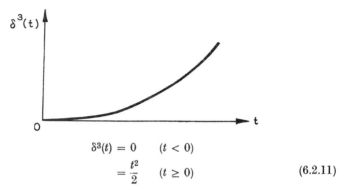

$$\delta^3(t) = 0 \qquad (t < 0)$$
$$= \frac{t^2}{2} \qquad (t \geq 0) \qquad\qquad (6.2.11)$$

Here even the discontinuity of the tangent is eliminated and only the curvature becomes discontinuous at $t = 0$. We now get by integrating by parts

$$\int_a^b f''(\xi)\delta^2(x - \xi)d\xi = -\left[f''(\xi)\delta^3(x - \xi)\right]_a^b + \int_a^b f'''(\xi)\delta^3(x - \xi)d\xi \quad (6.2.12)$$

and the resulting formula becomes

$$f(x) = f(a) + f'(a)(x - a) + \tfrac{1}{2}f''(a)(x - a)^2 + \int_a^b f'''(\xi)\delta^3(x - \xi)d\xi \quad (6.2.13)$$

Once more we have a *universal function* $\delta^3(t)$ which is shifted from point to point, multiplied by the proper weight factor and the sum formed. But now the base function is a *parabolic arc* which avoids discontinuity of either function or tangent. The resulting curve is so smooth that a small number

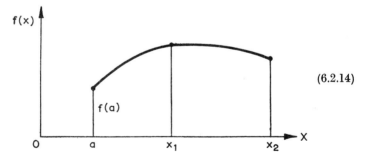

$$(6.2.14)$$

of parabolic arcs can cover rather large portions of the curve. Since the second derivative of $\delta^3(t)$ is a constant, our construction amounts to an approximation of $f(x)$ in which $f''(x)$ in a certain range is replaced by its average value and the same procedure is repeated from section to section.

In all these constructions the idea of a "building block" maintained its significance: we have a universal function which can be rigidly shifted from point to point, multiplied at each point with the proper weight factor and then the integral formed.

Problem 253. Show that the generation of a function in terms of parabolic arcs is equivalent to a solution of the differential equation

$$f''(x) = \gamma \qquad (6.2.15)$$

where γ is a constant which jumps from section to section.

Problem 254. Obtain the three parabolic building blocks for the generation of a function $f(x)$ defined as follows:

$$\begin{aligned}
f(x) &= 3x^2 &(0 \leq x \leq 1) \\
&= 6x - 3 &(1 \leq x \leq 2) \qquad (6.2.16)\\
&= \tfrac{3}{2}x^2 + 3 &(2 \leq x \leq 3)
\end{aligned}$$

[Answer: $6\delta^3$ at $x = 0$; $-6\delta^3$ at $x = 1$; $3\delta^3$ at $x = 2$.]

Problem 255. Approximate the function $f(x) = x^3$ in the range $[0, 2]$ with the help of two parabolic arcs in the interval $x = [0, 1]$ and $x = [1, 2]$, chosen in such a way that the constants of the differential equation (15) shall coincide with the second derivative of $f(x)$ at the *middle* of the respective intervals.

[Answer:

$$3\delta^3 \text{ at } x = 0, \quad 6\delta^3 \text{ at } x = 1$$

$$f^*(x) = \tfrac{3}{2}x^2 \text{ in } [0, 1]; \quad = \tfrac{9}{2}x^2 - 6x + 3 \text{ in } [1, 2].]$$

6.3. The step function response and higher order responses

Let us see how we can utilise these constructions in the problem of solving differential equations. In Chapter 5.5 the Green's function method was conceived as an application of the superposition principle. Then it was unnecessary to construct the adjoint equation. We obtain the solution of the differential equation

$$DG(x, \xi) = \delta(x, \xi) \qquad (6.3.1)$$

(considering x as the active variable). Then, making use of the generation of $\beta(x)$ in terms of $\delta(x, \xi)$:

$$\beta(x) = \int \beta(\xi)\delta(x, \xi)d\xi \qquad (6.3.2)$$

we have obtained the solution of the differential equation

$$Dv(x) = \beta(x) \qquad (6.3.3)$$

(augmented by the proper boundary conditions to make the solution unique), in the form

$$v(x) = \int G(x, \xi)\beta(\xi)d\xi \qquad (6.3.4)$$

Exactly the same principle holds if it so happens that a μ times differentiable function is generated in the form

$$\beta(x) = \int \delta^\mu(x, \xi)\beta^{(\mu)}(\xi)d\xi \qquad (6.3.5)$$

Then we construct the new Green's function by solving the equation

$$DG^\mu(x, \xi) = \delta^\mu(x, \xi) \qquad (6.3.6)$$

and obtain

$$v(x) = \int G^\mu(x, \xi)\beta^{(\mu)}(\xi)d\xi \qquad (6.3.7)$$

In the one-dimensional case (ordinary differential equations) we have the further advantage that the base functions $\delta^\mu(x, \xi)$ are in fact functions of *one variable* only, since they depend solely on the difference $x - \xi = t$:

$$\delta^\mu(x, \xi) = \delta^\mu(x - \xi) = \delta^\mu(t) \qquad (6.3.8)$$

Of particular interest is the case $\mu = 1$ which leads to Heaviside's method of obtaining the input–output relation of electric networks. Here the function $\delta^1(t)$ of Section 2 comes into operation which is in fact Heaviside's "unit step function". The associated Green's function is defined according to (6) and now the formula (2.5) yields the following representation of the solution of the given differential equation (3):

$$v(x) = \beta(a)G^1(x, a) + \int_a^b G^1(x, \xi)\beta'(\xi)d\xi \qquad (6.3.9)$$

From this solution we can return to the standard solution in terms of the Green's function $G(x, \xi)$, if we integrate by parts with respect to $\beta'(\xi)$:

$$\int_a^b G^1(x, \xi)\beta'(\xi)d\xi = \left[G^1(x, \xi)\beta(\xi)\right]_a^b - \int_a^b \frac{d}{d\xi}[G^1(x, \xi)]\beta(\xi)d\xi \qquad (6.3.10)$$

This shows the following relation between Heaviside's Green's function and the standard Green's function, which is the pulse response:

$$G(x, \xi) = -\frac{d}{d\xi}G^1(x, \xi) \qquad (6.3.11)$$

We observe, furthermore, the necessity of the boundary condition

$$G^1(x, b) = 0 \qquad (6.3.12)$$

which comes about in consequence of the fact that the right side of (6) vanishes throughout the given range if $\delta^1(x, \xi)$ moves out into the end point $\xi = b$.

Historically the Green's function defined with the help of the unit step function rather than the unit pulse played an important role in the theoretical researches of electrical engineering, since Heaviside, the ingenious originator of the Green's function in electrical engineering, used consistently the unit step function as input, instead of the unit pulse (i.e. the delta function). He thus established the formula (9) instead of the formula (4). In the later years of his life Heaviside became aware of the theoretical superiority of the pulse response compared with the unit step function response. However, the use of the unit step function became firmly established in engineering, although in recent years the pulse response gains more and more access into advanced engineering research.

From the practical standpoint the engineer's preference for the unit step function is well understandable. It means that at a certain time moment $t = 0$ the constant voltage 1 is applied to a certain network and the output observed. To imitate the unit pulse (in the sense of the delta function) with any degree of accuracy is physically much less realisable than to produce the unit step function and observe its effect on the physical system. The pulse response is a much more elusive and strictly speaking only theoretically available quantity.

There are situations, involving the motion of mechanical components, when even the step function response is experimentally unavailable because even the step function as input function is too unsmooth for practical operations. Let us consider for example a servo-mechanism installed on an aeroplane which coordinates the motion of a foot-pedal in the pilot's cockpit and the induced motion of the rudder at the rear of the aeroplane. The servo-mechanism involves hydraulic, mechanical, and electrical parts. To use the step function as input function would mean that the foot-pedal is pushed out suddenly into its extreme position. This is physically impossible since it would break the mechanism. Here we have to be satisfied with a Green's function which is one step still further removed from the traditional Green's function, by applying the linear input function (2.8). The foot-pedal is pushed out with *uniform speed* into its extreme position and the response observed. This function $G^2(x, \xi)$ is the negative integral of the step function response $G^1(x, \xi)$ considering ξ as the variable.

Problem 256. Obtain the solution of (3) in terms of $G^2(x, \xi)$ and $G^3(x, \xi)$, making use of the formulae (2.9) and (2.13).

[Answer:

$$v(x) = -\beta(a)\frac{d}{d\xi}G^2(x, a) + \beta'(a)G^2(x, a) + \int_a^b G^2(x, \xi)\beta''(\xi)d\xi \quad (6.3.13)$$

$$v(x) = \beta(a)\frac{d^2}{d\xi^2}G^3(x, a) - \beta'(a)\frac{d}{d\xi}G^3(x, a) + \beta''(a)G^3(x, a) + \int_a^b G^3(x, \xi)\beta'''(\xi)d\xi$$
$$(6.3.14)]$$

Problem 257. Obtain the relation of $G^2(x, \xi)$ and $G^3(x, \xi)$ to the standard Green's function $G(x, \xi)$.

[Answer:

$$G(x, \xi) = \frac{d^2}{d\xi^2}G^2(x, \xi)$$

$$= -\frac{d^3}{d\xi^3}G^3(x, \xi) \quad (6.3.15)$$

$$G^2(x, b) = \frac{d}{d\xi} G^2(x, b) = 0 \tag{6.3.16}$$

$$G^3(x, b) = \frac{d}{d\xi} G^3(x, b) = \frac{d^2}{d\xi^2} G^3(x, b) = 0 \tag{6.3.17}$$

6.4. The input–output relation of a galvanometer

Up to now we have not specified the nature of the ordinary differential operator, except for its linearity. It can involve derivatives of any order, the unknown function can be a scalar or a vector, and the coefficients may be any piecewise continuous functions of x. In electric network problems the further simplification occurs that the coefficients of the operator D are in fact *constants*. Furthermore, the given boundary conditions are all *initial conditions* because the given physical situation is such that at the time moment $x = 0$ all the functions involved have prescribed values. Under these conditions the mathematical problem is greatly simplified. The solution of the equation (3.6) is then reducible to the parameter-free equation

$$DG^\mu(t) = \delta^\mu(t) \tag{6.4.1}$$

because now the Green's function shares the property of the base function $\delta^\mu(x - \xi)$ to become a function of the single variable $t = x - \xi$ only:

$$G^\mu(x, \xi) = G^\mu(x - \xi) = \delta^\mu(t) \tag{6.4.2}$$

The solution of the differential equation (3.3)—apart from a boundary term—will now occur in the form

$$v(x) = \int_a^b G^\mu(x - \xi)\beta^{(\mu)}(\xi)d\xi \tag{6.4.3}$$

As a characteristic example we will discuss in this section the problem of the "galvanometer". The galvanometer is a recording instrument for the measurement of an electric current. A light mirror is suspended with the help of a very thin wire whose torsion provides the restoring force. Damping is provided by the motion in air or by electromagnetic damping. The differential equation of the galvanometer is identical with that of the vibrating spring with energy loss due to friction:

$$v''(x) + 2\alpha v'(x) + \rho^2 v(x) = \beta(x) \tag{6.4.4}$$

The constant α is called the "damping constant", while ρ is called the "stiffness constant"; the quantity $-\rho^2 v(x)$ is frequently referred to as the "restoring force". In the case of the galvanometer the variable x represents the time, $\beta(x)$ is the input current and $v(x)$ is the scale reading which may be recorded in a photographic way.

We are here interested in the galvanometer as a *recording instrument*. It represents a prototype of instruments which are used in many branches of physics. For example the pendulum used for the recording of earthquakes in seismographic research is a similar measuring device. Generally we speak of a "device of the galvanometer type" if an input–output relation is involved which is describable by a differential equation of the form (4). In our present investigation we shall be particularly interested in the galvanometer type of recording from the standpoint of comparing the output $v(x)$ with the input $\beta(x)$ and analysing the "fidelity" of the recording.

First of all we shall obtain the "Green's function" of our problem. This can be done according to the standard techniques discussed in Chapter 5. In fact, we have obtained the solution before, in Chapter 5.16 (cf. Problem 207, equation 5.16.33):

$$G(x, \xi) = e^{-\alpha(x-\xi)} \frac{\sin \sqrt{\rho^2 - \alpha^2}(x - \xi)}{\sqrt{\rho^2 - \alpha^2}} \qquad (\xi \leq x)$$
$$= 0 \qquad (\xi \geq x) \qquad (6.4.5)$$

The second equation tells us at once that the upper limit of integration will not be b but x since the integrand vanishes for all $\xi \geq x$. The lower limit of the integral is zero since we started our observations with the time moment $x = 0$.

For the sake of formal simplification we will introduce a natural time scale into the galvanometer problem by normalising the stiffness constant to 1. Furthermore, we will also normalise the output $v(x)$—the scale reading—by introducing a proper amplitude factor. In order to continue with our standard notations we will agree that the original x should be denoted by $\bar{x}$, the original $v(x)$ by $\bar{v}(x)$. Then we put

$$\bar{x} = \frac{x}{\rho}$$
$$\bar{v}(\bar{x}) = \frac{1}{\rho^2} v(x) \qquad (6.4.6)$$

With this transformation the original differential equation (4) (in which x is replaced by $\bar{x}$ and v by $\bar{v}$) appears now in the form

$$v''(x) + 2\kappa v'(x) + v(x) = \beta(x)$$
$$\kappa = \frac{\alpha}{\rho} \qquad (6.4.7)$$

Our problem depends now on one parameter only, namely on the "damping ratio" κ, for which we want to introduce an auxiliary angle γ, defined by

$$\kappa = \cos \gamma \qquad (6.4.8)$$

This angle is limited to the range $\pi/2$ to 0 as α increases from 0 to ρ. If α surpasses the critical value ρ, γ becomes purely imaginary, but our formulae do not lose their validity, although the sines and cosines now change to hyperbolic sines and cosines. The limiting value $\gamma = 0$ ($\kappa = 1$), which marks the transition from the periodic to the aperiodic range, is usually referred to as "critical damping".

If we write down the Green's function in the new variables, we obtain the expression

$$G(x, \xi) = G(x - \xi) = G(t)$$
$$= e^{-t \cos \gamma} \frac{\sin (t \sin \gamma)}{\sin \gamma} \qquad (t > 0) \tag{6.4.9}$$

which yields the output in the form of the integral

$$v(x) = \int_0^x G(x - \xi)\beta(\xi)d\xi$$
$$= \int_0^x \beta(x - t)e^{-t \cos \gamma} \frac{\sin (t \sin \gamma)}{\sin \gamma} \, dt \tag{6.4.10}$$

Problem 258.   Obtain the step function response $G^1(t)$ of the galvanometer problem.

[Answer:

$$G^1(t) = 1 - \frac{1}{\sin \gamma} e^{-t \cos \gamma} \sin (t \sin \gamma + \gamma) \tag{6.4.11}]$$

Problem 259.   Obtain the linear response $G^2(t)$ and the parabolic response $G^3(t)$ of the galvanometer problem.

[Answer:

$$G^2(t) = t - 2 \cos \gamma + \frac{1}{\sin \gamma} e^{-t \cos \gamma} \sin (t \sin \gamma + 2\gamma) \tag{6.4.12}$$

$$G^3(t) = \tfrac{1}{2}(t - 2 \cos \gamma)^2 + \cos 2\gamma - e^{-t \cos \gamma} [(1 + 2 \cos 2\gamma) \cos (t \sin \gamma)$$
$$- (\cot \gamma)(1 - 2 \cos 2\gamma) \sin (t \sin \gamma)] \tag{6.4.13}]$$

Problem 260.   Obtain $G(t)$ and $G^1(t)$ for critical damping.

[Answer:

$$G(t) = te^{-t} \tag{6.4.14}$$
$$G^1(t) = 1 - e^{-t}(1 + t) \tag{6.4.15}]$$

6.5. The fidelity problem of the galvanometer response

We will now investigate to what extent we may hope to find a close resemblance between the input function $\beta(x)$ and the output $v(x)$.   At first sight we can see no reason why there should be any resemblance between the two functions since the formula (4.10) shows that $v(x)$ is obtainable from $\beta(x)$ by the process of integration.   Hence $v(x)$ will not depend on the local value of $\beta(\xi)$ at the point $\xi = x$, but on *all* the values of $\beta(\xi)$ between 0 and x.   The galvanometer is an *integrating device* and what we get will be a certain *weighted average* of all the values of $\beta(\xi)$ between 0 and x.   However, this weight factor—which is in fact $G(x, \xi)$—has a strongly biased character.

Indeed, the weight factor

$$G(t) = e^{-t \cos \gamma} \frac{\sin (t \sin \gamma)}{\sin \gamma} \tag{6.5.1}$$

is such that it will emphasise the region around $t = 0$ while the region of very large values of t will be practically blotted out.   The galvanometer has a "memory" by retaining the earlier values $\beta(x - t)$ before the instantaneous value $\beta(x)$ but this memory is of short duration if the damping ratio is sufficiently large. For very small damping the memory will be so extended that the focusing power on small values of t is lost. In that case we cannot hope for any resemblance between $v(x)$ and $\beta(x)$.

Apart from these general, more qualitative results we do not obtain much information from our solution (4.10), based on a Green's function which was defined as the pulse response. We will now make our input function less extreme by using Heaviside's unit step function as input function. Then the response appeared—for the normalised form (4.7) of the differential equation—in the form (4.11). If we plot this function, we obtain a graph of the following character:

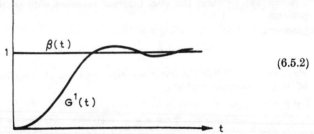

$$(6.5.2)$$

From this graph we learn that the output $G^1(t)$, although it does not resemble the input in the beginning, will *eventually* reproduce the input $\beta(t)$ with a gradually decreasing error. This shows that in our normalisation the proportionality factor between $v(x)$ and $\beta(x)$ will be 1. The original $v(x)$ of the general galvanometer equation (4.4) had to be multiplied by ρ^2 in order to give the new $v(x)$. Hence in the original form of the equation the proportionality factor of the output $v(x)$ will become ρ^2, in order to compare it with the input $\beta(x)$.

Still more information can be obtained if we use as input function the linear function $\beta(t) = t$ of Problem 259 (cf. 4.12). Here we obtain a graph of the following character:

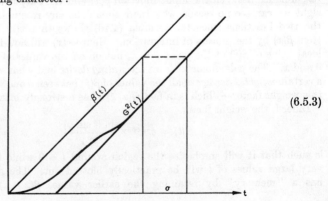

$$(6.5.3)$$

From this graph we learn that the output—apart from the initial disturbance —*follows* the input with a *constant time lag* of the amount

$$\sigma = 2 \cos \gamma = 2\kappa \tag{6.5.4}$$

If our input function is composed of straight line sections which follow each other in intervals which are long compared with the time $1/\rho$, i.e. the reciprocal stiffness constant of the galvanometer, this galvanometer will reproduce the curve with slight disturbances in the neighbourhood of the points where the sections meet. The constant time lag, with which the output follows the input, is in most cases not damaging and can be taken for granted. We then have to know that it is not $\rho^2 v(x)$ but $\rho^2 v(x + \sigma)$ which will correspond to $\beta(x)$.

Problem 261. Show that for no value of the damping constant α can the response curve (2) intersect the t-axis (apart from the point $t = 0$). Show also that the approach to the line $t = 1$ is monotonous for all values $\alpha \geq \rho$.

Problem 262. Show that for no value of the damping constant α can the response curve (3) intersect the t-axis (apart from the origin $t = 0$).

6.6. Fidelity damping

Even this information does not suffice to decide whether or not some particular value of the damping constant is preferable to other values in order to obtain maximum fidelity. We will now go one step further still and generate the input function with the help of *parabolic arcs*. This means that the input function (2.11) has to be used. The solution was obtained in Problem 259 (4.13), in the form of an expression which started with

$$G^3(t) = \tfrac{1}{2}(t - 2 \cos \gamma)^2 + \cos 2\gamma \tag{6.6.1}$$

plus further terms which go to zero with increasing t. We now see that there is indeed a distinguished value of γ, namely the value

$$\gamma = \frac{\pi}{4}, \quad \kappa = \cos \gamma = \frac{1}{\sqrt{2}}, \quad \cos 2\gamma = 0 \tag{6.6.2}$$

which will be of particular advantage from the standpoint of fidelity. With this choice of the damping ratio we obtain for the damping constant α the value

$$\alpha = \frac{\rho}{\sqrt{2}} = 0.707\rho \tag{6.6.3}$$

which is only 71% of the critical damping. Now we have fidelity for any curve of arbitrary *parabolic arcs*, which follow each other in intervals large compared with the time $1/\rho$ (since there is a quickly damped disturbance at the intersection of the arcs, due to the excitation of the eigen-vibration of the galvanometer). It is clear that under such conditions the fidelity of the galvanometer recording can be greatly increased since the parabolic arcs, which approximate a function, can be put much further apart than mere straight line sections which are too rigid for an effective approximation of a smooth function. Hence we will denote the choice (3) of the damping constant α as "fidelity damping".

Problem 263. Obtain the parabolic response $G^3(t)$ of the galvanometer for fidelity damping and demonstrate that at $t = 0$ function and derivatives vanish up to (and inclusive of) the third derivative.

[Answer:

$$G^3(t) = (t_1 - 1)^2 + e^{-t_1}(\cos t_1 - \sin t_1) \qquad (6.6.4)$$

(with $t_1 = t/\sqrt{2}$).]

6.7. The error of the galvanometer recording

It will be our aim to obtain a suitable error bound for the fidelity of galvanometer recording. We will start with the case of fidelity damping which assures the highest degree of fidelity recording. We have seen in the foregoing section that an input signal which can be sectionally represented by any polynomial of second order, will be faithfully reproduced—except for a constant time-lag—if the sections in which the representations hold, are sufficiently separated, since it is inevitable that with a jump in the second derivative the damped eigen-vibrations of the galvanometer will be excited, thus causing a short-lived disturbance at the points where the sections join.

In order to estimate the error of a galvanometer with fidelity damping, we will now employ as input signal the function

$$\beta(x) = (x + \sqrt{2})^3 \qquad (6.7.1)$$

The time-lag for fidelity damping is $2\kappa = \sqrt{2}$ and thus the differential equation

$$v'' + \sqrt{2}v' + v = \beta(x) \qquad (6.7.2)$$

would be satisfied by $v(x) = x^3$, if absolute fidelity could be expected. In actual fact the solution of our equation becomes

$$v(x) = x^3 + 2\sqrt{2} \qquad (6.7.3)$$

and this can be interpreted as

$$v(x) = \beta(x - \sqrt{2}) + \frac{\sqrt{2}}{3}\beta'''(x) \qquad (6.7.4)$$

We can make our error estimation still more accurate by allowing a certain time-lag at which the third derivative $\beta'''(x)$ is to be taken. Our error estimation can thus be made accurate for any polynomial of the order *four*. Let us use as input function

$$\beta(x) = (x + \sqrt{2})^4 \qquad (6.7.5)$$

Then the corresponding solution $v(x)$ becomes

$$v(x) = x^4 + 8\sqrt{2}x - 12 \qquad (6.7.6)$$

which can be interpreted as follows:

$$v(x) = \beta(x - \sqrt{2}) + \frac{\sqrt{2}}{3}\beta'''(x - \tfrac{7}{4}\sqrt{2}) \qquad (6.7.7)$$

The estimated error for critical damping is considerably higher since it is proportional to the *second* derivative of $\beta(x)$. If we use as input function

$$\beta(x) = (x + 2)^3 \qquad (6.7.8)$$

and solve the differential equation

$$v'' + 2v' + v = \beta(x) \qquad (6.7.9)$$

we obtain

$$v(x) = x^3 + 6x - 4 \qquad (6.7.10)$$

which can be interpreted in the form

$$v(x) = \beta(x - 2) + \beta''(x - \tfrac{8}{3}) \qquad (6.7.11)$$

Let us now return to the original formulation (4.4) of the galvanometer equation. Our results can now be summarised as follows:

Fidelity damping:

$$\beta(x) - \rho^2 v\left(x + \frac{\sqrt{2}}{\rho}\right) = \eta(x)$$

$$\eta(x) = -\frac{\sqrt{2}}{3\rho^3}\beta'''\left(x - \frac{3\sqrt{2}}{4\rho}\right) \qquad (6.7.12)$$

Critical damping:

$$\beta(x) - \rho^2 v\left(x + \frac{2}{\rho}\right) = \eta(x)$$

$$\eta(x) = -\frac{1}{\rho^2}\beta''\left(x - \frac{2}{3\rho}\right) \qquad (6.7.13)$$

Problem 264. Given the stiffness constants $\rho = 5$ and $\rho = 10$. Calculate the relative errors of the galvanometer response (disregarding the initial disturbance caused by the excitation of the eigen-vibration), for the input signal

$$\beta(x) = e^{-x} \qquad (6.7.14)$$

for both fidelity and critical damping. Compare these values with those predicted by the error formulae (12) and (13).

[Answer:

	$\rho = 5$	$\rho = 10$
fidelity:	0.004648	0.00052407
(formula:	0.004642)	(0.00052415)
critical:	-0.0474	-0.01078
(formula:	-0.0457	(-0.01069)]

$$(6.7.15)$$

Problem 265.   Apply the method of this section to the error estimation of the general case in which the damping ratio $\kappa = \alpha_/\rho$ is arbitrary (but not near to the critical value $1/\sqrt{2}$).

[Answer:

$$v(x) = \beta(x - 2\kappa) + (2\kappa^2 - 1)\beta''\left(x - \frac{4\kappa(5\kappa^2 - 3)}{3(2\kappa^2 - 1)}\right) \qquad (6.7.16)]$$

6.8. The input–output relation of linear communication devices

The galvanometer is the prototype of a group of much more elaborate mechanisms which have certain characteristics in common.   We can call them "communication devices" since their function is to transmit certain information from one place to another.   We may think for example of a broadcasting station which receives and transmits a speech given in acoustical signals and transformed into electric signals.   These electric signals are radiated out into space and are received again by a listener's radio set which transforms the electric signals once more into acoustic signals.   The microphone-receiver system of an ordinary telephone is another example of a device of the galvanometer type, the "input" being the speech which brings the microphone into vibration, the "output" being the air vibration generated by the membrane of the receiver and communicated to the listener.

Other situations may differ in appearance but belong mathematically to the same group of problems.   Consider for example the movements of an aeroplane pilot who operates certain mechanisms on the instrument-panel of his cockpit.   These movements are transmitted with the help of "servo-mechanisms" into a corresponding motion of the aileron, rudder, etc. The whole field of "servo-mechanism" can thus be considered as a special example of a communication device.

The mechanisms involved may be of an arbitrarily complicated type. They may involve mechanical or hydraulic or electric components.   They may consist of an arbitrary number of coupled electric networks.   They may contain a loudspeaker whose action is described by a partial differential equation and which is thus replaceable by an *infinity* of coupled vibrating springs.   In order to indicate that we are confronted with an unknown mechanism whose structure is left unspecified, we speak of a "black box" which contains some mechanism whose nature is not revealed to us.   However, there are two ends to this black box: at the one end some signal goes in as a function of the time t which in our notation will be called x.   At the other end some new function of the time t comes out which we will call the "output".   We will adhere to our previous notations and call the input signal $\beta(x)$, the output response $v(x)$.   In the case of the galvanometer $\beta(x)$ had the significance of the electric current which entered the galvanometer reading.   The coupling between these two functions was established by the differential equation of the damped vibrating spring.   In the general case we still have our two functions $\beta(x)$ and $v(x)$ as input and output but their significance may be totally different and the coupling between the two functions need not necessarily be established by a differential equation but can be of a much more general character.   Generally we cannot assume that there is necessarily a close resemblance between $\beta(x)$ and $v(x)$.   We may have the mathematical problem of restoring the input function $\beta(x)$ if $v(x)$ is given and this may lead to the solution of a certain *integral equation*. But frequently our problem is to investigate to what extent we can improve the resemblance of the output $v(x)$ to the input $\beta(x)$ by putting the proper mechanism inside the black box.

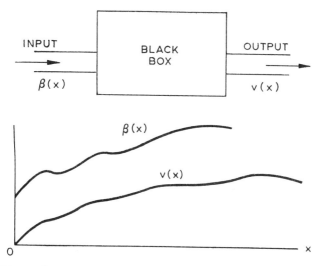

a) *Linear devices*. Although we have no direct knowledge of the happenings inside the black box, we can establish certain *general features* of this mechanism by experimenting with various types of input functions $\beta(x)$ and observing the corresponding outputs $v(x)$.   We will assume that by a number of preliminary experiments we establish the following two fundamental properties of the input–output relation :

1. Let us use an arbitrary input function $\beta(x)$ and observe the corresponding output $v(x)$.   We now change $\beta(x)$ to $\beta_1(x) = \alpha\beta(x)$, where α is an arbitrary constant.   Then we observe that the new output $v_1(x)$ becomes $v_1(x) = \alpha v(x)$.

2. Let $g_1(x)$ and $g_2(x)$ be two arbitrary input functions and $v_1(x)$, $v_2(x)$ be the corresponding outputs.   We now use $\beta_1(x) + \beta_2(x)$ as an input.   Then we observe that the output becomes $v_1(x) + v_2(x)$; this means that the *superposition principle* holds.   The simultaneous application of two inputs generates the sum of the corresponding outputs, *without any mutual interference*.

If our communication device satisfies these two fundamental conditions, we will say that our device is of the *linear type*.   As a consequence of the superposition principle and the proportionality principle we can state that the output which corresponds to the input function

$$\beta(x) = \alpha_1\beta_1(x) + \alpha_2\beta_2(x) + \ldots + \alpha_n\beta_n(x) \tag{6.8.1}$$

becomes

$$v(x) = \alpha_1 v_1(x) + \alpha_2 v_2(x) + \ldots + \alpha_n v_n(x) \tag{6.8.2}$$

where $v_1(x)$, $v_2(x)$, $\ldots$, $v_n(x)$ are the outputs which correspond to $\beta_1(x)$, $\beta_2(x)$, $\ldots$, $\beta_n(x)$ as inputs.

Now we have seen before (cf. equations (5.4.13–15)) that an arbitrary continuous function $f(x)$ can be considered as a linear superposition of delta functions. If now we obtain the response of the black box mechanism to the delta function as input function, we can obtain the response to an arbitrary input function $\beta(x)$, in a similar manner as the solution of a linear differential equation was obtained with the help of the Green's function $G(x, \xi)$:

$$v(x) = \int_a^b g(\xi)G(x, \xi)d\xi \tag{6.8.3}$$

We will assume that the lower limit of our variable x is $x = 0$ because we want to measure the time from the time moment $\xi = 0$, the start of our observations. The upper limit $\xi = b$ of our interval can be chosen arbitrarily but we can immediately add a further property of the function $G(x, \xi)$. As a consequence of the causality principle, the output has to *follow* the input. Hence $G(x, \xi)$ must vanish at any value of x which comes *before* the time of applying the unit pulse, that is before $x = \xi$:

$$G(x, \xi) = 0 \qquad (x < \xi) \tag{6.8.4}$$

We can equally say—considering ξ as the variable—that $G(x, \xi)$ vanishes for all values $\xi > x$. Hence the integration in (245) does not extend to $\xi = b$ but only to $\xi = x$:

$$v(x) = \int_0^x g(\xi)G(x, \xi)d\xi \tag{6.8.5}$$

b) *Time-independent mechanisms.* We will now add a further property of our communication device. We assume that the unknown components inside the black box *do not change in time.* They have physical characteristics which are *time independent.* If this is the case, it cannot make any difference, at what time moment ξ the unit pulse is applied, the output $v(x)$ will always be the same, except for a shift in the time scale. This additional property of the communication device can be mathematically formulated as follows. If $\beta(x)$ changes to $\beta(x + \alpha)$, $v(x)$ changes to $v(x + \alpha)$. Now the unit pulse applied at the time moment ξ is equal to the unit pulse applied at the time moment 0, with a shift of the time scale:

$$\delta(x, \xi) = \delta(x - \xi, 0) \tag{6.8.6}$$

Hence

$$G(x, \xi) = G(x - \xi, 0) \tag{6.8.7}$$

It is thus sufficient to observe the output $G(x)$ which follows the unit pulse input, applied at the time moment $x = 0$. This function is different from zero for *positive* values of x only while for all negative values $G(x)$ is zero:

$$G(x) = 0 \qquad (x < 0) \tag{6.8.8}$$

The function of *two* variables $G(x, \xi)$ is thus reducible to a function of a *single* variable $G(x)$ only and the relation (5) now becomes:

$$v(x) = \int_0^x \beta(\xi) G(x - \xi) d\xi \tag{6.8.9}$$

We will once more introduce the variable

$$x - \xi = t \tag{6.8.10}$$

as we have done before in the theory of the galvanometer (cf. (4.9–10)). We replace ξ by the new variable t, according to

$$\xi = x - t \tag{6.8.11}$$

and write the input–output relation (9) in the form

$$v(x) = \int_0^x \beta(x - t) G(t) dt \tag{6.8.12}$$

We see that the general theory of an arbitrary linear communication device is remarkably close to the theory of the galvanometer. The only difference is that in the case of the galvanometer the function $G(t)$ had a very definite form, given by (4.9). In the general case the function $G(t)$ will be determined by the more or less complex mechanism which is inside the black box.

c) *The dissipation of energy.* A mechanism of the following type may be considered. At the input end a switch is pulled. On the output an electric bulb is lighted. If we are not familiar with the content of the black box, we would think that a *perpetuum mobile* kind of device had been invented.

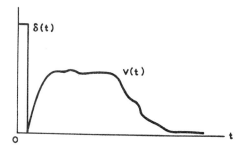

In actual fact the box contains an "internal generator" which constantly puts out energy. We want to assume that our device does *not* contain an internal generator. The input energy will be gradually dissipated by transforming it into heat. Hence the response to the unit pulse will be such that it will not last forever but disappear eventually. It is possible that strictly speaking $G(t)$ will never become *exactly* zero but approach zero *asymptotically*, as t grows to infinity:

$$\lim_{t \to \infty} G(t) = 0 \qquad (6.8.13)$$

This is the behaviour we have observed in the case of the galvanometer (cf. (4.9)), and the same will be true if the black box mechanism is composed of an arbitrary number of electric circuits or vibrating springs. The name "passive network" is applied to this kind of mechanism.

We see that we have gradually reduced the arbitrariness of the black box by the following restricting conditions which, however, are valid in most communication problems:

1. The device is linear.
2. The output follows the input.
3. The components of the device do not change with the time.
4. The device dissipates energy.

6.9. Frequency analysis

An entirely different kind of analysis is frequently of great practical and theoretical interest. We know from the theory of the Fourier series that a very large class of functions, defined in a finite or even infinite interval, can be resolved into purely periodic components. Instead of considering $\beta(x)$ as a superposition of pulses, as we have done before, we can equally consider $\beta(x)$ also as a superposition of periodic functions of the form $\cos \omega x$ and $\sin \omega x$. If we know how our communication device responds to the input function $\cos \omega x$ or $\sin \omega x$, we shall be able to generate the entire output as a superposition of these responses. In this analysis the basic building block for the generation of a function is not the pulse, shifted from point to point, but the periodic functions $\cos \omega x$ and $\sin \omega x$, with arbitrary values of ω.

For mathematical purposes it will be convenient to combine $\cos \omega x$ and $\sin \omega x$ in the complex form

$$e^{i\omega x} = \cos \omega x + i \sin \omega x \qquad (6.9.1)$$

If we know what the response of our device is to the complex input function (1), we shall immediately have the response to both $\cos \omega x$ and $\sin \omega x$ as input functions, by merely separating the real and imaginary parts of the response.

Before carrying out the computation we will make a small change in the formula (8.12). The limits of integration were 0 and x. The lower limit came about in view of $G(t)$ being zero for all negative values of t. The upper limit x came about since we have started our observation at the time moment $x = 0$. Now the input signal $\beta(x)$ did not exist before the time moment $x = 0$ which means that we can define $\beta(x)$ as zero for all negative values of x.

$$\beta(x) = 0 \qquad (x < 0) \tag{6.9.2}$$

But then we need not stop with the integration at $t = x$ but can continue up to $t = \infty$:

$$v(x) = \int_0^\infty \beta(x - t)G(t)dt \tag{6.9.3}$$

The condition (2) automatically reduces the integral (3) to the previous form (8.12). The new form has the advantage that we can now *drop* the condition (2) and assume that the input $\beta(x)$ started at an arbitrary time moment before or after the time moment $x = 0$.

This will be important for our present purposes because we shall assume that the periodic function (1) existed already for a very long—mathematically infinitely long—time. This will not lead to any difficulties in view of the practically finite memory time of our device.

We introduce now

$$\beta(x) = e^{i\omega x} \tag{6.9.4}$$

into our integral (3) and obtain

$$v(x) = \int_0^\infty e^{i\omega(x-t)}G(t)dt$$

$$= e^{i\omega x} \int_0^\infty e^{-i\omega t}G(t)dt \tag{6.9.5}$$

The factor of $e^{i\omega x}$ can be split into a real and imaginary part:

$$\int_0^\infty e^{-i\omega t}G(t)dt = A(\omega) - iB(\omega) \tag{6.9.6}$$

with

$$A(\omega) = \int_0^\infty G(t) \cos \omega t\, dt \tag{6.9.7}$$

$$B(\omega) = \int_0^\infty G(t) \sin \omega t\, dt \tag{6.9.8}$$

Moreover, we may write the complex number (6) in "polar form", with the "amplitude" $\rho(\omega)$ and the "argument" $\theta(\omega)$:

$$A(\omega) - iB(\omega) = \rho(\omega)e^{-i\theta(\omega)} \tag{6.9.9}$$

Then the output (5) becomes

$$v(x) = \rho(\omega)e^{i(\omega x - \theta(\omega))} \tag{6.9.10}$$

The significance of this relation can be formulated as follows. If we use a harmonic vibration $\cos \omega x$ or $\sin \omega x$ as input function, *the output will again be a harmonic vibration of the same frequency but modified amplitude and phase.*

The fact that a harmonic vibration remains a harmonic vibration with unchanged frequency, is characteristic for the *linearity* of a communication device with time-independent elements. We can in fact *test* our device for linearity by using a sine or cosine function of arbitrary frequency as input and observing the output. The necessary and sufficient condition for the linearity of the device is that a harmonic vibration remains a harmonic vibration with unchanged frequency, although with modified amplitude and a certain shift of the phase.

It may happen that in a large range of ω the amplitude factor $\rho(\omega)$ comes out as practically *independent* of ω. Moreover, the phase-shift $\theta(\omega)$ may come out as simply *proportional to* ω:

$$\rho(\omega) = \text{const.} = \rho_0$$
$$\theta(\omega) = \alpha\omega \tag{6.9.11}$$

In this case we can write for (10):

$$v(x) = \rho_0 e^{i\omega(x-a)} = \rho_0 u(x - \alpha) \tag{6.9.12}$$

Now the output follows the input with the time-lag α but reproduces the input with the proportionality factor ρ_0. If these conditions hold for a sufficiently large range of the frequency ω, the output will represent a *high-fidelity reproduction of the input*, apart from the constant time-lag α which for many purposes is not damaging.

6.10. The Laplace transform

The complex quantity (9.9) which characterises the amplitude factor and phase-shift of the response, is called the "transfer function". It is in a definite relation to the pulse-response $G(t)$ of our device and uniquely determined by that function. The relation between the two functions, viz. the pulse response and the transfer function, is intimately connected with a fundamental functional transformation, called the "Laplace transform", encountered earlier in Chapter 1.15. Let the function $G(t)$ be given in the interval $[0, \infty]$. We introduce a new function of the new variable p by putting

$$L(p) = \int_0^\infty e^{-pt} G(t) dt \tag{6.10.1}$$

This function $L(p)$ has generally no resemblance to the original function $G(t)$. In fact it is a much more regular function of p than $G(t)$ was of t. The function $G(t)$ need not be analytical at all. It can be prescribed freely as a generally continuous and absolutely integrable function, with a finite number of discontinuities and finite number of maxima and minima in any finite interval. The function $L(p)$, on the other hand, is an *analytical* function of the variable p, not only for real values of p but for arbitrary *complex* values of p as long as they lie in the *right half* of the complex plane (i.e. the real part of p is positive). The function $L(p)$ is thus an eminently regular and lawful function and we have no right to prescribe it freely, even in an arbitrarily small interval.

If now we consider the integral (9.6) which defines the transfer function

$$F(\omega) = \int_0^\infty e^{-i\omega t} G(t) dt \tag{6.10.2}$$

we see that by the definition of the Laplace transform we obtain the fundamental relation

$$F(\omega) = L(i\omega) \tag{6.10.3}$$

The transfer function $F(\omega)$ which determines the frequency response of our device, is thus obtained as the Laplace transform of the pulse response, taken along the imaginary axis.

Problem 266. Consider the pulse response of a galvanometer given by (4.9). Obtain the Laplace transform of this function and show that the singularity of $L(p)$ lies in the negative half plane, no matter what the value of the damping ratio $\kappa = \cos \gamma$ (between 0 and ∞) may be.

Problem 267. Obtain the transfer function (3) and study it from the standpoint of the fidelity problem. Explain the special role of the value $\kappa = 1/\sqrt{2}$ (fidelity damping).

[Answer:

$$F(\omega) = \frac{1}{1 - \omega^2 + 2i\kappa\omega} \tag{6.10.4}]$$

The amplitude response for the critical value becomes $(1 + \omega^4)^{-1/2}$ the distortion being of *fourth* instead of second order in ω.

Problem 268. Show that in the case of fidelity damping the maximum error of the galvanometer response for any ω between 0 and $1/2\sqrt{2}$ does not surpass 2% and that the error prediction on the basis of (7.12) in this frequency range holds with an accuracy of over 97.5%.

Problem 269. Show that from the standpoint of smallest phase distortion the value $\kappa = \sqrt{3}/2$ ($\gamma = \pi/6 = 30°$) represents the most advantageous damping ratio.

Problem 270. Assuming that ω varies between 0 and $1/4\kappa$ ($\kappa > \frac{1}{2}$), show that the maximum phase distortion of the galvanometer in that frequency range does not surpass the value

$$\Delta\theta = \frac{1}{32\kappa^2}\left(1 - \frac{4\kappa^2}{3}\right) \tag{6.10.5}$$

(excluding κ values which are near to the critical $\sqrt{3}/2$).

6.11. The memory time

We have mentioned before that our communication device absorbs the input energy, without creating energy of its own. Hence the response $G(t)$ to the input pulse will go asymptotically to zero as t goes to infinity. In actual fact it would be mere luxury to integrate out to infinity. Although $G(t)$ will disappear exactly in the theoretical sense only at $t = \infty$, it will be zero *practically* much sooner. No physical device can be taken with absolute mathematical exactitude since there are always disturbing accidental circumstances, called "noise", which interfere with the operation of the exact mathematical law by superimposing on our results some additional random phenomena which do not belong to the phenomenon we want to investigate. Hence we cannot aim at absolute accuracy. If we observe the function $G(t)$ which eventually becomes arbitrarily small but reaches zero strictly speaking at $t = \infty$, we come after a finite (and in practice usually very short) time T into a region which is of the same order of

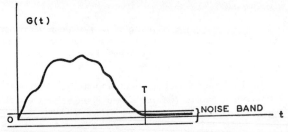

magnitude as the superimposed noise. We can then discard everything beyond $t = T$ as of no physical significance.

We see that under these circumstances a passive communication device possesses a definite "memory time" T which puts an upper limit to our integration. Instead of integrating out to infinity, it suffices to integrate only up to T:

$$v(x) = \int_0^T g(x - t)G(t)dt \tag{6.11.1}$$

Correspondingly the transfer function $F(\omega)$ will now become

$$F(\omega) = \int_0^T e^{-i\omega t}G(t)dt \tag{6.11.2}$$

The error thus induced in $F(\omega)$ can be estimated by the well-known integral theorem

$$\left| \int_a^b f(x)g(x)dx \right| \leq \int_a^b |f(x)||g(x)|dx \tag{6.11.3}$$

which in our case gives

$$|\eta| < \eta_0 \tag{6.11.4}$$

where

$$\eta_0 = \int_T^\infty |G(t)|dt \tag{6.11.5}$$

Although the "memory time" of an instrument is not an exact mathematical concept, to every given accuracy a definite memory time can be assigned.

Problem 271. Find the memory time of a galvanometer with critical damping if we desire that the error limit (11.5) shall be 3% of the total area under $G(t)$:

$$\eta_0 = 0.03 \int_0^\infty |G(t)| \, dt$$

[Answer: $T = 5.36$.]

Problem 272. Consider the Laplace transform $L(p)$ of the function

$$G(t) = e^{-t} \sin \beta t$$

Consider on the other hand the same transform $L_1(p)$ but integrating only up to $t = 4$. Show that the analytical behaviour of $L_1(p)$ is very different from that of $L(p)$ by being free of singularities in the entire complex plane, while $L(p)$ has a pole at the points $p = -1 \pm i\beta$. And yet, in the entire right half plane, including the imaginary axis $p = i\omega$, the relative error of $L_1(p)$ does not exceed 2%.

6.12. Steady state analysis of music and speech

The fact that every kind of input function $\beta(x)$ can be analysed in terms of pure sine and cosine vibrations, leads sometimes to wrong interpretations. It is an experimental fact that if a sustained vibration of the form

$$\sin(\omega x - \theta) \tag{6.12.1}$$

is presented to the human ear, the ear is not sensitive to the presence of the phase angle θ. It cannot differentiate between the input function $\sin \omega x$ and the input function $\sin(\omega x - \theta)$. From this fact the inference is drawn that the phase shift induced by the complex transfer function $F(\omega)$ is altogether of no importance. Since it makes no difference whether $v(x)$ is presented to the ear or that superposition of $\sin \omega x$ and $\cos \omega x$ functions which in their sum are equivalent to $v(x)$, it seems that we can completely discard the investigation of the phase-shift. The superposition principle holds, the ear does not respond to the phase of any of the components, hence it is altogether immaterial what phase angles are present in the output $v(x)$. According to this reasoning the two functions

$$v(x) = \sum b_k \sin(\omega_k x - \theta_k) \tag{6.12.2}$$

and

$$v_1(x) = \sum b_k \sin \omega_k x \tag{6.12.3}$$

are *equivalent* as far as our heading goes. The ear will receive the same impression whether $v(x)$ or $v_1(x)$ is presented to it.

This reasoning is in actual fact erroneous. The experimental fact that our perception is insensitive to the phase angle, θ_k, holds only if *sustained* musical sounds are involved. It is true that any combination of sine or cosine functions which are perceived as a continuous musical sound can be altered freely by adding arbitrary phase angles to every one of the components. But we have "noise" phenomena which are not of a periodic kind and which are not received by the ear as musical sounds. Such "noise" sequences can still be resolved into a superposition of steady state sine and cosine functions which have no beginning and no end. But the ear perceives these noises *as noises* and the steady state analysis is no longer adequate, although it is a mathematical possibility. Now it is no longer true that the periodic components into which the noise has been resolved can be altered by arbitrary phase shifts. Such an alteration would profoundly influence the noise and the perception of the noise. For noise phenomena which represent a succession of *transients* and cannot be perceived as a superposition of musical notes, the phase angle becomes of supreme importance. A phase shift which is not merely proportional to ω (giving a mere time-lag) will have no appreciable effect on the strictly "musical" portions of recorded music and speech but a very strong effect on the "transient" portions of the recording.

Even in pure music the transient phenomena are by no means of subordinate importance. The sustained tone of a violin has a beginning and an end. The ear receives the impression of a sustained tone after the first cycles of tone generation are over but the transition from one tone to the other represents a transient phenomenon which has to be considered separately. The same holds for any other instrument. It is the experience of many musically trained persons that they recognise the tone of a certain type of instrument much more by the *transitions* from tone to tone than by the sustained notes. The older acoustical investigations devoted the focus of attention almost completely to the distribution of "overtones" in a sustained note. Our ear perceives a musical sound under the aspects of "loudness", "pitch", and "tone quality". The loudness or intensity of the tone is determined by the amplitude of the air pressure vibrations which create in the ear the impression of a tone. The "pitch" of the tone is determined by the frequency of these vibrations. The "tone quality", i.e. the more or less pleasant or harmonious impression we get of a musical sound, is determined by the distribution of overtones which are present on account of the tone-producing mechanism. In strictly periodic tone excitement the overtones are in the frequency ratios $1:2:3:\ldots$, in view of the Fourier analysis to which a periodic function of time can be submitted. The presence of high overtones can give the sound a harsh and unpleasant quality. We must not forget, however, that generally the amplitude of the high overtones decreases rapidly and that in addition the sensitivity of our ear to high frequencies decreases rapidly. Hence it is improbable that in the higher frequency range anything beyond the third or fourth overtone is of actual musical significance. In the low notes we may perceive the influence of overtones up to the order 6 or 7. It would be a mistake to believe, however, that we actually perceive the overtones as separate tones since then the impression of the tone would be one of a *chord* rather than that of a single tone. It is the weakness of the overtones which prevents the ear from hearing a chord but their existence is nevertheless perceived and recorded as "tone quality".

In certain instruments which do not produce a sustained note, the value of an overtone analysis becomes doubtful since we do not have that periodicity which is a prerequisite for the application of the Fourier series to a physical phenomenon. The musical tone generated is here more a succession of transients than a steady state phenomenon. This is the case even with the piano tone which is the result of the sudden striking of a hammer and consists of damped vibrations. This is the reason that the piano tone does not blend well with the tone of string instruments; the hammer strikes the strings only once and produces a quickly damped vibration instead of a continuous tone. The damping is particularly strongly noticeable in the *upper* octaves. The transient is here more predominant than the steady state part of the tone, in fact one can hardly find a natural basis for a steady state analysis. Similar conditions hold for military percussion instruments and some jazz instruments.

Now the claim is often made in technical literature that the piano tone possesses "very high overtones". The tinny and flat sounds encountered in the piano recordings of older vintage have been explained on the basis that the amplitude response of the older recording instruments was unsatisfactory and failed to reproduce the very high overtones which accompany the tone generated by the piano. In actual fact these claims cannot be substantiated. Our ear is practically deaf to any overtones beyond about 8000 cycles per second. If the piano tone did possess such high overtones, they would still not be perceptible to our ear. The decisive factor is not the amplitude response but the *phase response* to which usually very little attention is paid. It was the strong phase distortion in the older recording instruments which interfered with the proper reproduction of transient phenomena, and this caused instruments of the piano type and all the other instruments with strong transient components to be poorly represented.

Similar phenomena are encountered in *speech recording*. Here the vowels represent the "sustained musical notes" while the consonants represent the noise part or transient part of speech. The construction of the larynx and the shape of the other organs which participate in sound production is such that the vowel sounds fall in a band of frequencies which vary between 300 and 3000 cycles per second, the most important range being that between 1000 and 2500 cycles. Our ordinary telephone is one of the most ingenious and most important communication devices. The ordinary telephone transmits frequencies between about 250 and 2750 cycles per second and is thus well qualified to the transmission of the steady part of our speech. At the same time the telephone is a poor device if it comes to the transmission of consonants as we can demonstrate by trying to telephone in a foreign language in which we are not perfectly at home. The good audibility of speech through the telephone is achieved through the method of "inferential extrapolation". We do not hear in the telephone what we believe we hear but we make up for the deficient information by reconstructing the distorted speech, because of familiarity with the language. If this familiarity is lacking, we are unable to use the method of inferential extrapolation and we are at a loss to understand the received communication. Consonants such as b, p, v, lose their identity if spoken in the telephone and thus in the spelling of unfamiliar words we have to take refuge to the method of "associative identification" by saying: "B for Billy, P fŏr Peter, V for Vincent."

The steady state parts of the speech, that is the vowels, do not suffer much in the process of telephone transmission and are easily recognisable. The reason is that these vowels are essentially composed of low-frequency trains of waves. Moreover, the phase distortion, which is a vital handicap in the transmission of noise, has practically no influence on the perception by the ear, as long as we stay within the boundaries of steady state analysis.

6.13. Transient analysis of noise phenomena

The resolution of signals into harmonic components is such a vital part of electrical engineering that it becomes second nature to the engineer. For him the signal $f(t)$ does not exist as $f(t)$ but as $F(\omega)$, that is its harmonic resolution into the components $e^{i\omega t}$, acted upon by the communication device. The communication device applies its own weight factor to each one of the harmonic components and after this weighting the component parts are put together again. Hence the input $f(t)$ is resolved into its harmonic components by Fourier *analysis*; the output $v(x)$, on the other hand, is assembled again out of its constituent parts, with the help of Fourier *synthesis*.

In this procedure the weighting of the *amplitudes* is traditionally considered as of vital importance. The amplitude response is technically much easier the subject of exact measurements than the phase response. The latter is frequently regarded as of negligible importance, in view of the fact that neither the perception of the sustained notes in music, nor the perception of the sustained elements of speech (i.e. primarily the vowel sounds) is sensitive to the presence or absence of arbitrary phase shifts. Our ear does not recognise the phase of a musical note and hence the question of phase fidelity is immaterial.

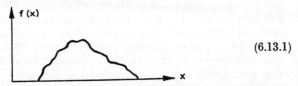

$$(6.13.1)$$

This picture changes, however, very considerably if we now come to the investigation of the second problem, viz. the reproduction of *transient phenomena*. A consonant such as p, q, r, and so on, presents itself primarily as a certain function of time which is of short duration and non-repetitive. We can speak of a certain "noise profile" by imagining that the air pressure generated by a consonant—or likewise by the onset of a musical tone—is plotted as a graph which characterises that particular element of speech or music. This graph can be of a rather irregular shape and the question arises how to obtain a faithful reproduction of it.

That this reproduction cannot be perfect, is clear from the outset. Perfect reproduction would mean that a delta function as input is recorded as a delta function as output, except for a certain constant time-lag. This is obviously impossible, because of the inertia of the mechanical components of the instrument. In fact we know that even the much more regular unit step function cannot be faithfully reproduced since no physical instrument can make a sudden jump. Nor is such an absolutely faithful reproduction necessary if we realise that our ear itself is not a perfect recording instrument. No recording instrument can respond to a definite local value of the input signal without a certain *averaging* over the neighbouring values. Our ear is likewise a recording instrument of the integrating type and thus unable to perceive an arbitrarily rugged $f(t)$. A certain amount of smoothing must characterise our acoustical perception, and it is this smoothing quality of the ear on which we can bank if we want to set up some reasonable standards in the fidelity analysis of noise.

We recognise from the beginning that the reproduction of a given rather complicated noise profile of short duration will be a much more exacting task than the recording of musical sounds of relatively low frequencies. But the question is how far have we to go in order to reproduce noise with a fidelity which is in harmony with the high, but nevertheless not arbitrarily high, capabilities of the human ear.

This seems to a certain degree a physiological question, but we can make some plausible assumptions towards its solution. We will not go far wrong if we assume that the ear smooths out the very rugged peaks of an input function by averaging. This averaging is in all probability very similar to that method of "local smoothing" that we have studied in the theory of the Fourier series and that we could employ so effectively towards an increased convergence of the Fourier series by cutting down the contribution of the terms of very high frequency. The method of local averaging attaches to the vibration $e^{i\omega t}$ the weight factor

$$\sigma(\omega) = \frac{\sin \omega \frac{\tau}{2}}{\omega \frac{\tau}{2}} = \frac{\sin \pi \nu \tau}{\pi \nu \tau} \tag{6.13.2}$$

(cf. 2.13.8), if we denote by τ the smoothing time. Our insensibility to higher overtones may easily be due to this smoothing operation of our ear. Since we do not perceive frequencies beyond 10,000 per second, we shall probably not go far wrong if we put the smoothing time of the human ear in the neighbourhood of $1/10{,}000 = 10^{-4}$ second. If we use the galvanometer response (4.9) as an operating model, we see from the formula (7.16) that the output does not give simply the local value $\beta(x - 2\kappa)$, but there is a correction term added which is of the same nature as that found in the theory of the Fourier series, caused by local smoothing (cf. 2.13.6). The smoothing time τ can thus be established by the relation

$$\frac{\tau^2}{\sigma} = 2\kappa^2 - 1 \tag{6.13.3}$$

or, returning to the general time scale of the galvanometer equation (4.4):

$$\tau = \sqrt{6}\,\frac{\sqrt{2\kappa^2 - 1}}{\rho} \tag{6.13.4}$$

With advancing age the smoothing time increases, partly due to an increase in the damping constant α, and partly due to a decrease in the stiffness constant ρ, considering the gradual relaxation of the elastic properties of living tissues. This explains the reduction of the frequency range to which the ear is sensitive with advancing age.

We often treat the phenomenon of "hard of hearing" as a decline of the hearing nerve in perceiving a certain sound *intensity*. The purpose of the hearing aid is to amplify the incoming sound, thus counteracting the reduction of sound intensity caused by the weakening of the hearing nerve. A second factor is often omitted, although it is of no smaller importance. This is the *reduced resolution power* of the ear in perceiving a given noise profile (1). As the smoothing time τ increases, the fine kinks of the noise profile (1) become more and more blurred. Hence it becomes increasingly more difficult to identify certain consonants, with their characteristic noise profiles. If in a large auditorium the lecturer gets the admonition "louder" from the back benches, he will not only raise his voice but he will instinctively talk slower and more distinctly. This makes it possible to recognise certain noise patterns which would otherwise be lost, due to smoothing. If we listen to a lecture delivered in a language with which we are not very familiar, we try to sit in the front rows. This has not merely the effect of higher tone intensity. It also contributes to the better resolution of noise patterns which in the back seats are blurred on account of the acoustical echo-effects of the auditorium. (Cockney English, however, cannot be resolved by this method.)

From this discussion certain consequences can be drawn concerning the fidelity analysis of noise patterns. In the literature of high fidelity sound reproducing instruments we encounter occasional claims of astonishing magnitude. One reads for example that the "flat top amplitude characteristics" (i.e. lack of any appreciable amplitude distortion) has been extended to 100,000 cycles per second. If this fact is technically possible, the question can be raised whether it is of any practical necessity (the distance travelled by the needle in the playing of a 12-in. record during the time of $1/100,000 = 10^{-5}$ second is not more than 0.005 mm). Taking into account the limited resolution power of the human ear, what are the actual fidelity requirements in the reproduction of noise patterns? Considering the inevitable smoothing operation of the ear, it is certainly unnecessary to try to reproduce all the kinks and irregularities of the given noise profile since the small details are obliterated anyway. In this situation we can to some extent relax the exact mathematical conditions usually observed in the generation of functions. In Section 2 we have recognised the unit step function as a fundamental building block in the generation of $f(x)$. The formula (2.5) obtained $f(x)$ by an integration over the base function $\delta^1(x - \xi)$. This integral could be conceived as the limit of the sum (2.6), reducing the Δx_i to smaller and smaller values. In the presence of smoothing,

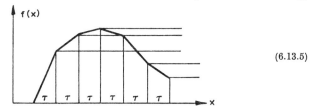

(6.13.5)

however, we can *leave this sum as a sum*, without going to a limit. We need not diminish the Δx_i below the smoothing time τ since smaller details will be blotted out by the ear. Under these circumstances we can generate the function $f(x)$ as a succession of "smoothed" unit step functions which do not jump up from zero to 1 but have a linear section between. The following figure demonstrates how such a function—except for the usual time-lag—can be faithfully reproduced by a galvanometer of critical damping.

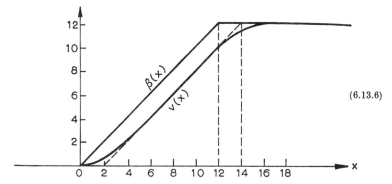

(6.13.6)

We have used 12 units for the horizontal part of the curve, in order to reduce the disturbance at the beginning and the end of the linear section to practically negligible amounts. Hence we have solved the problem of faithfully reproducing a noise pattern which is composed of straight line sections of the duration τ. Since the units used were normalised time units, the return to the general galvanometer equation (4.4) establishes a stiffness constant ρ which is in the following relation to the smoothing time τ:

$$\rho = \frac{12}{\tau} \qquad (6.13.7)$$

Now the transfer function (10.4) of a galvanometer demonstrates that we get a fairly undistorted amplitude and phase response—the amplitude remaining constant, the phase increasing linearly with ω—if we stay with ω within the limits 0 and $\frac{1}{2}$ (for κ near unity) and that means in the general time scale the range

$$\omega < \frac{\rho}{2} = \frac{6}{\tau} \tag{6.13.8}$$

If instead of the angular frequency ω we employ the number ν of cycles per second, we obtain as an upper bound of the faithfulness requirement

$$\nu < \frac{\rho}{4\pi} = \frac{3}{\pi\tau} \tag{6.13.9}$$

This result indicates that we need not extend the fidelity requirement in amplitude and phase beyond the limit $1/\tau$.

A similar result is deducible from a still different approach. We have seen in the treatment of the Fourier series that the infinite Fourier series could be replaced with a practically small error by a series which terminates with n terms, if we modify the Fourier coefficients by the sigma factors *and* at the same time replace $f(x)$ by $\bar{f}(x)$, obtained by local smoothing. The smoothing time of the local averaging process was

$$\tau = \frac{2\pi}{n} \tag{6.13.10}$$

or, expressing once more everything in terms of cycles per second, the last term of the finite Fourier series becomes $a_n \cos 2\pi\nu x + b_n \sin 2\pi\nu x$ while τ becomes $1/\nu$. We may equally reverse our argument and ask for the smoothing time τ which will make it possible to obtain $f(x)$ with sufficient accuracy by using frequencies which do not go beyond ν. The answer is $\tau = 1/\nu$.

Here again we come to the conclusion that *it is unnecessary to insist on the fidelity of amplitude and phase response beyond the upper limit $\nu = 1/\tau$.*

The result of our analysis is that the apparently very stringent fidelity requirements of a transient recording are in fact not as stringent as we thought in the first moment. If amplitude and phase distortion can be avoided up to about 10 or perhaps 15 thousand cycles per second, we can be pretty sure that we have attained everything that can be expected in the realm of high-fidelity sound reproduction. The great advances made in recent years in the field of high-fidelity equipment is not so much due to a spectacular extension of the amplitude fidelity to much higher frequencies but due to a *straightening out of the phase response* which is of vital importance for the high-fidelity reproduction of noise, although not demanded for the recording of sustained sounds. The older instruments suffered by strong phase distortion in the realm of high frequencies. To have eliminated this distortion up to frequencies of 10,000 cycles per second has improved the reproduction of the transients of music and speech to an admirable degree.

BIBLIOGRAPHY

[1] Bush, V., *Operational Circuit Analysis* (Wiley, New York, 1929)
[2] Carson, J. R., *Electric Circuit Theory and Operational Analysis* (McGraw-Hill, New York, 1920)
[3] Pipes, L. A., *Applied Mathematics for Physicists and Engineers* (McGraw-Hill, New York, 1946)

CHAPTER 7

STURM-LIOUVILLE PROBLEMS

Synopsis. The solution of ordinary second-order differential equations has played a fundamental role in the evolution of mathematical physics, starting with the eigenvibrations of a string, and culminating in the atomic vibrations of Schrödinger's wave equation. The separation of the fundamental differential operators of physics into functions of a single variable leads to a large class of important second order equations. While the solution of these equations cannot be given in closed form, except in special cases, it is possible to obtain an eminently useful *approximation* in terms of a mere quadrature. In the study of this method we encounter a certain refinement which permits us to deduce expressions in terms of elementary functions, which approximate some of the fundamentally important function classes of mathematical physics (such as the Bessel functions, and the Legendre, Hermite, and Laguerre type of polynomials), with a remarkably high degree of accuracy.

7.1. Introduction

There exists an infinite variety of differential equations which we may want to investigate. Certain differential equations came into focus during the evolution of mathematical theories owing to their vital importance in the description of physical phenomena. The "potential equation", which involved the Laplacian operator Δ, is in the foremost line among these equations. The separation of the Laplacian operator in various types of coordinates unearthed a wealth of material which demanded some kind of universal treatment. This was found during the nineteenth century, through the discovery of orthogonal expansions which generalised the outstanding properties of the Fourier series to a much wider class of functions. The early introduction of the astonishing hypergeometric series by Euler was one of the pivotal points of the development. Almost all the important function classes which came into use during the last two centuries are in some way related to the hypergeometric series. These function classes are characterised by ordinary differential equations of the second order. The importance of these function classes was first discovered by two French mathematicians, J. Ch. F. Sturm (1803–55) and J. Liouville (1809–82). These special types of differential operators are thus referred to as belonging to the "Sturm-Liouville type". Lord Rayleigh in his famous acoustical investigations (1898) came also near to the general theory of orthogonal expansions. The final unifying viewpoint came into being somewhat later, through the fundamentally new departure of I. Fredholm (1900), and its later development by D. Hilbert (1912). It will not be our aim here to go into any detailed study of this great branch of analysis which gave rise to a very extended literature. We will deal with the Sturm-Liouville type of differential equations primarily as an illustration of the general theory. Furthermore, we will pay detailed attention to a specific feature of second-order operators, namely that they are reducible to the solution of a first-order non-linear differential equation. This method of solution played a decisive role in the early development of quantum theory, characterised by Bohr's atom model. But certain refinements of the method are well adapted to a more exact approximation of the Sturm-Liouville type of function classes.

7.2. Differential equations of fundamental significance

The most general linear differential operator of second order can be written down in the following form:

$$Dv(x) = A(x)v''(x) + B(x)v'(x) + C(x)v(x) \qquad (7.2.1)$$

where $A(x)$, $B(x)$, $C(x)$ are given functions of x. We will assume that $A(x)$ does not go through zero in the domain of investigation since that would lead to a "singular point" of the differential equation in which generally the function $v(x)$ goes out of bound. It can happen, however, that $A(x)$ may become zero on the *boundary* of our domain.

We will enumerate a few of the particularly well investigated and significant differential equations of pure and applied analysis. The majority of the fundamental problems of mathematical physics are in one way or another related to these special types of ordinary second-order differential equations.

1. If

$$B(x) = A'(x) \qquad (7.2.2)$$

the differential equation (1) may be written in the form

$$Dv = \frac{d}{dx}[A(x)v'(x)] + C(x)v(x) \qquad (7.2.3)$$

and we have obtained the Sturm-Liouville type of differential operators which have given rise to very extensive investigations.

2. Bessel's differential equation:

$$v''(x) + \frac{1}{x}v'(x) + \left(1 - \frac{p^2}{x^2}\right)v(x) = 0 \qquad (7.2.4)$$

the solution of which are the "Bessel functions"

$$v(x) = J_p(x)$$

where the index p, called the "order of the Bessel function", may be an integer, or in general any real or even complex number.

3. Mathieu's differential equation

$$(1 - x^2)v''(x) - xv'(x) + [a + b(2x^2 - 1)]v(x) = 0 \qquad (7.2.5)$$

which depends on the two constants a and b.

4. The differential equation of Gauss:

$$x(1 - x)v''(x) + [\gamma - (\alpha + \beta + 1)x]v'(x) - \alpha\beta v(x) = 0 \qquad (7.2.6)$$

the solution of which is the hypergeometric series

$$F(\alpha, \beta, \gamma; x) = 1 + \frac{\alpha\beta}{\gamma \cdot 1} x + \frac{\alpha(\alpha + 1)\beta(\beta + 1)}{\gamma(\gamma + 1) \cdot 1 \cdot 2} x^2$$
$$+ \frac{\alpha(\alpha + 1)(\alpha + 2)\beta(\beta + 1)(\beta + 2)}{\gamma(\gamma + 1)(\gamma + 2)1 \cdot 2 \cdot 3} x^3 + \dots \qquad (7.2.7)$$

This differential equation embraces many of the others since almost all functions of mathematical physics are obtainable from the hypergeometric series by the proper specialisation of the constants α, β, γ, and the proper transformation of the variable x.

5. The differential equation of the Jacobi polynomials, obtained from the Gaussian differential equation by identifying α with a negative integer $-n$:

$$x(1 - x)v''(x) + (\gamma - \delta x)v'(x) + n(n + \delta - 1)v(x) = 0 \qquad (7.2.8)$$

$$v(x) = P_n^{(\gamma, \delta)}(x) = cF(-n, \delta + n - 1, \gamma; x) \qquad (7.2.9)$$

These polynomials depend on the two parameters γ and δ.

6. A special subclass of these polynomials are the "ultraspherical polynomials"

$$v(x) = P_n^{(\gamma)} = cF(-n, 2\gamma + n - 1, \gamma; x) \qquad (7.2.10)$$

which correspond to the choice $\delta = 2\gamma$ and thus depend on the single parameter γ. They are those Jacobi polynomials which possess left-right symmetry if x is transformed into $x_1 = 1 - 2x$. In this new variable we obtain the differential equation

$$(1 - x^2)v''(x) - 2\gamma xv'(x) + n(n + 2\gamma - 1)v(x) = 0 \qquad (7.2.11)$$

7. The choice $\gamma = 1$ leads to "Legendre's differential equation"

$$(1 - x^2)v''(x) - 2xv'(x) + n(n + 1)v(x) = 0 \qquad (7.2.12)$$

which defines the "Legendre polynomials"

$$v(x) = P_n(x) = F\left(-n, n + 1, 1; \frac{1 - x}{2}\right) \qquad (7.2.13)$$

8. The choice $\gamma = \frac{1}{2}$ leads to the "Chebyshev polynomials":

$$v(x) = T_n(x) = F\left(-n, n, \tfrac{1}{2}; \frac{1-x}{2}\right) \tag{7.2.14}$$

whose differential equation is

$$(1 - x^2)v''(x) - xv'(x) + n^2v(x) = 0 \tag{7.2.15}$$

9. Still another class of polynomials, associated with the range $[0, \infty]$, is established by the "differential equation of Laguerre":

$$xv''(x) + (1 - x)v'(x) + nv(x) = 0 \tag{7.2.16}$$

which defines the Laguerre polynomials $L_n(x)$. These polynomials can be conceived as limiting cases of the Jacobi polynomials, by choosing $\gamma = 1$ and letting δ go to infinity. This necessitates, however, a corresponding change of x, in the following sense:

$$x = \mu x_1, \quad \gamma = 1, \quad \delta = \frac{1}{\mu}$$

where μ goes to zero. We thus obtain:

$$v(x) = L_n(x) = c \lim_{\mu \to 0} F\left(-n, \frac{1}{\mu}, 1; \mu x\right) \tag{7.2.17}$$

10. The "differential equation of Hermite"

$$v''(x) - 2xv'(x) + 2nv(x) = 0 \tag{7.2.18}$$

is associated with the range $[-\infty, +\infty]$ and defines the "Hermitian polynomials" $H_n(x)$. These polynomials are limiting cases of the ultraspherical polynomials (10), by letting γ go to infinity and correspondingly changing x in the following sense:

$$x = \mu x_1, \quad \gamma = \frac{1}{\mu^2}$$

where μ goes once more to zero. We thus obtain:

$$v(x) = H_n(x) = c \lim_{\mu \to 0} F\left(-n, \frac{2}{\mu^2} + n - 1, \frac{1}{\mu^2}; \frac{1 - \mu x}{2}\right) \tag{7.2.19}$$

Problem 273. By substituting $x = \beta x_1$ in Bessel's differential equation, show that the function $J_p(\beta x)$ satisfies the following differential equation:

$$v'' + \frac{v'}{x} + \left(\beta^2 - \frac{p^2}{x^2}\right)v = 0 \tag{7.2.20}$$

Problem 274. By the added substitution $x = x_1{}^\alpha$ show that the function

$$v(x) = J_p(\beta x^\alpha)$$

satisfies the following differential equation:

$$v'' + \frac{v'}{x} + \alpha^2\left(\beta^2 x^{2\alpha-2} - \frac{p^2}{x^2}\right)v = 0 \tag{7.2.21}$$

Problem 275. By making the further substitution $v(x) = x^{-\gamma}v_1(x)$ show that the function

$$v(x) = x^{\gamma}J_p(\beta x^{\alpha}) \tag{7.2.22}$$

satisfies the following differential equation:*

$$v'' + \frac{1 - 2\gamma}{x}\,v' + \left(\alpha^2\beta^2 x^{2\alpha-2} + \frac{\gamma^2 - \alpha^2 p^2}{x^2}\right)v = 0 \tag{7.2.23}$$

Problem 276. Choose $\alpha = \tfrac{1}{2}$, $\gamma = -p/2$, $\beta = 1$. Show that the function

$$v(x) = x^{-p/2}J_p(\sqrt{x}) \tag{7.2.24}$$

satisfies the following differential equation:

$$xv'' + (p + 1)v' + \frac{v}{4} = 0 \tag{7.2.25}$$

Problem 277. By differentiating this equation n times, prove the following theorem: Let v_p be an arbitrary solution of Bessel's differential equation (4). Then an arbitrary solution of the same differential equation for $p_1 = p + n$ is obtainable as follows:

$$v_{p+n}(\sqrt{x}) = x^{(p+n)/2}(x^{-p/2}v_p(\sqrt{x}))^{(n)} \tag{7.2.26}$$

Problem 278. By substituting in (25) the value $p = -\tfrac{1}{2}$ show that Bessel's differential equation for $p = \tfrac{1}{2}$ is solvable by

$$v_{1/2}(x) = \frac{1}{\sqrt{x}}\,e^{\pm ix} \tag{7.2.27}$$

Problem 279. Obtain the general solution of Bessel's differential equation for the order $p = n - \tfrac{1}{2}$ (n an arbitrary integer) in terms of elementary functions as follows:

$$v_{n-1/2}(\sqrt{x}) = x^{n/2-1/4}(e^{i\sqrt{x}})^{(n)} \tag{7.2.28}$$

with the understanding that we take an arbitrary linear combination of the real and imaginary parts of this function.

7.3. The weighted Green's identity

In the matrix approach to the general theory of differential operators the independent variable plays a relatively minor role. We have represented the function $f(x)$ as a vector in a many-dimensional space (see Chapter 4.5). In this mode of representation the independent variable x provides merely the cardinal number of the axis on which the functional value $f(x)$ may be found if we project the vector on that particular axis. The matrix associated with a given linear differential operator D came about by breaking up the continuous variable x into a large but finite set of discrete points and replacing the derivatives by the corresponding difference coefficients. But

---

* This is a particularly useful generalised form of Bessel's differential equation. It is listed, together with many other differential equations which are solvable in terms of Bessel functions, in Jahnke-Emde (see Bibliography [2], pp. 214–15).

if we examine the procedure of Chapter 4.6, we shall notice that the correlation of the matrix A to the operator D is not unique. It depends on the method by which the continuum is atomised. For example the differential operator was translated into the matrix (4.6.11). In this translation the Δx_i of the atomisation process were considered as *constants* and put equal to ϵ. If these Δx_i *varied* from point to point, the associated matrix would be quite different. In particular, the originally symmetric matrix would now become non-symmetric. Now a variation of the Δx_i could also be conceived as keeping them once more as constants, but changing the independent variable x to some other variable t by the transformation

$$x = \varphi(t) \tag{7.3.1}$$

because then

$$\Delta x = \varphi'(t)\Delta t \tag{7.3.2}$$

A uniform change $\Delta t_i = \epsilon$ of the new variable t does *not* yield any longer a corresponding uniform change in x. Since the properties of the associated matrix A are of fundamental importance in the study of the differential operator D, we see that a transformation of the independent variable x into a new variable t according to (1) is not a trivial but in fact an essential operation. It happens to be of particular importance in the study of second order operators.

We have seen in Chapter 5.26 that the eigenfunctions $u_i(x)$, $v_i(x)$, which came about by the solution of the shifted eigenvalue problem (5.26.2), had the orthogonality property

$$\int u_i(x)u_k(x)dx = \delta_{ik}$$
$$\int v_i(x)v_k(x)dx = \delta_{ik} \tag{7.3.3}$$

If we introduce the new variable t, the same orthogonality changes its appearance, due to the substitution

$$dx = \varphi'(t)dt \tag{7.3.4}$$

and we now obtain

$$\int u_i(\varphi(t))u_k(\varphi(t))\varphi'(t)dt = \delta_{ik} \tag{7.3.5}$$

(With a similar relation for the functions v_i.) This modified orthogonality property can be avoided if we use x instead of t as our independent variable. But occasionally we have good reasons to operate with the variable t rather than x. For example the functions $u_i(x)$ may become *polynomials* if expressed in the variable t, while in the original variable x this property would be lost.

Under these circumstances it will be of advantage to insist on a complete freedom in choosing our independent variable x, allowing for an arbitrary transformation to any new variable t. For this purpose we can survey our previous results and every time we encounter a dx, replace it by $\varphi'(t)dt$. Since, however, we would like to adhere to our standard notation x for our independent variable, we will prefer to call the *original* variable t and the *transformed* variable x. Hence we prefer to re-write (4) in the form

$$t = \varphi(x), \quad dt = \varphi'(x)dx \tag{7.3.6}$$

Furthermore, we want to introduce a new symbol for $\varphi'(x)$ by putting

$$w(x) = \varphi'(x) \tag{7.3.7}$$

In order to preserve the one-to-one correspondence between the two variables x and t, we will demand that the sign of $w(x)$ shall not change within the interval of consideration $[a, b]$, although we will permit it to become zero at the *endpoints* of the range. We do not lose in generality if we normalise the sign of $w(x)$ by assuming that it is everywhere *positive* inside of our range.

This $w(x)$ yields a new degree of freedom which is particularly valuable in the study of second-order differential operators. Now the generalised orthogonality relation (5) can be written as follows:

$$\int_a^b u_i(x)u_k(x)w(x)dx = \delta_{ik} \tag{7.3.8}$$

We call it "weighted orthogonality", or "orthogonality with respect to the weight factor $w(x)$". We have encountered such a weighted orthogonality earlier, when dealing with the Laguerre polynomials (cf. 1.10.7), which were orthogonal with respect to the weight factor e^{-x}. The fundamental Green's identity appears now in the generalised variable x in the following form:

$$\int_a^b w(x)(uDv - v\tilde{D}u)dx = \text{boundary term} \tag{7.3.9}$$

which again leads to the following generalisation of the relation (4.12.5):

$$uDv - v\tilde{D}u = \frac{1}{w(x)}\frac{d}{dx}F(u, v) \tag{7.3.10}$$

We will call the identity (9), which defines the adjoint operator $\tilde{D}$ with respect to the weight factor $w(x)$, the "weighted Green's identity". We see that the definition of the adjoint operator is vitally influenced by the choice of the independent variable. If we transform the independent variable t to a new variable x, and thus express Dv in terms of x, we obtain the adjoint operator $\tilde{D}u$—likewise expressed in the new variable x—by defining this operator with the help of the weighted Green's identity (9).

The definition of the Green's function $G(x, \xi)$ is likewise influenced by the weight factor $w(x)$, because the delta function $\delta(x, \xi)$ is *not* an invariant of a coordinate transformation. The definition of $\delta(x - \xi) = \delta(t)$ demanded that

$$\int_{-\epsilon}^{+\epsilon} \delta(t)dt = 1 \tag{7.3.11}$$

If the independent variable t is changed to x according to the transformation (6), the condition (11) takes in the new variable the form

$$w(0) \int_{-\epsilon}^{+\epsilon} \delta(x)dx = 1 \tag{7.3.12}$$

Hence in the generalised variable the delta function has to be written in the form

$$\frac{1}{w(x)} \delta(x, \xi) \tag{7.3.13}$$

and the definition of the Green's function must occur on the basis of the equation (cf. 5.4.12)

$$\tilde{D}G(x, \xi) = \frac{1}{w(x)} \delta(x, \xi) \tag{7.3.14}$$

or, if we want to define the same function in terms of the original operator (cf. 5.12.15):

$$DG(x, \xi) = \frac{1}{w(\xi)} \delta(x, \xi) \tag{7.3.15}$$

The shifted eigenvalue problem

$$\begin{aligned} Dv &= \lambda u \\ \tilde{D}u &= \lambda v \end{aligned} \tag{7.3.16}$$

remains unchanged and the construction of the Green's function with the help of the bilinear expansion (5.27.7) is once more valid:

$$G(x, \xi) = \sum_{i=1}^{\infty} \frac{v_i(x)u_i(\xi)}{\lambda_i} \tag{7.3.17}$$

(We exclude zero eigenvalues since we assume that the given problem is complete and unconstrained.) However, in the solution with the help of the Green's function the weight factor $w(x)$ appears again:

$$v(x) = \int G(x, \xi)\beta(\xi)w(\xi)d\xi \tag{7.3.18}$$

$$u(x) = \int G(x, \xi)\gamma(\xi)w(\xi)d\xi \tag{7.3.19}$$

Problem 280.   Given the differential equation

$$v''(t) = \beta(t) \tag{7.3.20}$$

with the boundary conditions

$$v(0) = v'(0) = 0 \tag{7.3.21}$$

Transform t into the new variable x by putting

$$\sin t = x \tag{7.3.22}$$

Formulate the given problem in the new variable x and solve it with the help of the weight factor

$$w(x) = \frac{1}{\sqrt{1 - x^2}} \tag{7.3.23}$$

Demonstrate the equivalence of the results in t and in x.

[Answer:

$$\begin{aligned} Dv &= (1 - x^2)v'' - xv' \qquad x = [0, 1] \\ \tilde{D}v &= Dv \end{aligned} \tag{7.3.24}$$

$$\begin{aligned} G(x, \xi) &= \text{arc sin } x - \text{arc sin } \xi \qquad (\xi \le x) \\ &= 0 \qquad\qquad\qquad\qquad (\xi \ge x) \end{aligned} \tag{7.3.25}$$

$$v(x) = \int_0^x \beta(\xi) \frac{\text{arc sin } x - \text{arc sin } \xi}{\sqrt{1 - \xi^2}} d\xi \tag{7.3.26}]$$

7.4. Second-order operators in self-adjoint form

We have seen in the function space treatment of linear differential operators that the principal axis transformation of such an operator leads to two sets of orthogonal functions, operating in two different spaces. The one set can be used for the expansion of the right side, the other for the expansion of the solution. The relation between these two expansions was established by the eigenvalues associated with these two sets of eigenfunctions. But in the case where the operator is *self-adjoint*, a great simplification takes place because then these two spaces collapse into one. Then the functions on the left side and the right side can be analysed in the *same* set of orthogonal functions and the shifted eigenvalue problem becomes reduced to half the number of equations. The operator D does not now require pre-multiplication by $\tilde{D}$ to make it self-adjoint; it is self-adjoint in itself.

Now in the case of second-order differential operators we are in the fortunate position that the self-adjointness is an *automatic corollary* of the operator, provided that we employ the proper independent variable for the formulation of the problem, or else operate with an arbitrary variable but change from ordinary orthogonality to weighted orthogonality, in the sense of the previous section.

Let us consider the general form (2.1) of a linear differential operator. On the other hand, the most general *self-adjoint* operator has the form

$$Dv = \frac{d}{dt}(A_1 v') + Cv \tag{7.4.1}$$

If we change from the variable t to the variable x by means of the transformation (3.6), we obtain in the new variable:

$$Dv = \frac{1}{w(x)} \frac{d}{dx}(A_1 v') + Cv \tag{7.4.2}$$

We can dispose of the weight function $w(x)$ in such a way that the general operator (2.1) becomes transformed into the self-adjoint form (2). For this purpose we must define the functions A_1 and w according to the following conditions:

$$\frac{A_1}{w} = A, \quad \frac{A'_1}{w} = B \tag{7.4.3}$$

This yields for w the condition

$$(wA)' = B \tag{7.4.4}$$

With the substitution $wA = p$ we obtain

$$p' = \frac{B}{A}p \tag{7.4.5}$$

and thus

$$p = e^{\int (B/A)dx} \tag{7.4.6}$$

which means, if we return to the weight function $w(x)$:

$$w(x) = \frac{1}{A(x)} e^{\int (B/A)dx} \tag{7.4.7}$$

Here we have the weight factor which makes the general second order operator (2.1) self-adjoint.

The boundary term of the weighted Green's identity becomes

$$\left[A_1(uv' - vu')\right]_a^b = \left[wA(uv' - vu')\right]_a^b \tag{7.4.8}$$

We have a wide choice of boundary conditions which will insure self-adjointness. Let us write the two homogeneous boundary conditions imposed on $v(x)$ as follows:

$$\begin{aligned} v(b) &= p_1v(a) + p_2v'(a) \\ v'(b) &= q_1v(a) + q_2v'(a) \end{aligned} \tag{7.4.9}$$

Then the four constants of these relations can be chosen freely, except for the single condition:

$$w(b)A(b)(p_1q_2 - p_2q_1) = w(a)A(a) \tag{7.4.10}$$

Conditions which involve no interaction between the two boundary points, can be derived from (9) by a limit process. Let us put

$$\begin{aligned} p_2 &= \mu p_1 \\ q_2 &= \mu q_1 + \epsilon \end{aligned} \tag{7.4.11}$$

and let us go with ϵ toward zero. Then the quantity $p_1q_2 - p_2q_1$ is reduced to $p_1\epsilon$, while q_1 is freely at our disposal. Hence in the limit, as ϵ goes to zero, we obtain the boundary condition

$$v(b) + \nu v'(b) = 0 \tag{7.4.12}$$

where ν is arbitrary. This condition takes now the place of the second condition (9). At the same time, as ϵ goes to zero, p_1 must go to infinity. This implies that the first condition (9) becomes in the limit:

$$v(a) + \mu v'(a) = 0 \tag{7.4.13}$$

Hence the boundary conditions (12) and (13)—with arbitrary μ and ν—are *permissible self-adjoint boundary conditions.*

Our operator is now self-adjoint, with respect to the weight factor $w(x)$. Hence the shifted eigenvalue problem (3.16) becomes simplified to

$$Dv(x) - \lambda v(x) = 0 \tag{7.4.14}$$

in view of the fact that the functions $u_i(x)$ and $v_i(x)$ coincide. The resultant eigensolutions form an infinite set of ortho-normal functions, orthogonal with respect to the weight factor $w(x)$:

$$\int_a^b v_i(x)v_k(x)w(x)dx = \delta_{ik} \tag{7.4.15}$$

While generally the eigenvalue problem (14) need not have any solutions, and, even if the solutions exist, the λ_i will be generally complex numbers, the situation is quite different with second order (ordinary) differential operators, provided that the boundary conditions satisfy the condition (10). Here the eigenvalues are always real, the eigensolutions exist in infinite number and span the entire action space of the operator. The orthogonality of the eigenfunctions holds, however, with respect to the weight factor $w(x)$, defined by (7).

Problem 281. Obtain the weight factor $w(x)$ for Mathieu's differential operator (2.5) and perform the transformation of the independent variable explicitly.
[Answer:

$$w(x) = \frac{1}{\sqrt{1 - x^2}}, \quad x = -\cos t \tag{7.4.16}$$

$$v''(t) + (a + b \cos t)v(t) = 0 \tag{7.4.17}]$$

Problem 282. Consider Laguerre's differential equation (2.16). Obtain its weight factor $w(x)$ and show that for the realm $[0, \infty]$ the condition (10) is fulfilled for *any* choice of the four constants of (9).
[Answer:

$$w(x) = e^{-x} \tag{7.4.18}]$$

The boundary condition is nevertheless present by the demand that $v(x)$ must not grow to infinity stronger than $e^{x/2}$.

Problem 283. Show that boundary conditions involving the point a alone (or b alone) cannot be self-adjoint.

Problem 284. Find the condition, under which a self-adjoint periodic solution

$$
\begin{aligned}
v(b) &= v(a) \\
v'(b) &= v'(a)
\end{aligned}
\tag{7.4.19}
$$

becomes possible.

[Answer:

$$
\int_a^b \frac{B}{A}\, dx = 0
\tag{7.4.20]}
$$

7.5. Transformation of the dependent variable

There is still another method of securing the self-adjointness of a second order differential equation. Instead of transforming the independent variable x, we may transform the function $v(x)$. We do not want to lose the linearity of our problem and thus our hands are tied. The transformation we can consider is the multiplication of $v(x)$ by some function of x. This, however, is actually sufficient to make any second order problem self-adjoint.

We start with the weighted Green's identity

$$
\int_a^b w(uDv - vDu)dx = 0
\tag{7.5.1}
$$

Let us introduce the new functions $\bar{v}(x)$ and $\bar{u}(x)$ by putting

$$
\begin{aligned}
\bar{v}(x) &= \sqrt{w(x)}\,v(x) \\
\bar{u}(x) &= \sqrt{w(x)}\,u(x)
\end{aligned}
\tag{7.5.2}
$$

At the same time we transform the operator D to a new operator $\bar{D}$ by putting

$$
\bar{D}\bar{v} = \sqrt{w}\,D\,\frac{\bar{v}}{\sqrt{w}}
\tag{7.5.3}
$$

Then the transcription of (1) into the new variables yields:

$$
\int_a^b (\bar{u}\bar{D}\bar{v} - \bar{v}\bar{D}\bar{u})dx = 0
\tag{7.5.4}
$$

We are back at Green's identity *without* any weight factor. The transformation (2), (3) absorbed the weight factor $w(x)$ and the new operator has become self-adjoint without any weighting. The solution of the eigenvalue problem

$$
\bar{D}\bar{v}(x) = \lambda\bar{v}(x)
\tag{7.5.5}
$$

yields an ortho-normal function system for which the simple orthogonality condition holds:

$$
\int_a^b \bar{v}_i(x)\bar{v}_k(x)dx = \delta_{ik}
\tag{7.5.6}
$$

If we introduce this transformation into the general second order equation (2.1), we obtain the new self-adjoint operator $\bar{D}$ in the following form:

$$\bar{v}(x) = \frac{v(x)}{\sqrt{A(x)}}\, e^{1/2\int(B/A)dx}$$

$$\bar{D}\bar{v} = \frac{d}{dx}(A\bar{v}') + \left[C + \frac{A'' - B'}{2} - \frac{(A' - B)^2}{4A}\right]\bar{v} \qquad (7.5.7)$$

In summary we can say that the eigenvalue problem of a second order linear differential operator can be formulated in three different ways, in each case treating the problem as self-adjoint:

1. We operate with the given operator without any modification and switch from ordinary orthogonality to weighted orthogonality.

2. We transform the independent variable to make the operator self-adjoint, without weighting.

3. We transform the dependent variable and the operator to a self-adjoint operator, without weighting.

In all three cases the same eigenfunction system is obtained, in different interpretations, and the eigenvalues remain unchanged. There is, however, a fourth method by which unweighted self-adjointness can be achieved. We have written the general second-order operator in the form (4.2). Let us now multiply the given differential equation $Dv = \beta$ on both sides by $w(x)$. We then obtain

$$\frac{d}{dx}(wAv') + wCv = w\beta(x) \qquad (7.5.8)$$

and the new equation is automatically self-adjoint, although neither the independent nor the dependent variable has been transformed. The eigenfunctions and eigenvalues of the new problem are generally quite different from those of the original problem, since in fact we are now solving the eigenvalue problem

$$Dv = \frac{\lambda}{w(x)}v \qquad (7.5.9)$$

For the purpose of solving the differential equation, however, these eigenfunctions are equally applicable, and may result in a simpler solution method.

As an illustrative example we will consider the following differential equation:

$$v'' + \frac{v'}{x} - \frac{k^2 v}{x^2} = \beta(x) \qquad (7.5.10)$$

where k is an integer. This differential equation occurs in the problem of the loaded membrane (see Chapter 8.6), x signifying the polar coordinate r of the membrane, while $v(x)$ is the displacement, caused by the load density

$\beta(x)$. We assume as boundary condition that at $x = 1$—where the membrane is fixed—the displacement must vanish:

$$x = [0, 1], \quad v(1) = 0 \qquad (7.5.11)$$

The eigenvalue problem associated with this equation becomes

$$v'' + \frac{v'}{x} - \frac{k^2 v}{x^2} - \lambda v = 0 \qquad (7.5.12)$$

which is a special case of the differential equation (2.20), and thus solvable in the form

$$v(x) = c J_k(\sqrt{-\lambda} x) \qquad (7.5.13)$$

(We shall see in Section 12 that the second fundamental solution of Bessel's differential equation, viz. $J_k(x)$, becomes infinite at $x = 0$ and is thus ineligible as eigenfunction.) The boundary condition (11) demands

$$J_k(\sqrt{-\lambda}) = 0 \qquad (7.5.14)$$

and this means that $\sqrt{-\lambda}$ has to be identified with any of the zeros of the Bessel function of the order k (that is those x-values at which $J_k(x)$ vanishes). If these zeros are called x_m, we obtain for λ_m the selection principle

$$\lambda_m = -x_m{}^2 \qquad (7.5.15)$$

There is an infinity of such zeros, as we must expect from the fact that the eigenvalue problem of a self-adjoint differential operator (representing a symmetric matrix of infinite order) must possess an infinity of solutions.

The weight function $w(x)$ of our eigenvalue problem becomes (cf. 4.7)

$$w(x) = x \qquad (7.5.16)$$

and we have to normalise our functions (13) by a constant factor A_m, defined by

$$A_m{}^2 \int_0^1 J_k{}^2(x_m x) x \, dx = 1 \qquad (7.5.17)$$

The solution of the differential equation (10) can now occur by the standard method of expanding in eigenfunctions. First we expand the right side in our ortho-normal functions:

$$\beta(x) = \sum_{m=1}^{\infty} c_m A_m J_k(x_m x) \qquad (7.5.18)$$

where

$$c_m = A_m \int_0^1 \beta(\xi) J_k(x_m \xi) \xi \, d\xi \qquad (7.5.19)$$

Then we obtain the solution by a similar expansion, except for the factor $\lambda_m{}^{-1}$:

$$v(x) = \sum_{m=1}^{\infty} \frac{c_m A_m}{\lambda_m} J_k(x_m x) \qquad (7.5.20)$$

The same problem is solvable, however, in terms of *elementary* functions if we proceed in a slightly different manner. Making use of the transformation (2.22) with $\gamma = -\frac{1}{2}$ we first replace $v(x)$ by a new function $v_1(x)$, defined by

$$v_1(x) = x^{-1/2}v(x) \tag{7.5.21}$$

Our fundamental differential equation (10) appears now in the form

$$v_1'' + \frac{2}{x}v'_1 - \frac{k^2 - \frac{1}{4}}{x^2}v_1 = x^{-1/2}\beta(x) \tag{7.5.22}$$

The new weight factor is $w(x) = x^2$ and we will make our equation self-adjoint by multiplying through by x^2:

$$(x^2v'_1)' - (k^2 - \frac{1}{4})v_1 = x^{3/2}\beta(x) \tag{7.5.23}$$

In this new form the eigenvalue problem becomes

$$(x^2v'_1)' + (-\lambda - k^2 + \frac{1}{4})v_1 = 0 \tag{7.5.24}$$

This equation is solvable by putting

$$v = x^n \tag{7.5.25}$$

with the following condition for n:

$$(n + \frac{1}{2})^2 = -\mu^2 \tag{7.5.26}$$

where we have put

$$-\lambda = \mu^2 + k^2 \tag{7.5.27}$$

Then the general solution of (24), written in real form, becomes

$$v_1(x) = (A \cos \mu \log x + B \sin \mu \log x) \tag{7.5.28}$$

In contradistinction to the previous solution, in which the origin $x = 0$ remained a regular point since one of the solutions of Bessel's differential equation remained regular at $x = 0$, at present *both* solutions become singular and we are forced to exclude the point $x = 0$ from our domain, starting with the point $x = \epsilon$ and demanding at this point the boundary condition

$$v_1(\epsilon) = 0 \tag{7.5.29}$$

Since the differential equation (23) shows that in the neighbourhood of $x = 0$ the function $v_1(x)$ starts with the power $x^{3/2}$, we see that we do not commit a substantial error by demanding the boundary condition (29) at $x = \epsilon$ instead of $x = 0$, provided that we choose ϵ small enough.

Now we have the two boundary conditions

$$v_1(\epsilon) = v_1(1) = 0 \tag{7.5.30}$$

which determines $v_1(x)$ uniquely. We obtain for the normalised eigenfunctions

$$v_{1m}(x) = \sqrt{\frac{2}{-\log \epsilon}} \frac{\sin\left(\frac{m\pi}{-\log \epsilon} \log \frac{x}{\epsilon}\right)}{\sqrt{x}} \qquad (7.5.31)$$

and thus

$$\mu = \frac{m\pi}{-\log \epsilon} \qquad (7.5.32)$$

and

$$-\lambda = k^2 + \frac{\pi^2 m^2}{(\log \epsilon)^2} \qquad (7.5.33)$$

The new eigenfunctions oscillate with an even amplitude, which was not the case in our earlier solutions $J_k(x)$. In fact, we recognise in the new solution of the eigenvalue problem the *Fourier functions*, if we introduce instead of x a new variable t by putting

$$x = \epsilon e^t \qquad (7.5.34)$$

If, in addition, we change $\beta(x)$ to

$$\beta_1(x) = x^2 \beta(x) \qquad (7.5.35)$$

the expansion of the right side into eigenfunctions becomes a regular *Fourier sine analysis* of the function $\beta_1(\epsilon e^t)$.

The freedom of transforming the function $v(x)$ by multiplying it by a proper factor, plus the freedom of multiplying the given differential equation by a suitable factor, can thus become of great help in simplifying our task of solving a given second order differential equation. The eigenfunctions and eigenvalues are vitally influenced by these transformations, and we may wonder what eigenfunction system we may adopt as the "proper" system associated with a given differential operator. Mathematically the answer is not unique but in all problems of physical significance there is in fact a unique answer because in these problems—whether they occur in hydrodynamics, or elasticity, or atomic physics—the eigensolutions of a given physical system are determined by the separation of a time dependent differential operator *with respect to the time t*, thus reducing the problem to an eigenvalue problem in the space variables. We will discuss such problems in great detail in Chapter 8.

Problem 285. Transform Laguerre's differential equation (2.16) into a self-adjoint form, with the help of the transformation (2), (3).
[Answer:

$$v(x) = e^{-x/2} L_n(x) \qquad (7.5.36)$$

$$Dv = (xv')' + \left(n + \frac{1}{2} - \frac{x}{4}\right)v \qquad (7.5.37)]$$

Problem 286. Transform Hermite's differential equation (2.18) into a self-adjoint form.

[Answer:

$$v(x) = e^{-x^2/2}H_n(x) \tag{7.5.38}$$

$$Dv = v'' + (2n + 1 - x^2)v \tag{7.5.39}]$$

Problem 287. Transform Chebyshev's differential equation (2.15) into a self-adjoint form.

[Answer:

$$v(x) = \frac{T_n(x)}{\sqrt[4]{1 - x^2}} \tag{7.5.40}$$

$$Dv = [(1 - x^2)v']' + \left(n^2 - \frac{1}{4} - \frac{1}{4(1 - x^2)}\right)v \tag{7.5.41}]$$

Problem 288. Making use of the differential equation (2.23) find the normalised eigenfunctions and eigenvalues of the following differential operator:

$$-Dv = y'' + \frac{2y'}{x} \tag{7.5.42}$$

Range: $x = [0, a]$, boundary condition: $v'(a) = 0$.

[Answer:

$$v_m(x) = \sqrt{\frac{2}{a}}\frac{\sin\,(m + \tfrac{1}{2})\pi\dfrac{x}{a}}{x}, \quad \lambda_m = \frac{\pi^2}{a^2}(m + \tfrac{1}{2})^2 \tag{7.5.43}$$

$$w(x) = x^2]$$

7.6. The Green's function of the general second-order differential equation

We will now proceed to the construction of the Green's function for the general second-order differential operator (2.1). We have seen in (5.8) that a simple multiplication of the given differential equation

$$Dv(x) = \beta(x) \tag{7.6.1}$$

by $w(x)$ on both sides changes this equation to a self-adjoint equation. Hence it suffices to deal with our problem in the form

$$w(x)Dv(x) = w(x)\beta(x) \tag{7.6.2}$$

This means

$$wAv'' + wBv' + wCv = w\beta \tag{7.6.3}$$

for which we can also put

$$(wAv')' + wCv = w\beta \tag{7.6.4}$$

It will thus be sufficient to construct the Green's function for the self-adjoint equation (4).

Green's identity now becomes

$$\int_a^b (uDv - vDu)dx = \Big[A(uv' - vu')\Big]_a^b \tag{7.6.5}$$

As boundary conditions we prescribe the conditions

$$\alpha_1 v(a) + \alpha_2 v'(a) = 0, \quad \beta_1 v(b) + \beta_2 v'(b) = 0 \tag{7.6.6}$$

These are self-adjoint conditions, as we have seen in Section 4. Hence we have a self-adjoint operator with self-adjoint boundary conditions which makes our problem self-adjoint. The symmetry of the resulting Green's function is thus assured.

We proceed to the construction of the Green's function in the standard fashion, considering ξ as the variable and x as the fixed point. To the left of x we will have a solution of the homogeneous differential equation which satisfies the boundary condition at the point $\xi = a$. Now the homogeneous second-order differential equation has two fundamental solutions, let us say $\bar{v}_1(\xi)$ and $\bar{v}_2(\xi)$ and the general solution is a linear superposition of these two solutions:

$$v(\xi) = c_1 \bar{v}_1(\xi) + c_2 \bar{v}_2(\xi) \tag{7.6.7}$$

By satisfying the boundary condition at the point $\xi = a$ one of the constants can be reduced to the other and we obtain a certain linear combination of $\bar{v}_1(\xi)$ and $\bar{v}_2(\xi)$ (multiplied by an arbitrary constant), which satisfies not only the differential equation but also the boundary condition at the point $\xi = a$. We shall denote this linear combination by $v_1(\xi)$, and similarly the other linear combination which satisfies the boundary condition at $\xi = b$, by $v_2(\xi)$. Hence the two analytical expressions which give us the Green's function to the left and to the right of the point $\xi = x$, become

$$G_1(x, \xi) = C_1 v_1(\xi), \quad G_2(x, \xi) = C_2 v_2(\xi)$$
$$\xi = [a, x] \qquad \qquad \xi = [x, b] \tag{7.6.8}$$

with the two undetermined constants C_1 and C_2 which now have to be obtained from the conditions at the dividing point $\xi = x$. At this point we have *continuity* of the function. This gives

$$C_2 v_2(x) = C_1 v_1(x) \tag{7.6.9}$$

We have a jump, however, in the first derivative. The magnitude of this jump, going from the left to the right, is equal to 1 divided by the coefficient of $v''(\xi)$ at the point $\xi = x$. We thus get a second condition in the form

$$C_2 v'_2(x) = C_1 v'_1(x) + \frac{1}{w(x)A(x)} \tag{7.6.10}$$

The solution of the two simultaneous algebraic equations (9) and (10) for the constants C_1 and C_2 yields:

$$C_1 = \frac{v_2(x)}{w(x)A(x)} \frac{1}{v_1(x)v'_2(x) - v_2(x)v'_1(x)}$$

$$C_2 = \frac{v_1(x)}{w(x)A(x)} \frac{1}{v_1(x)v'_2(x) - v_2(x)v'_1(x)}$$

(7.6.11)

The quantity which appears in the denominator of the second factor is called the "Wronskian" of the two functions $v_1(x)$ and $v_2(x)$. We will denote it by $W(x)$:

$$W(x) = v_1(x)v'_2(x) - v_2(x)v'_1(x)$$

(7.6.12)

We thus obtain as final result:

$$G_1(x, \xi) = \frac{v_2(x)v_1(\xi)}{w(x)A(x)W(x)}$$

$$G_2(x, \xi) = \frac{v_1(x)v_2(\xi)}{w(x)A(x)W(x)}$$

(7.6.13)

Now the symmetry of the Green's function demands that an exchange of x and ξ in the first expression shall give the second expression. In the numerator we find that this is automatically fulfilled. In the denominator, however, we obtain the condition

$$w(x)A(x)W(x) = w(\xi)A(\xi)W(\xi)$$

(7.6.14)

Since the point ξ can be chosen arbitrarily, no matter how we have fixed the point x, we obtain the condition

$$w(\xi)A(\xi)W(\xi) = \text{const.} = \gamma$$

(7.6.15)

or, considering the expression (4.7) for $w(x)$:

$$W(x) = \gamma e^{-\int (B/A)dx}$$

(7.6.16)

We have obtained this result from the symmetry of the Green's function $G(x, \xi)$ but we can deduce it more directly from the weighted Green's identity

$$w(uDv - vDu) = \frac{d}{dx}[wA(uv' - vu')]$$

(7.6.17)

Since both v_1 and v_2 satisfy the homogeneous differential equation $Dv = 0$, we can identify v with v_2 and u with v_1. Then the relation (16) becomes

$$w(x)A(x)W(x) = \text{const.} = \gamma$$

(7.6.18)

which is exactly the relation we need for the symmetry of $G(x, \xi)$.

An important consequence of this result can be deduced if we consider the equation

$$v_1(x)v'_2(x) - v_2(x)v'_1(x) = \gamma e^{-\int (B/A)dx}$$

(7.6.19)

as a differential equation for $v_2(x)$, assuming that $v_1(x)$ is given. We can integrate this equation by the method of the "variation of constants". We put

$$v_2(x) = Cv_1(x) \tag{7.6.20}$$

Considering C as a function of x we obtain

$$C'v_1{}^2 = \gamma e^{-\int (B/A)dx} \tag{7.6.21}$$

and thus

$$C = \gamma \int \frac{1}{v_1{}^2} e^{-\int (B/A)dx} \tag{7.6.22}$$

and

$$v_2(x) = \gamma v_1(x) \int \frac{1}{v_1{}^2} e^{-\int (B/A)dx} dx \tag{7.6.23}$$

We see that a second-order differential equation has the remarkable property that *the second fundamental solution is obtainable by quadrature if the first fundamental solution is given.*

Problem 289. Consider the differential equation of the vibrating spring:

$$v'' + k^2 v = 0$$

Given the solution

$$v_1(x) = \cos kx$$

Obtain the second independent solution with the help of (23).

Problem 290. Consider Chebyshev's differential equation (2.14). Given the fundamental solution

$$v_1(x) = \cos k\theta \tag{7.6.24}$$

if

$$x = \cos \theta \tag{7.6.25}$$

Find the second fundamental solution with the help of (23).
[Answer:

$$v_2(x) = \sin k\theta \tag{7.6.26}$$]

Problem 291. Consider Legendre's differential equation (2.12) for $n = 2$. Given the solution

$$v_1(x) = \tfrac{1}{2}(3x^2 - 1) \tag{7.6.27}$$

Find the second fundamental solution by quadrature.
[Answer:

$$v_2(x) = (3x^2 - 1) \log \frac{1 + x}{1 - x} - 6x \tag{7.6.28}$$]

Problem 292. Consider Laguerre's differential equation (2.16) which defines the polynomials $L_n(x)$. Show that the second solution of the differential equation goes for large x to infinity with the strength e^x.

Problem 293. The proof of the symmetry of the Green's function, as discussed in the present section, seems to hold *universally* while we know that it holds only under self-adjoint boundary conditions. What part of the proof is invalidated by the not-self-adjoint nature of the boundary conditions?

[Answer: As far as $v_1(\xi)$ and $v_2(\xi)$ goes, they are always solutions of the homogeneous equation. But the dependence on x need not be of the form $f(x)v(\xi)$, as we found it on the basis of our self-adjoint boundary conditions.]

7.7. Normalisation of second order problems

The application of a weight factor $w(x)$ to a given second-order differential equation is a powerful tool in the investigation of the analytical properties of the solution and is often of great advantage in obtaining an approximate solution in cases which do not allow an explicit solution in terms of elementary functions. In the previous sections we have encountered the method of the weight factor in two different aspects. The one was to multiply the entire equation by a properly chosen weight factor $w(x)$ (cf. 6.2). If $w(x)$ is chosen according to (4.7), the left side of the equation is transformed into

$$\frac{d}{dx}(wAv') + wCv \tag{7.7.1}$$

Another method was to multiply $v(x)$ by $\sqrt{w(x)}$ and obtain a differential operator for

$$v_1(x) = \sqrt{w(x)}v(x) \tag{7.7.2}$$

But here we have not only made the substitution

$$v(x) = \frac{1}{\sqrt{w(x)}}v_1(x) \tag{7.7.3}$$

but also changed the operator D to D_1 (cf. 5.3). The new differential equation thus constructed became [cf. (5.7)]:

$$\frac{d}{dx}(Av'_1) + C_1v_1 = \beta(x) \tag{7.7.4}$$

where

$$C_1 = C + \frac{A'' - B'}{2} - \frac{(A' - B)^2}{4A} \tag{7.7.5}$$

We shall now combine these two methods in order to normalise the general second-order differential equation into a form which is particularly well suited to further studies. First of all we will divide the entire equation by the coefficient of $v''(x)$:

$$v'' + \frac{B}{A}v' + \frac{C}{A}v = \frac{\beta}{A} \tag{7.7.6}$$

We thus obtain two new functions which we will denote by $b(x)$ and $c(x)$:

$$\frac{B(x)}{A(x)} = b(x)$$

$$\frac{C(x)}{A(x)} = c(x) \tag{7.7.7}$$

Our equation to be solved has now the form

$$v'' + bv' + cv = \beta_1 \qquad (7.7.8)$$

Then we apply the transformation (3–5) which is now simplified due to the fact that $A(x) = 1$:

$$v(x) = e^{-1/2 \int b\,dx} v_1(x) \qquad (7.7.9)$$

$$v''_1 + \left(c - \frac{b'}{2} - \frac{b^2}{4} \right) v_1 = \beta_1 \qquad (7.7.10)$$

The new form of the differential operator has the conspicuous property that the *term with the first derivative is missing*. The new differential operator is of the Sturm-Liouville type (2.3) but with $A(x) = 1$. It is characterised by only *one* function which we will call $U(x)$:

$$U(x) = c - \frac{b'}{2} - \frac{b^2}{4} \qquad (7.7.11)$$

The homogeneous equation has now the form

$$v''_1 + U(x)v_1 = 0 \qquad (7.7.12)$$

If we know how to solve this differential equation, we have also the solution of an arbitrary second order equation since the solution of an arbitrary second order differential equation can be transformed into the form (12) which is often called the "normal form" of a linear homogeneous differential equation of second order.

Problem 294. Transform Bessel's differential equation (2.4) into the normal form (12).

[Answer:

$$v(x) = \frac{1}{\sqrt{x}} v_1(x)$$

$$U(x) = 1 - \frac{p^2 - \frac{1}{4}}{x^2} \qquad (7.7.13)]$$

Problem 295. Transform Laguerre's differential equation (2.16) into the normal form.

[Answer:

$$v(x) = \frac{1}{\sqrt{x}} e^{x/2} v_1(x)$$

$$U(x) = -\frac{1}{4} + \frac{n + \frac{1}{2}}{x} + \frac{1}{4x^2} \qquad (7.7.14)]$$

Problem 296. Transform Hermite's differential equation (2.18) into the normal form.

[Answer:

$$v(x) = e^{(1/2)x^2} v_1(x)$$
$$U(x) = 2n + 1 - x^2 \qquad (7.7.15)]$$

7.8. Riccati's differential equation

The following exponential transformation has played a decisive role in the development of wave mechanics and in many other problems of mathematical physics. We put

$$v(x) = e^{\varphi(x)} \tag{7.8.1}$$

and consider $\varphi(x)$ as a new function for which a differential equation is to be obtained.

We now have

$$v'(x) = \varphi'(x)e^{\varphi(x)}$$
$$v''(x) = [\varphi''(x) + \varphi'^2(x)]e^{\varphi(x)}$$

The differential equation

$$v'' + Uv = 0 \tag{7.8.2}$$

yields the following equation for $\varphi(x)$:

$$\varphi'' + \varphi'^2 + U = 0$$

If we put

$$\varphi'(x) = y(x) \tag{7.8.3}$$

we obtain for $y(x)$ the following first order equation:

$$y' + y^2 + U = 0 \tag{7.8.4}$$

This is a non-linear inhomogeneous differential equation of the first order for $y(x)$, called "Riccati's differential equation".

It is a fundamental property of this differential equation that the general solution is obtainable by quadratures if two independent particular solutions are given. Let y_1 and y_2 be two such solutions, then

$$\varphi_1(x) = \int y_1 dx$$
$$\varphi_2(x) = \int y_2 dx \tag{7.8.5}$$

yield two particular solutions of the homogeneous linear equation (2), in the form

$$v_1(x) = e^{\varphi_1(x)}$$
$$v_2(x) = e^{\varphi_2(x)} \tag{7.8.6}$$

But then the general solution of (2) becomes

$$v(x) = A_1 v_1(x) + A_2 v_2(x) \tag{7.8.7}$$

which gives for the associated function $\varphi(x)$:

$$\varphi(x) = \log\left(A_1 v_1 + A_2 v_2\right)$$

$$\varphi' = y = \frac{A_1 v'_1 + A_2 v'_2}{A_1 v_1 + A_2 v_2} = \frac{A_1 y_1 e^{\int y_1 dx} + A_2 y_2 e^{\int y_2 dx}}{A_1 e^{\int y_1 dx} + A_2 e^{\int y_2 dx}} \tag{7.8.8}$$

where A_1 and A_2 are two arbitrary constants.

Problem 297. In Riccati's differential equation (4) assume $U = \text{const.} = -C^2$. Then the differential equation is explicitly solvable by the method of the "separation of variables". Demonstrate the validity of the result (8), choosing as particular solutions: $y_1 = C$, $y_2 = -C$.

7.9. Periodic solutions

In a very large number of problems in which second order differential equations are involved, some kinds of *vibrations* occur. If the differential equation is brought into the form (8.2), we can see directly that we shall obtain two fundamentally different types of solutions, according to the positive or negative sign of $U(x)$. If $U(x)$ were a constant, the solution of the differential equation (8.2) would be of the form

$$v(x) = A \sin \sqrt{\overline{U}}x + B \cos \sqrt{\overline{U}}x \tag{7.9.1}$$

if U is positive and

$$v(x) = Ae^{\sqrt{-U}x} + Be^{-\sqrt{-U}x} \tag{7.9.2}$$

if U is negative. Although these solutions will not hold if $U(x)$ is a function of x, nevertheless, for a sufficiently small interval of x, $U(x)$ could still be considered as nearly constant and the general character of the solution will not be basically different from the two forms (1) or (2). We can say quite generally that the solution of the differential equation (8.2) has a *periodic* character if $U(x)$ is positive and an *exponential* character if $U(x)$ is negative. In most applied problems the case of a *positive* $U(x)$ is of primary interest. If $U(x)$ is negative and the solution becomes exponential, the exponentially increasing solution is usually excluded by the given physical situation while the exponentially decreasing solution exists only for a short interval, beyond which the function is practically zero. The case of positive $U(x)$ is thus of much more frequent occurrence.

Now in the case of positive $U(x)$ we usually look for solutions of Riccati's equation which are not real. A real solution would lead to an infinity of $\varphi(x)$ if $v(x)$ vanishes. Hence at every point at which a periodic oscillation goes to zero, the associated solution of the Riccati equation would become singular. This is not the case, however, if the solution is *complex*, i.e. if $y(x)$ is of the form

$$y(x) = \alpha(x) + i\beta(x) \tag{7.9.3}$$

Such a solution has the further advantage that it immediately supplies a *second* solution since we know in advance that also $\alpha(x) - i\beta(x)$ must be a solution in the case that $U(x)$ is real. And thus any complex solution of Riccati's equation (8.4) is equivalent to a *complete* solution of the associated equation (8.2).

Let us now introduce the complex quantity (3) in the equation (8.4). This equation now separates into the two real equations

$$\alpha' + \alpha^2 - \beta^2 + U = 0 \tag{7.9.4}$$

$$\beta' + 2\alpha\beta = 0 \tag{7.9.5}$$

The second equation is integrable and gives

$$\alpha = -\frac{1}{2}\frac{\beta'}{\beta} = (\log \beta^{-1/2})' \qquad (7.9.6)$$

thus

$$\int \alpha dx = \log \beta^{-1/2} + C \qquad (7.9.7)$$

But then

$$\varphi(x) = \log \beta^{-1/2} + i \int \beta dx \qquad (7.9.8)$$

and

$$v(x) = e^{\varphi(x)} = \frac{1}{\sqrt{\beta}} e^{i\int \beta dx} \qquad (7.9.9)$$

Since both the real and the imaginary part of this solution must be a solution of our equation (8.2) (assuming that $U(x)$ is real), we obtain the general solution in the form

$$v(x) = \frac{1}{\sqrt{\beta}} (A_1 \cos \int \beta dx + A_2 \sin \int \beta dx) \qquad (7.9.10)$$

for which we may also write

$$v(x) = \frac{A}{\sqrt{\beta}} \cos (\int \beta dx - \theta) \qquad (7.9.11)$$

where θ is an arbitrary phase angle. Since, however, the indefinite integral $\int \beta dx$ has already a constant of integration, it suffices to put

$$v(x) = \frac{A}{\sqrt{\beta}} \cos \int \beta dx \qquad (7.9.12)$$

This form of the solution leads to the remarkable consequence that the solution of an arbitrary linear homogeneous differential equation of second order for which the associated $U(x)$ is positive in a certain interval, may be conceived as a *periodic oscillation with a variable frequency and variable amplitude.* Ordinarily we think of a vibration in the sense of a function of the form

$$A \cos (\omega x - \theta) \qquad (7.9.13)$$

where A, ω, and θ are constants. If the amplitude A becomes a function of x, we can still conceive our function as a periodic vibration with a variable amplitude. But we can go still further and envisage that even the frequency ω is no longer a constant. Now if the argument of the sine function is no longer of the simple form ωx but some arbitrary function of x, we could introduce the concept of an "instantaneous frequency" which we can define as the *derivative* of the argument. But then we see that the solution (12) has the significance of an *oscillation with the instantaneous frequency* β. Furthermore, we come to the important conclusion that the amplitude and frequency of this oscillation are *necessarily coupled with each other*. If the frequency of the oscillation is a constant, the amplitude is also a constant. But if the frequency changes, the amplitude must also change according to a definite law. The amplitude of the vibration is always *inversely proportional to the square root of the instantaneous frequency*. If we study the distribution of zeros in the oscillations of the Bessel functions or the Jacobi polynomials or the Laguerre or Hermite type of polynomials, *the law of the zeros is not independent of the law according to which the maxima of the successive oscillations change*. The law of the amplitudes is uniquely related to the law of the zeros and vice versa.

This association of a certain vibration of varying amplitude and frequency with a solution of a second-order differential equation is not unique, however. The solution (11) contains two free constants of integration, viz. the phase constant θ and the amplitude constant A. This is all we need for the general solution of a second-order differential equation. And yet, if we consider the equations (4) and (5), which determine $\beta(x)$, we notice that we get for $\beta(x)$ a differential equation of second order, thus leaving two further constants free. This in itself is not so surprising, however, if we realise that we now have a *complex* solution of the given second-order differential equation, with the freedom of prescribing $v(x_0)$ and $v'(x_0)$ at a certain point $x = x_0$ as two *complex* values, which in fact means *four* constants of integration. But if we take the real part of the solution for itself:

$$v(x) = \frac{A}{\sqrt{\beta(x)}} \cos \left(\int \beta dx - \theta \right) \qquad (7.9.14)$$

then we see that the freedom of choosing $\beta(x_0)$ and $\beta'(x_0)$ freely must lead to a redundancy because to any given $v(x_0)$, $v'(x_0)$ we can determine the constants A and θ and, having done so, the further course of the function $v(x)$ is uniquely determined, no matter how $\beta(x)$ may behave. This means that the separation of our solution in amplitude and frequency cannot be unique but may occur in infinitely many ways.

Let us assume, for example, that at $x = x_0$ $v(x)$ *vanishes*. This means that, if the integral under the cosine starts from $x = x_0$, the phase angle becomes $\pi/2$. Now in this situation the choice of $\beta'(x_0)$ can have *no effect* on the resulting solution, while the choice of $\beta(x_0)$ can change only a factor of proportionality. And yet the course of $\beta(x)$—and thus the instantaneous frequency and the separation into amplitude and frequency—is profoundly influenced by these choices.

Problem 298. Investigate the differential equation

$$v'' + v = 0 \qquad (7.9.15)$$

with the boundary condition

$$v(0) = 0 \qquad (7.9.16)$$

which suggests a unique separation into a vibration of constant frequency and constant amplitude, while in actual fact this separation is not unique. The associated Riccati equation is here completely integrable (cf. Problem 297). Obtain the general solution of the problem in terms of instantaneous frequency and amplitude and show that the choice of $\beta'(0)$ has no effect on the solution, while the choice of $\beta(0)$ changes only a multiplicative factor.

[Answer:

$$\beta(x) = \frac{\rho}{\cos^2(\lambda - x) + \rho^2 \sin^2(\lambda - x)} \tag{7.9.17}$$

where ρ and λ are arbitrary constants.

$$v(x) = \frac{A}{\sqrt{\beta(x)}} \sin \int_0^x \beta(t)dt = \frac{A\sqrt{\rho}}{\sqrt{\cos^2\lambda + \rho^2\sin^2\lambda}} \sin x \tag{7.9.18}]$$

7.10. Approximate solution of a differential equation of second order

We have seen that an arbitrary linear homogeneous differential equation of the second order can be transformed into the form (8.2). That form, furthermore, could be transformed into the Riccati equation (8.4). We have no explicit method of solving this equation exactly. We have seen, however, that the solution is reducible, as far as the function $v(x)$ is concerned, to the form (9.11) and is thus obtainable by mere quadrature if $\beta(x)$ is at our disposal. This would demand an integration of the differential equation (9.4), expressing $\alpha(x)$ in terms of $\beta(x)$ according to the relation (9.5). We can make good use, however, of the redundancy of this differential equation which permits us to choose any *particular* solution of the differential equation and still obtain a complete solution of the original equation (8.2) While the equation for $\beta(x)$ is in fact a second-order differential equation, we shall handle it in a purely *algebraic* manner, neglecting altogether the terms with α' and α^2, thus putting simply

$$\beta(x) = \sqrt{U(x)} \tag{7.10.1}$$

The resulting solution for $v(x)$:

$$v(x) = \frac{1}{\sqrt[4]{U}} e^{i(\int\sqrt{U}dx-\theta)} \tag{7.10.2}$$

is of great value in many problems of atomic physics and gives very satisfactory approximations for many important function classes, arising out of the Sturm-Liouville type of problems. It is called the "Kramers-Wentzel-Brillouin approximation", or briefly the "KWB solution". The corresponding solution in the range of *negative* U becomes

$$v(x) = \frac{1}{\sqrt[4]{-U}} (A_1 e^{\int\sqrt{-U}dx} + A_2 e^{-\int\sqrt{-U}dx}) \tag{7.10.3}$$

The condition for the successful applicability of the simplified solution (1) is that U *must be sufficiently large.* Our solution will certainly fail in the neighbourhood of $U = 0$. It so happens, however, that in a large class of problems U ascends rather steeply from the value $U = 0$ and thus the range in which the KWB approximation fails, is usually limited to a relatively small neighbourhood of the point at which $U(x)$ vanishes.

In order to estimate the accuracy of the KWB solution, we shall substitute in Riccati's differential equation

$$y = y_0 + \eta \tag{7.10.4}$$

where y_0 is the KWB solution, namely

$$y_0 = i\sqrt{U} - \frac{1}{4}\frac{U'}{U} \tag{7.10.5}$$

Then we obtain for $\eta(x)$ the differential equation

$$\eta' + \eta^2 + 2y_0\eta + [\tfrac{1}{4}(\log U)']^2 - \tfrac{1}{4}(\log U)'' = 0 \tag{7.10.6}$$

Since η is small, η^2 is negligible. Moreover, we expect that η will be small relative to y_0. This makes even η' negligible in comparison to $2y_0\eta$. Hence we can put with sufficient accuracy for estimation purposes:

$$\eta = \frac{(\log U)'' - \tfrac{1}{4}[(\log U)']^2}{8i\sqrt{U}} \tag{7.10.7}$$

(We have replaced in the denominator y_0 by its leading term.)

Here we have an estimation of the error of the solution of Riccati's differential equation if we accept the simplified solution (1) for $\beta(x)$ (and correspondingly the solution (2) for $v(x)$). This is the quantity which will decide how closely the KWB solution approximates the true solution. *The KWB solution will only be applicable in a domain in which the quantity* (7) *is sufficiently small.*

So far we have only obtained the error of y. We have to go further in order to estimate the error of $v(x)$. Here the solution of Riccati's differential equation appears in the exponent and requires an additional quadrature. Hence we need the quantity

$$\int_a^x \eta(t)dt \tag{7.10.8}$$

if our realm starts with $x = a$ and we assume that $v(x)$ is at that point adjusted to the proper value $v(a)$ in amplitude and phase. To carry through an exact quadrature with the help of (7) as integrand will seldom be possible. But an approximate estimation is still possible if we realise that the second term in the numerator of (7) is of *second order* and can thus be considered as small. If, in addition, we replace in the denominator $\sqrt{U}$ by its minimum value between a and x, we obtain the following estimation of the *relative error* of the KWB approximation:

$$\left|\frac{\Delta v(x)}{v(x)}\right| < \frac{1}{8\sqrt{|U|_{\min}}}\left|\frac{U'(x)}{U(x)} - \frac{U'(a)}{U(a)}\right| \tag{7.10.9}$$

(We have assumed that $(\log U)''$ does not go through zero in our domain, otherwise we have to sectionalise the error estimation.)

Problem 299. Given the differential equation

$$v'' - (1 + x^2)v = 0 \qquad (7.10.10)$$

In this example one of the fundamental solutions is explicitly available in exact form:

$$v(x) = e^{x^2/2} \qquad (7.10.11)$$

which makes an exact error analysis possible. Obtain the solution by the KWB method and compare it with the exact solution in the realm $x = [3, \infty]$. Choosing $a = \infty$, estimate the maximum error of $y(x)$ and $v(x)$ on the basis of the formulae (7) and (9) and compare them with the actual errors.

[Answer:

Maximum error of $y(x)$ (which occurs at $x = 3$): $\eta(3) = -0.0123$

(formula (7) gives -0.0107)

Maximum relative error of $v(x)$: $\dfrac{\Delta v}{v}(3) = 0.0195$

(formula (9) gives 0.0237)]

Problem 300. For what choice of $U(x)$ will the KWB approximation become accurate? Obtain the solution for this case.

[Answer:

$$U(x) = \frac{1}{(ax + b)^4} \qquad (7.10.12)$$

where a and b are arbitrary constants.

$$v(x) = (ax + b)e^{\pm ix/[b(ax+b)]} \qquad (7.10.13)]$$

Problem 301. Obtain the KWB-approximation for Bessel's differential equation, transformed to the normal form.

[Answer:

$$v(x) = \frac{1}{\sqrt[4]{x^2 - p_1^2}} e^{\pm i[\sqrt{x^2 - p_1^2} + p_1 \text{ arc sin } p_1/x]} \qquad (p_1 = \sqrt{p^2 - \tfrac{1}{4}}) \qquad (7.10.14)]$$

Problem 302. Obtain the KWB-approximation for Hermite's differential equation (cf. 7.15).

[Answer:

$$v(x) = \frac{1}{\sqrt[4]{2n + 1 - x^2}} e^{x^2/2 \pm i/2[x\sqrt{2n+1-x^2} + (2n+1) \text{ arc sin } (x/\sqrt{2n+1})]} \qquad (7.10.15)]$$

7.11. The joining of regions

In many important applications of second-order differential equations, in particular in the eigenvalue problems of atomic physics, the situation is encountered that the function $U(x)$, which appears in the normal form of a differential equation of second order, *changes its sign* in the domain under consideration. This change of sign does not lead to any singularity but it does have a profound effect on the general character of the solution since the solution has a *periodic* character if $U(x)$ is positive and an *exponential* character if $U(x)$ is negative. The question arises how we can continue our solution from the one side to the other, in view of the changed behaviour of the function. The KWB approximation is often of inestimable value in giving a good overall picture of the solution. The accuracy is not excessive but an error of a few per cent can often be tolerated and the KWB method has frequently an accuracy of this order of magnitude. The method fails, however, in the neighbourhood of $U(x) = 0$ and in this interval a different approach will be demanded. Frequently the transitory region is of limited extension because $U(x)$ has a certain steepness in changing from the negative to the positive domain (or vice versa) and the interval in which $U(x)$

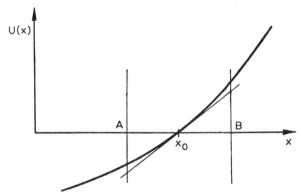

becomes too small for an effective approximation by the KWB method, is often sufficiently reduced to allow an approximation of a different kind. If in the figure the KWB approximation fails between the points A and B, we see that on the other hand $U(x)$ changes in this interval gently enough to permit a *linear* approximation of the form $a(x - x_0)$. Since we can always transfer the origin of our reference system into the point $x = x_0$, we will pay particular attention to the differential equation

$$v'' + axv = 0 \qquad (7.11.1)$$

If we have the solution of this particular differential equation, we shall have the link we need in order to bridge the gap between the exponential and the periodic domains. This differential equation can be conceived as a special case of (2.23) which is solvable in terms of Bessel functions. For this purpose we have to choose the constants α, β, γ, p as follows:

$$\alpha = \tfrac{3}{2}, \quad \gamma = \tfrac{1}{2}, \quad p = \tfrac{1}{3}, \quad \alpha^2\beta^2 = a \qquad (7.11.2)$$

We shall choose $\beta = 1$ and thus put $a = 9/4$. The differential equation

$$v'' + \tfrac{9}{4}xv = 0 \qquad (7.11.3)$$

can thus be solved in the following form:

$$v(x) = \sqrt{x}[A_1 J_{1/3}(x^{3/2}) + A_2 J_{-1/3}(x^{3/2})] \qquad (7.11.4)$$

In order to study the behaviour of this solution for both positive and negative values of x, a brief outline of the basic analytical properties of the Bessel functions will be required.

7.12. Bessel functions and the hypergeometric series

The hypergeometric series (2.7) is closely related to the Bessel functions. Let us put $\alpha = \beta$ and $x = \alpha^{-2}x_1$. We will thus consider the infinite expansion

$$F\left(\alpha, \alpha, \gamma; \frac{x}{\alpha^2}\right) = 1 + \frac{x}{\gamma} + \frac{[\alpha(\alpha + 1)]^2}{1 \cdot 2 \cdot \gamma(\gamma + 1)} \frac{x^2}{\alpha^4} + \cdots$$

$$= 1 + \frac{x}{\gamma} + \frac{x^2}{1 \cdot 2 \cdot \gamma(\gamma + 1)} \left(1 + \frac{1}{\alpha}\right)^2$$

$$+ \frac{x^3}{1 \cdot 2 \cdot 3 \cdot \gamma(\gamma + 1)(\gamma + 2)} \left[\left(1 + \frac{1}{\alpha}\right)\left(1 + \frac{2}{\alpha}\right)\right]^2 + \cdots$$

$$(7.12.1)$$

If in this expansion we let α go to infinity, we obtain a definite limit:

$$\lim_{a \to \infty} F\left(\alpha, \alpha, \gamma; \frac{x}{\alpha^2}\right) = F(\gamma; x)$$

$$= 1 + \frac{x}{\gamma} + \frac{x^2}{2! \gamma(\gamma + 1)} + \frac{x^3}{3! \gamma(\gamma + 1)(\gamma + 2)} + \cdots \quad (7.12.2)$$

This series defines a function of x which is in fact an "entire function", i.e. it has no singular point anywhere in the entire complex plane. Indeed, the only singular point of the hypergeometric function $F(\alpha, \beta, \gamma; x)$ is the point $x = 1$ but the substitution x/α^2 for x pushes the singular point out into infinity and makes the series (2) converge for all real or complex values of x. The parameter γ is still at our disposal.

Now the substitution $\alpha = \beta$, and

$$x = \frac{x_1}{\alpha^2} \qquad (7.12.3)$$

transforms the Gaussian differential equation (2.6) as follows:

$$x_1\left(1 - \frac{x_1}{\alpha^2}\right)v'' + \left(\gamma - (2\alpha + 1)\frac{x_1}{\alpha^2}\right)v' - v = 0 \qquad (7.12.4)$$

If we now go with α to infinity, we obtain in the limit for the function $F(\gamma; x)$, defined by the expansion (2), the following differential equation (dividing through by x_1 which we denote once more by x):

$$v'' + \frac{\gamma}{x} v' - \frac{v}{x} = 0 \qquad (7.12.5)$$

This differential equation can again be conceived as a special case of (2.23). For this purpose the following choices have to be made (replacing the γ of (2.23) by γ_1 since the present γ refers to a different quantity):

$$1 - 2\gamma_1 = \gamma, \quad \gamma_1 = -\alpha p, \quad 2\alpha - 2 = -1, \quad \alpha\beta = i$$

This means:

$$\alpha = \tfrac{1}{2}, \quad \beta = 2i, \quad \gamma_1 = -\frac{p}{2} \tag{7.12.6}$$

and thus we see that the entire function $F(\gamma; x)$ satisfies Bessel's differential equation if we make the following correlation:

$$v_p(2i\sqrt{x}) = cx^{p/2}F(p + 1; x) \tag{7.12.7}$$

The same correlation may be written in the following form:

$$v_p(x) = c\left(\frac{x}{2}\right)^p F\left(p + 1; -\left(\frac{x}{2}\right)^2\right) \tag{7.12.8}$$

Since, furthermore, the constant p appears in (2.4) solely in the form p^2, we obtain the general solution of Bessel's differential equation as follows:

$$v_p(x) = A_1\left(\frac{x}{2}\right)^p F\left(p + 1; -\left(\frac{x}{2}\right)^2\right) + A_2\left(\frac{x}{2}\right)^{-p} F\left(-p + 1; -\left(\frac{x}{2}\right)^2\right) \tag{7.12.9}$$

(The second solution is invalidated, however, if p is an integer n, since the γ of the hypergeometric series must not be zero or a negative integer.) Bessel's function of the order p is specifically defined as follows:

$$J_p(x) = \frac{1}{p!} \left(\frac{x}{2}\right)^p F\left(p + 1; -\left(\frac{x}{2}\right)^2\right)$$

$$= \frac{1}{p!} \left(\frac{x}{2}\right)^p \left[1 - \frac{1}{p+1}\left(\frac{x}{2}\right)^2 + \frac{1}{(p+1)(p+2)\cdot 2!}\left(\frac{x}{2}\right)^4 - \ldots\right] \tag{7.12.10}$$

We will define the following entire function of x^2:

$$\Lambda_p(x^2) = \frac{1}{p!}\frac{1}{2^p} F\left(p + 1; -\left(\frac{x}{2}\right)^2\right)$$

$$= \frac{1}{p!} 2^{-p}\left[1 - \frac{1}{p+1}\left(\frac{x}{2}\right)^2 + \frac{1}{2!(p+1)(p+2)}\left(\frac{x}{2}\right)^4 - \ldots\right] \tag{7.12.11}$$

Then we can put

$$J_p(x) = x^p\Lambda_p(x^2) \tag{7.12.12}$$

If p happens to be a (positive) integer n, the function $J_p(x)$ is in itself an "entire function", i.e. a function which is analytical throughout the complex plane $z = x + iy$. But if p is not an integer, the factor x^p will interfere with the analytical nature of $J_p(x)$ and require that along some half-ray of the complex plane, between $r = 0$ and ∞, a "cut" is made and we must not pass from one border to the other. However, this cut is unnecessary if we stay with x in the *right* complex half plane:

$$z = x + iy \qquad (x \geq 0) \tag{7.12.13}$$

It is in fact unnecessary to pass to the negative half plane. The function $J_p(z)$ is—as we see from the definition (10)—reducible to the basic function $\Lambda_p(z^2)$ which assumes the same values for $\pm z$. Hence $J_p(-z)$ is reducible to $J_p(z)$. Moreover, the Taylor expansion (10) of $J_p(z)$, having real coefficients, demonstrates that $J_p(x - iy)$ is simply the complex conjugate of $J_p(x + iy)$. Hence it suffices to study the analytical behaviour of $J_p(z)$ in the *right upper quarter* of the complex plane, letting the polar angle θ of the complex number $z = re^{i\theta}$ vary only between 0 and $\pi/2$.

Problem 303. Show the following relation on the basis of the definition (10) of $J_p(z)$:

$$J_p(-z) = e^{-ip\pi} J_p(z) \tag{7.12.14}$$

7.13. Asymptotic properties of $J_p(z)$ in the complex domain

The behaviour of the Bessel functions in the outer regions of the complex plane is of considerable interest. We need this behaviour for our problem of joining the periodic and exponential solutions of a differential equation of second order. We will not go into the profound function-theoretical investigations of K. Neumann and H. Hankel which have shed so much light on the remarkable analytical properties of the Bessel and related functions.* Our aim will be to stay more closely to the defining differential equation and draw our conclusions accordingly. In particular, we will use the KWB method of solving this differential equation in the complex, for a simple derivation of the asymptotic properties of the function $J_p(z)$ for large complex values of the argument z.

For this purpose we write Bessel's differential equation once more in the normal form:

$$v'' + \left(1 - \frac{p^2 - \frac{1}{4}}{z^2}\right)v = 0 \tag{7.13.1}$$

with

$$J_p(z) = \frac{1}{\sqrt{z}}\,v(z) \tag{7.13.2}$$

We now introduce the complex variable z in polar form:

$$z = re^{i\theta} \tag{7.13.3}$$

---

* Cf. the chapters on Bessel functions in the fundamental literature: Courant-Hilbert, Whittaker-Watson, Watson, quoted in the Bibliography.

We want to move along a large circle with the radius $r = r_0$, the angle θ changing between 0 and $\pi/2$. This demands the substitution

$$dz = ir_0e^{i\theta}d\theta \qquad (7.13.4)$$

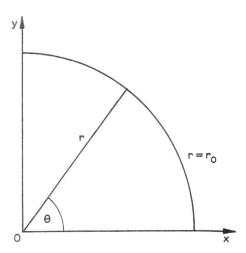

In the new variable θ our differential equation (1) becomes

$$v'' - iv' - (r_0^2e^{2i\theta} - p^2 + \tfrac{1}{4})v = 0 \qquad (7.13.5)$$

Now we have to make the substitution (cf. 7.9)

$$v(x) = e^{i(\theta/2)}v_1(x) \qquad (7.13.6)$$

thus bringing (5) into the normal form

$$v''_1 - (r_0^2e^{2i\theta} - p^2)v_1 = 0 \qquad (7.13.7)$$

Now the function

$$U(x) = -r_0^2e^{2i\theta} + p^2 \qquad (7.13.8)$$

is large, in view of the largeness of r_0, and is thus amenable to the KWB solution. In fact, p^2 is negligible in comparison to the first term (except if the order of the Bessel function is very large), which shows that the asymptotic behaviour of the Bessel functions will be similar for all orders p.

The KWB solution (10.3) becomes in our case (considering that r_0 is a constant which can be united with the constants A_1 and A_2):

$$v_1(\theta) = \frac{1}{e^{i(\theta/2)}} \left[A_1e^{ir_0(\cos\theta + i\sin\theta)} + A_2e^{-ir_0(\cos\theta + i\sin\theta)} \right] \qquad (7.13.9)$$

Returning to the original $v(\theta)$ (cf. 13.6) and writing our result in terms of z we obtain

$$v(z) = A_1e^{iz} + A_2e^{-iz} \qquad (7.13.10)$$

and finally, in view of (2):

$$J_p(z) = \frac{A_1 e^{iz} + A_2 e^{-iz}}{\sqrt{z}} \tag{7.13.11}$$

The undetermined constants A_1 and A_2 have to be adjusted to the initial values, as they exist at the point $\theta = 0$, that is on the *real axis*:

$$J_p(x) = \frac{A_1 e^{ix} + A_2 e^{-ix}}{\sqrt{x}} \tag{7.13.12}$$

If we know how $J_p(x)$ behaves for large *real* values of the argument, the equation (11) will tell us how it behaves for large *complex* values. Our problem is thus reduced to the investigation of $J_p(x)$ for large real values of x.

7.14. Asymptotic expression of $J_p(x)$ for large values of x

In our investigation of the interpolation problem of the Bessel functions (cf. Chapter 1.22), we encountered the following integral representation of the Bessel functions which holds for any *positive p*:

$$J_p(x) = \frac{2}{\sqrt{\pi}} \frac{(\tfrac{1}{2}x)^p}{(p - \tfrac{1}{2})!} \int_0^{\pi/2} \cos(x \sin \varphi) \cos^{2p} \varphi \, d\varphi \tag{7.14.1}$$

With the transformation $\sin \varphi = t$ the same integral may be written in the form

$$\int_0^1 \cos tx \sqrt{1 - t^2}^{\,2p-1} dt \tag{7.14.2}$$

Moreover, since the integrand is an even function, the same integral may also be written in the complex form:

$$\frac{1}{2} \int_{-1}^{+1} e^{-itx} (1 - t^2)^{p-(1/2)} dt \tag{7.14.3}$$

Now let x be large. Then we will modify the path of integration of the variable t as follows:

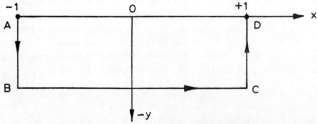

We will first investigate the contribution of the path CD. Let us put

$$t = 1 - i\tau \tag{7.14.4}$$

Then the function e^{-ixt} becomes

$$e^{-ix}e^{-x\tau} \tag{7.14.5}$$

The first factor is a mere constant with respect to the integration in τ. The second factor diminishes very rapidly to zero, due to the largeness of x. Hence only the immediate neighbourhood of $\tau = 0$ contributes to the value of the integral. But if τ is small, we can put

$$1 - t^2 = 1 - 1 + 2i\tau = 2i\tau$$

and

$$(1 - t^2)^{p-(1/2)} = (2i\tau)^{p-(1/2)} = 2^{p-(1/2)}e^{(\pi/2)i[p-(1/2)]}\tau^{p-(1/2)} \tag{7.14.6}$$

The result of the integration can be written down as the constant factor

$$\tfrac{1}{2}2^{p-(1/2)}ie^{-i\{x-(\pi/2)[p-(1/2)]\}} \tag{7.14.7}$$

multiplied by the integral

$$\int_0^\infty e^{-x\tau}\,\tau^{p-(1/2)}d\tau \tag{7.14.8}$$

(The limit of integration can be extended to infinity because the largeness of x blots out everything beyond a small neighbourhood of $\tau = 0$.) This integral becomes Euler's integral of the factorial function if we introduce $x\tau = \xi$ as a new variable:

$$\frac{1}{x^{p+(1/2)}}\int_0^\infty e^{-\xi}\xi^{p-(1/2)}d\xi = \frac{(p - \tfrac{1}{2})!}{x^{p+(1/2)}} \tag{7.14.9}$$

The path AB contributes the same, except that all i have to be changed to $-i$. The path BC contributes nothing since here the integrand becomes arbitrarily small. The final result of our calculation is that the integral (3) becomes

$$\frac{2^{p-(1/2)}(p - \tfrac{1}{2})!}{x^{p+(1/2)}} \sin\left(x - \frac{\pi}{2}p + \frac{\pi}{4}\right) \tag{7.14.10}$$

and thus we obtain the following asymptotic representation of $J_p(x)$ for large positive values of x:

$$J_p(x) = \sqrt{\frac{2}{\pi x}} \sin\left(x - \frac{\pi}{2}p + \frac{\pi}{4}\right)$$

$$= \sqrt{\frac{2}{\pi x}} \cos\left(x - \frac{\pi}{2}(p + \tfrac{1}{2})\right) \tag{7.14.11}$$

Although we have proved this asymptotic behaviour only for *positive* p, it holds in fact for *all* p, as we can see from the recurrence relation

$$J_{p-1}(x) - \frac{2p}{x}J_p(x) + J_{p+1}(x) = 0 \tag{7.14.12}$$

which holds for negative orders as well and which carries the asymptotic relation (11) over into the realm of negative p.

7.15. Behaviour of $J_p(z)$ along the imaginary axis

The asymptotic relation (13.11) shows that $J_p(z)$ for purely *imaginary* values of z will have an exponentially increasing or decreasing character. Whenever A_2 is present, the exponentially increasing part will swamp the first term and the function will go exponentially to infinity for large values of $z = iy$. Only for one particular linear combination of the two Bessel functions $J_p(x)$ and $J_{-p}(x)$ can we obtain an exponentially *decreasing* course. This will happen if the constant A_2 is obliterated and that again demands— as we can see from (13.12)—that the function shall go to infinity for real values of x like $e^{ix}/\sqrt{x}$, without any intermingling of $e^{-ix}/\sqrt{x}$. Now, if we write (11) in complex form:

$$J_p(x) = \frac{1}{\sqrt{2\pi x}} \left[e^{ix}e^{-i(\pi/2)[p+(1/2)]} + e^{-ix}e^{i(\pi/2)[p+(1/2)]} \right] \qquad (7.15.1)$$

we notice that the e^{-ix} part of $J_p(x)$ can only be obliterated, if we choose the following linear combination of $J_p(x)$ and $J_{-p}(x)$:

$$K_p(x) = -e^{-i(\pi/2)p}J_p(x) + e^{i(\pi/2)p}J_{-p}(x) \qquad (7.15.2)$$

This function becomes for large values of x:

$$i\sqrt{\frac{2}{\pi x}} \sin \pi p \; e^{i[x-(\pi/4)]} \qquad (7.15.3)$$

Accordingly in the formula (13.11) the constant A_2 drops out and we obtain along the imaginary axis $z = iy$:

$$K_p(iy) = i\sqrt{\frac{2}{\pi i y}} \sin \pi p \; e^{-i(\pi/4)}e^{-y}$$

$$= \sqrt{\frac{2}{\pi y}} e^{-y} \sin \pi p \qquad (7.15.4)$$

Apart from an arbitrary complex constant, the function $K_p(z)$ is the only linear combination of Bessel functions which decreases exponentially along the positive imaginary axis.

Problem 304. Show on the basis of (12.10) that the function $K_p(z)$ is real everywhere along the imaginary axis.

Problem 305. Obtain the asymptotic value of the function

$$M_p(z) = e^{i(\pi/2)p}J_p(z) - e^{-i(\pi/2)p}J_{-p}(z) \qquad (7.15.5)$$

along the imaginary axis.
[Answer:

$$M_p(iy) = i\sqrt{\frac{2}{\pi y}} e^y \sin \pi p \qquad (7.15.6)]$$

Problem 306.   Obtain the asymptotic value of $M_p(-iy)$ (cf. 12.14).
[Answer:

$$M_p(-iy) = -\sqrt{\frac{2}{\pi y}}\, e^{-y} \sin \pi p \qquad (7.15.7)]$$

Problem 307.   Obtain the asymptotic value of the combination

$$I_p(x) = e^{-i(\pi/2)p} J_p(x) + e^{i(\pi/2)p} J_{-p}(x) \qquad (7.15.8)$$

along the real and imaginary axes.
[Answer:

$$I_p(x) = \sqrt{\frac{2}{\pi x}} \left[e^{i[x-(\pi/4)]} \cos \frac{\pi}{2} p + e^{-i[x-(\pi/4)]} \right] \qquad (7.15.9)$$

$$I_p(iy) = \sqrt{\frac{2}{\pi y}}\, e^{y} \qquad (7.15.10)]$$

7.16. The Bessel functions of the order $\frac{1}{3}$

We are now sufficiently prepared to return to the problem of Section 11. We wanted to study the solution of the special differential equation (11.3) which will enable us to join the periodic and exponential branches of the KWB solution in the case that the function $U(x)$ changes its sign. We have found the solution in the form (11.4). Now we must dispose of the constants A_1 and A_2 properly.

First of all we notice on the basis of (12.10–12) that our solution is in the following relation to the entire function $\Lambda_p(x^2)$:

$$v(x) = \sqrt{x}[A_1 x^{1/2} \Lambda_{1/3}(x^3) + A_2 x^{-1/2} \Lambda_{-1/3}(x^3)]$$
$$= A_1 x \Lambda_{1/3}(x^3) + A_2 \Lambda_{-1/3}(x^3) \qquad (7.16.1)$$

We thus see that $v(x)$ is for any choice of the constants A_1 and A_2 an *entire analytical function* of the complex variable z, regular throughout the complex plane. The cuts needed in the study of the Bessel functions disappear completely in the resulting function (11.4).

We will now choose as fundamental solutions of our differential equation the following combinations of Bessel functions:

$$f(x) = \sqrt{x}[J_{1/3}(x^{3/2}) + J_{-1/3}(x^{3/2})]$$
$$g(x) = \sqrt{x}[-J_{1/3}(x^{3/2}) + J_{-1/3}(x^{3/2})] \qquad (7.16.2)$$

The asymptotic behaviour of these two functions can be predicted on the basis of the general asymptotic behaviour of the Bessel functions. First of all we will move along the *positive* x-axis. Then we obtain for large values of x:

$$f(x) = \sqrt{x}\,\sqrt{\frac{2}{\pi x^{3/2}}} \left[\cos \left(x^{3/2} - \frac{\pi}{4} - \frac{\pi}{6} \right) + \cos \left(x^{3/2} - \frac{\pi}{4} + \frac{\pi}{6} \right) \right]$$
$$= \sqrt{\frac{2}{\pi}} \frac{2 \cos \pi/6}{\sqrt[4]{x}} \cos \left(x^{3/2} - \frac{\pi}{4} \right) \qquad (7.16.3)$$

$$g(x) = \sqrt{x}\sqrt{\frac{2}{\pi x^{3/2}}}\left[-\cos\left(x^{3/2} - \frac{\pi}{4} - \frac{\pi}{6}\right) + \cos\left(x^{3/2} - \frac{\pi}{4} + \frac{\pi}{6}\right)\right]$$

$$= \sqrt{\frac{2}{\pi}}\frac{-2\sin\pi/6}{\sqrt[4]{x}}\sin\left(x^{3/2} - \frac{\pi}{4}\right) \tag{7.16.4}$$

We see that in the positive range of x the functions $f(x)$ and $g(x)$ represent two oscillations (of variable amplitude and frequency) which have the constant phase shift of $\pi/2$ relative to one another.

We now come to the study of the *negative* range of x, starting with the function $f(x)$. As we see from (1) (considering x as a positive quantity):

$$f(x) = -x\Lambda_{1/3}(-x^3) + \Lambda_{-1/3}(-x^3) \tag{7.16.5}$$

We will now make the correlation

$$x^3 = t^2 \tag{7.16.6}$$

Then by definition:

$$\begin{aligned}\Lambda_{1/3}(-t^2) &= (it)^{-1/3}J_{1/3}(it)\\ \Lambda_{-1/3}(-t^2) &= (it)^{1/3}J_{-1/3}(it)\end{aligned} \tag{7.16.7}$$

and thus

$$\begin{aligned}f(-x) &= -t^{1/3}i^{-1/3}J_{1/3}(it) + t^{1/3}i^{1/3}J_{-1/3}(it)\\ &= t^{1/3}[-e^{-i(\pi/6)}J_{1/3}(it) + e^{i(\pi/6)}J_{-1/3}(it)]\\ &= t^{1/3}K_{1/3}(it)\end{aligned} \tag{7.16.8}$$

But then, making use of the asymptotic behaviour of $K_p(iy)$ (see 15.4), we obtain for large values of x:

$$\begin{aligned}f(-x) &= t^{1/3}\sqrt{\frac{2}{\pi t}}e^{-t}\sin\frac{\pi}{3}\\ &= \sqrt{\frac{2}{\pi}}\frac{\sin\pi/3}{\sqrt[4]{x}}e^{-x^{3/2}}\end{aligned} \tag{7.16.9}$$

We proceed quite similarly for $g(-x)$:

$$\begin{aligned}g(-x) &= x\Lambda_{1/3}(-x^3) + \Lambda_{-1/3}(-x^3)\\ &= t^{1/3}[e^{-i(\pi/6)}J_{1/3}(it) + e^{i(\pi/6)}J_{-1/3}(it)]\end{aligned} \tag{7.16.10}$$

For large values of x we obtain (on the basis of the asymptotic behaviour of the function $I_p(x)$ of Problem 307 (cf. 15.10)):

$$g(-x) = \sqrt{\frac{2}{\pi}}\frac{1}{\sqrt[4]{x}}e^{x^{3/2}} \tag{7.16.11}$$

7.17. Jump conditions for the transition "exponential–periodic"

The differential equation (11.3), on which we have concentrated, is characterised by

$$U(x) = \tfrac{9}{4}x \tag{7.17.1}$$

Hence

$$\sqrt{U(x)} = \tfrac{3}{2}\sqrt{x} \tag{7.17.2}$$

$$\int_0^x \sqrt{U(\xi)}\,d\xi = x^{3/2} \tag{7.17.3}$$

In the negative range we will put

$$x = -t \tag{7.17.4}$$

$$\sqrt{-U(x)} = \tfrac{3}{2}\sqrt{t} \tag{7.17.5}$$

$$\int_0^{-x} \sqrt{-U(\xi)}\,d\xi = -\int_0^t \tfrac{3}{2}\sqrt{\xi}\,d\xi = -t^{3/2} \tag{7.17.6}$$

The general KWB approximation in the negative range can be written as follows. We select the point $x = x_0$ (in which $U(x_0) = 0$) as our point of reference and write

$$v(x) = \frac{1}{\sqrt[4]{-U}}\left[A_1 e^{\int_{x_0}^x \sqrt{-U}\,d\xi} + A_2 e^{-\int_{x_0}^x \sqrt{-U}\,d\xi}\right] \tag{7.17.7}$$

In the positive range, on the other hand, we put

$$v(x) = \frac{1}{\sqrt[4]{U}}\left[A'_1 e^{i\int_{x_0}^x \sqrt{U}\,d\xi} + A'_2 e^{-i\int_{x_0}^x \sqrt{U}\,d\xi}\right] \tag{7.17.8}$$

The four constants of these solutions cannot be independent of each other. If we start with the exponential region and know from certain boundary conditions the values of A_1 and A_2, the transition to the periodic region is uniquely established, and thus A'_1 and A'_2 must be expressible in terms of A_1 and A_2. On the other hand, if we start from the periodic region and know on the basis of some boundary conditions the values of A'_1 and A'_2, the transition back to the exponential range is once more uniquely determined and thus A_1 and A_2 must be expressible in terms of A'_1, A'_2.

If we could neglect the singular character of the point $U(x_0) = 0$, we could argue that we should have

$$A_1 = \sqrt[4]{-1}\,A'_1, \quad A_2 = \sqrt[4]{-1}\,A'_2 \tag{7.17.9}$$

since in the exponents $\sqrt{-U}$ and $i\sqrt{U}$ are identical expressions. But this argument is in fact wrong, because there is a gulf between the two types of solutions which cannot be bridged without the proper precautions. We have to use our function $f(x)$ as a test function for the coefficient A_1 and $g(x)$ as a test function for the coefficient A_2. The comparison of the formulae

(16.9) and (16.3)—the latter written in complex form—yields the following relation:

$$A'_1 = e^{-(\pi/4)i}A_1$$
$$A'_2 = e^{(\pi/4)i}A_1 \tag{7.17.10}$$

On the other hand, the comparison of the formulae (16.11) and (16.4)—the latter written in complex form—yields:

$$A'_1 = \frac{i}{2} e^{-(\pi/4)i}A_2$$
$$A'_2 = -\frac{i}{2} e^{(\pi/4)i}A_2 \tag{7.17.11}$$

and thus the complete relation between the two pairs of constants becomes

$$A'_1 = e^{-(\pi/4)i}\left(A_1 + \frac{i}{2} A_2\right)$$
$$A'_2 = e^{(\pi/4)i}\left(A_1 - \frac{i}{2} A_2\right) \tag{7.17.12}$$

which means

$$A'_1 = \frac{1}{\sqrt{2}} [(1 - i)A_1 + \tfrac{1}{2}(1 + i)A_2]$$
$$A'_2 = \frac{1}{\sqrt{2}} [(1 + i)A_1 + \tfrac{1}{2}(1 - i)A_2] \tag{7.17.13}$$

The inverse relations become

$$A_1 = \frac{1}{2\sqrt{2}} [(1 + i)A'_1 + (1 - i)A'_2]$$
$$A_2 = \frac{1}{\sqrt{2}} [(1 - i)A'_1 + (1 + i)A'_2] \tag{7.17.14}$$

7.18. Jump conditions for the transition "periodic–exponential"

If it so happens that $U(x)$ changes its sign from plus to minus during its transition through the critical point $x_0 = 0$, we can write the approximate solutions once more in the form (17.7) and (17.8), but using now the formula (17.8) for $x < x_0$ and the formula (17.7) for $x > x_0$. Once more we can ask for the relation between the two pairs of constants A_1, A_2 on the one hand and A'_1, A'_2, on the other hand. The new situation is reducible to the previous one by changing x to $-x$. This means that A_1 and A_2, and similarly A'_1 and A'_2, become interchanged. Accordingly, the relations (17.13) and (17.14) have now to be formulated as follows:

$$A'_1 = \frac{1}{\sqrt{2}} [\tfrac{1}{2}(1 - i)A_1 + (1 + i)A_2]$$
$$A'_2 = \frac{1}{\sqrt{2}} [\tfrac{1}{2}(1 + i)A_1 + (1 - i)A_1] \tag{7.18.1}$$

while the reciprocal relations become

$$A_1 = \frac{1}{\sqrt{2}}[(1 + i)A'_1 + (1 - i)A'_2]$$

$$A_2 = \frac{1}{2\sqrt{2}}[(1 - i)A'_1 + (1 + i)A'_2]$$

(7.18.2)

7.19. Amplitude and phase in the periodic domain

In the exponential domain the two branches of the KWB approximation define two completely different analytical functions whose separation is often of vital importance. The one is an exponentially *increasing*, the other an exponentially *decreasing* function. In the periodic domain, however, the two branches of the solution represent two analytical functions of the same order of magnitude whose separation is not advocated on natural grounds. In most practical applications we deal with *real* solutions while the two branches of the KWB approximation operate with complex quantities Hence it is preferable to combine the two complex branches into one real branch by writing the solution in the form

$$v(x) = \frac{C}{\sqrt[4]{U}} \cos \left[\int_{x_0}^{x} \sqrt{U} d\xi - \theta \right]$$

(7.19.1)

The solution appears in the form of a vibration of variable amplitude and frequency. The two constants of integration A'_1 and A'_2 are now replaced by the amplitude factor C and the phase angle θ. The relation between the two types of constants is obtained if we write (1) in complex form, obtaining

$$A'_1 = \frac{C}{2} e^{-i\theta}$$

$$A'_2 = \frac{C}{2} e^{i\theta}$$

(7.19.2)

The relations (17.14) which hold for the transition from the exponential to the periodic domain, now become:

$$A_1 = \frac{C}{2\sqrt{2}} (\cos \theta + \sin \theta) = \frac{C}{2} \cos \left(\theta - \frac{\pi}{4} \right)$$

$$A_2 = \frac{C}{\sqrt{2}} (\cos \theta - \sin \theta) = C \cos \left(\theta + \frac{\pi}{4} \right) = -C \sin \left(\theta - \frac{\pi}{4} \right)$$

(7.19.3)

and thus

$$\tan \left(\theta - \frac{\pi}{4} \right) = -\frac{A_2}{2A_1}$$

$$C = \sqrt{4A_1^2 + A_2^2}$$

(7.19.4)

These are the formulae by which the constants of the *exponential* domain are determined if the constants of the *periodic* domain are given, and vice versa. The transition occurs in the sequence : exponential–periodic. If the sequence is the reverse, viz. periodic–exponential, we have to utilize the formulae (18.2) which now give

$$A_1 = \frac{C}{\sqrt{2}} (\cos \theta + \sin \theta) = C \cos \left(\theta - \frac{\pi}{4} \right)$$

$$A_2 = \frac{C}{2\sqrt{2}} (\cos \theta - \sin \theta) = \frac{C}{2} \cos \left(\theta + \frac{\pi}{4} \right) = - \frac{C}{2} \sin \left(\theta - \frac{\pi}{4} \right)$$

(7.19.5)

and

$$\tan \left(\theta - \frac{\pi}{4} \right) = - \frac{2A_2}{A_1}$$

$$C = \sqrt{A_1{}^2 + 4A_2{}^2}$$

(7.19.6)

7.20. Eigenvalue problems

In the application of differential equations to the problems of atomic physics we have often problems characterised by certain homogeneous boundary conditions. We demand for example that the function shall *vanish* at the two end-points of the domain. Or we may have the situation encountered in the differential equation of Gauss where the boundary points are singular points of the domain and we demand that the function shall remain *bounded* at the two end-points of our range. Generally such conditions cannot be met without the freedom of a certain parameter, called the "eigenvalue" of the differential equation. In the case of the hypergeometric series, for example, the two fundamental solutions go out of bound at either the lower or the upper boundary. To demand that they shall remain finite at *both* boundaries means that a certain restriction has to be fulfilled by the parameters of the solution. Generally, if

$$Dv = 0 \qquad (7.20.1)$$

is a homogeneous linear differential operator and we add the proper number of homogeneous boundary conditions, we shall get as the only possible solution the trivial solution $u = 0$, because the "right side" is missing which excites the various eigenfunctions of the operator. These eigenfunctions are in the case of a self-adjoint operator (and all second-order operators are self-adjoint, if we include the proper weight factor), defined by the modified differential equation

$$Dv + \lambda v = 0 \qquad (7.20.2)$$

which is once more homogeneous but now contains the parameter λ, adjustable in such a way that the given homogeneous boundary conditions shall be satisfied. That this is indeed possible, and with an infinity of *real* λ_i, follows from the general matrix treatment of linear differential operators (as we have seen in Chapters 4 and 5). In fact, the solution of the equation (2) yields those "principal axes" of the operator which establish an orthogonal frame of reference in function space and make an expansion of both right side and solution into eigenfunctions possible.

The very same eigenfunctions have, however, direct *physical* significance in all vibration problems. It is the time dependent part of the differential operators of mathematical physics which provides the λv-term of the differential equation (2) and thus yields those "characteristic vibrations", which the physical system is capable of performing. In atomic physics the eigen*values* have the significance of the various *energy values* associated with the possible states of the atom, while the eigen*solutions* represent the various "quantum states" in which the atom may maintain itself.

Before the advent of Schrödinger's wave-equation, in the epoch of Bohr's atomic model, the various quantum states were interpreted on the basis of certain "quantum conditions" which led to definite stable configurations of the atom, associated with definite energy values. These "quantum conditions" are closely related to the KWB approximation applicable to Schrödinger's wave equation—although this connection was not known before the discovery of wave-mechanics. In particular, the selection principle which brings about the quantum conditions within the framework of the KWB approximation lies in the condition that in the exponential domain only the exponentially *decreasing* solution is permitted, since the other solution would increase to infinity and is void of physical significance.

The close agreement of the older results of Bohr's atomic theory with the later eigenvalue theory demonstrates the usefulness of the KWB method in atomic problems. Although the wave-mechanical solution replaced the approximative treatment by the exact solution and the exact eigenvalues, the KWB method retains its great usefulness. Many of the fundamentally important function classes of mathematical physics arise from Sturm-Liouville problems, since the separation of the wave-equation in various systems of coordinates leads to ordinary second-order differential equations (as we will see in Chapter 8). Although the hypergeometric series provides us with the exact eigenvalues and exact eigenfunctions in all these cases, this is of little help if our aim is to study the actual course of the function since the sum of too many terms would be demanded for this purpose (except for the eigenfunctions of lowest order). We fare much better if we possess a solution in *closed form* which, although only approximative, can be handled in explicit terms. In the following sections we will see how some of the fundamental function classes of mathematical physics can be represented in good approximation with the help of elementary functions, on the basis of the KWB solution.

7.21. Hermite's differential equation

The differential equation (2.18) of Hermite, if transformed in the normal form, has a $U(x)$ given by (5.39). It is this differential operator which describes in wave-mechanics an "atomic oscillator". The differential operator is here simply

$$Dv = v'' - x^2 v \tag{7.21.1}$$

and the associated eigenvalue problem becomes

$$v'' + (\lambda - x^2)v = 0 \tag{7.21.2}$$

The domain of our solution is the infinite range $x = [-\infty, +\infty]$ and we demand that the function shall not go to infinity as we approach the two end-points $x = \pm \infty$.

Now this eigenvalue problem is solvable with the help of the hypergeometric series, after the proper transformations. The solution is well known in terms of the "Hermitian polynomials" $H_n(x)$ (cf. 2.19). But let us assume that this transformation had escaped us and we would tackle our problem by the KWB method. We see that if x stays within the limits $\pm \sqrt{\lambda}$, we have a *periodic*, outside of those limits an *exponential* domain. The KWB method requires the following integration:

$$F = \int \sqrt{U}\, dx = \int \sqrt{\lambda - x^2}\, dx \tag{7.21.3}$$

We put

$$\lambda = \beta^2, \quad x = \beta \sin t$$

$$F(x) = \beta^2 \int \cos^2 t\, dt = \beta^2 \left(\frac{t}{2} + \frac{\sin 2t}{4} \right)$$

$$= \frac{\beta^2}{2} \left(\arcsin \frac{x}{\beta} + \frac{x}{\beta} \sqrt{1 - \left(\frac{x}{\beta} \right)^2} \right) \tag{7.21.4}$$

For the exponential range we need the integral

$$K = \int \sqrt{-U}\, dx = \int \sqrt{x^2 - \lambda}\, dx$$

Here we make the substitution

$$\lambda = \beta^2, \quad x = \beta \cosh t$$

$$K = \beta^2 \int \sinh^2 t\, dt = \beta^2 \left(-\frac{t}{2} + \frac{\sinh 2t}{4} \right)$$

$$= \frac{\beta^2}{2} \left(\frac{x}{\beta} \sqrt{\left(\frac{x}{\beta} \right)^2 - 1} - \log \left[\frac{x}{\beta} + \sqrt{\left(\frac{x}{\beta} \right)^2 - 1} \right] \right) \tag{7.21.5}$$

Now the exponential solution associated with the constant A_1 *increases to infinity*, while the exponential solution associated with A_2 *decreases to zero*. It is this second solution which we admit but not the first one. Hence we obtain a very definite *selection principle* by demanding that the constant A_1 *must become zero*. This condition imposes a definite demand on the phase constant θ which exists in the periodic range. We have now (for $x > 0$) the sequence "periodic–exponential" and have to make use of the relations (19.5). The requirement $A_1 = 0$ means

$$\cos\left(\theta - \frac{\pi}{4}\right) = 0$$

or

$$\theta = -\frac{\pi}{4} \pm k\pi \tag{7.21.6}$$

where k is an arbitrary integer. This means that the periodic solution must arrive at the critical point $U = 0$ *with a definite phase angle*.

The solution appears in the periodic range according to (19.1) in the following form:

$$v(x) = \frac{C \cos\left[F(x) - F(x_0) - \theta\right]}{4\sqrt{U}} \tag{7.21.7}$$

So far only *one* of the constants, namely θ, has been restricted, but we still have the constant C at our disposal and thus a solution seems possible for all λ. We have to realise, however, that x assumes both positive and negative values and the transition to the exponential domain occurs at both points $x = \pm\sqrt{\lambda}$. The second point adds its own condition, except if $F(x)$ is either an even or an odd function, in which case the two conditions on the left and on the right collapse into one, on account of the left–right symmetry of the given differential operator. In fact, this is the *only* chance of satisfying both conditions. Now the solution (7) yields in the numerator

$$\cos F(x) \cos\left[F(x_0) - \theta\right]$$
$$+ \sin F(x) \sin\left[F(x_0) - \theta\right] \tag{7.21.8}$$

The condition that our function shall become *even* demands the vanishing of the second term, which means

$$F(x_0) - \theta = m\pi \tag{7.21.9}$$

and thus

$$F(x_0) = -\frac{\pi}{4} + k\pi \tag{7.21.10}$$

Similarly, the condition that our function shall become *odd* demands the vanishing of the first term, which means

$$F(x_0) - \theta = \frac{\pi}{2} + m\pi \tag{7.21.11}$$

and thus

$$F(x_0) = \frac{\pi}{4} + k\pi \tag{7.21.12}$$

where k is an arbitrary integer. But then, since

$$F(x_0) = F(\sqrt{\lambda}) = \lambda \frac{\pi}{4} \qquad (7.21.13)$$

we obtain for λ the following selection rules:

$$\begin{aligned} \lambda &= 4k - 1 \qquad \text{(even function)} \\ &= 4k + 1 \qquad \text{(odd function)} \end{aligned} \qquad (7.21.14)$$

Both conditions are included if we put

$$\lambda = 2n + 1 \qquad (7.21.15)$$

where an even n belongs to the even, an odd n to the odd eigenfunctions.

A comparison with the exact treatment shows that in the present example we obtained an *exact* result which is more than we should have expected, in view of the purely *approximative* character of the KWB method. The solution $v(x)$ is in the present problem obtainable in terms of the "Hermitian polynomials" $H_n(x)$, in the form

$$v(x) = e^{-x^2/2} H_n(x) \qquad (7.21.16)$$

The eigenvalue of Hermite's differential equation is *exactly* $2n + 1$, in agreement with our result (15).

7.22. Bessel's differential equation

Another interesting example is provided by the celebrated differential equation (2.4) which defines the Bessel functions. We bring this differential equation in the normal form (7.13) and have thus

$$U(x) = 1 - \frac{p_1^2}{x^2}, \quad p_1 = \sqrt{p^2 - \tfrac{1}{4}} \qquad (7.22.1)$$

We consider the order p of the Bessel function to be an arbitrary real number which, however, shall not be too small (less than about 4), since otherwise our approximation procedure becomes too inaccurate (the Bessel functions of low order can be made amenable to the KWB method if we first transform the defining differential equation into the more general form (2.23). An example was provided by the Bessel functions of the order 1/3).

First of all we carry through the necessary integrations:

$$F = \int \sqrt{1 - \frac{p_1^2}{x^2}}\, dx = x \sqrt{1 - \frac{p_1^2}{x^2}} - \int \frac{p_1^2}{x^2} \frac{1}{\sqrt{1 - \frac{p_1^2}{x^2}}}\, dx$$

The second integral becomes manageable by the substitution

$$\frac{p_1}{x} = t, \quad -\frac{p_1 dx}{x^2} = dt$$

and we obtain

$$F = x\sqrt{1 - \frac{p_1^2}{x^2}} + p_1 \arc \sin \frac{p_1}{x} \tag{7.22.2}$$

This holds in the *periodic* domain $x \geq p_1$. In the *exponential* domain $x \leq p_1$ we obtain similarly

$$K = \int \sqrt{\frac{p_1^2}{x^2} - 1} \; dx = x\sqrt{\frac{p_1^2}{x^2} - 1} + \int \frac{p_1^2}{x^2} \frac{dx}{\sqrt{\frac{p_1^2}{x^2} - 1}}$$

The substitution

$$\frac{p_1}{x} = \cosh t, \quad -\frac{p_1}{x^2} dx = \sinh t \, dt$$

reduces the second integral to

$$-p_1 t = -p_1 \log \left(\frac{p_1}{x} + \sqrt{\frac{p_1^2}{x^2} - 1} \right)$$

and the complete result, valid in the exponential domain, may be written in the following form

$$K = p_1 \left[\sqrt{1 - \frac{x^2}{p_1^2}} + \log x - \log \left(1 + \sqrt{1 - \frac{x^2}{p_1^2}} \right) \right] \tag{7.22.3}$$

The two fundamental KWB approximations in the exponential region can thus be put in the following form:

$$v(x) = \frac{B}{\sqrt[4]{\frac{p_1^2}{x^2} - 1}} e^{\pm p_1 [\sqrt{1 - (x^2/p_1^2)} + \log x/(1 + \sqrt{1 - (x^2/p_1^2)})]} \tag{7.22.4}$$

The general solution has two free constants B_1 and B_2, associated with the $\pm$ signs in the exponent.

The point $x = 0$ is a singular point of our differential equation in which $U(x)$ becomes infinite. However, our approximation does not fail badly in this neighbourhood. If we let x go towards zero, we find

$$v(x) = \frac{Be^{\pm p_1}}{\sqrt{p_1}} \left(\frac{x}{2} \right)^{\pm p_1} \sqrt{x} \tag{7.22.5}$$

We must remember that our function $v(x)$ is *not* the Bessel function $J_p(x)$ but $J_p(x)\sqrt{x}$. Moreover, the Bessel functions $J_p(x)$, respectively $J_{-p}(x)$, assume by definition in the vicinity of zero the values

$$\frac{1}{(\pm p)!} \left(\frac{x}{2} \right)^{\pm p} \tag{7.22.6}$$

Hence we conclude that the two branches of our approximation belong to the two Bessel functions $J_p(x)$ and $J_{-p}(x)$. If B_1 alone is present, we shall obtain $J_p(x)$, if B_2 alone is present, we shall obtain $J_{-p}(x)$.* The error of our approximation is very small in this neighbourhood since p and p_1 (for not too small p) are very nearly equal (for example for $p = 5$ the relative error is only $\frac{1}{2}\%$). The values of B_1 and B_2 can be determined by the initial values (6), obtaining

$$B_1 = \frac{\sqrt{p_1}}{p!} e^{-p_1} \quad \text{(for } J_p(x)\text{)}$$

$$B_2 = \frac{\sqrt{p_1}}{(-p)!} e^{p_1} \quad \text{(for } J_{-p}(x)\text{)}$$

(7.22.7)

We will now see what happens if we come to the end of the exponential domain and enter the *periodic* domain. In order to make the transition, we must put our solution in the form (17.7). But let the upper branch with the $+$ sign in the exponent be given in the more general form

$$v(x) = \frac{B_1}{\sqrt[4]{-U}} e^{K(x)} \tag{7.22.8}$$

Then the form (17.7) demands that we shall write this solution as follows:

$$\frac{A_1}{\sqrt[4]{-U}} e^{K(x)-K(x_0)} \tag{7.22.9}$$

which gives

$$A_1 = B_1 e^{K(x_0)} \tag{7.22.10}$$

Similarly

$$A_2 = B_2 e^{-K(x_0)} \tag{7.22.11}$$

Now in our problem the point x_0 in which $U(x)$ vanishes becomes the point $x = p_1$. The value of $K(x_0)$ can be taken from the form (3) of our solution:

$$K(p_1) = p_1 \log p_1 \tag{7.22.12}$$

Let us first consider the case of $J_p(x)$. Here only B_1 is present and we obtain

$$A_1 = \frac{\sqrt{p_1}}{p!} e^{-p_1} p_1^{p_1}$$

$$A_2 = 0$$

(7.22.13)

The transition to the periodic range occurs according to the formulae (19.4) which now gives

$$\theta = \frac{\pi}{4}$$

$$C = 2A_1$$

(7.22.14)

---

* That we were rash in this identification will be seen a little later.

Hence the representation of $J_p(x)\sqrt{x}$ beyond the point $x = p_1$ becomes, in view of (2):

$$v(x) = \frac{C}{\sqrt[4]{1 - \dfrac{p_1^2}{x^2}}} \cos \left[\sqrt{x^2 - p_1^2} + p_1 \arcsin \frac{p_1}{x} - p_1 \frac{\pi}{2} - \frac{\pi}{4} \right] \quad (7.22.15)$$

It is of interest to compare this result with the previously obtained asymptotic estimation (14.11) of the Bessel functions for large values of the argument. If x goes to infinity, the argument of the cosine function in (15) approaches

$$x - \left(p_1 + \frac{1}{2} \right) \frac{\pi}{2} \quad (7.22.16)$$

which differs from the corresponding quantity in (14.11) only by the fact that p is replaced by p_1. This involves a very small error, as we have seen before.

Another change can be noticed in the amplitude constant C. In the traditional estimation C should have the value

$$C_1 = \sqrt{\frac{2}{\pi}} \quad (7.22.17)$$

while our result is

$$C = \frac{2\sqrt{p_1}}{p!} \, p_1^{p_1} e^{-p_1} \quad (7.22.18)$$

Now p_1 is very near to p and $p!$ can be estimated on the basis of Stirling's formula:

$$p! = \sqrt{2\pi p} \, p^p e^{-p} \quad (7.22.19)$$

Hence our C is near to

$$\frac{2}{\sqrt{2\pi}} = \sqrt{\frac{2}{\pi}} \quad (7.22.20)$$

which is in agreement with the correct asymptotic value. Here again the error is fairly small, e.g. for $n = 5$ not more than 5.7%. It is frequently more advisable, however, to make the amplitude factor C correct in the *periodic* range and transfer the amplitude error to the exponential domain. In this case the approximation of $J_p(x)$ will be given as follows:

for $x > p_1$:

$$J_p(x) = \sqrt{\frac{2}{\pi}} \frac{1}{\sqrt[4]{x^2 - p_1^2}} \cos \left[\sqrt{x^2 - p_1^2} + p_1 \arcsin \frac{p_1}{x} - p_1 \frac{\pi}{2} - \frac{\pi}{4} \right] \quad (7.22.21)$$

for $x < p_1$:

$$J_p(x) = \sqrt{\frac{2}{\pi}} \frac{1}{\sqrt[4]{p_1^2 - x^2}} \left[\frac{x}{p_1 + \sqrt{p_1^2 - x^2}} \right]^{p_1} e^{\sqrt{p_1^2 - x^2}} \quad (7.22.22)$$

We now come to the discussion of $J_{-p}(x)$. Here we obtain instead of (13) the new constants

$$A_1 = 0$$

$$A_2 = \frac{\sqrt{p_1}}{(-p)!} e^{p_1} p_1^{-p_1} \tag{7.22.23}$$

The transition to the periodic range occurs once more on the basis of the formulae (19.4) which now gives

$$\theta = -\frac{\pi}{4} \tag{7.22.24}$$

$$C = A_2$$

The representation of $J_{-p}(x)\sqrt{x}$ beyond the point $x = p_1$ becomes

$$v(x) = \frac{C}{\sqrt[4]{1 - \frac{p_1^2}{x^2}}} \cos\left[\sqrt{x^2 - p_1^2} + p_1 \arcsin\frac{p_1}{x} - p_1\frac{\pi}{2} + \frac{\pi}{4}\right] \tag{7.22.25}$$

For large values of x the cosine part of the function becomes

$$\cos\left[x - \left(p_1 - \frac{1}{2}\right)\frac{\pi}{2}\right]$$

which is replaceable by

$$-\sin\left[x - \left(p_1 + \frac{1}{2}\right)\frac{\pi}{2}\right] \tag{7.22.26}$$

while in actual fact the periodicity factor of $J_{-p}(x)$ should come out as

$$\cos\left[x + \left(p_1 - \frac{1}{2}\right)\frac{\pi}{2}\right]$$

Hence we have arrived in the periodic range with the *wrong phase*.

Let us investigate the value of the constant C. Disregarding the small difference which exists between p and p_1 and making use of Stirling's formula (19) we obtain

$$C = \frac{\sqrt{2\pi p}}{(-p)!\,p!} = \frac{\sqrt{2\pi}}{(-p)!\,(p-1)!} \tag{7.22.27}$$

Now we can take advantage of the reflection theorem of the Gamma function which gives

$$(p-1)!\,(-p)! = \frac{\pi}{\sin \pi p} \tag{7.22.28}$$

Hence

$$C = \sqrt{\frac{2}{\pi}} \sin \pi p \tag{7.22.29}$$

Here again our result is erroneous since the amplitude factor of $\sqrt{x}J_{-p}(x)$ in infinity is $\sqrt{2/\pi}$, *without* the factor $\sin \pi p$.

The phenomenon here encountered is of considerable interest. We have tried to identify a certain solution of Bessel's differential equation by starting from a point where the solution went to infinity. But the differential equation has *two* solutions, the one remaining finite (in fact going to zero with the power x^p) the other going to infinity. Now the solution which remains finite at $x = 0$ allows a *unique identification* since the condition of finiteness automatically excludes the second solution. If, however, we try to identify the second solution by fitting it in the neighbourhood of the singular point $x = 0$, we cannot be sure that we have in fact obtained the right solution since *any admixture of the regular solution would remain undetected*. The solution which goes out of bound swamps the regular solution.

What we have obtained by our approximation, is thus not necessarily $J_{-p}(x)$ but

$$v(x) = J_{-p}(x) + \alpha J_p(x) \tag{7.22.30}$$

where α is an arbitrary constant. The behaviour of $v(x)$ in the periodic range shows that it is the *Neumann function* $N_p(x)$ to which our solution is proportional:

$$N_p(x) = \frac{1}{\sin p\pi} [J_p(x) \cos p\pi - J_{-p}(x)]$$
$$J_{-p}(x) - J_p(x) \cos p\pi = -(\sin p\pi) N_p(x) \tag{7.22.31}$$

The constant α in (30) is thus identified as $-\cos p\pi$. The Neumann function $N_p(x)$ has the property that for large x its periodicity is given by

$$\sin\left[x - \left(p + \frac{1}{2}\right)\frac{\pi}{2}\right] = -\cos\left[x - \left(p - \frac{1}{2}\right)\frac{\pi}{2}\right] \tag{7.22.32}$$

Hence the function $-N_p(x)$ has a periodicity which agrees with (26). Moreover, the amplitude factor (29) is explained by the fact that it is not $-N_p(x)$ itself but $-(\sin p\pi) N_p(x)$ that has been represented by our approximation.

If again we agree that the amplitude factor C shall become correct in the periodic region and only approximately correct in the exponential region, we obtain the following approximate representation of the Neumann function $N_p(x)$:

for $x > p_1$:

$$N_p(x) = \sqrt{\frac{2}{\pi}} \frac{1}{\sqrt[4]{x^2 - p_1^2}} \sin\left(\sqrt{x^2 - p_1^2} + p_1 \arcsin\frac{p_1}{x} - p_1\frac{\pi}{2} - \frac{\pi}{4}\right) \tag{7.22.33}$$

for $x < p_1$:

$$N_p(x) = -\sqrt{\frac{2}{\pi}} \frac{1}{\sqrt[4]{p_1^2 - x^2}} \left[\frac{x}{p_1 + \sqrt{p_1^2 - x^2}}\right]^{-p_1} e^{-\sqrt{p_1^2 - x^2}} \tag{7.22.34}$$

7.23. The substitute functions in the transitory range

While we have obtained a useful approximative solution of a second order differential equation (after we have transformed it in its normal form), there is nevertheless the difficulty that our solution loses its significance if we come near to the point $U(x_0) = 0$ where the amplitude factor becomes infinite. This difficulty can be overcome, however, by the following artifice. Before $U(x_0)$ vanishes, there is a certain finite region in which $U(x)$ allows a linear approximation. In this region we were able to solve the resulting differential equation (11.1) with the help of Bessel functions of the order 1/3. But if this is so, we can proceed as follows. Before our approximate solution goes out of hand, we have a KWB approximation which can be matched to the KWB approximation of the differential equation (11.1). Having performed the matching we can now *discard* our KWB approximation and substitute for it the solution of the differential equation (11.1). This substitute solution is now valid in the transitory region and with this solution we have no difficulty, even if we come to the critical point $x = x_0$ and go beyond. It will thus be sufficient to tabulate *four auxiliary functions* which we have to substitute in the place of the KWB approximation in the vicinity of the critical point $x = x_0$. We need *four* such functions because we have in the exponential range the constants A_1 and A_2 which are multiplied by the two branches of the KWB solution and we have to know what functions we have to substitute as factors of A_1 and A_2. We will call these functions $\varphi_1^e(x)$, respectively $\varphi_2^e(x)$. But then we have in addition the solution in the *periodic* range, of the form

$$\frac{C}{\sqrt[4]{U}} \cos [F(x) - F(x_0) - \theta] \tag{7.23.1}$$

For this we may put

$$\frac{C \cos \theta}{\sqrt[4]{U}} \cos [F(x) - F(x_0)]$$
$$+ \frac{C \sin \theta}{\sqrt[4]{U}} \sin [F(x) - F(x_0)] \tag{7.23.2}$$

and thus we have to tabulate the two substitute functions which will become the factors of $C \cos \theta$ and $C \sin \theta$ in the transitory region. These latter functions shall be called $\varphi_1^p(x)$ and $\varphi_2^p(x)$.

Let us first normalise the constant a of the differential equation (11.1) to 9/4, as we have done in (11.3). We have chosen the two functions (16.2) as the two fundamental solutions of our differential equation. Moreover, we have seen by the formula (16.9) that the upper branch of the KWB approximation will go with $f(-x)$. However, the asymptotic solution should become

$$\frac{1}{\sqrt[4]{\frac{9}{4}x}} e^{-x^{3/2}} \tag{7.23.3}$$

The comparison with (16.9) shows that the function $\varphi_1{}^e(x)$ should become identified with

$$\sqrt{\frac{2\pi}{3}}\;\sqrt[4]{\frac{4}{9}}\,f(-x) \tag{7.23.4}$$

It will be more natural, however, to normalise the constant a of the differential equation (11.1) to 1. This change of the constant is equivalent to the transformation

$$x = \sqrt[3]{\frac{4}{9}}\,x_1 \tag{7.23.5}$$

Furthermore, in order to arrive at the proper amplitude in the asymptotic range, we have to multiply our function by $(2/3)^{-1/3}$ and the final expression for $\varphi_1{}^e(x)$ becomes

$$\varphi_1{}^e(x) = \sqrt{\frac{2\pi}{3}}\left(\frac{2}{3}\right)^{1/6} f\!\left(-\sqrt[3]{\frac{4}{9}}\,x\right) \tag{7.23.6}$$

In a similar fashion we obtain with the help of (16.11):

$$\varphi_2{}^e(x) = \sqrt{\frac{\pi}{2}}\left(\frac{2}{3}\right)^{1/6} g\!\left(-\sqrt[3]{\frac{4}{9}}\,x\right) \tag{7.23.7}$$

In the periodic range the formulae (16.3) and (16.4) come in operation and our final result becomes:

$$\varphi_1{}^p(x) = \frac{\sqrt{\pi}}{2}\left(\frac{2}{3}\right)^{1/6}\left[\frac{f\!\left(\sqrt[3]{\frac{4}{9}}\,x\right)}{\sqrt{3}} + g\!\left(\sqrt[3]{\frac{4}{9}}\,x\right)\right] \tag{7.23.8}$$

$$\varphi_2{}^p(x) = \frac{\sqrt{\pi}}{2}\left(\frac{2}{3}\right)^{1/6}\left[\frac{f\!\left(\sqrt[3]{\frac{4}{9}}\,x\right)}{\sqrt{3}} - g\!\left(\sqrt[3]{\frac{4}{9}}\,x\right)\right] \tag{7.23.9}$$

The method of computing these four functions will be given in the following section.

Generally the differential equation which is valid in the transitory range will be of the form (11.1) with a constant a which is not 1 but $U'(x_0)$. The transition to the general case means that our previous x has to be replaced by

$$\bar{x} = \sqrt[3]{U'(x_0)}(x - x_0) \tag{7.23.10}$$

Hence in the general case, where the value of x_0 and $U'(x_0)$ is arbitrary, the substitute functions have to be taken with the argument $\sqrt[3]{U'(x_0)}(x - x_0)$. Moreover, a constant factor has to be applied in order to bring the asymptotic representation of these functions in harmony with the KWB approximation. This factor is $|U'(x_0)|^{-1/6}$.

Let us first assume that $U'(x_0) > 0$. We then have the transition *exponential to periodic.* We can now obtain the substitute functions of the transitory region according to the following table:

1. factor of A_1:

$$v(x) = \frac{1}{\sqrt[6]{U'(x_0)}} \, \varphi_1^e[\sqrt[3]{U'(x_0)}(x - x_0)] \qquad (x \leq x_0)$$

2. factor of A_2:

$$v(x) = \frac{1}{\sqrt[6]{U'(x_0)}} \, \varphi_2^e[\sqrt[3]{U'(x_0)}(x - x_0)] \qquad (x \leq x_0)$$

(7.23.11)

3. factor of $C \cos \theta$:

$$v(x) = \frac{1}{\sqrt[6]{U'(x_0)}} \, \varphi_1^p[\sqrt[3]{U'(x_0)}(x - x_0)] \qquad (x \geq x_0)$$

4. factor of $C \sin \theta$:

$$v(x) = \frac{1}{\sqrt[6]{U'(x_0)}} \, \varphi_2^p[\sqrt[3]{U'(x_0)}(x - x_0)] \qquad (x \geq x_0)$$

Let us assume, on the other hand, that the transition occurs in the sequence *periodic to exponential.* In this case the correlation occurs as follows:

1. factor of A_1:

$$v(x) = \frac{1}{\sqrt[6]{|U'(x_0)|}} \, \varphi_2^e[\sqrt[3]{|U'(x_0)|}(x_0 - x)] \qquad (x \geq x_0)$$

2. factor of A_2:

$$v(x) = \frac{1}{\sqrt[6]{|U'(x_0)|}} \, \varphi_1^e[\sqrt[3]{|U'(x_0)|}(x_0 - x)] \qquad (x \geq x_0)$$

(7.23.12)

3. factor of $C \cos \theta$:

$$v(x) = \frac{1}{\sqrt[6]{|U'(x_0)|}} \, \varphi_1^p[\sqrt[3]{|U'(x_0)|}(x_0 - x)] \qquad (x \leq x_0)$$

4. factor of $C \sin \theta$:

$$v(x) = \frac{-1}{\sqrt[6]{|U'(x_0)|}} \, \varphi_2^p[\sqrt[3]{|U'(x_0)|}(x_0 - x)] \qquad (x \leq x_0)$$

As an example let us consider the case of the Bessel functions $J_p(x)$ (with positive p), studied before. We want to determine the value of $J_p(p_1)$, or still better the value of the function $v(x) = \sqrt{x} J_p(x)$, at the transition point $x = x_0$. Here we have

$$U(x) = 1 - \frac{p_1^2}{x^2}$$

(7.23.13)

$$U'(x) = \frac{2p_1^2}{x^3}, \quad U'(p_1) = \frac{2}{p_1}$$

Furthermore, the definition of the functions $f(x)$ and $g(x)$ according to (16.2) yields

$$f(0) = g(0) = \frac{\sqrt[3]{2}}{(-\tfrac{1}{3})!} \tag{7.23.14}$$

We substitute these values in (8) and (9):

$$\varphi_1{}^p(0) = \frac{\sqrt{\pi}}{2} \left(\frac{2}{3}\right)^{1/6} \frac{\sqrt[3]{2}}{(-\tfrac{1}{3})!} \frac{1+\sqrt{3}}{\sqrt{3}} = 1.21566$$

$$\varphi_2{}^p(0) = \frac{\sqrt{\pi}}{2} \left(\frac{2}{3}\right)^{1/6} \frac{\sqrt[3]{2}}{(-\tfrac{1}{3})!} \frac{1-\sqrt{3}}{\sqrt{3}} = -0.32573 \tag{7.23.15}$$

Now in our problem we have according to (22.21):

$$C = \frac{1}{\sqrt{\dfrac{\pi}{2}}} \qquad \theta = \frac{\pi}{4}$$

and thus

$$C \cos \theta = C \sin \theta = \frac{1}{\sqrt{\pi}}$$

We have to make use of the table (11) which now gives

$$v(p_1) = p_1{}^{1/6} \left(\frac{2}{9}\right)^{1/3} \frac{1}{(-\tfrac{1}{3})!}$$

and going back to $J_p(x)$:

$$J_p(p_1) = \left(\frac{2}{9}\right)^{1/3} \frac{1}{(-\tfrac{1}{3})!} p_1{}^{-1/3} = 0.44730732 p_1{}^{-1/3} \tag{7.23.16}$$

If, on the other hand, the Neumann function $N_p(x)$ is in question, we have (cf. (22.33))

$$C = \frac{1}{\sqrt{\dfrac{\pi}{2}}} \qquad\qquad \theta = \frac{3\pi}{4}$$

$$C \cos \theta = -\frac{1}{\sqrt{\pi}}, \qquad C \sin \theta = \frac{1}{\sqrt{\pi}}$$

and the table (11) gives (in view of (15)):

$$v(p_1) = -p_1{}^{1/6} \left(\frac{2}{9}\right)^{1/3} \frac{\sqrt{3}}{(-\tfrac{1}{3})!}$$

$$N_p(p_1) = -\left(\frac{4}{3}\right)^{1/6} \frac{1}{(-\tfrac{1}{3})!} p_1{}^{-1/3} = -0.77475900 p_1{}^{-1/3} \tag{7.23.17}$$

Example. As an example we substitute for p the values 4, 6, 8, 10, and 12, and make the comparison of our approximation with the corresponding exact values of the Bessel functions $J_p(x)$, taken at the point p_1. The values of p_1 now become:

$$3.9686, \quad 5.9791, \quad 7.9843, \quad 9.9875, \quad 11.990$$

Substitution in the formula (16) yields

$p = 4$	6	8	10	12	
$J_p(p_1) =$ 0.2825,	0.2464,	0.2238,	0.2077,	0.1954	(7.23.18)
Exact v_1: 0.2765,	0.2434,	0.2219,	0.2064,	0.1945	

7.24. Tabulation of the four substitute functions

The numerical evaluation of the four functions $\varphi_1^e(x)$, $\varphi_2^e(x)$, $\varphi_1^p(x)$, $\varphi_2^p(x)$ causes no difficulties since all these functions are entire functions and possess a Taylor expansion which in the region concerned has satisfactory convergence. Certain linear combinations of these four functions exist, however, in tabulated form* and we can make use of these tables in order to obtain our functions. The actually tabulated functions have a real and an imaginary part, denoted by $R(h_1)$ and $I(h_1)$. We have to take the following linear combinations of the tabulated functions in order to obtain our functions $\varphi(x)$:

$$\varphi_1^e(x) = \sqrt{\frac{\pi}{2}} \left(\frac{2}{3}\right)^{1/6} (\sqrt{3}R - I) \qquad (x \leq 0)$$

$$\varphi_2^e(x) = \frac{1}{2}\sqrt{\frac{\pi}{2}} \left(\frac{2}{3}\right)^{1/6} (-R - \sqrt{3}I) \tag{7.24.1}$$

$$\varphi_1^p(x) = \frac{\sqrt{\pi}}{4} \left(\frac{2}{3}\right)^{1/6} [(\sqrt{3} - 1)R - (\sqrt{3} + 1)I] \qquad (x \geq 0)$$

$$\varphi_1^p(x) = \frac{\sqrt{\pi}}{4} \left(\frac{2}{3}\right)^{1/6} [(\sqrt{3} + 1)R + (\sqrt{3} - 1)I] \tag{7.24.2}$$

This means in numerical terms:

$$x \leq 0$$
$$\varphi_1^e = 2.028953167R - 1.171416657I$$
$$\varphi_2^e = -0.5857083285R - 1.014476584I \tag{7.24.3}$$

$$x \geq 0$$
$$\varphi_1^p = 0.3031849410R - 1.131501602I$$
$$\varphi_2^p = 1.131501602R + 0.3031849410I \tag{7.24.4}$$

The Table III of the Appendix gives the numerical values of these four

* Tables of the modified Hankel functions of order one-third and of their derivatives (Harvard University Press, Cambridge, Mass., 1945); cf. in particular the case $y = 0$ on pp. 2 and 3.

functions, in intervals of 0.1, for the range $x = [0, -3]$, respectively [0, 3]. Beyond this range no substitution is demanded since the KWB approximation becomes sufficiently accurate.

Problem 308. Obtain with the help of the tables the values of $J_5(4.7)$, $J_5(5)$, $J_5(5.2)$ and likewise $J_{10}(9.5)$, $J_{10}(10)$, $J_{10}(10.5)$.

[Answer:

$J_5(4.7) = 0.2307$	$J_5(5) = 0.2661$	$J_5(5.2) = 0.2977$
exact: (0.2213)	(0.2611)	(0.2865)
$J_{10}(9.5) = 0.1695$	$J_{10}(10) = 0.2086$	$J_{10}(10.5) = 0.2459$
(0.1650)	(0.2075)	(0.2477)]

Problem 309. The x-value at which $\varphi_1{}^p + \varphi_2{}^p$, respectively $\varphi_1{}^p - \varphi_2{}^p$ first vanishes, is $x = 2.3381$, resp 1.1737. Find accordingly the approximate position of the first zero of $J_p(x)$ and $N_p(x)$. In the case of $J_p(x)$ compare the result with the exact position of the first zero for $p = 5, 7, 9$, and 10.

[Answer:

$$J_p(x) = 0: \quad x_1 = p_1 + 1.8557 \sqrt[3]{p_1}$$
$$N_p(x) = 0: \quad x_1 = p_1 + 0.9316 \sqrt[3]{p_1} \tag{7.24.5}$$

$p =$	5	7	9	10
$x_1 =$	8.14	10.53	12.844	13.98
exact:	(8.771)	(11.09)	(13.35)	(14.48)

The relatively poor agreement indicates the presence of a systematic error which will be investigated in the next section.]

7.25. Increased accuracy in the transition domain

In using the substitution functions $\varphi(x)$ in the vicinity of the point x_0 we observe that we obtain good results only in a very narrow interval around x_0. As soon as we depart from the transition point $x = x_0$ to a substantial degree, our results deviate considerably from the true functional values. For example the position of the first zero of the Bessel functions involves the x-value 2.34 of the functions $\varphi_1{}^p(x)$ and $\varphi_2{}^p(x)$ but the answers do not agree well with the actual position of the zeros.

The reason for this phenomenon is that the purely linear approximation of $U(x)$ in the neighbourhood of $x = x_0$ holds only within a very *small* interval. The question arises whether we could not extend this interval to a somewhat larger domain, without losing too much in simplicity.

If we expand $U(x)$ in the neighbourhood of $U(x_0) = 0$, we obtain

$$U(x) = U'(x_0)(x - x_0) + \tfrac{1}{2}U''(x_0)(x - x_0)^2 \tag{7.25.1}$$

The higher order terms shall be neglected but not the quadratic term. We will briefly put

$$U'(x_0) = U'_0$$
$$U''(x_0) = U''_0 \tag{7.25.2}$$
$$x - x_0 = \xi$$

Then the given differential equation becomes in the transitory region:

$$v''(\xi) + (U'_0\xi + \tfrac{1}{2}U''_0\xi^2)v(\xi) = 0 \tag{7.25.3}$$

On the other hand, our four functions $\varphi(x)$ satisfied the differential equation

$$v''(x) + xv(x) = 0 \tag{7.25.4}$$

It will be our aim to bring the two differential equations (3) and (4) in harmony with each other. For this purpose we establish a relation between the variables x and ξ. While before we have assumed that x and ξ are simply *proportional* to each other, we will now apply a correction term and put

$$x = A\xi + \epsilon B\xi^2 \tag{7.25.5}$$

considering ϵ as a small quantity the square of which may be neglected.

Let us first investigate the transformation of the independent variable in general terms. The transformation

$$x = x(\xi) \tag{7.25.6}$$

means that the operation of differentiating has to be changed as follows:

$$\frac{d}{dx} = \frac{1}{x'}\frac{d}{d\xi}$$

$$\frac{d^2}{dx^2} = \frac{1}{(x')^2}\frac{d^2}{d\xi^2} - \frac{x''}{(x')^3}\frac{d}{d\xi}$$

Our differential equation

$$v'' + Uv = 0 \tag{7.25.7}$$

thus becomes in the new variable ξ:

$$\frac{1}{(x')^2}v'' - \frac{x''}{(x')^3}v' + Uv = 0 \tag{7.25.8}$$

which gives

$$v'' - \frac{x''}{x'}v' + (x')^2Uv = 0 \tag{7.25.9}$$

Now we make use of the method of taking out a proper factor in order to obliterate the first order term (cf. 7.9–10). We put

$$v(\xi) = e^{(1/2)\int(x''/x')d\xi}v_1(\xi) \tag{7.25.10}$$

and obtain for $v_1(\xi)$ the new equation

$$v_1''(\xi) + \left[(x')^2U - \frac{b'}{2} - \frac{b^2}{4}\right]v = 0 \tag{7.25.11}$$

where

$$b = -\frac{x''}{x'}$$

We obtain from (10):

$$v_1(\xi) = \frac{v(\xi)}{\sqrt{x'(\xi)}} \qquad (7.25.12)$$

Moreover, our postulated relation (5) gives

$$x'(\xi) = A + 2\epsilon B\xi$$
$$x''(\xi) = 2\epsilon B \qquad (7.25.13)$$

and thus, neglecting quantities of second order in ϵ:

$$b = -\frac{x''}{x'} = -\frac{2\epsilon B}{A}$$

Since b is a constant and thus $b' = 0$ (and since b^2 is negligible, being of second order in ϵ), the differential equation (11) becomes:

$$v''_1(\xi) + (A + 2\epsilon B\xi)^2 (A\xi + \epsilon B\xi^2)v(\xi) = 0 \qquad (7.25.14)$$

The factor of $v(\xi)$ becomes, neglecting quantities of second order:

$$A^2\left(1 + 4\epsilon \frac{B}{A}\xi\right)A\xi\left(1 + \epsilon \frac{B}{A}\xi\right) = A^3\xi\left(1 + 5\epsilon \frac{B}{A}\xi\right)$$

The corresponding factor of the differential equation (3) may be written in the form

$$U'_0\xi\left(1 + \frac{1}{2}\frac{U''_0}{U'_0}\xi\right)$$

and we obtain agreement by the choice

$$A = \sqrt[3]{U'_0}$$
$$\epsilon \frac{B}{A} = \frac{1}{10}\frac{U''_0}{U'_0} \qquad (7.25.15)$$

Finally, going back to the relation (5):

$$x = \sqrt[3]{U'_0}\xi\left(1 + \frac{1}{10}\frac{U''_0}{U'_0}\xi\right)$$
$$x' = \sqrt[3]{U'_0}\left(1 + \frac{1}{5}\frac{U''_0}{U'_0}\xi\right) \qquad (7.25.16)$$

Considering (12) the final result becomes that in the transition domain we have to use the φ-functions in the following manner:

$$v(x) = \frac{\varphi\left[\sqrt[3]{U'_0}(x - x_0)\left(1 + \frac{1}{10}\frac{U''_0}{U'_0}(x - x_0)\right)\right]}{\sqrt[6]{U'_0}\left[1 + \frac{1}{10}\frac{U''_0}{U'_0}(x - x_0)\right]} \qquad (7.25.17)$$

For the sake of increased accuracy the tables (23.11) and (23.12) have to be modified according to this correction. The correction is in fact quite effective. Let us obtain for example once more the first zero of $J_p(x)$. This demands the first zero of the function $\varphi_1^p + \varphi_2^p$ which is at the point $x = 2.3381$. Now in the present case

$$\frac{U''}{U'} = -\frac{3}{x} = -\frac{3}{p_1}$$

and we obtain the condition

$$\sqrt[3]{\frac{2}{p_1}}\,(x - p_1)\left[1 - \frac{3}{10p_1}(x - p_1)\right] = 2.3381$$

which yields with sufficient accuracy

$$x - p_1 = 1.8557\sqrt[3]{p_1}\left(1 + \frac{3}{10p_1}\,1.8557\sqrt[3]{p_1}\right)$$

$$x = p_1 + 1.8557\sqrt[3]{p_1} + \frac{1.0331}{\sqrt[3]{p_1}} \tag{7.25.18}$$

The same condition for the Neumann functions $N_p(x)$ becomes

$$\sqrt[3]{\frac{2}{p_1}}\,(x - p_1)\left(1 - \frac{3}{10p_1}(x - p_1)\right) = 1.1737$$

$$x - p_1 = 0.9316\sqrt[3]{p_1}\left(1 + \frac{3}{10p_1}\,0.9316\sqrt[3]{p_1}\right)$$

$$x = p_1 + 0.9316\sqrt[3]{p_1} + \frac{0.2604}{\sqrt[3]{p_1}} \tag{7.25.19}$$

The last term is the correction which has to be added to our previous formulae (24.5). The corrected values of the zeros obtained in Problem 309 now become

$p =$	5	7	9	10
$x_1 =$	8.748	11.069	13.341	14.463
	(8.771)	(11.086)	(13.354)	(14.475)

The agreement is now quite satisfactory.

Problem 310. Obtain with the help of the corrected formula the values of $J_5(4.5)$ and $J_5(5.5)$ and likewise the values of $J_{10}(9)$, $J_{10}(9.5)$, $J_{10}(10.5)$, $J_{10}(11)$.
[Answer:

$$J_5(4.5) = 0.1993 \qquad J_5(5.5) = 0.3118$$
$$(0.1947) \qquad\qquad (0.3209)$$

$$J_{10}(9) = 0.1262, \quad J_{10}(9.5) = 0.1664, \quad J_{10}(10.5) = 0.2493, \quad J_{10}(11) = 0.2817$$
$$(0.1247) \qquad\qquad (0.1650) \qquad\qquad (0.2477) \qquad\qquad (0.2804)\;]$$

7.26. Eigensolutions reducible to the hypergeometric series

The differential operator associated with the hypergeometric series (2.7) is given as follows:

$$Dv = x(1 - x)v'' + (a - bx)v' \qquad (7.26.1)$$

where we have put

$$a = \gamma, \quad b = \alpha + \beta + 1 \qquad (7.26.2)$$

The factor $A(x)$ of $v''(x)$ *vanishes* at the two points $x = 0$ and 1. These points are thus natural boundary points of the operator which limits the range of x to $[0, 1]$. According to the general theory the weight factor $w(x)$ of our operator becomes (cf. 4.7)

$$w(x) = \frac{1}{x(1 - x)} e^{\int [(a-bx)/x(1-x)] dx}$$

$$= x^{a-1}(1 - x)^{b-a-1} \qquad (7.26.3)$$

The eigenvalue equation associated with our operator assumes the form

$$Dv + \lambda v = 0 \qquad (7.26.4)$$

(if we agree that the eigenvalues are denoted by $-\lambda$ in order to make λ positive). According to the differential equation of Gauss (cf. 2.6) this equation is solvable in terms of the hypergeometric function $F(\alpha, b - \alpha - 1, a; x)$ if the following identification is made:

$$\lambda = \alpha(\alpha + 1 - b) \qquad (7.26.5)$$

The boundary term of Green's identity associated with our operator becomes

$$\left[x^a (1 - x)^{b-a} (uv' - vu') \right]_{-1}^{+1} \qquad (7.26.6)$$

This term has the peculiarity that it vanishes without any imposed conditions on u and v, due to the vanishing of the first factor. In fact, however, this implies that $v(x)$ and $v'(x)$ remain *finite* at the points $x = \pm 1$. Since the points $x = \pm 1$ are *singular points* of our differential operator where the solution goes out of bound if no special precautions are taken, the very condition that $v(x)$ and $v'(x)$ must remain *finite* at the two end-points of the range, represents two homogeneous boundary conditions of our differential operator which selects its eigenvalues and eigenfunctions. In particular, the hypergeometric function $F(\alpha, \beta, \gamma; x)$ goes to infinity at the point $x = 1$, *except* in the special case that the series *terminates* automatically after a finite number of terms. This happens if the parameter α (or equally β but this does not give anything new since F is completely symmetric in α and β) is equated to a *negative integer* $-n$. Then the eigenvalues λ_n become (according to (5)):

$$\lambda_n = n(n + b - 1) \qquad (7.26.7)$$

Hence the eigenfunctions of our operator become simply *polynomials* of the order n:

$$v_n(x) = p_n(x) = cF(-n, n + b - 1, a; x) \qquad (7.26.8)$$

The integer n can assume all values between 0 and infinity:

$$n = 0, 1, 2, 3, \ldots \qquad (7.26.9)$$

The polynomials thus generated are called "Jacobi polynomials".

Certain choices of the parameters a and b of the operator (1) lead to particularly interesting function classes. For example the choice $a = 1$ brings about the weight factor

$$w(x) = (1 - x)^{b-2} \qquad (7.26.10)$$

This weighting has the peculiarity that it puts the emphasis on the neighbourhood of the origin $x = 0$ (for $b > 2$). With increasing b we obtain a weight factor which is practically exponential since for large b we obtain practically

$$w(x) = e^{-(b-2)x}$$

This kind of weighting occurs in the Laguerre polynomials which we will consider in Section 29.

Problem 311. By substituting in the differential equation (2.6) of Gauss

$$v(x) = x^{1-\gamma}v_1(x) \qquad (7.26.11)$$

show that a second solution is obtainable in the form

$$v(x) = x^{1-\gamma}F(\alpha + 1 - \gamma, \beta + 1 - \gamma, 2 - \gamma; x) \qquad (7.26.12)$$

Problem 312. Let in (10) b go to infinity like $1/\mu$ but at the same time transform x to μx_1. Show that in the limit the weight factor $w(x)$ becomes e^{-x}.

7.27. The ultraspherical polynomials

Let us put the origin of our reference system in the *middle* of the range by putting $x = (1/2) + \xi$. Now the weight factor (26.3) changes to

$$(\tfrac{1}{2} + \xi)^{a-1}(\tfrac{1}{2} - \xi)^{b-a-1} \qquad (7.27.1)$$

A particularly interesting mode of weighting arises if we maintain *left–right symmetry*. This demands the condition

$$a - 1 = b - a - 1$$

which means

$$b = 2a \qquad (7.27.2)$$

or in terms of the original parameters (26.2):

$$\alpha + \beta + 1 = 2\gamma \qquad (7.27.3)$$

It is more convenient to change the range $\pm \tfrac{1}{2}$ of ξ to ± 1 and that means that our original x is transformed into

$$x_1 = 1 - 2x \qquad (7.27.4)$$

Now the operator (1) becomes in the new variable and under the condition (3):

$$Dv = (1 - x^2)v'' - 2\gamma xv' \qquad (7.27.5)$$

The weight factor $w(x)$ is now

$$w(x) = (1 - x^2)^{\gamma - 1} \qquad (7.27.6)$$

and the eigensolutions are given by the polynomials

$$P_n^{(\gamma)}(x) = cF(-n, n + 2\gamma - 1, \gamma; x) \qquad (7.27.7)$$

with the eigenvalues

$$\lambda_n = n(n + 2\gamma - 1) \qquad (7.27.8)$$

This special class of Jacobi polynomials received the name "ultraspherical".

These polynomials are related to each other by many interesting analytical properties but from the standpoint of applications two special cases are of particular importance. The one corresponds to the choice $\gamma = \frac{1}{2}$ (and thus the weight factor $(1 - x^2)^{-1/2}$):

$$T_n(x) = F\left(-n, n, \tfrac{1}{2}; \frac{1 - x}{2}\right) \qquad (7.27.9)$$

They are called "Chebyshev polynomials". They have many outstanding advantages in approximation problems* and have the further advantage that they are representable in terms of elementary functions. If we put $x = \cos \theta$, these polynomials become simply

$$T_n(x) = \cos n\theta \qquad (7.27.10)$$

thus establishing a relation between polynomial expansions and the Fourier cosine series.

The other fundamentally important choice is $\gamma = 1$, in which case the weight factor $w(x)$ becomes 1 and we obtain a set of polynomials which are orthogonal in the range ± 1 *without any weighting*. They are called "Legendre polynomials"

$$P_n(x) = F\left(-n, n + 1, 1; \frac{1 - x}{2}\right) \qquad (7.27.11)$$

These polynomials have important applications in potential theory, in least square problems, in Gaussian quadrature, in statistical investigations, and hence deserve special attention. We will apply the KWB method to the defining differential equation and thus obtain an approximation of these important functions which sheds interesting light on their general analytic behaviour. The fact that these functions are obtainable in polynomial form, does not mean that we can handle them easily since a polynomial of high or even medium order is not directly amenable to analytical studies.

* Cf. A. A., Chapter 7.

7.28. The Legendre polynomials

We shall not assume in advance that the eigenvalues of Legendre's differential equation are given. We shall prefer to consider the equation in the form

$$\frac{d}{dx}[(1 - x^2)v'(x)] + \alpha^2 v(x) = 0 \qquad (7.28.1)$$

(replacing the undetermined constant λ by α^2) and find the eigenvalues in the course of our investigation. Our first aim will be to transform our equation into the normal form in which the first derivative is missing. We could do that by splitting away a proper factor as we have seen it in the discussion of Hermite's differential equation and likewise in dealing with Bessel's differential equation. However, we will prefer to follow a slightly different course. Instead of splitting away a proper factor we want to *change the independent variable*. Let us multiply the entire equation by $(1 - x^2)$:

$$(1 - x^2)\frac{d}{dx}[(1 - x^2)v'(x)] + \alpha^2(1 - x^2)v(x) = 0 \qquad (7.28.2)$$

If we now introduce a new variable ξ by putting

$$\frac{dx}{1 - x^2} = d\xi$$

$$\xi = \tfrac{1}{2}\log\frac{1 + x}{1 - x} \qquad (7.28.3)$$

then our equation in ξ becomes

$$\frac{d^2 v}{d\xi^2} + \alpha^2(1 - x^2)v = 0 \qquad (7.28.4)$$

It will be convenient to introduce an additional angle variable θ by putting

$$x = \cos\theta$$

Then

$$e^\xi = \cot\frac{\theta}{2} \qquad (7.28.5)$$

Moreover,

$$1 - x^2 = \frac{4e^{2\xi}}{(e^{2\xi} + 1)^2} \qquad (7.28.6)$$

$$\int \sqrt{U}\,d\xi = 2\alpha\int\frac{e^\xi d\xi}{e^{2\xi} + 1} = -\alpha\int d\theta$$

$$= -\alpha\theta + \beta \qquad (7.28.7)$$

The constant β can be determined if we consider that an increase of x from 0 to 1 means that θ decreases from $\pi/2$ to 0. If we want to integrate from

$x = 0$, we have to put $\beta = \alpha(\pi/2)$ and we obtain our approximate solution in the form

$$\frac{c}{\sqrt{\sin \theta}} \left.\begin{matrix} \cos \\ \sin \end{matrix}\right\} \alpha\left(\frac{\pi}{2} - \theta\right) \qquad (7.28.8)$$

where the constant c is arbitrary and the cosine has to be chosen if an *even*, the sine if an *odd* function of x is involved.

Up to now we see no reason why α should be restricted to some exceptional values. On the other hand, we do not know yet, how our solution will behave if we approach the critical point $\theta = 0$. Here our approximation goes out of hand because we come to the point x_0 in which $U(x_0)$ vanishes. Generally this point causes no difficulty and leads to no singularity. We merely replace our solution by the substitute functions $\varphi(x)$ which we have studied in the previous sections. At present, however, this method is not available since the vanishing of

$$U(\xi) = \left(\frac{2\alpha e^{\xi}}{e^{2\xi} + 1}\right)^2 \qquad (7.28.9)$$

occurs at the point $\xi = \infty$ and thus recedes to infinity. We have to find some other method to study our solution as ξ becomes large.

Now for large ξ we have practically

$$U(\xi) = 4\alpha^2 e^{-2\xi} \qquad (7.28.10)$$

since the 1 becomes negligible in the denominator of (9) if ξ grows large. Let us then put

$$e^{-\xi} = t$$

$$\frac{d}{d\xi} = -e^{-\xi}\frac{d}{dt}$$

$$\frac{d^2}{d\xi^2} = e^{-2\xi}\frac{d^2}{dt^2} + e^{-\xi}\frac{d}{dt}$$

Hence in the new variable t our differential equation (for large ξ) becomes

$$t^2 v'' + t v' + 4\alpha^2 t^2 v = 0$$
$$v'' + \frac{v'}{t} + 4\alpha^2 v = 0 \qquad (7.28.11)$$

But this is *Bessel's differential equation* for the order $n = 0$, taken at the point $2\alpha t$ (cf. 2.20):

$$v(t) = J_0(2\alpha t)$$

$$= J_0(2\alpha e^{-\xi}) = J_0\left(2\alpha \tan \frac{\theta}{2}\right) \qquad (7.28.12)$$

Generally a differential equation of second order has *two* solutions and in actual fact the differential equation (11) has a second solution $N_0(t)$ which, however, goes logarithmically to infinity at $t = 0$, i.e. $\xi = \infty$. The demand that our solution shall remain *finite* at $x = 1$—i.e. $\xi = \infty$—*excludes* the second solution and restricts the function $v(t)$ to $J_0(2\alpha t)$, except for a factor of proportionality. But this factor is in our case equal to 1 because the Legendre polynomials $P_n(x)$ are normalised by the condition

$$P_n(1) = 1 \tag{7.28.13}$$

This now means that $v(0)$ must become 1, but then the factor of proportionality of J_0 must become 1 in view of the fact that $J_p(x)$ starts with the value

$$\frac{1}{p!} \left(\frac{x}{2}\right)^p$$

which becomes 1 for the case of $J_0(x)$.

Now $J_0(x)$ assumes for an x which is not too small, the asymptotic law

$$\frac{\cos\left(x - \frac{\pi}{4}\right)}{\sqrt{\frac{\pi}{2} x}} \tag{7.28.14}$$

which in our case becomes

$$\frac{\cos\left(2\alpha \tan\frac{\theta}{2} - \frac{\pi}{4}\right)}{\sqrt{\frac{\pi}{2} 2\alpha \tan\frac{\theta}{2}}}$$

Since for small values of θ the tangent becomes practically the angle itself, we obtain for small θ:

$$v(\theta) = \frac{\cos\left(\alpha\theta - \frac{\pi}{4}\right)}{\sqrt{\frac{\pi}{2} \alpha\theta}} \tag{7.28.15}$$

This solution can be linked to the solution (8) which is valid in the domain of larger θ. Let us first assume the case of an *even* function. Then the matching of the cosine factors demands the condition

$$\alpha\theta - \alpha\frac{\pi}{2} = \alpha\theta - \frac{\pi}{4} - k\pi \tag{7.28.16}$$

where k is an arbitrary integer. This gives

$$\alpha = 2k + \tfrac{1}{2} \tag{7.28.17}$$

The constant c is then determined to

$$c = \frac{(-1)^k}{\sqrt{\frac{2}{\pi} \alpha}} \tag{7.28.18}$$

Let us now assume the case of an *odd* function. The solution can now be written in the form

$$\frac{c \cos\left(\frac{\pi}{2} + \alpha\theta - \alpha\frac{\pi}{2}\right)}{\sqrt{\sin\theta}}$$

and the matching of the cosine factor yields the condition

$$\frac{\pi}{2} - \alpha\frac{\pi}{2} = -\frac{\pi}{4} - k\pi \qquad (7.28.19)$$

which gives

$$\alpha = \tfrac{3}{2} + 2k = 2k + 1 + \tfrac{1}{2} \qquad (7.28.20)$$

while c becomes once more

$$c = \frac{(-1)^k}{\sqrt{\frac{\pi}{2}\alpha}} \qquad (7.28.21)$$

The case of both even and odd functions is included in the selection rule

$$\alpha = n + \tfrac{1}{2} \qquad (7.28.22)$$

where n is an integer, with the understanding that all *even* n demand the choice *cosine* and all *odd* n the choice *sine* in the general expression (8).

The determination of the eigenvalues α according to (22) is not exact but *very close*. The exact law for α is

$$\lambda = \alpha^2 = n(n+1) = (n+\tfrac{1}{2})^2 - \tfrac{1}{4}$$
$$\alpha = \sqrt{(n+\tfrac{1}{2})^2 - \tfrac{1}{4}} \qquad (7.28.23)$$

Hence the approximation (22) is quite satisfactory.

Of particular interest is the distribution of the *zeros* of $P_n(x)$. In order to study this distribution and make a comparison with the zeros of the Chebyshev polynomials, it will be appropriate to change θ to the complementary angle φ by putting

$$\frac{\pi}{2} - \theta = \varphi$$
$$x = \sin\varphi \qquad (7.28.24)$$

Then the expression for the polynomials of even order becomes

$$P_{2\mu}(x) = \frac{(-1)^\mu}{\sqrt{\frac{\pi}{2}\left(2\mu + \frac{1}{2}\right)\cos\varphi}} \cos\left(2\mu + \frac{1}{2}\right)\varphi \qquad (7.28.25)$$

while the polynomials of odd order become

$$P_{2\mu+1} = \frac{(-1)^\mu}{\sqrt{\frac{\pi}{2}\left(2\mu + \frac{3}{2}\right)\cos\varphi}} \sin\left(2\mu + \frac{3}{2}\right)\varphi \qquad (7.28.26)$$

The corresponding expressions for the Chebyshev polynomials become (not in approximation but exactly):

$$T_{2\mu}(x) = (-1)^\mu \cos 2\mu\varphi$$
$$T_{2\mu+1}(x) = (-1)^\mu \sin (2\mu + 1)\varphi$$

(7.28.27)

The zeros of the Chebyshev polynomials of *odd* order follow the law

$$x_k = \pm \sin \frac{k\pi}{2\mu + 1} \qquad (k = 0, 1, 2, \ldots, \mu) \qquad (7.28.28)$$

while the Gaussian zeros follow (in close approximation) the law

$$x_k = \pm \sin \frac{k\pi}{2\mu + \frac{3}{2}} \qquad (k = 0, 1, 2, \ldots, \mu) \qquad (7.28.29)$$

The difference is that in the Chebyshev case the full circle is divided into $4\mu + 2 = 2n$ equal parts and the points projected down on the diameter. In the Gaussian case the full circle is divided into $4\mu + 3 = 2n + 1$ equal parts and again the points projected down on the diameter.

In the case of the polynomials of *even* order the zeros of the Chebyshev polynomials become

$$x_k = \pm \sin \frac{\pi}{4\mu} (1 + 2k) \qquad (k = 0, 1, 2, \ldots, n - 1) \qquad (7.28.30)$$

while the Gaussian zeros become

$$x_k = \pm \sin \frac{\pi}{4\mu + 1} (1 + 2k) \qquad (k = 0, 1, 2, \ldots, n - 1) \qquad (7.28.31)$$

Now the *half circle* is divided into $4\mu = 2n$, respectively $4\mu + 1 = 2n + 1$ equal parts and the points are projected down on the diameter, but skipping all points of even order 2, 4, $\ldots$ and keeping only the points of the order $1, 3, 5, \ldots, n - 1$.

The asymptotic law of the zeros is remarkably well represented even for small n. We obtain for example for $n = 4$, 5, 6, 7, 8 the following distribution of zeros, as compared with the exact Gaussian zeros:

$n = 4$		$n = 5$	
0.3420	(0.3400)	0.5406	(0.5385)
0.8660	(0.8611)	0.9096	(0.9062)

$n = 6$		$n = 7$		$n = 8$	
0.2393	(0.2386)	0.4067	(0.4058)	0.1837	(0.1834)
0.6631	(0.6612)	0.7431	(0.7415)	0.5265	(0.5255)
0.9350	(0.9325)	0.9510	(0.9491)	0.7980	(0.7967)
				0.9618	(0.9608)

In order to test the law of the amplitudes we will substitute $x = 0$. Then we obtain for the polynomials of even order $n = 2\mu$ the following starting values:

$$P_n(0) = \frac{(-1)^\mu}{\sqrt{\pi(\mu + \frac{1}{4})}} \qquad (7.28.32)$$

while the derivative of the polynomials of odd order $n = 2\mu + 1$ at the point $x = 0$ becomes:

$$P'_n(0) = (-1)^\mu \sqrt{\frac{4\mu + 3}{\pi}} \qquad (7.28.33)$$

The Legendre polynomials have the property that they are derivable from a "generating function" in the following way:

$$P\alpha(x) = \frac{1}{2^n \cdot n!} (x^2 - 1)^{(n)} \qquad (7.28.34)$$

From this property of $P_n(x)$ we derive the following exact values of $P_{2\mu}(0)$ and $P'_{2\mu+1}(0)$:

$$P_{2\mu}(0) = \frac{(-1)^\mu}{2^{2\mu}} \binom{2\mu}{\mu}$$

$$P'_{2\mu+1}(0) = \frac{(-1)^\mu}{2^{2(\mu+1)}} \binom{2\mu + 2}{\mu + 1} 2(\mu + 1) \qquad (7.28.35)$$

If we make use of Stirling's formula (22.19) which approximates $n!$ remarkably well even in the realm of fairly small n, we obtain

$$\frac{1}{2^\mu} \binom{2\mu}{\mu} = \frac{1}{\sqrt{\pi\mu}} \qquad (7.28.36)$$

Within this accuracy the value of $P_{2\mu}(0)$ coincides with that given in (32), except that $\mu + \frac{1}{4}$ is replaced by μ, while in the case of $P'_{2\mu+1}(0)$ we find that $4\mu + 3$ is replaced by $4\mu + 4$.

The numerical comparison shows that in the realm of $n = 5$ to $n = 10$ we obtain on the basis of the asymptotic formula the following initial values, respectively initial derivatives (the numbers in parenthesis give the corresponding exact values):

	$n = 6$	$n = 8$	$n = 10$
$P_n(0) =$	-0.3130	0.2737	-0.2462
	(-0.3125)	(0.2734)	(-0.2461)

	$n = 5$	$n = 7$	$n = 9$
$P'_n(0) =$	1.8712	-2.1851	2.4592
	(1.8750)	(-2.1875)	(2.4609)

We see that the asymptotic law gives very good results even in the realm of small n (starting from $n = 5$).

7.29. The Laguerre polynomials

Another eminently important class of polynomials which is closely related to the theory of the Laplace transform, was introduced in analytical research by the French mathematician Laguerre. The Laguerre polynomials are associated with the range $x = [0, \infty]$ and they are orthogonal with respect to the weight factor e^{-x}. The Laguerre *functions* are thus the functions

$$\Psi_k(x) = e^{-x/2}L_k(x) \qquad (7.29.1)$$

They are orthogonal in themselves, without any weight factor:

$$\int_0^\infty \Psi_i(x)\Psi_k(x)dx = 0 \qquad (i \neq k) \qquad (7.29.2)$$

Moreover, they have the remarkable property that their "norm" becomes 1:

$$\int_0^\infty \Psi_k{}^2(x)dx = 1 \qquad (7.29.3)$$

if we prescribe the following simple initial condition for the polynomials $L_k(x)$:

$$L_k(0) = 1 \qquad (7.29.4)$$

We have encountered these polynomials earlier, in Chapter 1, when dealing with the Gregory-Newton type of equidistant interpolation, on account of their outstanding properties in the calculus of finite differences. At that time we could draw our conclusions without any reference to the differential equation that these polynomials satisfy. In our present discussions we consider them as a special case of the eigenfunctions which arise from the hypergeometric series and will carry through the same kind of treatment which we applied to a close approximation of the other kinds of hypergeometric polynomials.

The Laguerre polynomials are in many respects counterpart to the Legendre polynomials. Any given *finite* range $x = [a, b]$ can be normalised to the range $[-1, +1]$ by a simple linear transformation of x. For an *infinite* range, however, such a transformation is not possible. It is still possible, however, to change any range $x = [a, \infty]$ into the range $[0, \infty]$ by a proper linear transformation of x. Since the powers of x are unbounded and grow to infinity if x approaches infinity, we cannot use them for approximation purposes without a proper weight factor which will cut down the contribution from the domain which is very far out. The most natural weighting for this purpose is the factor $e^{-\alpha x}$ and since α can always be normalised to 1 by a proper scaling of x, we come automatically to the definition of the Laguerre polynomials and the associated Laguerre functions (1). We have seen in (7.14) that the function

$$v(x) = \sqrt{x}\Psi_n(x) \qquad (7.29.5)$$

satisfies the following differential equation:

$$v'' + \left(-\frac{1}{4} + \frac{2n + 1}{2x} + \frac{1}{4x^2}\right)v = 0 \qquad (7.29.6)$$

We prefer once more to write this equation in the form of an eigenvalue problem:

$$v'' + \left(-\frac{1}{4} + \frac{1}{4x^2} + \frac{\lambda}{2x}\right)v = 0 \qquad (7.29.7)$$

leaving the eigenvalue λ for the first undetermined.

In the present problem

$$U(x) = \frac{1}{4x^2} + \frac{\lambda}{2x} - \frac{1}{4} \qquad (7.29.8)$$

and we see that for small values of x we are in the *periodic*, for large values in the *exponential* domain. The dividing point $U(x_0) = 0$ is determined by the root of a quadratic equation. We will simplify our task, however, by the observation that for a sufficiently large λ the first term of $U(x)$ will quickly become negligible. We fare better if we do not neglect the first term but combine it with the second term in the form

$$\frac{\lambda}{2x - \mu}, \quad \mu = \frac{1}{\lambda}$$

Then, if x is not too small, we have practically

$$\frac{\lambda}{2x\left(1 - \frac{1}{2\lambda x}\right)} = \frac{\lambda}{2x}\left(1 + \frac{1}{2\lambda x}\right) = \frac{\lambda}{2x} + \frac{1}{4x^2} \qquad (7.29.9)$$

and we can consider

$$U(x) = \frac{\lambda}{2x - \mu} - \frac{1}{4} \qquad (7.29.10)$$

as a sufficiently close representation of our problem, except in the realm of very small x. Here, however, another simplification takes place. In the range of very small x the constant term $\frac{1}{4}$ becomes negligible and we can put

$$U(x) = \frac{1}{4x^2} + \frac{\lambda}{2x} \qquad (7.29.11)$$

But then we have a differential equation which belongs to a recognisable class. It is of the form (2.23) which is solvable in terms of Bessel functions. In order to reduce (11) to the form (2.23), the following choice of the constants has to be made:

$$\gamma = \frac{1}{2}, \quad 2\alpha - 2 = -1, \quad \gamma^2 - \alpha^2 p^2 = \frac{1}{4}, \quad \alpha^2\beta^2 = \frac{\gamma}{2}$$

This gives

$$\alpha = \tfrac{1}{2}, \quad p = 0, \quad \gamma = \tfrac{1}{2}, \quad \beta = \sqrt{2\gamma} \qquad (7.29.12)$$

and our solution becomes

$$v(x) = \sqrt{x}J_0(\sqrt{2\lambda x}) \qquad (7.29.13)$$

The second solution, $\sqrt{x}N_0(\sqrt{\lambda x})$, has to be rejected since we know that $x^{-1/2}v(x)$ must remain finite at $x = 0$. Moreover, the solution (13) is already properly normalised in view of the condition (4) which holds for all Laguerre polynomials. We have thus a situation similar to that encountered in the study of the Legendre polynomials around the point $x = 1$.

We now come to the $U(x)$ given by (10) and the KWB approximation associated with it. We will put

$$x = \frac{\mu}{2} + 2\lambda \sin^2 \varphi \qquad (7.29.14)$$

$$U(x) = \tfrac{1}{4} \cot^2 \varphi \qquad (7.29.15)$$

Then

$$\int \sqrt{U}\,dx = 2\lambda \int \cos^2 \varphi\,d\varphi = \lambda\!\left(\varphi + \frac{\sin 2\varphi}{2}\right) \qquad (7.29.16)$$

and the KWB solution becomes

$$v(x) = C\sqrt{2\tan\varphi}\,\cos\left[\lambda\!\left(\varphi + \frac{\sin 2\varphi}{2}\right) - \delta\right] \qquad (7.29.17)$$

with the two constants of integration C and δ.

On the other hand, the asymptotic law of the Bessel functions gives on account of (13):

$$v(x) = \sqrt{\frac{2x}{\pi}}\,\frac{\cos\left(\sqrt{2\lambda x} - \dfrac{\pi}{4}\right)}{\sqrt[4]{2\lambda x}} \qquad (7.29.18)$$

and we have to link the solution (17) to (18) for sufficiently small values of x.

Now small values of x mean small values of φ. But in the realm of small φ the argument of the cosine function in (17) is replaceable by $2\lambda\varphi - \delta$ and this quantity becomes, if we go back to the original variable x on the basis of (14)—neglecting in this realm the small constant $\mu/2$:

$$2\lambda\varphi - \delta = \sqrt{2\lambda x} - \delta$$

The comparison with (17) shows that we must put

$$\delta = \frac{\pi}{4} \qquad (7.29.19)$$

On the other hand, the amplitude factor becomes according to (18):

$$\sqrt{\frac{\pi}{2}}\,\sqrt[4]{\frac{x}{2\lambda}} = \sqrt{\frac{\pi}{2}}\,\sqrt{\sin\varphi}$$

This determines the factor C of (17) to

$$C = \frac{1}{\sqrt{\pi}} \qquad (7.29.20)$$

and we obtain the solution

$$v(x) = \sqrt{\frac{2 \tan \varphi}{\pi}} \cos \left[\lambda \left(\varphi + \frac{\sin 2\varphi}{2} \right) - \frac{\pi}{4} \right] \qquad (7.29.21)$$

Now the transition point x_0 at which $U(x_0)$ becomes zero, belongs to the value $\varphi = \pi/2$. Hence the phase angle θ becomes, according to (21.7):

$$\theta = \frac{\pi}{4} - \lambda \frac{\pi}{2} \qquad (7.29.22)$$

The transition into the exponential domain must be such that the branch with the positive exponent *vanishes* since otherwise our solution would go exponentially to infinity which is prohibited since we know that our solution must go exponentially to zero as x grows to infinity. This means $A_1 = 0$ and we see from (19.5) that this condition demands

$$\theta = -\frac{\pi}{4} - n\pi \qquad (7.29.23)$$

where n is an arbitrary integer. The comparison of (22) and (23) gives the selection rule

$$\lambda = 2n + 1 \qquad (7.29.24)$$

and this is the *exact eigenvalue of Laguerre's differential equation* (6). Once more, as in the solution of Hermite's differential equation, the KWB method leads to an *exact* determination of the eigenvalue which is generally not to be expected, in view of the *approximate* nature of our integration procedure.

The constant of the negative branch of the exponential domain becomes, according to (19.5):

$$A_2 = \frac{(-1)^n}{2\sqrt{\pi}} \qquad (7.29.25)$$

The transition occurs at the point

$$x_0 = 2\lambda + \frac{1}{2\lambda} = 4n + 2 + \frac{1}{4n + 2} \qquad (7.29.26)$$

For the exponential range we will make the substitution

$$x = \frac{\mu}{2} + 2\lambda \cosh^2 \varphi$$

$$U(x) = -\tfrac{1}{4} \coth^2 \varphi \qquad (7.29.27)$$

The solution in the exponential range becomes

$$v(x) = (-1)^n \sqrt{\frac{1}{2\pi}} \coth \varphi \; e^{-\lambda[(\sinh 2\varphi)/2 - \varphi]} \qquad (7.29.28)$$

In order to obtain the Laguerre functions $\Psi_n(x)$, we have to divide by $\sqrt{x}$. The final solution can thus be written down as follows:

$$x < x_0:$$

$$\Psi_n(x) = \sqrt{\frac{2 \tan \varphi}{\pi x}} \cos \left[(2n + 1)\left(\varphi + \frac{\sin 2\varphi}{2} \right) - \frac{\pi}{4} \right] \qquad (7.29.29)$$

$$x > x_0:$$

$$\Psi_n(x) = (-1)^n \sqrt{\frac{\coth \varphi}{2\pi x}} \; e^{-(2n+1)[(\sinh 2\varphi)/2 - \varphi]}$$

Problem 313.   Obtain the value of $\Psi_n(x)$ at the transition point x_0.
[Answer:

$$\Psi_n(x_0) = (-1)^n \cdot 0.50207 \; \frac{1}{\sqrt{x_0}} \sqrt[6]{2n + 1}, \quad x_0 = 4n + 2 + \frac{1}{4n + 2} \qquad (7.29.30)]$$

Problem 314.   Obtain the position of the last zero of $\Psi_n(x)$.
[Answer:

$$x = x_0 - 4.6762 \sqrt[3]{2n + 1} + \frac{2.1867}{\sqrt[3]{2n + 1}} \qquad (7.29.31)]$$

7.30. The exact amplitude equation

The representation of the solution of a second-order linear differential equation in terms of an oscillation of variable frequency and variable amplitude was based on a definite approximation procedure, called the KWB method. We may ask whether this separation into a purely oscillatory part and a variable amplitude part could not be put on a *rigorous* basis, omitting all reference to an approximation. This is indeed possible and can be achieved as follows. We start again with the normal form of our differential equation:

$$v'' + Uv = 0 \qquad (7.30.1)$$

(omitting any specific reference to the argument x which we take for granted). We shall take out a proper amplitude factor $C(x)$ by putting

$$v(x) = C(x)v_1(x) \qquad (7.30.2)$$

Then

$$v'' = C''v_1 + 2C'v'_1 + Cv''_1$$

and our differential equation becomes, if we divide through by C:

$$v''_1 + 2\frac{C'}{C} v'_1 + \left(\frac{C''}{C} + U \right)v_1 = 0 \qquad (7.30.3)$$

Let us put

$$U + \frac{C''}{C} = U_1 \qquad (7.30.4)$$

We will now again make use of the exponential transformation

$$v_1 = e^\varphi, \qquad \varphi' = y$$
$$v'_1 = y e^\varphi$$
$$v''_1 = (y' + y^2) e^\varphi \qquad (7.30.5)$$

obtaining once more a Riccati type of differential equation:

$$y' + y^2 + 2 \frac{C'}{C} y + U_1 = 0 \qquad (7.30.6)$$

But now we can assume that φ is *purely imaginary* since we have split away the amplitude $C(x)$ and what remains is a *pure oscillation with constant amplitude.* But then we can put

$$y = i y_1 \qquad (7.30.7)$$

considering y_1 as a real quantity. Then the differential equation (6) separates into a real and imaginary part and splits into the two equations

$$U_1 - y_1^2 = 0$$
$$y'_1 + 2 \frac{C'}{C} y_1 = 0 \qquad (7.30.8)$$

The second equation is integrable in the form

$$\frac{C'}{C} = -\frac{y'_1}{2y_1} = -\frac{U'_1}{4U_1}$$
$$C = \frac{c_0}{\sqrt[4]{U_1}} \qquad (7.30.9)$$

where c_0 is an arbitrary constant.

We have thus obtained the following solution:

$$v(x) = \frac{c_0}{\sqrt[4]{U_1}} e^{i\int \sqrt{U_1} dx} \qquad (7.30.10)$$

which we may write in the real form

$$v(x) = \frac{c_0}{\sqrt[4]{U_1}} \cos\left(\int \sqrt{U_1} dx - \theta\right) \qquad (7.30.11)$$

with the two constants of integration c_0 and θ

We recognise the solution (11) once more as tne KWB solution of the given differential equation, if we are in the periodic range $U_1 > 0$. The difference is, however, that the function $U(x)$ is replaced by $U_1(x)$, defined according to (4). The solution (11) is no longer an approximate but an *exact* solution of the given problem. In order to obtain this solution we must first obtain the function $C(x)$ by solving the non-linear second-order differential equation

$$U + \frac{C''}{C} = \left(\frac{c_0}{C}\right)^4 \qquad (7.30.12)$$

The customary KWB approximation results if we neglect C''/C in comparison to U. But if we do not neglect this term, we obtain a tool by which we can study with complete accuracy the change of the amplitude of our vibration in a given limited range. Once more we see that the separation of our solution into amplitude and frequency is not unique. The differential equation (12) allows the prescription of the initial value of C and C' at the point of start. In addition we have the value of c_0 at our disposal but this constant is irrelevant since it can be absorbed into the initial value of C. Hence we can put $c_0 = 1$ without loss of generality. But the initial value of C and C' is still arbitrary.

While it will generally not be possible to solve the differential equation (12), we may draw important conclusions from it concerning the law according to which the amplitude $C(x)$ changes. We may want to know for example whether the ratio of a maximum amplitude to the next maximum is smaller or larger than 1. For this purpose we can start at a point x_1 at which the cosine factor of the solution (11) assumes its maximum value 1. This means that the integration of $\sqrt{U_1(x)}$ will start from that particular point $x = x_1$:

$$v(x) = \frac{1}{\sqrt[4]{U_1(x)}} \cos \int_{x_1}^{x} \sqrt{U} d\xi$$

We insure a maximum of $v(x)$ by choosing the initial values of $C(x)$ as follows:

$$C(x_1) = \frac{1}{\sqrt[4]{U(x_1)}}$$

$$C'(x_1) = 0$$

Then also $C''(x_1) = 0$ and we see that $C(x)$ will change *very slowly* in this neighbourhood. We may be able to proceed to the next maximum on the basis of a local expansion of a few terms and estimate the increase or decrease of the maximum without actually integrating the amplitude equation (12).

Problem 315. In the discussion of the differential equation (9.15) of the vibrating spring a solution was obtained which could be interpreted as an oscillation of variable amplitude $C(x)$ and frequency $\beta(x)$. Show that the variable amplitude $[\beta(x)]^{-1/2}$ satisfies the exact amplitude equation (12).

7.31. Sturm-Liouville problems and the calculus of variations

We have seen that all linear second-order differential operators are self-adjoint if the proper weight factor is applied. On the other hand, self-adjoint differential equations are derivable from a variational principle. This means that the given differential equation can be conceived as the solution of the problem of making a certain definite integral to a minimum, or at least to a stationary value. The existence of such a minimum principle is of great advantage because it often enables us to establish the existence of a minimum by tools which do not require the solution of a differential equation. Furthermore, the actual numerical solution of a differential equation is frequently greatly facilitated by trying to minimise the variational integral *directly*, paying no attention to the differential equation which is the consequence of the variational problem.

In the case of a second-order operator the integrand of the variational principle is particularly simple. We have here

$$Q = \int_a^b L dx, \quad L = \tfrac{1}{2}[A(x)\dot{v}^2 - C(x)v^2] + \beta(x)v \qquad (7.31.1)$$

Making use of the standard technique discussed in Chapter 5.9 (see particularly the second equation of the system (5.9.4)), we arrive at the differential equation

$$\frac{d}{dx}(Av') + Cv = \beta \qquad (7.31.2)$$

which coincides with the standard form (4.1) of a self-adjoint second-order differential equation.

This in itself, however, does not establish the full solution of the given variational problem. During the process an integration by parts takes place which gives rise to the following boundary term:

$$A(b)v'(b)\delta v(b) - A(a)v'(a)\delta v(a) \qquad (7.31.3)$$

The differential equation (2) is a necessary but not sufficient condition for the stationary value of the variational integral. A further condition is required by the vanishing of the boundary term. This establishes the *boundary conditions* without which our problem is not fully determined.

Now it is possible that our variational problem is constrained by certain conditions which have to be observed during the process of variation. For example we may demand the minimum of the integral

$$Q = \int_a^b L dx \qquad (7.31.4)$$

but under the restricting condition that some definite boundary values are prescribed for $v(x)$:

$$v(a) = \alpha, \quad v(b) = \beta \qquad (7.31.5)$$

In this case the variation of $v(x)$ *vanishes* on the boundary:

$$\delta v(a) = \delta v(b) = 0 \qquad (7.31.6)$$

because $v(x)$ remains fixed at the two end-points during the process of variation. Then the boundary term (3) vanishes automatically and we get no further boundary conditions through the variational procedure.

It is equally possible, however, that we have *no* prescribed boundary conditions, that is, the variational integral (4) is free of further constraints. In that case $\delta v(x)$ is not constrained on the boundary but can be chosen freely. Here the vanishing of the boundary term (3) introduces two "natural boundary conditions" (caused by the variational principle itself, without outside interference), namely—assuming that $A(x)$ does not vanish on the boundary:

$$v'(b) = v'(a) = 0 \qquad (7.31.7)$$

We may also consider the case that the problem is *partly constrained*, by prescribing *one* boundary condition, for example

$$v(a) = \alpha \qquad (7.31.8)$$

Then $\delta v(x)$ vanishes on the lower boundary but is free on the upper boundary, and the variational principle provides the second boundary condition in the form

$$v'(b) = 0 \qquad (7.31.9)$$

In Section 4 we dealt in full generality with the question of self-adjoint boundary conditions and came to the conclusion that a wide class of boundary conditions satisfy the condition of self-adjointness. Assuming any two linear relations between the four quantities $v(b)$, $v'(b)$, $v(a)$, $v'(a)$, we could eliminate $v(b)$ and $v'(b)$ in terms of $v(a)$, $v'(a)$; then the condition (4.10) demanded only *one single condition* between the four constants of these relations. In view of the equivalence between self-adjointness and existence of a variational principle we must be able to prove the same result on a variational basis.

In Chapter 5 we have found that any variational problem can be transformed into the Hamiltonian normal or "canonical" form (cf. particularly Chapter 5.10). In our problem we bring about the canonical form by adding $v'(x)$ as surplus variable $w(x)$, by the auxiliary condition

$$v'(x) - w(x) = 0 \qquad (7.31.10)$$

Then our variational integrand becomes

$$L = \tfrac{1}{2}(Aw^2 - Cv^2) + \beta v \qquad (7.31.11)$$

with the auxiliary condition (10). This condition can be united, however, with the Lagrangian by the method of the Lagrangian multiplier (cf. Chapter 5.9), which we want to denote by p:

$$L' = \tfrac{1}{2}(Aw^2 - Cv^2) + p(v' - w) + \beta v \qquad (7.31.12)$$

This adds to our original problem the two new variables p and w, but w is purely algebraic and can be eliminated, by the equation

$$\frac{\partial L'}{\partial w} = Aw - p = 0 \tag{7.31.13}$$

This yields

$$w = \frac{p}{A} \tag{7.31.14}$$

and

$$L' = pv' - H \tag{7.31.15}$$

where

$$H = \frac{1}{2}\left(\frac{p^2}{A} + Cv^2\right) - \beta v \tag{7.31.16}$$

The resulting canonical equations

$$\begin{aligned} v' &= \frac{\partial H}{\partial p} = \frac{p}{A} \\ p' &= -\frac{\partial H}{\partial v} = -Cv + \beta \end{aligned} \tag{7.31.17}$$

are equivalent to the single equation (2).

These equations are not influenced, however, by adding to L' a complete derivative of an arbitrary function since such a term is equivalent to a mere *boundary term* in Q which has no effect on any inside point of the range. If we choose the quantity $-\frac{1}{2}(pv)'$ as such a term, our L' is changed to the equivalent

$$L_1 = \frac{1}{2}(pv' - vp') - H \tag{7.31.18}$$

With this L_1 the boundary term of the process of variation becomes

$$\frac{1}{2}\left[p\delta v - v\delta p\right]_a^b \tag{7.31.19}$$

and if we assume the boundary conditions

$$\begin{aligned} p(b) &= \alpha_1 p(a) + \alpha_2 v(a) + \gamma_1 \\ v(b) &= \beta_1 p(a) + \beta_2 v(a) + \gamma_2 \end{aligned} \tag{7.31.20}$$

(which demands the same conditions for the variations $\delta v(x)$ and $\delta p(x)$, omitting the constants γ_1 and γ_2 on the right side), we see that the boundary term (19) vanishes if the single condition

$$\alpha_1\beta_2 - \alpha_2\beta_1 = 1 \tag{7.31.21}$$

is fulfilled. Considering the relation $p = Av'$ we see that this result is in perfect agreement with our previous result, except for the notations.

The canonical equations (17) lead to an interesting conclusion. In view of these equations the Lagrangian L_1 becomes for the actual solution

$$L_1 = \frac{1}{2}\left(p\,\frac{\partial H}{\partial p} + v\,\frac{\partial H}{\partial v}\right) - H \qquad (7.31.22)$$

We will consider the case of the *homogeneous* differential equation and thus put $\beta(x) = 0$. Then the Hamiltonian function (16) becomes a homogeneous algebraic form of the second order in the variables p and v, and thus by Euler's formula of homogeneous forms:

$$H = \frac{1}{2}\left(p\,\frac{\partial H}{\partial p} + v\,\frac{\partial H}{\partial v}\right) \qquad (7.31.23)$$

which in view of (22) yields

$$L_1 = 0 \qquad (7.31.24)$$

and also

$$Q = Q_1 = \int_a^b L_1 dx = 0 \qquad (7.31.25)$$

Hence the variational integral becomes *zero* for the actual solution. It is worth remarking that this result is *independent of any boundary conditions*. It is a mere consequence of the canonical equations which in themselves do not guarantee a stationary value of Q.

Let us now consider the eigenvalue problem associated with our operator:

$$Dv - \lambda v = 0 \qquad (7.31.26)$$

(In the boundary conditions (20) we now have to replace γ_1 and γ_2 by zero.) This equation can be conceived as consequence of our variational principle if in our Lagrangian (1) we put $\beta(x) = 0$ and replace $C(x)$ by $C(x) + \lambda$. But another interpretation is equally possible. We normalise the solution $v(x)$ by the condition

$$\int_a^b v^2(x)dx = 1 \qquad (7.31.27)$$

which is considered as a given constraint during the process of variation, without changing our L. Then the Lagrangian multiplier method of Chapter 5.9 gives again a modification of L and, if the multiplier is denoted by $-\frac{1}{2}\lambda$, the new Lagrangian becomes

$$L' = \frac{1}{2}[A\dot{v}^2 - (C + \lambda)v^2] \qquad (7.31.28)$$

which is entirely equivalent to a modification of $C(x)$ by the constant λ. Now we make use of the result (25) which has shown that as the result of the variational equations the variational integral vanishes for the actual solution:

$$\int_a^b Ldx - \frac{\lambda}{2}\int_a^b v^2 dx = 0 \qquad (7.31.29)$$

This again yields in view of the constraint (27):

$$\lambda = 2 \int_a^b L dx = 2Q \qquad (7.31.30)$$

The significance of this result is as follows. The eigenvalues and eigenfunctions associated with our self-adjoint operator D can be obtained by the solution of the following variational problem: "Make the variational integral Q stationary, under the normalisation condition (27)." Any solution of this problem yields one of the eigenfunctions of the operator D. Moreover, the eigenvalue λ itself is equal to twice the stationary value of the given integral Q.

More definite statements can be made if it so happens that not only is $A(x)$ everywhere positive inside the range $[a, b]$ but $C(x)$ is everywhere *negative*. Then we can put $C(x) = -K(x)$ and write our Lagrangian in the following form:

$$L = \tfrac{1}{2}(Av^2 + Kv^2) \qquad (7.31.31)$$

This expression is now a *positive definite quadratic form* of v and v' which cannot take negative values for *any* choice of $v(x)$. The same holds then of the integral Q and—in view of (30)—of the eigenvalues λ_i. Hence *the eigenvalues associated with a positive definite Lagrangian can only be positive numbers*.

But we can go further and speak of the *absolute minimum* obtainable for Q under the constraint (27). This absolute minimum (apart from the factor 2) will give us the *smallest eigenvalue* of our eigenvalue spectrum. This process can be continued. After obtaining the lowest eigenvalue and the associated eigenfunction $v_1(x)$, we now minimise once more the integral Q with the auxiliary condition (27), but now we add the further constraint that we move *orthogonally* to the first eigenfunction $v_1(x)$:

$$\int_a^b v(x)v_1(x)dx = 0 \qquad (7.31.32)$$

The absolute minimum of this new variational problem (which excludes the previous solution $v_1(x)$ due to the constraint (32)) yields the *second* lowest eigenvalue λ_2 and its eigenfunction $v_2(x)$. We can continue this process, always keeping all the previous constraints, plus one new constraint, namely the orthogonality to the last eigenfunction obtained. In this fashion we activate more and more dimensions of the function space and *the eigenvalues enter automatically in the natural arithmetical order*.

We will now give an example which seems to contradict our previous result by yielding a negative eigenvalue, in spite of a positive definite Lagrangian. We make the simple choice

$$A(x) = 1, \quad C(x) = 0, \quad x = [0, \pi] \qquad (7.31.33)$$

and prescribe the self-adjoint boundary conditions

$$v'(0) + \alpha v(0) = 0, \quad v'(\pi) + \alpha v(\pi) = 0 \qquad (7.31.34)$$

The solution

$$v(x) = ce^{-\alpha x} \tag{7.31.35}$$

obviously satisfies the given boundary conditions, together with the differential equation

$$v'' + \lambda v = 0 \tag{7.31.36}$$

where λ has the *negative* value $\lambda = -\alpha^2$.

The explanation of this apparent paradox lies in the fact that the minimisation of our Lagrangian integral (1) cannot lead to boundary conditions of the form (34), in spite of their self-adjoint character. Only such boundary conditions can be prescribed which make the boundary term

$$A(b)v(b)v'(b) - A(a)v(a)v'(a) \tag{7.31.37}$$

to zero. For boundary conditions of a more general type we had to subtract the term $\frac{1}{2}(pv)'$ from L, that is our variational integral becomes now

$$\bar{Q} = \int_a^b [L - \frac{1}{2}(Avv')']dx \tag{7.31.38}$$

The added term amounts to a mere boundary term but this term is *no longer positive definite*, in fact it may become negative and counteract the first positive term to such an extent that the resulting $\bar{Q}$ becomes *negative*. This is what actually happens in the example (33–34).

Problem 316. Obtain the complete ortho-normal system associated with the problem (33), (34).

[Answer:

$$v_0(x) = \sqrt{\frac{2\alpha}{1 - e^{-2\pi\alpha}}}\, e^{-\alpha x}, \quad \lambda_0 = -\alpha^2$$

$$v_k(x) = \sqrt{\frac{\pi}{2}} \cos(kx + \theta_k), \quad \lambda_k = k^2 \tag{7.31.39}$$

$$\tan \theta_k = \frac{\alpha}{k}, \quad k = 1, 2, 3, \ldots\,]$$

Problem 317. Obtain the eigenfunctions and eigenvalues which belong to the Lagrangian

$$L = \frac{1}{2}x^\mu \dot{v}^2, \quad (0 \le \mu \le 2)$$
$$x = [0, 1], \quad v(1) = 0 \tag{7.31.40}$$

[Answer:

Added boundary condition: $[x^\mu \dot{v}]_{x=0} = 0$

$$v_k(x) = c_k x^{(1-\mu)/2} J_p\left(\frac{\sqrt{\lambda}}{1 - \frac{\mu}{2}} x^{1-(\mu/2)}\right), \quad p = \frac{\mu - 1}{2 - \mu}, \quad \lambda_k = \left(1 - \frac{\mu}{2}\right)^2 \xi_k^2$$

$$\tag{7.31.41}$$

where $J_p(\xi_k) = 0.]$

Problem 318.   Show that for any $\mu > 2$ the integral Q can be made arbitrarily small by a function of the type $e^{-\alpha t}$ where α is very large.   Hence the minimum problem cannot lead to a discrete smallest eigenvalue.

Problem 319.   Obtain the eigenfunctions and eigenvalues for the limiting case $\mu = 2$.

[Answer: The eigenvalues become continuous since only the boundary condition at $v(1)$ gives a selection principle:

$$\lambda = \tfrac{1}{4} + \rho^2, \quad v(x) = \frac{1}{\sqrt{x}} \sin \rho(\log x - 1) \tag{7.31.42}$$

The eigenfunctions cannot be normalised since they are not square-integrable. However, by cutting out an arbitrarily small neighbourhood around the singular point $x = 0$, the continuous spectrum changes to a dense line spectrum and the functions become once more normalisable.]

Problem 320.   Find the solution of the variational problem associated with the integral

$$Q = \frac{1}{2} \int_0^\pi \dot{v}^2 dx \tag{7.31.43}$$

with the constraints (27) and (34).   (Although this problem seems to coincide with Problem 316, this is in fact not the case because now the given Lagrangian is positive definite and the boundary conditions (34) must be treated as *constraints*. A negative eigenvalue is not possible under these conditions.)

[Answer: The method of the Lagrangian multiplier yields the differential equation

$$v'' + \lambda v = \alpha v(0)\delta(x, 0) - \alpha v(\pi)\delta(x, \pi) \tag{7.31.44}$$

with the following interpretation. *A strict minimum is not possible under the given constraints.* The minimum comes arbitrarily near to the solution of the eigenvalue problem which belongs to the boundary conditions $v'(0) = v'(\pi) = 0$, without actually reaching it.

This example shows that *a variational problem does not allow any tampering with its inherent boundary conditions* (which demand the vanishing of the boundary term (37)). If constraints are prescribed which do not harmonise with the inherent boundary conditions, the resulting differential equation is put out of action at the end-points, in order to allow the fulfilment of the inherent boundary conditions which must be rigidly maintained.]

BIBLIOGRAPHY

[1] Cf. {1}, pp. 82–97, 324–36, 466–510, 522–35

[2] Cf. {12}, Chapters XIV–XVII (pp. 281–385)

[3] Jahnke, E., and F. Emde, *Tables of Functions with Formulae and Curves* (Dover, New York, 1943)

[4] MacLachlan, N. W., *Bessel Functions for Engineers* (Clarendon Press, Oxford, 1934)

[5] Magnus, W., and F. Oberhettinger, *Formulas and Theorems of the Special Functions of Mathematical Physics* (Chelsea, New York, 1949)

[6] Szegö, G., *Orthogonal Polynomials* (Am. Math. Soc. Colloq. Pub., 23, 1939)

[7] Watson, G. N., *A Treatise on the Theory of Bessel Functions* (Cambridge University Press, 1944)

CHAPTER 8

BOUNDARY VALUE PROBLEMS

Synopsis. For irregular boundaries the boundary value problem of even simple differential operators becomes practically unmanageable, if our aim is to arrive at an analytical solution. For certain regular boundaries, however, the fundamental partial differential equations of mathematical physics become solvable by the method of the "separation of variables". While the number of explicitly solvable boundary value problems is thus very restricted, the study of these problems has had a profound impact on the general theory of partial differential equations, and given rise to a large class of auxiliary function systems which have a wide field of application. The present chapter discusses the standard type of boundary value problems which played such a decisive role in the understanding of the general theory of second order operators. We enlarge, however, our field of interest by the addition of certain unconventional types of boundary value problems which are not less solvable than the conventional types, although they do not submit to the customary conditions of a "well-posed" problem. We encounter the "parasitic spectrum" with its ensuing consequences. Finally we arrive at a perturbation method which transforms all non-conformist problems into the traditional "well-posed" ones and obtains the solution by a certain limit process.

8.1. Introduction

In all the previous chapters we were primarily concerned with the general theory of linear differential operators which were characterised by *homogeneous* boundary conditions; that is, certain linear combinations of the unknown function $v(x)$ and its derivatives on the boundary were prescribed as *zero* on the boundary, while, on the other hand, the "right side" of the differential equation was prescribed as some given function $\beta(x)$ of the domain. Historically, another problem received much more elaborate attention. The given differential equation itself is homogeneous by having zero on the right side (hence $\beta(x) = 0$). On the other hand, some linear combinations of the unknown function and its partial derivatives are now prescribed as given values, generally different from zero. Instead of an inhomogeneous differential equation with homogeneous boundary conditions we now have a homogeneous differential equation with inhomogeneous

432

boundary conditions. Problems of this type occur particularly frequently in mathematical physics and in fact the entire theory of boundary value problems took its departure from the exploration of natural phenomena. For example one of the best investigated partial differential equations of mathematics, the "Laplacian equation", or "potential equation":

$$\Delta v = \frac{\partial^2 v}{\partial x^2} + \frac{\partial^2 v}{\partial y^2} + \frac{\partial^2 v}{\partial z^2} = 0 \qquad (8.1.1)$$

originated from the Newtonian theory of gravitation, but it occurred equally in hydrodynamics, in elasticity, in electrostatics, in heat conduction. Again, it was the problem of sound propagation which induced Riemann (around 1860) to discover a completely different method for the investigation of another type of differential equation which later received the name "hyperbolic". Based on earlier results of G. Monge (1795), P. du Bois-Raymond introduced (in 1889) the classification of second-order differential operators into the "elliptic", "parabolic", and "hyperbolic" types. The potential equation is characteristic for the "elliptic" type. The heat flow equation

$$\frac{\partial v}{\partial t} - \frac{\partial^2 v}{\partial x^2} = 0 \qquad (8.1.2)$$

or more generally

$$\frac{\partial v}{\partial t} - \Delta v = 0 \qquad (8.1.3)$$

represents the "parabolic" type, while the equation of the vibrating string:

$$\frac{\partial^2 v}{\partial t^2} - \frac{\partial^2 v}{\partial x^2} = 0 \qquad (8.1.4)$$

or more generally

$$\frac{\partial^2 v}{\partial t^2} - \left(\frac{\partial^2 v}{\partial x^2} + \frac{\partial^2 v}{\partial y^2} \right) = 0 \qquad (8.1.5)$$

(the problem of the vibrating membrane) and

$$\frac{\partial^2 v}{\partial t^2} - \Delta v = 0 \qquad (8.1.6)$$

(called the "wave equation"), belong to the "hyperbolic" type. With the advent of wave-mechanics the Schrödinger equation

$$\frac{\partial v}{i\partial t} - \Delta v + V(r)v = 0 \qquad (8.1.7)$$

and many allied equations came in the focus of interest. Here the question of boundary values is often of subordinate importance since the entire space is the domain of integration. But in atomic scattering problems we encounter once more the same kind of boundary value problems which occur in optical and electro-magnetic diffraction phenomena, associated with the Maxwellian equations.

In the investigation of this particular class of second-order differential equations the observation was made that the analytical nature of the solution differed widely according to the "type" of the given differential operator. Furthermore, each type required its own kind of boundary conditions. For example the elliptic type of differential equations required conditions all along the boundary,* while the hyperbolic type required either initial or end conditions, the parabolic type initial conditions, excluding end conditions. Under such circumstances a truly universal approach to the theory of boundary value problems seemed out of the question. In every single case one had to establish first, what kind of boundary conditions are appropriate to the given problem, in order to make the problem analytically feasible. Hadamard in his famous Lectures on the Cauchy Problem (cf. Chapter Bibliography [5]), defined the conditions which a given boundary value problem had to satisfy in order to be admitted as a "well-posed" problem, with the implication that problems not satisfying these conditions are analytically inadmissible.

In marked contrast with these views we find that the fundamental "matrix decomposition theorem", which we have encountered in Chapter 3.9—and which in proper interpretation carries over into the field of arbitrary linear differential operators, with arbitrarily given boundary conditions—provides a *universal platform* for the theory of boundary value problems, irrespective of the "type" to which the differential operator belongs. In this treatment the subclass of "well-posed" problems is distinguished by a very definite property of the eigenvalue spectrum associated with the given operator. This property is that the eigenvalue $\lambda = 0$ is *excluded* from the eigenvalue spectrum, both with respect to the U-space (thus excluding over-determination) and the V-space (thus excluding under-determination). Then the operator is activated in the entire function space and not only in a restricted sub-space of that function space. But even this condition is not enough. The further condition has to be made that the eigenvalue $\lambda = 0$ *must not be a limit point of the eigenvalue spectrum*, that is the spectrum of the eigenvalues must start with a definite finite eigenvalue $\lambda_1 > \epsilon$, while the segment between $\lambda = 0$ and $\lambda = \epsilon$ must remain free of eigenvalues.

From the standpoint of the numerical solution of a given boundary value problem Hadamard's "well-posed" condition is well justified, since problems which do not satisfy this condition, lead to numerical instability, which is an undesirable situation. From the *analytical* standpoint, however, it is hardly advantageous to restrict our investigation by a condition which does not harmonise with the nature of a linear operator. Many fundamentally important operators of mathematical physics—as we have seen in Chapter 4—are *not* activated in the entire function space. They may omit axes in either the U or the V space, or in both. Accordingly our data

---

* Such conditions we will call "peripheral conditions", in order to avoid collision with the more general term "boundary conditions" which includes all classes of boundary data.

cannot be chosen freely but are subject to a finite or infinite number of compatibility conditions. Moreover, the solution of the given problem may not be unique, the uncertainty occurring in a finite or even infinite number of dimensions of the full V-space. But these apparent deficiencies are fully compensated by the fact that the given operator is *within its own field of activation both complete and unconstrained.* Our difficulties arise solely by going *outside* the field of operation which is inherently allotted to the operator. If we stay within that field, all our difficulties disappear and neither over-determination nor under-determination comes into existence.

What remains is the possibility of the eigenvalue $\lambda = 0$ as a limit point. This happens if boundary conditions are given which are not "according to type", that is initial (or end) conditions for the elliptic type of equations, or peripheral conditions for the hyperbolic type, or end-conditions for the parabolic type of equations. But if we have admitted the previous restriction of the given data to a definite subspace of the U-space, then we cannot exclude this new class of problems from our considerations either, because they represent a natural modification of the previous compatibility conditions. If before we required our data to have no projections in the "forbidden" (that is inactive) dimensions of the function space, now we have to demand a *less* stringent condition, because in those dimensions which belong to arbitrarily small eigenvalues, the projections of the data need not be zero, but only *sufficiently weak.*

With these restrictions *any* kind of boundary value problem, whose data are properly given, has a solution, and in fact a unique solution. Finally we shall encounter a further fact which tends to disprove the unique position of the "well-posed" problems. By a finite but arbitrarily weak perturbation we can wipe out the eigenvalue $\lambda = 0$ *and its infinitesimal neighbourhood,* thus transforming any arbitrarily "ill-posed" problem to a "well-posed" one (cf. Section 18).

8.2. Inhomogeneous boundary conditions

Although we shall deal specifically in this chapter with solution methods adapted to the case of inhomogeneous boundary conditions associated with a homogeneous differential equation, it is of greatest theoretical importance that our previous theory which assumed an inhomogeneous differential equation with homogeneous boundary conditions, *includes* the solution of the present problem.

We have recognised the expansion into eigenfunctions as a particularly powerful and universal method of solving linear differential equations. The presence of inhomogeneous boundary conditions seems to put this method out of action since the eigenfunctions cannot satisfy anything but homogeneous boundary conditions and the same holds for any linear combination of them. By the following detour, however, we get round the difficulty. First of all we find a function $v_0(x)$ which need not satisfy any differential equation but is chosen in such a way that it shall satisfy the given boundary conditions. We require a certain smoothness of this function $v_0(x)$, because the operator $Dv_0(x)$ must be able to operate on it and thus it must be differentiable to the proper degree. This may require a certain smoothness of the boundary data which is more than necessary for the existence of a solution but puts us on the safe side. Now we will put

$$v(x) = v_0(x) + V(x) \tag{8.2.1}$$

Then, if the given problem is to solve the homogeneous differential equation

$$Dv = 0 \tag{8.2.2}$$

with inhomogeneous boundary conditions, the substitution of (1) in our equation yields for the function $V(x)$ the inhomogeneous equation

$$DV(x) = -Dv_0(x) \tag{8.2.3}$$

but the given inhomogeneous boundary conditions are absorbed by $v_0(x)$, with the consequence that $V(x)$ must satisfy the same kind of boundary conditions which have been prescribed for $v(x)$, but *with zero on the right side*. And thus we have succeeded in transforming our originally given homogeneous differential equation with inhomogeneous boundary conditions into an inhomogeneous differential equation with homogeneous boundary conditions.

Let us see how the method of the eigenfunctions would operate under these circumstances. We assume that we possess the complete system of eigenfunctions and eigenvalues associated with the eigenvalue problem

$$\begin{aligned} Dv &= \lambda u \\ \check{D}u &= \lambda v \end{aligned} \tag{8.2.4}$$

Then $V(x)$ can be expanded into the eigenfunctions $v_i(x)$:

$$V(x) = \sum_{i=1}^{\infty} c_i v_i(x) \tag{8.2.5}$$

The coefficients of this expansion are obtainable in terms of the expansion coefficients of the right side:

$$c_i = \frac{\beta_i}{\lambda_i} \tag{8.2.6}$$

where

$$\beta_i = -\int u_i(x) Dv_0(x) dx \tag{8.2.7}$$

Now we make use of Green's identity (4.17.4):

$$\int u_i(x) Dv_0(x) dx - \int v_0(x) \check{D}u_i(x) dx = \int \sum_{\alpha} F_{\alpha}(v_0, u_i) v_{\alpha} d\sigma \tag{8.2.8}$$

which yields

$$\beta_i = -\lambda_i \int v_0(x) v_i(x) dx - \int \sum_{\alpha} F_{\alpha}(v_0, u_i) v_{\alpha} d\sigma \tag{8.2.9}$$

The second term is extended over the boundary surface σ and involves the boundary values of $u_i(\sigma)$ and its partial derivatives, together with the boundary values of $v_0(x)$, which in fact coincide with the given boundary values for $v(x)$.

If we separate the first term on the right side of (9) and examine its contribution to the function, we find the infinite sum

$$- \sum_{i=1}^{\infty} \int v_0(\xi) v_i(\xi) d\xi v_i(x)$$

and it seems that we have simply obtained $-v_0(x)$ which compensates the $v_0(x)$ we find on the right side of the expression (1). It seems, therefore, that the whole detour of separating the preliminary function $v_0(x)$ is unnecessary. But in fact the function $v_0(x)$ does not belong to the functions which can be expanded into the eigenfunctions $v_i(x)$, since it does not satisfy the necessary homogeneous boundary conditions. Moreover, the separation of this term from the sum (9) might make the remaining series divergent. In spite of the highly arbitrary nature of $v_0(x)$ and the independence of the final solution (2) of the choice of $v_0(x)$, this separation is nevertheless necessary if we want to make use of the expansion of the solution into a convergent series of eigenfunctions $v_i(x)$.

The method of the Green's function is likewise applicable to our problem and here we do not have to split away an auxiliary function but can apply the given inhomogeneous boundary values directly, in spite of the fact that the Green's function is defined in terms of homogeneous boundary conditions. We have defined the Green's function by the differential equation

$$\tilde{D}G(x, \xi) = \delta(x, \xi) \tag{8.2.10}$$

(the operator $\tilde{D}$ includes the adjoint homogeneous boundary conditions; moreover, the right side has to be modified, if necessary, by the proper constraints, as we have seen in (5.22.9)). While, however, in our previous problem (when $v(x)$ was characterised by homogeneous boundary conditions), Green's identity appeared in the form

$$\int [GDv - v\tilde{D}G]dx = 0 \tag{8.2.11}$$

and led to the solution

$$v(x) = \int G(x, \xi)\beta(\xi)d\xi \tag{8.2.12}$$

now this volume integral is zero, due to the vanishing of $\beta(x)$. On the other hand, we now have to make use of the "extended Green's identity" (4.17.4):

$$\int [GDv - v\tilde{D}G]dx = \int \sum_{\alpha} F_\alpha(G(x, \sigma), v(\sigma))\nu_\alpha d\sigma \tag{8.2.13}$$

and obtain $v(x)$ in the form of an integral extended over the boundary surface σ:

$$v(x) = - \int \sum_{\alpha} F_\alpha(G(x, \sigma), v(\sigma))\nu_\alpha d\sigma \tag{8.2.14}$$

Here the functions $F_\alpha(u, v)$ have automatically the property that they blot out all those boundary values which have *not* been prescribed. What remains is a surface integral involving all the "given right sides" of the boundary conditions. We may denote them by

$$f_1(\sigma), f_2(\sigma), \ldots, f_p(\sigma) \qquad (8.2.15)$$

Then the general form of the solution can be written as follows:

$$v(x) = \int [G_1(x, \sigma)f_1(\sigma) + \ldots + G_p(x, \sigma)f_p(\sigma)]d\sigma \qquad (8.2.16)$$

where the auxiliary functions $G_1(x, \sigma), \ldots, G_p(x, \sigma)$ are formed with the help of the Green's function $G(x, \xi)$ and its partial derivatives, applied to the boundary surface σ.

The unique solution thus obtained is characterised by the following properties. If our operator is incomplete by allowing solutions $v^i(x)$ of the homogeneous equation

$$Dv^i(x) = 0 \qquad (8.2.17)$$

our solution is made *orthogonal* to all these solutions. Furthermore, if the adjoint homogeneous equation possesses non-zero solutions:

$$\tilde{D}u^j(x) = 0 \qquad (8.2.18)$$

we assume that the given boundary conditions $f_\alpha(\sigma)$ are such that the integral conditions

$$\int \sum_\alpha F_\alpha(u^j(\sigma), v(\sigma))\nu_\alpha d\sigma = 0 \qquad (8.2.19)$$

are automatically fulfilled since otherwise the boundary data are incompatible and the given problem is unsolvable.

Problem 321. Assume the existence of non-zero solutions of (18) and the fulfilment of the required compatibility conditions (19). Now apply the method of transforming the given inhomogeneous boundary value problem into a homogeneous boundary value problem with inhomogeneous differential equation. Show that now the orthogonality of the right side to the "forbidden" axes u^j is automatically satisfied.

8.3. The method of the "separation of variables"

Although the general theory of the Green's function and the associated double set of eigenfunctions is of great value for the general analytical investigation of boundary value problems, the actual solution of such problems can often be accomplished by simpler tools. The solution of an inhomogeneous differential equation involves a right side which is given in an n-dimensional domain, while the boundary data of a homogeneous differential equation belong to the boundary surface, i.e. a domain of only $n - 1$ dimensions. Hence we can expect that under the proper circumstances a boundary value problem is solvable without the full knowledge of the complete set of eigenfunctions which are associated with the given differential operator.

In many of the particularly important differential operators of mathematical physics a fortunate circumstance exists, without which our knowledge concerning the nature of boundary value problems would be much more restricted. It consists of an artifice first employed by D. Bernoulli (1775) which has retained its fundamental importance to our day. We try to reduce the given partial differential equation to the solution of a set of *ordinary* differential equations, which depend on a single variable only. We do that by trying a solution which is set up as a product of functions of one single variable only:

$$v(x) = V_1(x_1)V_2(x_2) \ldots V_n(x_n) \tag{8.3.1}$$

That such an experiment succeeds is by no means self-evident. It is merely a fortunate circumstance that most of the basic differential operators of mathematical physics actually allow such a separation in the variables, in fact in many cases a separation is possible in a great variety of coordinates (for example the Laplacian operator Δ can be separated in rectangular, polar, cylindrical, parabolic, and many other coordinates). Furthermore, in most cases in which the separation succeeds, we obtain an infinite set of particular solutions and by a linear superposition of all these particular solutions the *complete* solution of the given differential equation can be accomplished, inside of a properly chosen domain. The coefficients of this linear expansion are obtainable in terms of the given boundary values, by integrating over the boundary surface.

The drawback of this method is only that the domain of the validity of these expansions is restricted to boundary surfaces of great regularity, such as a sphere, a cylinder, a parallelepiped—and their counterparts in two dimensions—occasionally also surfaces or curves of second order. For boundaries of irregular shape the method loses its applicability and in such cases we are frequently forced to take recourse to purely numerical methods.

8.4. The potential equation of the plane

The Laplacian equation in two dimensions:

$$\Delta V = \frac{\partial^2 V}{\partial x^2} + \frac{\partial^2 V}{\partial y^2} = 0 \tag{8.4.1}$$

has many exceptional properties, due to its close relation to the celebrated "Cauchy-Riemann differential equations", which are at the foundation of the theory of analytical functions. A function of the complex variable $z = x + iy$:

$$f(z) = f(x + iy) = u + iv \tag{8.4.2}$$

has the property that its real and imaginary parts are related to each other
by the two partial differential equations

$$\frac{\partial u}{\partial x} - \frac{\partial v}{\partial y} = 0$$

$$\frac{\partial v}{\partial x} + \frac{\partial u}{\partial y} = 0$$

(8.4.3)

these are the differential equations (first discovered independently by
Cauchy and by Riemann), which have the consequence that both the real
and imaginary part of $f(z)$ satisfy the potential equation (1):

$$\Delta u = 0$$

$$\Delta v = 0$$

(8.4.4)

Hence an arbitrary function $f(z)$, if separated into real and imaginary parts,
solves the potential equation of the plane. Furthermore, the Cauchy-
Riemann equations (3) can be conceived as a "conformal mapping" of the
plane on itself, that is a mapping which preserves angles but not lengths.
Such a mapping can be used to transform an irregular closed boundary into
a simple boundary, in particular into a circle. The Cauchy-Riemann
equations—and thus also the potential equation—are invariants of such a
mapping. Hence in principle—as it was shown by Riemann—any simply
connected domain of not too great irregularity can be mapped into the
inside of a circle, although unfortunately we do not possess the tools for the
explicit construction of such a mapping, except in a few simple cases.

For a circle it is advantageous to change from the rectangular coordinates
x, y, to polar coordinates r, θ, by the transformation

$$x + iy = re^{i\theta}$$

(8.4.5)

In these coordinates the equations (3) appear in the form

$$r\frac{\partial u}{\partial r} - \frac{\partial v}{\partial \theta} = 0$$

$$r\frac{\partial v}{\partial r} + \frac{\partial u}{\partial \theta} = 0$$

(8.4.6)

Now we try separation by putting

$$u(r, \theta) = U_1(r)U_2(\theta)$$
$$v(r, \theta) = V_1(r)V_2(\theta)$$

(8.4.7)

which yields the conditions

$$\frac{rU'_1(r)}{V_1(r)} = \frac{V'_2(\theta)}{U_2(\theta)}$$

$$\frac{rV'_1(r)}{U_1(r)} = -\frac{U'_2(\theta)}{V_2(\theta)}$$

(8.4.8)

But a function of r can only be a function of θ if in fact both functions are reduced to mere *constants*. Hence the equations (8) separate into the ordinary differential equations

$$rU'_1 = \alpha V_1 \quad V'_2 = \alpha U_2$$
$$rV'_1 = \beta U_1 \quad U'_2 = -\beta V_2$$

(8.4.9)

The second set of equations yields

$$V''_2 + \alpha\beta V_2 = 0$$
$$U''_2 + \alpha\beta U_2 = 0$$

(8.4.10)

solvable by exponential functions:

$$U_2 = A_2 e^{\sqrt{-\alpha\beta}\,\theta} + B_2 e^{-\sqrt{-\alpha\beta}\,\theta}$$

(8.4.11)

(with a similar solution for V_2). But now we have to demand that our solution be *periodic* in θ with the period 2π, since two points which belong to the angles θ and $\theta + 2\pi$, in fact *coincide*, and without the required periodicity our solution would not be single-valued. This condition restricts the possible values of the product $\alpha\beta$ to k^2, where k is an arbitrary positive integer, or zero:

$$\alpha\beta = k^2$$

(8.4.12)

We thus obtain the solutions

$$U_2 = A \cos k\theta + B \sin k\theta$$
$$V_2 = -\frac{\alpha}{k^2} U'_2 = \frac{\alpha}{k} (A \sin k\theta - B \cos k\theta)$$

(8.4.13)

We now come to the solution of the first set of conditions (9). Here we obtain for U_1 alone the differential equation

$$r(rU'_1)' = k^2 U_1$$

(8.4.14)

which has the solution

$$U_1 = ar^k + br^{-k}$$

(8.4.15)

But the second term becomes infinite at $r = 0$ and has to be rejected. Moreover, the arbitrary constant a does not add anything new to the free constants A and B of (13), and may be normalised to 1. Hence the method of separation, applied to polar coordinates, yields the following class of particular solutions of the problem (6):

$$u(r, \theta) = r^k(A_k \cos k\theta + B_k \sin k\theta)$$

(8.4.16)

while the first of the conditions (9), combined with (13), gives

$$v(r, \theta) = r^k(A_k \sin k\theta - B_k \cos k\theta)$$

(8.4.17)

From these particular solutions we proceed to form by linear superposition the infinite sum

$$u(r, \theta) + iv(r, \theta) = \sum_{k=0}^{\infty} c_k r^k e^{ik\theta}$$

$$= \sum_{k=0}^{\infty} c_k z^k \qquad (8.4.18)$$

where the complex constant c_k stands for

$$c_k = A_k - iB_k \qquad (8.4.19)$$

It is shown in the theory of analytical functions that this is indeed the *general* solution of the Cauchy-Riemann differential equations, inside of a circle with the radius r.

Let us now consider the function $u(r, \theta)$ alone, without reference to $v(r, \theta)$. It can be considered as a solution of the Laplacean equation (1) which in polar coordinates becomes

$$\frac{\partial r \dfrac{\partial V}{\partial r}}{r \partial r} + \frac{1}{r^2} \frac{\partial^2 V}{\partial \theta^2} = 0 \qquad (8.4.20)$$

The infinite sum

$$V(r, \theta) = \sum_{k=0}^{\infty} r^k (a_k \cos k\theta + b_k \sin k\theta) \qquad (8.4.21)$$

represents once more the *complete solution* of the equation (20), if we stay inside a circle within which the equation holds. Let us normalise the radius of that circle to $r = 1$ and prescribe on the periphery of the circle the boundary values

$$V(1, \theta) = \varphi(\theta) \qquad (8.4.22)$$

Then our problem is to satisfy the equation

$$\sum_{k=0}^{\infty} (a_k \cos k\theta + b_k \sin k\theta) = \varphi(\theta) \qquad (8.4.23)$$

which is equivalent to the problem of expanding a given function into a trigonometric series. We have solved this problem in Chapter 2.2, with the following result:

$$a_0 = \frac{1}{2\pi} \int_{-\pi}^{+\pi} \varphi(t)dt, \quad a_k = \int_{-\pi}^{+\pi} \varphi(\theta) \cos k\theta \, d\theta$$

$$b_k = \int_{-\pi}^{+\pi} \varphi(\theta) \sin k\theta \, d\theta \qquad (8.4.24)$$

The entire solution can be combined into the single equation

$$V(r, \theta) = \frac{1}{\pi} \sum_{k=0}^{\infty}{}' \int_{-\pi}^{+\pi} r^k \cos k(t - \theta)\varphi(t)dt \qquad (8.4.25)$$

We can arrive at a Green's function type of solution if we succeed in getting a closed expression for the infinite sum

$$G_1(r, t - \theta) = \frac{1}{\pi} \sum_{k=0}^{\infty}{}' r^k \cos k(t - \theta) \qquad (8.4.26)$$

This can be done in the present case since we have the real part of an infinite series which is summable by the formula of the geometrical series:

$$\sum_{k=0}^{\infty}{}' r^k e^{iks} = \frac{1}{1 - re^{is}} - \frac{1}{2} \qquad (8.4.27)$$

and thus, putting $t - \theta = s$:

$$G_1(r, s) = \frac{1}{2\pi} \frac{1 - r^2}{1 - 2r \cos s + r^2} \qquad (8.4.28)$$

This Green's function is not identical with the full Green's function $G(x, \xi)$ of the potential equation, but the two functions are closely related to each other. We have seen in (2.16) that the auxiliary functions $G_\alpha(x, \sigma)$ are expressible in terms of the Green's function $G(x, \xi)$ and its partial derivatives, *taken on the boundary surface* σ, which in our case of two dimensions is reduced to a boundary *curve*. But in our simple problem of a circle we have no difficulty in constructing even the full Green's function $G(x, \xi)$ which satisfies the differential equation

$$\Delta G(x, \xi) = \delta(x, \xi) \qquad (8.4.29)$$

together with the homogeneous boundary condition

$$G(x, \sigma) = 0 \qquad (8.4.30)$$

The high symmetry of the Laplacian operator permits us to study the Green's function of the potential operator in much more detail than that of an arbitrary operator, and for a few sufficiently regular domains we can actually construct the Green's function in explicit form.

The following properties of the Laplacian operator

$$\Delta v = \frac{\partial^2 V}{\partial x_1^2} + \frac{\partial^2 V}{\partial x_2^2} + \ldots + \frac{\partial^2 V}{\partial x_n^2} \qquad (8.4.31)$$

hold in spaces of arbitrary dimensions, although the spaces of two, three, and four dimensions are of primary interest from the applied standpoint:

1. The operator remains invariant with respect to arbitrary translations and rotations. Hence any solution of the Laplacian equation remains a

solution if we translate it to any other point of space, or rotate it rigidly around an arbitrary axis by an arbitrary angle.

2. The Laplacian operator is separable in polar coordinates and a solution exists which is a function of r only. For this solution the defining differential equation becomes

$$\frac{d(r^{n-1}V')}{dr} = 0 \tag{8.4.32}$$

whose solution is

$$V(r) = a + br^{-n+2} \tag{8.4.33}$$

In two dimensions the corresponding solution becomes

$$V(r) = \alpha + \beta \log r \tag{8.4.34}$$

3. If we apply to (31) the Gaussian integral transformation

$$\int \operatorname{div} F d\tau = \int \frac{\partial F}{\partial \nu} d\sigma \tag{8.4.35}$$

we obtain that in any "analytical" domain (in which the differential equation (31) is satisfied without any singularities) for any closed surface σ the fundamental relation holds:

$$\int \frac{\partial V}{\partial \nu} d\sigma = 0 \tag{8.4.36}$$

On the other hand, if we identify V with that negative power of r which satisfies the Laplacian equation (in the plane we have to choose $\log r$), there is a point of singularity at $r = 0$ which has to be excluded from our analytical domain. Then the theorem (36) still holds, but we now have an inner and outer boundary, the integral extended over the inner *plus* the outer boundary is zero, which means that the integral extended over the outer boundary alone becomes a *constant*. We can evaluate this constant by integrating over a sphere (in the plane a circle) of the radius r. For example, choosing $V = \log r$, and integrating over a circle of the radius r, we obtain for the integral (36):

$$\frac{2\pi r}{r} = 2\pi \tag{8.4.37}$$

while in three dimensions, choosing $V = r^{-1}$, we obtain

$$-\frac{4\pi r^2}{r^2} = -4\pi \tag{8.4.38}$$

4. Let us consider the solution of the inhomogeneous equation

$$\Delta v = \beta \tag{8.4.39}$$

and let us apply the Gaussian integral transformation to this equation. We now obtain the integral theorem (36) in the more general form:

$$\int \frac{\partial V}{\partial \nu} \, d\sigma = \int \beta d\tau \tag{8.4.40}$$

Let us assume in particular that $\beta(x)$ is a pure function of r. Then a solution of (39) can be found which is likewise a function of r only, while the general solution will be this particular $V_1(r)$, plus a solution of the homogeneous equation

$$\Delta V = 0 \tag{8.4.41}$$

5. We will assume that $\beta(r)$ is different from zero only within a certain (n-dimensional) sphere of the radius $r = \epsilon$. Then our particular solution $V_1(r)$ outside of this sphere must be of the form (33) and the constant b will be determined by the application of the relation (40), integrating over the inner sphere:

$$V'_1(r) \int d\sigma = B = \int_0^\epsilon \beta(\rho)d\tau \quad (r \geq \epsilon) \tag{8.4.42}$$

For example in two dimensions we get

$$V_1(r) = \frac{B}{2\pi} \log r \tag{8.4.43}$$

in three dimensions

$$V_1(r) = -\frac{B}{4\pi}\frac{1}{r} \tag{8.4.44}$$

in four dimensions

$$V_1(r) = -\frac{B}{4\pi^2}\frac{1}{r^2} \tag{8.4.45}$$

and in a space of the arbitrary dimensionality n, depending on the even or odd character of n:

for $n = 2k$:

$$V_1(r) = -\frac{B}{4\pi^k}\frac{(k-2)!}{r^{2k-2}}$$

for $n = 2k + 1$:

$$V_1(r) = -\frac{B}{(4\pi)^k}\frac{(2k-2)!}{(k-1)!}\frac{1}{r^{2k-1}} \tag{8.4.46}$$

6. Let us now investigate the solution of the equation

$$\Delta V = \delta(0, \xi) \tag{8.4.47}$$

By the definition of the delta function the integral over the right side becomes 1. The delta function can be assumed to be spherically symmetric and concentrated in a small sphere—whose centre is at the origin—with the radius ϵ which shrinks to zero. Hence the constant B is now 1 and the validity of our solution (43–46) applies in the limit to any point outside the point $r = 0$. We thus obtain for the Green's function $G(x, \xi)$:

in two dimensions:

$$G(x, \xi) = \frac{1}{2\pi} \log r_{x\xi} + V(\xi) \qquad (8.4.48)$$

in three dimensions:

$$G(x, \xi) = -\frac{1}{4\pi} \frac{1}{r_{x\xi}} + V(\xi) \qquad (8.4.49)$$

in four dimensions:

$$G(x, \xi) = -\frac{1}{4\pi^2} \frac{1}{r_{x\xi}^2} + V(\xi) \qquad (8.4.50)$$

where $V(\xi)$ is a solution of the Laplacian equation (41) which is regular throughout the given domain. This part of the Green's function will be uniquely determined by the homogeneous boundary conditions prescribed for $G(x, \xi)$.

With this excursion into the general properties of the Green's function associated with a Laplacian equation in arbitrary dimensions, we now return to the boundary value problem (22) which we have previously solved by the method of the separation of variables. Now we will solve the same problem on the basis of the Green's function. For this purpose we make use of the geometrical property of the so-called "conjugate points" of a circle or sphere. The "conjugate" of a point x with respect to the unit circle lies on the same radius vector but has the reciprocal distance $1/r$ from the origin (in Figure (51) $x = P$, $x' = P'$, while the running point is $\xi = Q$).

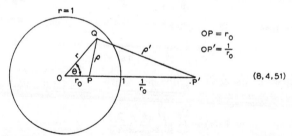

(8, 4, 51)

The contribution of the point P is fixed (according to (48)) to $(1/2\pi) \log \rho$. A similar contribution from the point P' satisfies the condition of the additional function $V(\xi)$, since the singularity of this function is outside the circle and thus does not violate the condition that $V(\xi)$ has to be analytical everywhere inside the circle.

By the laws of geometry we have

$$\rho = \sqrt{r^2 + r_0^2 - 2rr_0 \cos \theta}$$
$$\rho' = \frac{1}{r_0} \sqrt{r_0^2 r^2 + 1 - 2rr_0 \cos \theta} \qquad (8.4.52)$$

and we observe that ρ and $r_0\rho'$ become equal on the unit circle. Hence the linear superposition

$$\frac{1}{2\pi}[\log \rho - \log (r_0\rho')] \tag{8.4.53}$$

yields a solution of the Laplacian equation which vanishes on the boundary and satisfies all the other conditions of the Green's function. The same holds in three dimensions if we choose the solution

$$-\frac{1}{4\pi}\left(\frac{1}{\rho} - \frac{1}{r_0\rho'}\right) \tag{8.4.54}$$

Hence we have obtained the Green's function of the "first boundary value problem of potential theory" (when the values of $V(\sigma)$ are prescribed on the boundary), for the case that the boundary is the unit circle or the unit sphere. The result is for the case of two dimensions:

$$G(x, \xi) = \frac{1}{4\pi}\log\frac{r^2 + r_0{}^2 - 2rr_0\cos\theta}{r_0{}^2r^2 + 1 - 2rr_0\cos\theta} \tag{8.4.55}$$

while in three dimensions we obtain the expression

$$G(x, \xi) = -\frac{1}{4\pi}\left[\frac{1}{\sqrt{r^2 + r_0{}^2 - 2rr_0\cos\theta}} - \frac{1}{\sqrt{r^2r_0{}^2 + 1 - 2rr_0\cos\theta}}\right] \tag{8.4.56}$$

Furthermore, in our problem the extended Green's identity becomes (8.4.56)

$$\int (G\varDelta V - V\varDelta G)d\tau = \int\left(G\frac{\partial V}{\partial r} - V\frac{\partial G}{\partial r}\right)d\sigma \tag{8.4.57}$$

which leads to the solution

$$V(x) = \int V(\sigma)\frac{\partial G(x, \sigma)}{\partial r}\,d\sigma \tag{8.4.58}$$

If—for the case of the circle—we substitute in this formula the expression (55) (σ corresponds to $r = 1$), we obtain

$$V(r_0, \theta) = \frac{1}{2\pi}\int_{-\pi}^{+\pi}\varphi(\theta)\frac{1 - r_0{}^2}{1 + r_0{}^2 - 2r_0\cos\theta}\,d\theta \tag{8.4.59}$$

Finally, if the point x does not have the coordinates r_0, 0 but r, θ, we can reduce this problem to the previous one by a mere rotation of our reference system. The final formula becomes, if the integration variable is denoted by t:

$$V(r, \theta) = \frac{1}{2\pi}\int_{-\pi}^{+\pi}\varphi(t)\frac{1 - r^2}{1 + r^2 - 2r\cos(t - \theta)}\,dt \tag{8.4.60}$$

which is in full agreement with our previous result (28).

Problem 322. By a method which corresponds to that employed in the derivation of the equations (23–28), find the solution of the "second boundary value problem of potential theory" (the "Neumann problem"), in which the prescribed boundary values—in our case specified to the unit circle—belong to $\partial V/\partial v$ instead of V, that is

$$\frac{\partial V}{\partial r}(1, \theta) = g(\theta) \tag{8.4.61}$$

[Answer:

$$V(r, \theta) = \frac{1}{\pi} \sum_{k=1}^{\infty} \frac{r^k \cos k(t - \theta)}{k} g(t)dt \tag{8.4.62}$$

$$G_1(r, s) = -\frac{1}{2\pi} \log(1 - 2r \cos s + r^2) \tag{8.4.63}$$

Constraint:

$$\int_{-\pi}^{+\pi} g(\theta)d\theta = 0 \tag{8.4.64}]$$

Problem 323. Construct the complete Green's function $G(x, \xi)$ of this problem (constrained on account of (64)) and show that the solution (63) is identical with the solution obtained on the basis of the Green's function method. Demonstrate the symmetry of $G(x, \xi)$.

(*Hint*: Use again a proper linear combination of the contributions of the point x and its conjugate x'.)

[Answer:

Definition of $G(x, \xi)$:

$$\Delta G(x, \xi) = \delta(x, \xi) - \frac{1}{\pi}$$

$$\frac{\partial G}{\partial r}(1, \theta) = 0 \tag{8.4.65}$$

$$V(x) = -\int G(x, \sigma)g(\sigma)d\sigma \;]$$

Solution:

$$G(x, \xi) = G(\xi, x) = \frac{1}{4\pi}[\log(r_0^2 + r^2 - 2rr_0 \cos s)$$

$$+ \log(r_0^2 r^2 + 1 - 2rr_0 \cos s) + \tfrac{3}{4} - r_0^2 - r^2] \tag{8.4.66}$$

Problem 324. Obtain special solutions of the potential equation (20) by assuming that $V(r, \theta)$ is the *sum* of a function of r and a function of θ:

$$V(r, \theta) = P(r) + Q(\theta) \tag{8.4.67}$$

Formulate the result as the real part of a function $f(z)$ of the complex variable $z = re^{i\theta}$.

[Answer:

$$f(z) = \alpha \log^2 z + (\beta + i\gamma) \log z + \delta \tag{8.4.68}]$$

8.5. The potential equation in three dimensions

The Laplacian equation in three dimensions:

$$\Delta V = \frac{\partial^2 V}{\partial x^2} + \frac{\partial^2 V}{\partial y^2} + \frac{\partial^2 V}{\partial z^2} \tag{8.5.1}$$

leads to the definition of a number of important function classes. It can be separated in a great variety of coordinates, such as rectangular, polar, cylindrical, elliptic, and parabolic coordinates, every one of these separations occurring in actual physical situations. We will restrict ourselves to the case of polar coordinates. In these coordinates r, θ, ϕ the Laplacian equation appears in the following form*:

$$\frac{\partial r^2 \dfrac{\partial V}{\partial r}}{r^2 \partial r} + \frac{1}{r^2}\left[\frac{\partial \sin \theta \dfrac{\partial V}{\partial \theta}}{\sin \theta\, \partial \theta} + \frac{1}{\sin^2 \theta}\frac{\partial^2 V}{\partial \phi^2}\right] = 0 \qquad (8.5.2)$$

We separate first of all in the variable r by writing V in the form

$$V(r, \theta, \phi) = R(r)Y(\theta, \phi) \qquad (8.5.3)$$

This leads to the ordinary differential equation

$$(r^2 R')' = \alpha R \qquad (8.5.4)$$

and the following partial differential equation for the function $Y(\theta, \phi)$:

$$\frac{\partial \sin \theta \dfrac{\partial Y}{\partial \theta}}{\sin \theta \partial \theta} + \frac{1}{\sin^2 \theta}\frac{\partial^2 Y}{\partial \phi^2} = -\alpha Y \qquad (8.5.5)$$

This differential equation (5) has an independent significance of its own. Its left side represents the Laplacian operator ΔY, written down for the surface of a *sphere*, instead of a plane. The radius of this sphere is 1. The entire equation expresses the *eigenvalue problem* associated with the self-adjoint operator $-\Delta Y$, α being the eigenvalue. But we know from the general theory that the eigenvalue spectrum of a finite domain must be a *discrete* spectrum (except for the case of singular operators which lead to non-normalisable eigenfunctions). We can thus state in advance that the separation constant α must be restricted to an infinity of *discrete* values.

For the purpose of studying our eigenvalue problem in more detail we will apply the method of separating the variables once more:

$$Y(\theta, \phi) = Q(\theta)S(\phi) \qquad (8.5.6)$$

obtaining the two ordinary differential equations

$$S'' + \beta S = 0 \qquad (8.5.7)$$

$$\frac{d\left(\sin \theta \dfrac{dQ}{d\theta}\right)}{\sin \theta\, d\theta} - \frac{\beta}{\sin^2 \theta} Q + \alpha Q = 0 \qquad (8.5.8)$$

---

* The formulation of invariant differential operators in arbitrary curvilinear coordinates is the subject matter of "tensor calculus", or "absolute calculus" (cf., e.g., the books [7], [9], [10] of the Chapter Bibliography). For a brief introduction into the principles of tensor calculus see the author's article "Tensor Calculus" in the Handbook of Physics (Condon and Odeshaw) (McGraw-Hill, 1958), Part 1, pp. 111–122.

Now the new constant β is also an eigenvalue, belonging to the operator $-S''$. Exactly as in Section 4, the only possible values of β become

$$\beta = m^2 \qquad (m = 0, 1, 2, \ldots) \tag{8.5.9}$$

in view of the fact that $S(\phi)$ must become a *periodic* function of ϕ. The associated eigenfunctions are

$$S(\phi) = A \cos m\phi + B \sin m\phi \tag{8.5.10}$$

We start with $m = 0$. Then the equation (8) becomes, in the new variable

$$\cos \theta = x \tag{8.5.11}$$

identical with *Legendre's differential equation* (cf. (7.2.12)):

$$\frac{d(1 - x^2)Q'}{dx} + \alpha Q = 0 \tag{8.5.12}$$

We have encountered this differential equation earlier as a special case of the hypergeometric differential equation, and seen that the singularity at $x = \pm 1$ can only be avoided if the hypergeometric series *terminates* after a finite number of terms. This requires the selection principle

$$\alpha = n(n + 1) \qquad (n = 0, 1, 2, \ldots) \tag{8.5.13}$$

The functions $Q_n(x)$ then become the *Legendre polynomials*, expressed in $x = \cos \theta$.

We have now obtained a special class of spherical harmonics which are independent of the azimuth angle ϕ:

$$Y(\theta) = c_n P_n(\cos \theta) \tag{8.5.14}$$

and we will return to the equation (4), in order to obtain the full solution of $V(r, \theta, \phi)$:

$$(r^2 R')' = n(n + 1)R \tag{8.5.15}$$

The two solutions of this differential equation are obtainable by putting

$$R = r^\mu \tag{8.5.16}$$

which yields for μ the determining equation

$$\mu(\mu + 1) = n(n + 1) \tag{8.5.17}$$

with the two solutions

$$\mu = n, \quad \mu = -(n + 1) \tag{8.5.18}$$

The second solution has to be rejected because it leads to a point of infinity at $r = 0$. On the other hand, if our aim is to solve the potential equation *outside* of a certain sphere, then *only* the second solution must be kept and the first one dropped, since r^n goes to infinity with increasing r.

We can now obtain by superposition the complete solution of the Laplacian equation inside of a certain sphere with the radius $r = a$, if that solution has *cylindrical symmetry* by not depending on the azimuth angle ϕ:

$$V(r, \theta) = \sum_{n=0}^{\infty} c_n r^n P_n(\cos \theta) \qquad (8.5.19)$$

Legendre's differential equation gives valuable clues even toward the general problem of obtaining the spherical harmonics which do depend on the azimuth angle ϕ. Let us differentiate Legendre's differential equation

$$(1 - x^2)v'' - 2xv' + \lambda v = 0 \qquad (8.5.20)$$

m times. We obtain then for the function $y = v^{(m)}(x)$ the following differential equation

$$(1 - x^2)y'' - (2 + 2m)xy' + (\lambda - m(m + 1))y = 0 \qquad (8.5.21)$$

Now the substitution

$$y(x) = u(x)w(x) \qquad (8.5.22)$$

yields for $w(x)$ the following differential equation:

$$(1 - x^2)w'' - \left[(2 + 2m)x - 2\frac{u'}{u}(1 - x^2)\right]w'$$

$$+ \left[\lambda - m(m + 1) + (1 - x^2)\frac{u''}{u} - (2 + 2m)x\frac{u'}{u}\right]w = 0 \quad (8.5.23)$$

and we will dispose of $u(x)$ in such manner that the factor of w' shall remain $-2x$. For this purpose we have to put

$$\frac{u'}{u} = \frac{mx}{1 - x^2}$$

$$u = (1 - x^2)^{-m/2} \qquad (8.5.24)$$

The factor of w in (23) now becomes

$$\lambda - \frac{m^2}{1 - x^2} \qquad (8.5.25)$$

and we obtain for

$$w(x) = y(x)(1 - x^2)^{m/2} = P_n^{(m)}(\cos \theta)(\sin \theta)^m \qquad (8.5.26)$$

the following differential equation:

$$(1 - x^2)w'' - 2xw' + \left[n(n + 1) - \frac{m^2}{1 - x^2}\right]w = 0 \qquad (8.5.27)$$

But this is exactly the differential equation (8) for $\beta = m^2$ and $\alpha = n(n + 1)$. We have thus obtained the following particular solutions of the Laplacian differential equation:

$$(r, \theta, \phi) = cr^n P_n^{(m)}(\cos \theta) \sin^m \theta e^{\pm im\phi} \qquad (8.5.28)$$

Now a polynomial of the order n cannot be differentiated more than n times (the higher derivatives vanishing identically). Hence the integer m in (28) can only assume the values $0, 1, 2, \ldots, n$. With the exception of $m = 0$ every one of these values leads to *two* solutions, in view of the $\pm$ sign of the last factor. Hence *the total multiplicity of the eigenvalue n is $2m + 1$.*

Problem 325. Show that to any solution $V(r, \theta, \phi)$ of the Laplacian equation a second solution can be constructed by putting

$$V_1 = \frac{1}{r} V\left(\frac{1}{r}, \theta, \phi\right) \tag{8.5.29}$$

(excluding the point $r = 0$).

Problem 326. Show that the particular solutions (28) are all polynomials in the rectangular coordinates x, y, z, where

$$\begin{align} x + iy &= r \sin \theta e^{i\phi} \\ z &= r \cos \theta \end{align} \tag{8.5.30}$$

(The previous notation x for $\cos \theta$ is here discarded.)

Problem 327. Show the orthogonality of the function system (26), for fixed m and variable n.

Problem 328. Assume that $V(r, \theta, \phi)$ has cylindrical symmetry (i.e. ϕ independent). Given are the boundary values of $V(r, \theta)$ on the unit sphere:

$$V(1, \theta) = g(\theta) \tag{8.5.31}$$

Obtain the coefficients c_n of the expansion (19) in terms of the given boundary values.

[Answer:

$$c_n = \frac{2n + 1}{2} \int_0^\pi g(\theta) P_n(\cos \theta) \sin \theta d\theta \tag{8.5.32}]$$

Problem 329. Obtain particular solutions of the potential equation (2) by assuming a $V(r, \theta)$ which is a *sum* of a function of r and a function of θ:

$$V(r, \theta) = A(r) + B(\theta) \tag{8.5.33}$$

Answer:

$$V(r, \theta) = \alpha \log r + (\alpha + \beta) \log \sin \theta - \beta \log (1 + \cos \theta) + \frac{a}{r} + b \tag{8.5.34}$$

In particular for $\beta = \mp \alpha$ we obtain the solution

$$\begin{align} V(r, \theta) &= \alpha \log r(1 \pm \cos \theta) \\ &= \alpha \log (r \pm z) \end{align} \tag{8.5.35}]$$

Problem 330. Choose the upper sign in (35) and demonstrate the following property of this solution. We select the point $r = a$, $\theta = 0$ on the positive z-axis and construct a sphere of the radius $b < a$ around the point $(a, 0)$ as centre. Then the normal derivative of $V(r, \theta)$ along this sphere becomes

$$\frac{\partial V}{\partial \nu} = \frac{1}{b}\left(1 - \frac{a}{r}\right) \tag{8.5.36}$$

Problem 331. The Green's function for the case of a unit sphere was obtained in (4.56) as far as the "first boundary value problem of potential theory" is concerned. Solve the same problem for the "second boundary value problem" ("Neumann problem"). *Hint:* the operation with the conjugate points is not enough, but we succeed if the result (36) is taken into account.

[Answer:

Definition of Green's function:

$$\Delta G(x, \xi) = \delta(x, \xi) - \frac{3}{4\pi}, \quad \frac{\partial G}{\partial r}(1, \theta, \phi) = 0$$

$$V(x) = -\int G(x, \sigma)g(\sigma)d\sigma$$

$$(8.5.37)$$

Constraint:

$$\int g(\sigma)d\sigma = 0 \qquad (8.5.38)$$

Solution (cf. Fig. (4.51)):

$$G(x, \xi) = -\frac{1}{4\pi}\left[\frac{1}{\sqrt{r^2 + r_0^2 - 2rr_0\cos\theta}} + \frac{1}{\sqrt{r_0^2r^2 + 1 - 2rr_0\cos\theta}}\right.$$
$$- \log\left(\sqrt{r_0^2r^2 + 1 - 2rr_0\cos\theta} + 1 - r_0r\cos\theta\right)$$
$$\left.+ \frac{r^2 + r_0^2}{2} - \frac{8}{5} + \log 2\right] \qquad (8.5.39)$$

$$G(r_0, 0, 0; 1, \theta, \phi) = -\frac{1}{4\pi}\left[\frac{2}{\sqrt{1 + r_0^2 - 2r_0\cos\theta}}\right.$$
$$\left.- \log\left(\sqrt{1 + r_0^2 - 2r_0\cos\theta} + 1 - r_0\cos\theta\right)\right] \qquad (8.5.40)$$

(The additional constant has no effect on the integration and may be omitted.)]

Problem 332. Show that the solution thus obtained automatically satisfies the conditions

$$\begin{aligned} a) && V(0, \theta, \phi) &= 0 \\ b) && \int V(\tau)d\tau &= 0 \end{aligned} \qquad (8.5.41)$$

provided that the compatibility condition (38) is satisfied. Explain the origin of the condition a).

[Answer: the vanishing of the coefficient c_0 in the expansion (19).]

Problem 333. Consider the problem of minimising the integral

$$\frac{1}{2}\int|\text{grad } V|^2d\tau = \frac{1}{2}\int\left[\left(\frac{\partial V}{\partial x}\right)^2 + \left(\frac{\partial V}{\partial y}\right)^2 + \left(\frac{\partial V}{\partial z}\right)^2\right]d\tau \qquad (8.5.42)$$

inside of a certain domain τ, with prescribed values of $V(\sigma)$ on a certain portion C of the boundary surface σ, while no restrictions are imposed on the complementary boundary C'. Show that the solution of this variational problem is the following boundary value problem:

$$\begin{aligned} \Delta V &= 0 && (\text{in } \tau) \\ V(\sigma) &= g(\sigma) && (\text{on } C) \\ \frac{\partial V}{\partial \nu} &= 0 && (\text{on } C') \end{aligned} \qquad (8.5.43)$$

("mixed boundary value problem").

Problem 334. Given the boundary value problem

$$-\Delta V = 0 \qquad (8.5.44)$$

$$\frac{\partial V}{\partial \nu}(\sigma) + \gamma(\sigma) V(\sigma) = g(\sigma) \qquad (8.5.45)$$

where $\gamma(\sigma)$ is a given function on the boundary surface σ.

a) Show that this problem is self-adjoint and thus deducible from a variational problem.

b) Find the boundary integral which has to be added to the volume integral (42) in order to obtain the boundary condition (45) as the inherent boundary condition of the variational problem.

c) Consider the eigenvalue problem of (44), putting $g(\sigma) = 0$, and show that for $\gamma(\sigma) \geq 0$ the eigenvalues become all positive, with the possible exception of $\lambda = 0$ in the extreme case that $\gamma(\sigma)$ vanishes identically.

[Answer:

Added boundary term:

$$\int \left[\tfrac{1}{2}\gamma(\sigma) V^2(\sigma) - \left(\frac{\partial V}{\partial \nu} + \gamma(\sigma) V(\sigma) \right) g(\sigma) \right] d\sigma \qquad (8.5.46)]$$

8.6. Vibration problems

Partial differential equations which involve the time t, are frequently of the following structure. Space and time are separated. We have an operator Dv which involves the space variables only and whose coefficients are independent of time. This operator Dv originates by the minimisation of a certain space integral which is *positive definite*. Hence Dv is a self-adjoint operator (with self-adjoint boundary conditions), whose eigenvalues are all positive. Moreover, the time enters only by an added term which contains the second derivative of v with respect to t:

$$Dv + \frac{\partial^2 v}{\partial t^2} = 0 \qquad (8.6.1)$$

We will assume that we possess the complete function system associated with the eigenvalue problem

$$Dv_i = \lambda_i v_i \qquad (\lambda_i > 0) \qquad (8.6.2)$$

Now we expand $v(x, t)$ into the complete ortho-normal eigenfunction system v_i, with coefficients which are functions of t:

$$v(x, t) = \sum_{i=1}^{\infty} c_i(t) v_i(x) \qquad (8.6.3)$$

Our differential equation (1) now separates into the ordinary differential equation

$$c''_i(t) + \lambda_i c_i(t) = 0 \qquad (8.6.4)$$

which is solved by

$$c_i(t) = a_i \cos \sqrt{\lambda_i} t + b_i \sin \sqrt{\lambda_i} t \qquad (8.6.5)$$

Our problem is thus reduced to the determination of the coefficients a_i, b_i. This can be done by prescribing the proper *initial conditions* at the time moment $t = 0$. Let us assume that we possess the data

$$v(x, 0) = f(x) \tag{8.6.6}$$

$$\frac{\partial v}{\partial t}(x, 0) = g(x) \tag{8.6.7}$$

This means that at the time moment $t = 0$ the initial *displacements* and the initial *velocities* of the vibrating medium are given. Then we can expand both $f(x)$ and $g(x)$ into the eigenfunction system $v_i(x)$:

$$f(x) = \sum_{i=1}^{\infty} c_i v_i(x)$$
$$g(x) = \sum_{i=1}^{\infty} c'_i v_i(x) \tag{8.6.8}$$

with the coefficients

$$c_i = c_i(0) = \int f(\xi) v_i(\xi) d\xi$$
$$c'_i = c'_i(0) = \int g(\xi) v_i(\xi) d\xi \tag{8.6.9}$$

But then the comparison with (5) shows that we have in fact obtained the coefficients a_i, b_i explicitly:

$$a_i = c_i$$
$$b_i = \frac{1}{\sqrt{\lambda}} c'_i \tag{8.6.10}$$

and we can consider our problem as solved.

Problem 335.   Let us assume that the homogeneous equation

$$Dv^j = 0 \tag{8.6.11}$$

possesses non-zero solutions under the given homogeneous boundary conditions. Show that $f(x)$ can still be prescribed freely, while $g(x)$ is constrained by the orthogonality to the homogeneous solutions:

$$\int g(\xi) v^j(\xi) d\xi = 0 \tag{8.6.12}$$

Problem 336.   Show that the Green's function $G(x, t; \xi, \tau)$ of the problem (1), with $f(x) = g(x) = 0$, can be constructed as follows:

$$G(x, t; \xi, \tau) = 0 \qquad (t \leq \tau)$$
$$G(x, t; \xi, \tau) = \sum_{i=1}^{\infty} \frac{\sin \sqrt{\lambda_i}(t - \tau)}{\sqrt{\lambda_i}} v_i(x) v_i(\xi) \qquad (t \geq \tau) \tag{8.6.13}$$

Problem 337.   In the case of a vibrating membrane the differential operator D becomes the negative Laplacian operator of the plane:

$$Dv = -\Delta v = -\left(\frac{\partial^2}{\partial x^2} + \frac{\partial^2}{\partial y^2}\right)v \qquad (8.6.14)$$

Assume a circular membrane of the radius 1 and determine the eigenfrequencies and vibrational modes of the membrane, under the boundary condition that the membrane is fixed at $r = 1$:

$$v(1, \theta) = 0 \qquad (8.6.15)$$

[Answer:

$$v(r, \theta) = v(r)e^{\pm ik\theta}, \quad \frac{1}{r}(rv')' - \frac{k^2 v}{r^2} + \lambda v = 0 \qquad (8.6.16)$$

$$v_i(r, \theta) = c_{km}J_k(\sqrt{\lambda_{km}}r)e^{\pm ik\theta} \qquad (8.6.17)$$

where the eigenvalues λ_{km} are determined by the condition

$$\lambda_{km} = \xi_{km}{}^2 \qquad (8.6.18)$$

if ξ_{km} denotes the (infinitely many) zeros of the Bessel function of the order k:

$$J_k(\xi_{km}) = 0 \qquad (m = 1, 2, \ldots) \qquad (8.6.19)]$$

Problem 338.   Show that a vibration problem with inhomogeneous but time-independent boundary conditions can be solved as follows.   We first solve the given inhomogeneous boundary value problem for the differential equation

$$Dv_0 = 0 \qquad (8.6.20)$$

Then we replace the given inhomogeneous boundary conditions by the corresponding homogeneous conditions and follow the previous procedure, merely replacing the initial displacement function $f(x)$ by $f(x) - v_0(x)$.   The resulting solution of our problem becomes

$$v_0(x) + v(x, t) \qquad (8.6.21)$$

8.7. The problem of the vibrating string

The problem of the "vibrating string" (which, in fact, as we will see, is far from "vibrating"), is one of the historically most interesting examples of a boundary value problem.   It was in connection with this problem that D. Bernoulli discovered the method of eigensolutions and their super-position, thus obtaining a complete solution of a given inhomogeneous boundary value problem.   The same problem brought about the remarkable controversy concerning the nature of the trigonometric series, which involved D'Alembert, Lagrange, Euler, Bernoulli, and finally Fourier.

This problem is once more associated with the Laplacian operator, which, however, is here reduced to a single dimension.   The differential equation

$$\frac{\partial^2 v}{\partial t^2} - \Delta v = 0 \qquad (8.7.1)$$

leads to physically important applications, whether the Laplacian operator

Δ involves one, two, or three rectangular space coordinates. The simplest case of one single coordinate leads (in proper normalisation of units) to the differential equation of the vibrating string:

$$\frac{\partial^2 v}{\partial t^2} - \frac{\partial^2 v}{\partial x^2} = 0 \qquad (8.7.2)$$

The string is fixed at the points $x = 0$ and $x = l$, which imposes the boundary conditions

$$v(0, t) = 0, \quad v(l, t) = 0 \qquad (8.7.3)$$

According to (6.13) we can proceed immediately to the construction of the Green's function of our problem. For this purpose we have to solve the eigenvalue problem (6.2) which in our case becomes

$$v''_k(x) + \lambda_k v_k(x) = 0 \qquad (8.7.4)$$

In view of the boundary conditions (3) we obtain the normalised eigensolutions

$$v_k(x) = \sqrt{\frac{2}{l}} \sin k \frac{\pi}{l} x \qquad (8.7.5)$$

with

$$\lambda_k = \frac{\pi^2}{l^2} k^2 \qquad (k = 1, 2, \ldots) \qquad (8.7.6)$$

Then, in view of (6.13), we obtain for the Green's function $G(x, t; \xi, \tau)$ which we prefer to denote by $G(x, \xi; t, \tau)$:

$$G(x, \xi; t, \tau) = 0 \qquad (t \le \tau)$$

$$= 2 \sum_{k=1}^{\infty} \frac{\sin k \frac{\pi}{l} (t - \tau)}{k \pi} \sin k \frac{\pi}{l} x \sin k \frac{\pi}{l} \xi \qquad (t \ge \tau) \quad (8.7.7)$$

If we denote

$$t - \tau = s \qquad (8.7.8)$$

and replace the product of two sine functions by the difference of two cosine functions, we obtain the general term of the expansion (7) in the following form:

$$\frac{1}{k \pi} \sin k \frac{\pi}{l} s \left[\cos k \frac{\pi}{l} (x - \xi) - \cos k \frac{\pi}{l} (x + \xi) \right]$$

$$= \frac{1}{2 \pi k} \left[\sin k \frac{\pi}{l} (s + x - \xi) + \sin k \frac{\pi}{l} (s -- x + \xi) \right.$$

$$\left. - \sin k \frac{\pi}{l} (s + x + \xi) - \sin k \frac{\pi}{l} (s - x - \xi) \right] \qquad (8.7.9)$$

Hence it suffices to evaluate *one* universal function $F(p)$, defined by

$$F(p) = \frac{1}{2\pi}\sum_{k=1}^{\infty} \frac{\sin k\frac{\pi}{l}p}{k} \tag{8.7.10}$$

In terms of this function the resulting Green's function becomes:

$$G(x, \xi; s) = F(s + x - \xi) + F(s - x + \xi)$$
$$- F(s + x + \xi) - F(s - x - \xi) \tag{8.7.11}$$

The Green's function of our problem is not only a mathematically important function, but has a very definite and simple physical significance. The delta function on the right side of the defining differential equation means in physical terms that at the point $x = \xi$ and the time moment $t = \tau$ an infinitely sharp *hammer blow* is applied, of infinitely short duration. This hammer blow conveys to the particles of the string a mechanical momentum of the magnitude 1, localised to the immediate neighbourhood of the point $x = \xi$. Then the string is left alone and starts to perform its motion on the basis of the principles of mechanics—more precisely the "principle of least action", which demands that the time integral of $T - V$ (where T is the kinetic energy of the string and V its potential energy), shall be made a minimum, or at least a stationary value:

$$\delta \int_0^{t_1} (T - V)dt = 0 \tag{8.7.12}$$

where

$$T = \frac{1}{2} \int_0^l \left(\frac{\partial v}{\partial t}\right)^2 dx, \quad V = \frac{1}{2} \int_0^l \left(\frac{\partial v}{\partial x}\right)^2 dx \tag{8.7.13}$$

The solution of this variational problem is the differential equation (2), with the boundary conditions (3), caused by the constraints which are imposed on the string by the forces which prevent it from moving at the two fixed ends.

The Green's function itself is the displacement of the string, as a function of x and t, after the hammer blow is over. The expression (7) shows at once an interesting property of the motion of the string, observed by the early masters of acoustical research: "an overtone which has a nodal point at the point where the hammer strikes, cannot be present in the harmonic spectrum of the vibrating string". Indeed, the sum (7) represents a harmonic resolution of the motion of any particle of the string, if we consider x as fixed and describe the motion as a function of time. The overtones have frequencies which are integer multiples of the fundamental frequency $1/(2l)$ and any overtone receives the weight zero if the last factor has a nodal point at the point $x = \xi$.

It is this harmonic analysis in time which leads to the notion that the string performs some kind of "vibration", as the name "vibrating string" indicates. That such vibrations are possible is clear from the mathematical form of the eigenfunctions, if they are taken separately. But this does not mean that under the ordinary conditions of bringing the string into motion, some kind of vibration will occur. It is a curious fact that our "physical intuition" can easily mislead us if we try to make predictions without the aid of the exact mathematical theory. What happens if we strike the string with a hammer? How will the disturbance propagate? The answer frequently given is that some kind of "wave" will propagate along the string, similar to the waves observed in a pond if a stone is dropped in the water. Another guess may be made from the way in which an electromagnetic disturbance is propagated in space: it spreads out on an ever-increasing sphere at the velocity of light, giving a short disturbance at the points swept over by this expanding sphere. The picture derived from this analogy would be a narrow but sharp "hump" on the wire which will propagate with constant velocity (in fact with the velocity 1 due to the normalisation of our differential equation), both to the right and to the left from the point of excitation.

Both guesses are in fact wrong. In order to study the phenomenon in greater detail, we have to find first of all the significance of the sum (10). We have encountered the same sum earlier (cf. 2.2.10), and making use of the result there obtained we get

$$F(p) = \frac{1}{4} - \frac{p}{4l} \qquad (8.7.14)$$

together with the two further conditions
$$F(-p) = -F(p)$$
$$F(l + p) = -F(l - p) \qquad (8.7.15)$$
which implies

$$F(-p) - F(-p + l) = -[F(p) - F(p + l)] \qquad (8.7.16)$$

Let us now apply the hammer blow first of all at the *centre* of the string, that is the point $\xi = l/2$ and examine the displacement of the string, given by (11), at the points $x = l/2 \pm x_1$. By the principle of symmetry we know in advance that the disturbance must spread symmetrically to the right and to the left, and thus the displacement $v(s \pm x_1)$ which belongs to these points, must be the same. If we substitute the result (14) in the general formula (11), we get

$$v(x_1, s) = F(s + x_1) - F(s + x_1 + l) + F(s - x_1) - F(s - x_1 + l)$$
$$= \tfrac{1}{2} \quad (s > x_1)$$
$$= 0 \quad (s < x_1) \qquad (8.7.17)$$

The picture we derive from this formula can be characterised as follows. The disturbance propagates from the point of striking in the form of a *hill of constant height* $\tfrac{1}{2}$, which spreads out with the velocity 1 to larger and larger portions of the string. After arriving at the end-points 0 and l, the hill *recedes* with the same velocity and finally collapses to zero after the time l, when it jumps over to the height $-\frac{1}{2}$ and repeats the same cycle over once more, on the negative side. After the time of $2l$ the full cycle is repeated in identical terms.

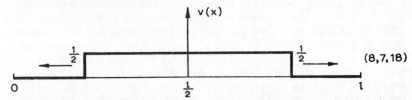

$$(8,7,18)$$

The remarkable feature of this result is that with the limited kinetic energy of the initial blow larger and larger portions of the string are excited, which seems to contradict the conservation law of energy. In actual fact the expression (13) shows that the accumulated potential energy is zero because the hill is of constant height and thus $\partial v/\partial x = 0$. The local kinetic energy is likewise zero because the points of the hill, after rising to the constant height of $\frac{1}{2}$ (or $-\frac{1}{2}$) remain *perfectly still*, their velocity dropping to zero. The exchange between potential and kinetic energy takes place solely at the *end* of the hill which travels out more and more but repeats the same phenomenon in identical terms.

If the hammer blow is applied away from the centre $\xi = l/2$, the only change is that the reflection of the hill at the two end-points now occurs at different time moments and thus the receding of the hill starts at one end at a time when the hill is still moving forward on the other side. The collapsing and reversal of sign now occurs at the mirror image of the point of excitation, but again after the time $2l$ has passed from the time moment of excitation.

The actual motion of the particles is far from a harmonic oscillation. It

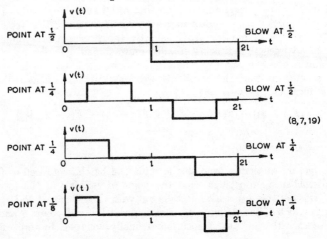

$$(8,7,19)$$

consists in a sudden rise from the equilibrium position into a maximum position, staying there for a while and then falling back once more into the equilibrium position, with a repetition of the same motion in reversed sequence and reversed sign, until after the time $2l$ the entire cycle is completed and the play starts again. In Figure (19) a few characteristic motion forms are graphed, with the hammer striking in the middle, and half way from the middle.

Under these circumstances we may wonder how a hammer-blow instrument such as a piano can serve as a musical instrument at all. We have to remember, however, that any periodic disturbance of the air will be recorded by the ear as a musical tone of definite pitch, while the more or less regular shape of the disturbance influences merely the "tone quality". Furthermore, the sounding board and the resonating cavities of the piano body put weight factors to the partial harmonics with the result that the air particles which bring the ear membrane into forced vibrations perform a much more regular motion than the jerky motions illustrated in Figure (19).

Another method of exciting the string is by "plucking", as practised in instruments such as the harp, the guitar, the lute, and many others: the string is pulled out with the finger, giving it a nearly triangular shape; then it is released. Here the motion starts with a certain shape $f(x)$ at the time moment $t = 0$, while the velocity of the particles at $t = 0$ can be considered as zero:

$$v(x, 0) = f(x)$$

$$\frac{\partial v}{\partial t}(x, 0) = 0 \tag{8.7.20}$$

Now the method of separating the variables yields the solution

$$v(x, t) = \sum_{k=1}^{\infty} c_k \cos k \frac{\pi}{l} t \sin k \frac{\pi}{l} x$$

$$= \tfrac{1}{2} \sum_{k=1}^{\infty} c_k \left[\sin k \frac{\pi}{l}(x + t) + \sin k \frac{\pi}{l}(x - t) \right]$$

$$= \tfrac{1}{2}[f(x + t) + f(x - t)] \tag{8.7.21}$$

This formula can be interpreted as follows. The point $P = (x, t)$ is

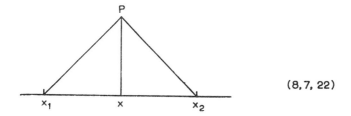

$$(8, 7, 22)$$

projected down on the X-axis by drawing two straight lines at the angles of $-45°$ downward. We arrive at the two points x_1 and x_2. Then we take the arithmetic mean of the two values $f(x_1)$ and $f(x_2)$ as the solution $v(x, t)$. If $f(x)$ is plotted geometrically, we can obtain the solution by a purely geometrical construction.

If we pluck the string at the centre, the shape of the string at $t = 0$ will be an isosceles triangle. How will this triangle move, if we release the string? Will the entire triangle move up and down as a unit, vibrating in unison? This is not the case. The geometrical construction according to Figure (22) demonstrates that *the outer contour of the triangle remains at rest* but a straight line moves down with uniform speed, truncating the triangle to a quadrangle of diminishing height, until the figure collapses into the axis OL. Then the same phenomenon is repeated downward in reversed

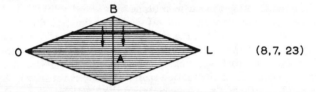

$$(8, 7, 23)$$

sequence, building up the triangle gradually, until the mirror image of the original triangle is restored. We are now at the time moment $t = l$ and the half-cycle of the entire period is accomplished. The second half-cycle repeats the same motion with opposite sign.

If the plucking occurs at a point away from the centre, the descending straight line will not be horizontal. Furthermore, the triangle which develops below the zero-line is the result of two reflections, about the X *and* about the Y axes. In other respects the phenomenon is quite similar to the previous case.

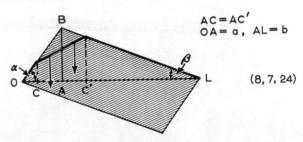

$$AC = AC'$$
$$OA = a, \quad AL = b$$

$$(8, 7, 24)$$

If we follow the motion of a single particle of the string as a function of time, we observe that the motion is now less jerky than it was in (19) but still far from a regular vibration. The following figure plots a few characteristic motion patterns for central and non-central plucking.

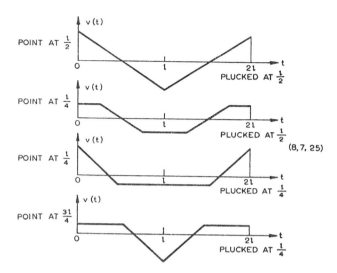

Problem 339. Obtain the solution of the boundary value problem (2), (3), with the initial conditions

$$v(x, 0) = f(x)$$
$$\frac{\partial v}{\partial t}(x, 0) = g(x) \tag{8.7.26}$$

in terms of the Green's function $G(x, \xi; t)$ $(\tau = 0)$.

[Answer:

$$v(x, t) = \int_0^l \left[f(\xi) \frac{\partial G(x, \xi; t)}{\partial t} + g(\xi) G(x, \xi; t) \right] d\xi \tag{8.7.27}$$

Problem 340. a) By making the transformation

$$\eta_1 = x + t$$
$$\eta_2 = x - t \tag{8.7.28}$$

obtain the general solution of the differential equation (2) in the form $P(x + t) + Q(x - t)$ which may also be written in the form

$$\tfrac{1}{2}[A(x + t) + A(x - t)] + \tfrac{1}{2}[B(x + t) - B(x - t)] \tag{8.7.29}$$

b) Show that the boundary conditions (3) demand the following extension of the functions $A(p)$ and $B(p)$ beyond the original range $[0, l]$:

$$A(-p) = -A(p) \quad A(l + p) = -A(l - p)$$
$$B(-p) = B(p) \quad B(l + p) = B(l - p) \tag{8.7.30}$$

c) Obtain the solutions (17) and (21) on the basis of this method, without any eigenfunction analysis.

Problem 341. Show geometrically that the conservation law of energy:

$$T + V = E = \text{const.} \tag{8.7.31}$$

is satisfied for the plucked string (24).

[Answer:

Putting $OA = a$, $AL = b$,

$$V = \tan^2 \alpha (a \cos \alpha - \xi) + \tan^2 \beta (b \cos \beta - \xi) + \left[\frac{\tan \alpha - \tan \beta}{2} \right]^2 2\xi$$

$$= \text{const} - \frac{\xi}{2} (\tan \alpha + \tan \beta)^2 \tag{8.7.32}$$

$$T = \left[\frac{\tan \alpha + \tan \beta}{2} \right]^2 2\xi \tag{8.7.33}]$$

8.8. The analytical nature of hyperbolic differential operators

If we examine the solutions obtained in the previous sections, we can derive some conclusions of principal significance. The differential equation (7.1) shows a characteristic property. The term in which the time t appears enters the differential equation with the *opposite* sign, compared with the other terms. Whether the Laplacian operator Δ contains only one space variable (vibrating string), or two (vibrating membrane) or three coordinates (the wave equation), in all cases the space terms enter the equation with a negative, the time with a positive sign. It is this change of sign which has a profound effect on the analytical character of the solutions. When we were dealing with the solution of the potential equation (in Sections 4 and 5), the boundary values on the surface of a sphere (or in two dimensions on a circle) were analysed in terms of functions of the Fourier type. The expansion of the boundary values into these orthogonal functions does not require any great regularity. If the boundary data are *sectionally continuous* functions, with arbitrary discontinuities at the common boundaries between the continuous regions, this is entirely sufficient to guarantee the convergence of the expansion. If then we investigate the solution *inside* the domain, we find that the potential function shows a surprisingly high degree of regularity. Although the boundary data could not even be differentiated since the first derivatives went to infinity at the points of discontinuity, yet at any inside point the function $V(x)$ is *differentiable any number of times*, with respect to any of the coordinates. The integration with the help of the Green's function has this remarkable smoothing effect as its consequence. We have found explicit expressions for this Green's function and these expressions show that $G(x, \xi)$, considered as a function of x, is an *analytical* function of x, which can be differentiated partially any number of times. The only point of singularity is the point $x = \xi$, but this equality can never occur, since x is in the inside, ξ on the boundary of the given domain.

The solution of a hyperbolic type of differential equation behaves very differently. Here the singularities *propagate from the boundary into the inner domain.* The propagation occurs along the so-called "character-istics". For example in the case of the vibrating string the characteristics are two straight lines at 45° which emanate from the singular point. We can demonstrate this behaviour if we examine the example of the plucked string. Initially the string had the shape (23), with a singular point at B in which the function $f(x)$ is continuous but its tangent becomes dis-continuous. With increasing time this point moved with constant velocity to the left and to the right, changing the contour of the string to a quadrangle. The *discontinuity in the tangent remained.*

Let us now pay attention to the fact that the operator on the left side of (7.2) requires the formation of the second derivative with respect to t and x. This means that not only must $v(x, t)$ be a continuous function of x and t, but *even the first derivative must be everywhere continuous* since otherwise the second derivative would become infinite at certain points and our operator would lose its significance. Hence the shape of the string as pictured in (24) is not a mathematically or physically acceptable initial condition. We have to subject the initial displacement $v(x, 0)$ to the constraint that it must be a *continuous and differentiable function*, with a piecewise continuous second derivative. Hence we must assume that the sharp corner shown in the figure is in fact replaced by a *round* corner which changes the tangent sharply but not instantly. This smoothing can occur along a microscopically small portion of the string and causes a merely local disturbance which has no effect in the distance. But without this smoothing the nature of the differential equation is violated.

The same objection has to be raised even more pointedly against the form (7.18) of the Green's function. Here the *function itself* becomes dis-continuous at two points which spread to the right and to the left with the velocity 1. We should remember, however, that the Green's function is not more than an auxiliary function with the help of which the solution is obtained. The right side $\delta(x, \xi)$ of the defining differential equation prevents the function $G(x, \xi)$ from satisfying the differential equation at the critical point $x = \xi$. But we would expect that it does satisfy the given differential equation everywhere else, in view of the local nature of the delta function. And this is indeed the case with the Green's functions (4.55) and (4.56) of the potential equation. But if we make a comparison with the Green's function (7.18), we observe an important difference. The previous Green's function remained a regular solution of the homogeneous differential equation everywhere *except* at the point $x = \xi$, where the function went out of bound. The new Green's function fails to satisfy the differential equation not only at the source $x = \xi$, but all along the two characteristics which emanate from the point $x = \xi$. And even more can be said. The delta function is *not* a legitimate input function, as we have emphasized at several occasions. It is the *limit* of a legitimate input function. An arbitrarily small local smoothing changed the extreme nature of the delta function to a legitimate function. Then we had a right side which was generally zero but jumped from zero to the high constant value $1/\epsilon$ in the vicinity ϵ of the point

$x = \xi$. In our present problem, if such a function is applied as input function, this means in physical terms that a sharp hammer blow of short (but not instantaneous) duration hits a small (but finite) portion of the string, in the neighbourhood of the point $x = \xi$. The resulting solution becomes quite similar to the previous figure (7.18), but the two ends of the hill descend now with a *finite* tangent and the discontinuity of the function is avoided. This, however, is *not enough*. We have just seen that the operator (7.2) requires that the solution must have a continuous first derivative in both x and t. Hence the solution (7.18) is not acceptable as the solution of the given differential equation, even if the ends are slanted. We have to demand that the sharp corners of the figure are rounded off. Hence the Green's function is not *one* but *two* smoothings away from a permissible solution of the differential equation and that again means that the right side of a hyperbolic type of differential equation *cannot be prescribed with the same freedom* that is permitted in the case of an elliptic (or, as we will see later, a parabolic) equation. There the right side need not be more than piecewise continuous. Now we have to require the right side to be a continuous *and even differentiable* function. Without this restriction our problem becomes unsolvable because, while we should obtain a solution $v(x, t)$ which is in itself continuous, the first partial derivative v_x or v_t, or both, would go out of bound.

This phenomenon must give us a certain disquietude if we look at it from the standpoint of eigenfunction expansion. A piecewise continuous function can certainly be expanded into a convergent series of orthogonal functions and now, if we construct the solution by dividing by the eigenvalues λ_i, the convergence should become *better* rather than worse, since the eigenfunctions of high order—which may cause divergence—are divided by numbers which grow to infinity. Now it is true that the difficulty is not caused by the function $v(x, t)$ itself, but by the partial derivatives v_x and v_t which are less smooth than v itself. But a closer analysis reveals that we cannot escape by this argument. We have seen in Chapter 5.11 that partial differential equations can be reduced to a "canonical form" in which no higher than first derivatives appear. This means in our problem that the partial derivatives v_x and v_t can be added to the function v as new independent variables, with the result that we get a coupled system of *three* equations instead of the previous single equation. In this system we need no differentiation any more to get the quantities v_x and v_t; they are fully-fledged components of the solution, which now consists of the three quantities v, v_x, v_t.

Let us carry out the actual procedure according to the general technique discussed in Chapter 5.11. We introduce the two partial derivatives as new variables:

$$v_x - w_1 = 0$$
$$v_t - w_2 = 0 \qquad (8.8.1)$$

(making use of the simplified notation v_x for $\partial v/\partial x$, v_t for $\partial v/\partial t$). Now our

Lagrangian becomes, according to (7.12), if we multiply by -1 for the sake of convenience:

$$L = \tfrac{1}{2}(v_x{}^2 - v_t{}^2)$$
$$= \tfrac{1}{2}(w_1{}^2 - w_2{}^2) \tag{8.8.2}$$

We modify the Lagrangian according to the method of the Lagrangian multiplier:

$$L' = \tfrac{1}{2}(w_1{}^2 - w_2{}^2) + p_1(v_x - w_1) + p_2(v_t - w_2)$$
$$= p_1 v_x + p_2 v_t - H \tag{8.8.3}$$

with

$$H = p_1 w_1 + p_2 w_2 - \tfrac{1}{2}(w_1{}^2 - w_2{}^2) \tag{8.8.4}$$

But the algebraic variables w_1 and w_2 can be eliminated which reduces H to

$$H = \tfrac{1}{2}(p_1{}^2 - p_2{}^2) \tag{8.8.5}$$

and thus our final Lagrangian becomes

$$L'' = p_1 v_x + p_2 v_t - \tfrac{1}{2}(p_1{}^2 - p_2{}^2) \tag{8.8.6}$$

Variation with respect to p_1, p_2, and v yields the system

$$v_x - p_1 = 0$$
$$v_t + p_2 = 0 \tag{8.8.7}$$
$$-p_{1x} - p_{2t} = 0$$

with the boundary conditions

$$v(0, t) = v(l, t) = 0 \tag{8.8.8}$$

and the initial conditions

$$v(x, 0) = p_2(x, 0) = 0 \tag{8.8.9}$$

These latter conditions arise from the fact that we want to employ the method discussed in Section 2 which transforms the inhomogeneous boundary value problem (in our case initial value problem) with a homogeneous differential equation into an inhomogeneous differential equation with homogeneous boundary conditions. The "right side" $\beta(x, t)$ of this inhomogeneous equation has to be put in the new formulation (7) in the place of the zero of the third equation, while the right sides of the first and the second equation remain zero. If we eliminate p_1 and p_2 from these two equations and substitute in the third equation, we are back at the single equation

$$\frac{\partial^2 v}{\partial t^2} - \frac{\partial^2 v}{\partial x^2} = \beta(x, t) \tag{8.8.10}$$

We will now place the paradox in sharp focus by formulating it in exact quantitative terms. We assume that we have solved the eigenvalue problem associated with our system (7) (whose differential operator is self-adjoint but not the boundary conditions (9)). We will call the orthogonal eigenfunctions of our problem $U_i(x)$ and $V_i(x)$, each one of these functions representing in fact a *vector*, with the three components

$$V_i = (p_{1i}, p_{2i}, v_i)$$
$$U_i = (q_{1i}, q_{2i}, u_i)$$

(8.8.11)

We prescribe $\beta(x, t)$ as a finite, sectionally continuous function of bounded variation, with a finite number of discontinuities in the variable x, if we freeze t to a constant value (the same could be done with the variable t, considering x as a constant). We expand the vector of the right side $(0, 0, \beta)$ into the eigenfunctions $U_i(x, t)$:

$$(0, 0, \beta) = \sum_{i=1}^{\infty} c_i U_i(x, t)$$

(8.8.12)

with

$$c_i = \int \beta(\xi, \tau) U_i^{(3)}(\xi, \tau) d\xi d\tau$$

(8.8.13)

Correspondingly the vector of the solution can be expanded—in accordance with the general theory—into the following series:

$$(p_1, p_2, v) = \sum_{i=1}^{\infty} \frac{c_i V_i(x, t)}{\lambda_i}$$

(8.8.14)

Now, by the properties of orthogonal expansions we obtain for the "length" or "norm" of the function $\beta(x, t)$ (cf. Chapter 4.7):

$$\beta^2 = \int \beta^2(x, t) dx dt = \sum_{i=1}^{\infty} c_i^2$$

(8.8.15)

while the corresponding norm of the solution becomes

$$\int (p_1^2 + p_2^2 + p_3^2) dx dt = \sum_{i=1}^{\infty} \frac{c_i^2}{\lambda_i^2}$$

(8.8.16)

This quantity, however, *cannot converge* under the given circumstances. We have found in our previous discussion that a discontinuity of $\beta(x, t)$ with respect to x causes that $p_1 = v_x$ becomes *infinite* at that point and the solution loses its quadratic integrability. Hence the sum (16) must go to infinity, although the sum (15) converged. But how can it happen that the solution is *less* convergent than the right side, when in fact we divide the c_i by the λ_i which with increasing i go to infinity?

We leave the resolution of this apparent contradiction to Section 16, where we will be able to understand it as a special case of a much more general phenomenon.

8.9. The heat flow equation

The differential equation which describes the conduction of heat, is formally very similar to the wave-equation in one, two, or three dimensions. The only difference is that the second derivative with respect to t is replaced by the first derivative. The fundamental differential equation thus becomes

$$\frac{\partial v}{\partial t} - \Delta v = 0 \tag{8.9.1}$$

with the proper boundary and initial conditions. If we once more solve the time-independent eigenvalue problem

$$\Delta v_i + \lambda_i v_i = 0 \tag{8.9.2}$$

we can once more expand $v(\tau, t)$—where τ symbolises the coordinates of space—in the following form:

$$v_i(\tau, t) = \sum_{i=1}^{\infty} c_i(t) v_i(\tau) \tag{8.9.3}$$

obtaining for $c(t)$ the ordinary differential equation

$$c'_i + \lambda_i c_i = 0 \tag{8.9.4}$$

which has the solution

$$c_i(t) = c_i e^{-\lambda_i t} \tag{8.9.5}$$

Hence the complete solution becomes

$$v(\tau, t) = \sum_{i=1}^{\infty} c_i e^{-\lambda_i t} v_i(\tau) \tag{8.9.6}$$

The undetermined coefficients c_i can be obtained by prescribing the initial condition

$$v(\tau, 0) = f(\tau) \tag{8.9.7}$$

Then

$$c_i = \int f(\tau) v_i(\tau) d\tau \tag{8.9.8}$$

Although formally our solution is very similar to that of the vibrating string (or membrane, or space), there is in fact a fundamental difference between the two solutions, because the appearance of an *exponential* in place of the earlier *periodic* function in time changes the analytical nature of the solution profoundly. We have observed in the problem of the vibrating string that any discontinuity in the initial position or its tangent propagated into the inside of the domain and maintained its character unchanged. Such behaviour is now out of the question. The infinite sum (6) has exceedingly good convergence because for large values of i also λ_i becomes large and, as i goes to infinity, the factor $e^{-\lambda_i t}$ cuts down so effectively the contribution of the high order terms that the sum (6) *and even all its derivatives of arbitrary order* (with respect to x or t) remain convergent, for all values $t > 0$. Hence an initial discontinuity in function or derivative is immediately *smoothed out* by the phenomenon of heat conduction and the initially irregular function $f(\tau)$ is transformed into a function $v(\tau, t)$ which is *analytical* in all its derivatives. In fact, Weierstrass made use of this property of the phenomenon of heat conduction to demonstrate that an arbitrarily non-analytical, although single-valued and sectionally continuous function can be approximated to any degree by strictly analytical functions (for example polynomials).

If we consider a one-dimensional heat flow in a rod whose ends at $x = 0$ and $x = l$ are kept at zero temperature, we have a problem which is quite analogous to the motion of the vibrating string. The function $v(x, t)$ satisfies once more the boundary conditions

$$v(0, t) = v(l, t) = 0 \tag{8.9.9}$$

and the differential equation

$$\frac{\partial v}{\partial t} - \frac{\partial^2 v}{\partial x^2} = 0 \tag{8.9.10}$$

is once more solvable by a similar separation of variables as employed in (7.21):

$$v(x, t) = \sum_{k=1}^{\infty} c_k e^{-\pi^2 l^{-2} k^2 t} \sin \frac{k\pi}{l} x \tag{8.9.11}$$

with

$$c_k = \frac{2}{l} \int_0^l f(\xi) \sin k \frac{\pi}{l} \xi \, d\xi \tag{8.9.12}$$

This solution may also be written in terms of a Green's function $G(x, \xi; t)$:

$$v(x, t) = \int_0^l f(\xi) G(x, \xi; t) d\xi \tag{8.9.13}$$

where

$$G(x, \xi; t) = \frac{2}{l} \sum_{k=1}^{\infty} e^{-\pi^2 l^2 k^2 t} \sin k \frac{\pi}{l} x \sin k \frac{\pi}{l} \xi$$

$$= F(x - \xi, t) - F(x + \xi, t) \tag{8.9.14}$$

if we define the following function of two variables:

$$F(p, t) = \frac{1}{l} \sum_{k=1}^{\infty} \cos k \frac{\pi}{l} p e^{-\pi^2 l^{-2} k^2 t} \tag{8.9.15}$$

The significance of the Green's function (14) is the heat flow generated by a heat source of the intensity 1, applied during an infinitesimal time at $t = 0$, and in the infinitesimal neighbourhood of the point $x = \xi$. Such a heat flow would occur even if the body extended on both sides to infinity

We can thus ask for the limiting value of the Green's function if l goes to infinity. Accordingly, we will place the point ξ at the *midpoint* of the rod: $\xi = l/2$ and put once more (as we have done in the problem of the vibrating string):

$$x = \frac{l}{2} + x_1 \tag{8.9.16}$$

Then the function we want to get becomes

$$F(x_1, t) - F(l + x_1, t) \tag{8.9.17}$$

letting l go to infinity. But then the sum on the right side of (15) becomes more and more an integral; in the limit, as l grows to infinity, we obtain

$$\int_0^\infty \cos \pi p z e^{-\pi^2 t z^2} dz \tag{8.9.18}$$

The integrand is the real part of

$$e^{-\pi^2 t z^2 + i \pi p z} = e^{-(\pi z \sqrt{t} - i(p/2)\sqrt{t})^2} e^{-p^2/4t} \tag{8.9.19}$$

The integration is thus reducible to the definite integral

$$\int_0^\infty e^{-\pi^2 t z^2} dz = \frac{1}{\pi \sqrt{t}} \int_0^\infty e^{-z^2} dz = \frac{1}{2\sqrt{\pi t}} \tag{8.9.20}$$

and we obtain

$$F_\infty(p, t) = \frac{1}{2\sqrt{\pi t}} e^{-p^2/4t} \tag{8.9.21}$$

Thus the Green's function of the heat flow equation in one dimension for an infinitely extended medium becomes

$$G(x, \xi; t - \tau) = 0 \qquad (t < \tau)$$
$$= \frac{1}{2\sqrt{\pi}\sqrt{t - \tau}} e^{-(x-\xi)^2/4(t-\tau)} \quad (t > \tau) \tag{8.9.22}$$

This is the fundamental solution of the heat conduction equation, comparable to the solution $-(4\pi r)^{-1}$ in the case of the three-dimensional potential equation.

Problem 342. Define the Green's function of the heat flow equation by the standard technique (making use of the adjoint equation which does not coincide with the original one) and show that the Green's function $G(x, \xi; t, \tau)$ of the problem (7), (9), (10) is in the following relation to the Green's function (14):

$$G(x, \xi; t, \tau) = G(x, \xi; t - \tau) \qquad (t > \tau)$$
$$= 0 \qquad (t < \tau) \tag{8.9.23}$$

Problem 343. Change the boundary conditions (9) to

$$v_x(0, t) = v_x(l, t) = 0 \tag{8.9.24}$$

which hold if the two ends of the rod are insulated against heat losses.

a) Show that any solution of the heat flow equation under these boundary conditions satisfies the condition

$$\int_0^l v(x, t)dx = \text{const.} \tag{8.9.25}$$

b) Show that, as t increases to infinity, we get in the limit

$$v(x, \infty) = C \tag{8.9.26}$$

where

$$C = \frac{1}{l}\int_0^l f(\xi)d\xi \tag{8.9.27}$$

c) Demonstrate that the solution (22), if integrated with respect to x between $\pm\infty$, gives 1.

8.10. Minimum problems with constraints

Many of the previously considered boundary value problems were characterised by self-adjoint operators (and boundary conditions) and were thus derivable by minimising a certain integral. This is particularly the case if the equilibrium position of a mechanical system is to be found, composed of continuously distributed masses. The potential energy of such a system is given by a definite integral, extended over the domain of the masses. For example the equilibrium of a stretched membrane requires the minimisation of the integral

$$Q = \frac{1}{2}\int\left[\left(\frac{\partial v}{\partial x}\right)^2 + \left(\frac{\partial v}{\partial y}\right)^2\right]dxdy \tag{8.10.1}$$

with prescribed boundary values, determined by the closed space curve which terminates the membrane. We will consider the simple case of a plane circular membrane of unit radius whose frame is kept at the constant distance a from the horizontal plane. Here the solution

$$v(r, \theta) = \text{const.} = a \tag{8.10.2}$$

makes the potential energy to a minimum, namely zero.

We will now restrict the movability of the membrane by a vertical peg of circular cross-section which pins the membrane down to the horizontal plane. Let ϵ be the radius of the peg and let us make ϵ smaller and smaller.

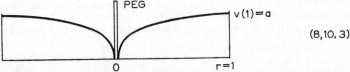

(8,10,3)

We then approach the limit in which only *one point* of the membrane is pinned down. For the sake of simplicity we assume that the peg is centrally applied.

Our problem is a "minimum problem with constraints" since the potential energy has to be minimised under the auxiliary condition that $v(r, \theta)$ is fixed not only at $r = 1$ but also at $r = \epsilon$. The differential equation of the free membrane (4.20)

$$\Delta v = \frac{\partial r \frac{\partial v}{\partial r}}{r \partial r} + \frac{1}{r^2} \frac{\partial^2 v}{\partial \theta^2} = 0$$

will hold between $r = [\epsilon, 1]$ while between $r = 0$ and ϵ we have to put

$$\Delta v = \beta(r) \tag{8.10.4}$$

where the right side is proportional to the force density required for the pinning down of the membrane.

Now we have seen that the only possible solution of (3) under circular symmetry is

$$v(r) = A \log r + B \tag{8.10.5}$$

and since the constraints demand the two conditions

$$v(\epsilon) = 0, \quad v(1) = a \tag{8.10.6}$$

we obtain the solution

$$v(r) = a\left(1 - \frac{\log r}{\log \epsilon}\right) \tag{8.10.7}$$

Let us now multiply the inhomogeneous equation (4) by the area element $2\pi r dr$ and integrate between 0 and ϵ. We then obtain

$$2\pi \int_0^\epsilon \beta(r) r dr = - \frac{2\pi}{\log \epsilon} \tag{8.10.8}$$

The quantity on the left side is proportional to the *total force* required for pinning the membrane down. We see that this force is becoming smaller and smaller as the radius ϵ of the peg decreases. At the same time the

(8,10,9)

solution (7) shows that the indention caused by the peg becomes more and more *local* since for very small $\epsilon/v(r)$ becomes practically $v = a$, except in the immediate neighbourhood of $r = \epsilon$. In the limit, as ϵ recedes to zero, we obtain the following solution: the membrane is everywhere horizontal but it is pinned down at $r = 0$. This means that $v(r)$ assumes everywhere the constant value $v = a$, except at $r = 0$, where $v(0) = 0$.

While this solution exists *as a limit*, it is *not a legitimate solution* because it is a function which cannot be differentiated at $r = 0$ and hence does not belong to that class of functions which are demanded in the process of minimising the integral (1). We thus encounter the peculiar situation that if we require the minimisation of the integral (1) with the boundary condition $v(1) = a$ and the inside condition

$$v(0) = 0 \qquad (8.10.10)$$

this problem has *no solution*. We can make the given integral *as small as we wish but not zero*, and thus no definite minimum can be found under the given conditions.

The situation is quite different, however, if the membrane is not pinned down at a *point* but along a *line*. A line has the same dimensionality as the boundary and a constraint along a line can in fact be considered as a *boundary condition*, if we add the line of constraint to the outer boundary. Constraints of this type are of frequent occurrence in physical problems. We may consider for example a three-dimensional flow problem of a fluid which is forced to flow around an obstacle which is given in the form of a surface. Or we may have a problem in electrostatics in which the potential along an inner surface is given as zero, since the surface is earthed. Again, in a diffraction problem which requires the solution of the differential equation

$$\Delta V + \omega^2 V = 0 \qquad (8.10.11)$$

the function V along an absorbing screen may be prescribed as zero.

In problems of this type it is often desirable to consider the added conditions as constraints, rather than as parts of the boundary conditions. But then we have to take into consideration that *the maintaining of a constraint demands an external force*—corresponding to the appearance of a "right side" of the given differential equation—and this is equivalent of saying that *at the points of constraint the differential equation will be violated*. The differential equation now to be solved can be written in the form

$$Dv(x) = \beta(x) \qquad (8.10.12)$$

where $\beta(x)$ is not zero in the domain of the constraint and the symbolic notation x refers again to an arbitrary point of an n-dimensional manifold. The solution of our new problem will be once more a function $v_0(x)$ which satisfies the given inhomogeneous boundary data, together with the homogeneous differential equation, but to this solution we now have to add the solution of the inhomogeneous equation (12), with homogeneous boundary conditions. This can be done in terms of the Green's function of our problem and thus the complete solution of our problem can be given as follows:

$$v(x) = v_0(x) + \int G(x, \xi)\beta(\xi)d\xi \qquad (8.10.13)$$

But now we must make use of the fact that the given constraint exists in a certain *sub*-domain of our space, more precisely on a given inner *surface* which we want to denote by σ', in distinction to the boundary surface σ. Accordingly we have to rewrite the equation (13) as follows:

$$v(x) = v(x_0) + \int G(x, \sigma')\beta(\sigma')d\sigma' \qquad (8.10.14)$$

The quantity $\beta(\sigma')$ is proportional to the density of the surface force which is required for the fulfilment of the given constraint. But this force is not a given quantity. What is given is the *constraint*, which demands that $v(x)$ becomes zero on the surface σ', or more generally that $v(x)$ becomes some *prescribed function* on this surface. The force needed for the maintenance of this condition adjusts itself in such a way that the constraint is satisfied. Let us express this physical situation in mathematical terms. We will denote by s' an arbitrarily selected point of the inner surface σ'. Then our constraint demands that the following equation shall be satisfied:

$$v(s') = v_0(s') + \int G(s', \sigma')\beta(\sigma')d\sigma' \tag{8.10.15}$$

The peculiar feature of this equation—called an "integral equation"—is that it is not $\beta(s')$ that is given to us (in order to obtain $v(s')$ by the process of integration), but $\beta(s')$ is the *unknown function* and the left side $v(s')$ is the *given function* (together with $v_0(s')$ which can be transferred to the left side and combined with $v(s')$). The function $G(s', \sigma')$ is called the "kernel" of the integral equation. The general form of an integral equation is thus

$$\int K(x, \xi)f(\xi)d\xi = g(x) \tag{8.10.16}$$

where $f(\xi)$ is the unknown and $g(x)$ the given function.

Generally integration is a smoothing operation and the function on the left and on the right side of an integral equation cannot belong to the same class of functions. If the kernel function is everywhere *bounded*, the result of the integration on the left side of (16) is that we obtain a continuous and even *differentiable* function. Hence the given function $g(x)$ must be prescribed as a function of sufficient regularity to make the integral equation solvable. But in the potential problem discussed in the beginning of this section the situation is quite different. Our kernel $K(x, \xi)$ is here the Green's function $G(x, \xi)$ of the potential equation which goes out of bound at the point $x = \xi$. For example in two dimensions the Green's function goes to infinity at the critical point with the strength $\log r_{x\xi}$, in three dimensions with the strength $r_{x\xi}^{-1}$. Furthermore, the integration is restricted to a lower dimensional manifold—in two dimensions to a curve, in three dimensions to a surface—which increases the strength of the singularity. Integral equations with such kernels are called "singular integral equations". The singularity of the kernel counterbalances the usual discrepancy which exists between the smoothness of the functions $f(x)$ and $g(x)$.

Problem 344. Consider the minimum problem (1) in three dimensions, with the boundary condition $v = a$ on the sphere $r = 1$, and the added constraint $v(0) = 0$. Show that the minimum can be made as small as we wish, but not zero.
[Answer:
Choose

$$v(r) = 0, \quad (r \le \epsilon); \quad v(r) = \frac{a}{1 - \epsilon}\left(1 - \frac{\epsilon}{r}\right) \quad (r \ge \epsilon)$$

Then

$$Q = \frac{2\pi a^2 \epsilon}{1 - \epsilon} \qquad (8.10.17)]$$

8.11. Integral equations in the service of boundary value problems

The method of the "separation of variables" is of eminent theoretical importance since it yields almost all the fundamental function classes of mathematical physics. As a tool for solving boundary value problems it is of limited applicability because it is restricted to boundaries of simple shape. Moreover, even in the case of boundaries of high symmetry the kind of boundary conditions prescribed must be of considerable simplicity. For example in the case of the potential equation we have succeeded in solving the boundary value problem of the first and the second kind. But if we consider the more general boundary condition (5.45), we do not succeed with the separation in polar coordinates because we do not obtain any explicit expression for the determination of the expansion coefficients.

Under these circumstances it is of great advantage that the theory of "integral equations" can help solve boundary value problems. Any boundary value problem can be formulated as an integral equation and thus the methods for solving integral equations are applicable to the solution of boundary value problems.

We have seen in the previous section how certain constraints on an inner boundary could be solved in two ways. We could extend the outer boundary by the inner boundary and consider the entire problem as one single boundary value problem; the constraints then appear as boundary data. But we could also use a more direct approach, replacing the given constraints on the inner boundary by an integral equation. We can now extend this method to the outer boundary and reduce our entire problem to the solution of an integral equation.

We will illustrate the method by the example of the potential equation, although it is equally applicable to boundary value problems of the parabolic or hyperbolic type. Let our problem be to solve the potential equation in a given closed domain in which the Laplacian equation is to be solved. It would be enough to prescribe the boundary values of the potential function, or the normal derivatives, but we want to over-determine the problem by giving both sets of values. This cannot be done freely, of course. If we prescribe improper boundary values, the forces of constraint will come into operation, alter the differential equation in the vicinity of the boundary, and we shall not obtain what we want. If, however, the prescribed surplus data are the *correct* data, we have done no harm to the given problem. The question is merely from where to take these surplus data, but we will find a solution to this problem.

The giving of these surplus data has the following fortunate effect. We can solve our problem in terms of the Green's function. But in the original formulation the construction of the Green's function is not an easy task. We have that part of the Green's function which becomes singular at the point $x = \xi$, but to this part we had to add a regular solution of the potential equation (cf. (4.47–49)), chosen in such a way that on the boundary σ the required homogeneous boundary conditions of the Green's function are satisfied. This is now quite different in our present problem. The adjoint equation—which defines the Green's function—is now strongly under-determined and in fact we obtain *no boundary conditions* of any kind for the function $G(x, \xi)$. Hence we can choose the added function $V(x)$ as *any* solution of the potential equation, even as *zero*. Hence the over-determination of the problem has the fortunate consequence that the Green's function can be explicitly given in the form of a simple power of the distance $r_{x\xi}$ (times a constant), in particular in two dimensions

$$G(x, \xi) = \frac{1}{2\pi} \log r_{x\xi} \qquad (8.11.1)$$

and in three dimensions

$$G(x, \xi) = -\frac{1}{4\pi} \frac{1}{r_{x\xi}} \qquad (8.11.2)$$

Then the solution of our over-determined problem appears in the following form (making use of the extended Green's identity (4.57), but generalised to any boundary surface instead of a sphere):

$$V(x) = -\frac{1}{4\pi} \int \left[V(\sigma) \frac{\partial}{\partial \nu} \frac{1}{r_{\sigma x}} - \frac{1}{r_{\sigma x}} \frac{\partial V(\sigma)}{\partial \nu} \right] d\sigma \qquad (8.11.3)$$

Now the fact that our data have been properly given has the following consequence. If we approach with the inside point x any point s on the boundary, we actually approach the given boundary value $V(s)$. Hence we can consider as the criterion of properly given boundary values that the equation (3) remains valid even in the limit, when the point x coincides with the boundary point s. Then we get the following integral relation, valid for any point s of the boundary surface σ:

$$V(s) = -\frac{1}{4\pi} \int \left[V(\sigma) \frac{\partial}{\partial \nu} \frac{1}{r_{\sigma s}} - \frac{1}{r_{\sigma s}} \frac{\partial V(\sigma)}{\partial \nu} \right] d\sigma \qquad (8.11.4)$$

Now, instead of using this relation as a check on the prescribed boundary data, we can use it for the *determination* of the surplus data. Let us assume for example that the data corresponding to the Neumann problem are given. Then the second integral on the right side of (4) is at our disposal and we obtain for the data $V(s)$ on the boundary the following integral equation:

$$V(s) + \frac{1}{4\pi} \int V(\sigma) \frac{\partial}{\partial \nu} \frac{1}{r_{\sigma s}} d\sigma = \frac{1}{4\pi} \int \frac{1}{r_{\sigma s}} \frac{\partial V(\sigma)}{\partial \nu} d\sigma \qquad (8.11.5)$$

This integral equation is of the following general form:

$$f(x) - \lambda \int K(x, \xi) f(\xi) d\xi = g(x) \qquad (8.11.6)$$

to be solved for $f(x)$, with given $g(x)$.

The general theory of integral equations of this type was developed by
I. Fredholm (1900) and subsequently the same subject gave rise to a very
extensive literature.* All the methods which have been designed for the
solution of the Fredholm type of integral equations, are immediately
applicable to the solution of boundary value problems, under much more
general conditions than those under which a solution is obtainable by the
separation of variables, or by an explicit construction of the Green's function.
The basic method may be characterised as follows. It is often possible to
give the solution of the differential equation

$$\check{D}G(x, \xi) = \delta(x, \xi) \tag{8.11.7}$$

if we do not demand any additional boundary conditions. This can be
achieved by over-determining the original problem by the addition of
surplus data. These data are obtained by solving an integral equation for
the boundary surface.

We add one more example by considering the general boundary value
problem (5.45) for an arbitrary boundary surface σ. For this purpose we
write the integral relation (4) as follows:

$$V(s) = -\frac{1}{4\pi} \int \left[V(\sigma)\left(\frac{\partial}{\partial \nu}\frac{1}{r_{\sigma s}} + \frac{\gamma(\sigma)}{r_{\sigma s}}\right) - \frac{1}{r_{\sigma s}}\left(\frac{\partial V(\sigma)}{\partial \nu} + \gamma(\sigma)V(\sigma)\right)\right] d\sigma \tag{8.11.8}$$

or, putting the given data to the right side:

$$V(s) + \frac{1}{4\pi}\int V(\sigma)\left(\frac{\partial}{\partial \nu}\frac{1}{r_{\sigma s}} + \frac{\gamma(\sigma)}{r_{\sigma s}}\right) = \int \frac{g(\sigma)}{r_{\sigma s}} d\sigma \tag{8.11.9}$$

This is once more a Fredholm type of integral equation, only the kernel
$K(x, \xi)$ has changed, compared with the previous problem (5).

From the standpoint of obtaining an explicit solution in numerical terms
we may fare better if we avoid the solution of an integral equation whose
kernel goes out of bound at the point $s = \sigma$. The surplus data are also
obtainable by making use of the *compatibility conditions* which have to be
satisfied by our data. We then have a greater flexibility at our disposal
because the compatibility conditions appear in the form

$$\int \left[u(\sigma)\frac{\partial V(\sigma)}{\partial \nu} - V(\sigma)\frac{\partial u(\sigma)}{\partial \nu}\right] d\sigma = 0 \tag{8.11.10}$$

where $u(\sigma)$ can be chosen as any function which satisfies the homogeneous
equation

$$\Delta u = 0 \tag{8.11.11}$$

and is free of any singularities inside the given domain. We can once
more choose as our $u(\sigma)$ the reciprocal distance $r_{\sigma x}^{-1}$, provided that the fixed
point x is chosen as any point *outside* the boundary surface. By putting the
point x sufficiently near to the surface, yet not directly *on* the surface we

---

* For a more thorough study of the theory of integral equations, cf. [8] and [11] of
the Chapter Bibliography.

avoid the singularity of the kernel and reduce the determination of the surplus data numerically to the solution of a well-conditioned large-scale system of ordinary linear equations.

Problem 345. Show on the basis of (3) that the potential function $V(\tau)$ is everywhere inside the domain τ an *analytical function* of the rectangular coordinates (x, y, z) (that is the partial derivatives of all orders exist), although the boundary values themselves need not be analytical.

8.12. The conservation laws of mechanics

The boundary values associated with a given homogeneous differential equation are not always freely at our disposal. We have just seen that in a potential problem, if we prescribe both the function and its normal derivative on the boundary, we have strongly over-determined our problem and accordingly we have to satisfy an infinity of compatibility conditions. But even without over-determination our data may be subject to constraints. We have seen for example that the data of the Neumann problem had to satisfy the condition (5.38) which for an arbitrary boundary surface σ becomes

$$\int \frac{\partial V}{\partial \nu} \, d\sigma = 0 \tag{8.12.1}$$

Under all circumstances we have an unfailing method by which we can decide how much or how little the given data are constrained. The decision lies with the *adjoint homogeneous equation* (cf. Section 8.2, particularly (2.18)). To every independent non-zero solution of the adjoint homogeneous equation belongs a definite compatibility condition, and vice versa, these are *all* the compatibility conditions that our data have to satisfy. In physical problems these compatibility conditions have frequently an important significance, as they express the *conservation* of some physical quantity. Of particularly fundamental importance are the *conservation laws of mechanics*, which in the case of continuously distributed masses are the consequence of certain compatiblity conditions of partial differential equations.

The physical state of such masses is characterised by a fundamental set of quantities, called the "matter tensor". The components of this tensor form a matrix and are thus characterised by two subscripts. We can conceive the components of the matter tensor as an $n \times n$ matrix of the n-dimensional space, whose components are continuous and differentiable functions of the coordinates. We will denote this tensor by T_{ik}, with the understanding that the subscripts i and k assume independently the values 1, 2 in two dimensions, 1, 2, 3 in three dimensions, and 1, 2, 3, 4 in four dimensions (the fourth dimension is in close relation to the time t). This matter tensor has two fundamental properties. First of all it has the algebraic property that the components of the matter tensor form a matrix which is *symmetric*:

$$T_{ik} = T_{ki} \tag{8.12.2}$$

and for this reason we call the matter tensor a "symmetric tensor". Hence the number of independent components is reduced from n^2 to $n(n + 1)/2$, which means in 2, 3, and 4 dimensions respectively 3, 6, and 10 independent components. A further fundamental property of the matter tensor is that *its divergence vanishes at all points*:

$$\sum_{\alpha=1}^{n} \frac{\partial T_{i\alpha}}{\partial x_{\alpha}} = 0 \qquad (8.12.3)$$

This represents a vectorial system of n homogeneous partial differential equations of first order. Since only n equations are prescribed for $n(n + 1)/2$ quantities, our system is obviously strongly under-determined, and accordingly the adjoint system strongly over-determined. This, however, does not mean that the adjoint system does not possess non-zero solutions. In fact such solutions exist and each one of them yields a condition between the boundary values which has an important physical significance.

Making use of the usual technique of obtaining the adjoint equation by multiplying the given system by an undetermined factor and then "liberating" the original functions by the method of integrating by parts, we now have to apply as undetermined factor a *vector* Φ_i of n components. We obtain

$$\int \sum_{i,\alpha=1}^{n} \Phi_i \frac{\partial T_{i\alpha}}{\partial x_{\alpha}} d\tau + \frac{1}{2} \int \sum_{i,\alpha=1}^{n} T_{i\alpha}\left(\frac{\partial \Phi_i}{\partial x_{\alpha}} + \frac{\partial \Phi_{\alpha}}{\partial x_i}\right) = \int \sum_{i,\alpha=1}^{n} T_{i\alpha}\Phi_i\nu_{\alpha}d\sigma \quad (8.12.4)$$

Hence the adjoint homogeneous equation becomes

$$\frac{\partial \Phi_i}{\partial x_k} + \frac{\partial \Phi_k}{\partial x_i} = 0 \qquad (8.12.5)$$

To every independent solution of this equation a condition between the boundary values is obtained, of the form

$$\int \sum_{i,\alpha=1}^{n} T_{i\alpha}\Phi_i\nu_{\alpha}d\sigma = 0 \qquad (8.12.6)$$

The integration is extended over the boundary surface σ which encloses the domain τ in which the equations (3) hold.

Now the equations (5) do not possess many solutions, as we can imagine if we realise that the vector Φ_i of n components is subjected to $n(n + 1)/2$ conditions. First of all we have the solutions

$$\Phi_i = \text{const.} \qquad (8.12.7)$$

These solutions can be systematised by putting first $\Phi_1 = 1$, all other $\Phi_i = 0$, then $\Phi_2 = 1$, all other $\Phi_i = 0$, ..., finally $\Phi_n = 1$, all other

$\Phi_i = 0$. Accordingly the boundary conditions (6) for these special solutions become

$$\int \sum_{\alpha=1}^{n} T_{i\alpha}\nu_\alpha d\sigma = 0 \qquad (i = 1, 2, \ldots, n) \tag{8.12.8}$$

If the matter tensor were not symmetric, these would be all the adjoint solutions since in that case the operator on the left side of (5) would be replaced by $\partial \Phi_i / \partial x_k$ alone. The symmetry of the matter tensor has, however, another class of solutions in its wake. We choose an arbitrary pair of subscripts, for example i and k, and put

$$\Phi_i = x_k, \quad \Phi_k = -x_i \tag{8.12.9}$$

while all the other Φ_α are equated to zero. These solutions, whose total number is $n(n-1)/2$, give us an additional set of boundary conditions, namely

$$\int \sum_{\alpha=1}^{n} (T_{i\alpha}x_k - T_{k\alpha}x_i)\nu_\alpha d\sigma = 0 \tag{8.12.10}$$

The total number of independent boundary conditions is thus $n(n+1)/2$.

We now come to the discussion of the physical significance of these conditions. We begin with the case $n = 3$. This means that we consider a physical system *in equilibrium* because the omission of the fourth coordinate means that everything is *time-independent*, in other words, the masses are *in rest*. Let us now surround a mass at rest with the surface which terminates the mass distribution, and let us apply to this surface σ the condition (8). The coordinates x_1, x_2, x_3 have the significance of rectangular coordinates, usually denoted by x, y, z. The components ν_1, ν_2, ν_3 are the three components of the outward normal ν. Furthermore, the vector

$$F_i = \sum_{\alpha=1}^{3} T_{i\alpha}\nu_\alpha \tag{8.12.11}$$

has the following physical significance. In the field theoretical description of events it represents the *external force* which is impressed from the outside on the material body, per unit surface. The integration over the entire surface represents accordingly the *resulting force* of all the forces which act on the body, from the surrounding field.

Now the boundary condition

$$\int \sum_{\alpha=1}^{3} T_{i\alpha}\nu_\alpha d\sigma = \int F_i d\sigma = \overline{F}_i = 0 \tag{8.12.12}$$

expresses the fact that *a material body can be in equilibrium only if the resultant of the external forces acting on it is zero.*

Let us now turn to the second class of boundary conditions of the type

(10), which involve a pair of indices i and k. In three dimensions we have only the three possible combinations $2,3$; $3,1$; $1,2$. Moreover, the three quantities

$$M_1 = F_2 x_3 - F_3 x_2, \quad M_2 = F_3 x_1 - F_1 x_3, \quad M_3 = F_1 x_2 - F_2 x_1 \quad (8.12.13)$$

form the components of a vector, called the "moment" M of the force F. Hence the second set of boundary conditions:

$$\int (F_i x_k - F_k x_i) d\sigma = 0 \qquad (i, k = 1, 2, 3) \qquad (8.12.14)$$

obtain the following physical significance:

$$\int M_i d\sigma = \overline{M}_i = 0 \qquad (8.12.15)$$

which means that *a material body can be in equilibrium only if the resultant moment of the external forces acting on the body is zero*. The conditions (12) and (15) are fundamental in the statics of rigid or any other kind of bodies. We have obtained them as the compatibility conditions of a partial differential equation, namely the equation which expresses the divergence-free nature of the matter tensor. Earlier, in Chapter 4.15, when dealing with an elastic bar which is free at the two end-points, we found that the differential equation of the elastic displacement was only solvable if two compatibility conditions are satisfied: the sum of the forces and the sum of the moments of the forces had to be zero. At that time we had an ordinary differential equation of fourth order; now we have a system of three partial differential equations of first order which leads in a more general setting to similar compatibility conditions.

We will now leave the realm of statics and enter the realm of dynamics. Einstein in his celebrated "Theory of Relativity" has shown that space and time belong inseparably together by forming a single manifold. Minkowski demonstrated that the separation of the physical world into space and time is purely accidental. All the equations of mathematical physics can be written down in a form in which not merely the three space variables x_1, x_2, x_3 (corresponding to the three rectangular coordinates x, y, z) play an equivalent role, but these three coordinates are supplemented by the fourth coordinate x_4, which in physical interpretation corresponds to the product ict, where c is the velocity of light:

$$x_4 = ict \qquad (8.12.16)$$

The extension of the matter tensor to four dimensions introduces a fourth row and column $T_{i4} = T_{4i}$, which means four new quantities. In view of the imaginary character of x_4 we must assume that the three components T_{i4} ($i = 1, 2, 3$) are *purely imaginary*, while T_{44} is real.

The change from statics to dynamics means that the divergence condition of the matter tensor is extended from $n = 3$ to $n = 4$. Instead of equilibrium conditions we shall now get principles which govern the *motion* of material bodies. We shall have to interpret the four boundary conditions of the type (8), and the $4 \times 3/2 = 6$ conditions of the type (10).

The operations in the space-time world of Relativity require some special experiences which we will not assume at this phase of our discussion. We shall prefer to formulate our results in the usual fashion which separates space and time, although in the basic equations they play a similar role. In relativistic deductions the index i runs from 1 to 4, while we will restrict i to the values 1, 2, 3, and write down separately the terms which belong to the dimension "time" (in our formalism x_4). Similarly the Gaussian integral transformation will be restricted to a volume of the ordinary three-dimensional space and not to a four-dimensional volume. The "boundary surface" σ thus remains a surface of our ordinary space, although now no longer in equilibrium but in some form of motion.

Our fundamental equations are the four equations

$$\sum_{\alpha=1}^{3} \frac{\partial T_{i\alpha}}{\partial x_\alpha} + \frac{\partial T_{i4}}{\partial x_4} = 0$$

$$\sum_{\alpha=1}^{3} \frac{\partial T_{4\alpha}}{\partial x_\alpha} + \frac{\partial T_{44}}{\partial x_4} = 0$$

(8.12.17)

with the six symmetry conditions

$$T_{ik} = T_{ki}$$
$$T_{i4} = T_{4i}$$

(8.12.18)

First of all we consider the four conditions (8). In view of the added terms we have to complement the previous surface integrals by further integrals which are extended over the entire volume of the masses. We will introduce the following four quantities, three of whom correspond to the three components of a vector and the fourth to a scalar:

$$\int T_{i4}d\tau = P_i$$
$$\int T_{44}d\tau = E$$

(8.12.19)

Since we have integrated over the total volume of our domain, these four quantities are no longer functions of x_1, x_2, x_3, but they are still functions of x_4. Now the equation (12) appears in the following more general form:

$$\int F_i d\sigma + \frac{dP_i}{dx_4} = 0$$

(8.12.20)

to which we have to add a fourth equation in the form

$$\int \sum_{\alpha=1}^{3} T_{4\alpha}v_\alpha d\sigma + \frac{dE}{dx_4} = 0$$

(8.12.21)

Let us now remember that the first term of (20) was physically interpreted as the "total force" exerted on the body by the outside forces. Since Newton's law of motion states that "the time rate of change of the total momentum is equal to the external force", we come in harmony with this law if we interpret

$$-\frac{1}{ic} P_i = p_i \tag{8.12.22}$$

as the "total momentum" contained in the volume τ. But the surface integral of the first term of (20) allows the interpretation that it is the "momentum flux" through the boundary surface σ and in this interpretation the equation (20) can be conceived as the *conservation law of momentum*. But then the equation (21) must also have the significance of a conservation law since in Relativity a vector has *four* instead of three components and the three equations (20) and the equation (21) form an inseparable unity. In analogy to (11) we have to define as the "fourth component of the external force" the quantity

$$F_4 = \int \sum_{\alpha=1}^{3} T_{4\alpha} \nu_\alpha d\sigma \tag{8.12.23}$$

and now write (21) in the form

$$F_4 + \frac{dE}{dx_4} = 0 \tag{8.12.24}$$

And since it is shown in Relativity that momentum and energy go inseparably together, we must interpret the equation (21) as the *conservation law of energy*. Accordingly we must interpret the quantity E as the *total energy* of the body (or the material system), while the first term of (21)—if multiplied by ic—represents the *energy flux* through the boundary surface. But now we make use of the fundamental symmetry of the matter tensor which has the consequence that the components of the energy flux become identical with the components of the momentum density—namely iT_{i4}/c—multiplied by c^2. And, since the *conservation of mass* is expressed by exactly the same equation, except that E is replaced by mc^2, we arrive at the monumental *identification of mass and energy*, according to the celebrated equation of Einstein:

$$E = mc^2 \tag{8.12.25}$$

Up to now the "momentum" p_i had no specific significance. We have called it "momentum" but a motion law will only result if we succeed in interpreting this quantity in kinematic terms. Now we still have six more conservation laws, corresponding to the second set (9) of the solutions of the adjoint homogeneous equation. We will in particular choose $\Phi_4 = x_k$ and thus multiply the fourth equation of the system (17) by x_k. Then this equation may be written as follows:

$$\sum_{\alpha=1}^{3} \frac{\partial(x_k T_{4\alpha})}{\partial x_\alpha} - T_{4k} + \frac{\partial(x_k T_{44})}{\partial x_4} = 0 \tag{8.12.26}$$

Now we define the "centre of energy" or "centre of mass" of our mechanical system by putting

$$\xi_k = \frac{\int x_k T_{44} d\tau}{\int T_{44} d\tau} \qquad (8.12.27)$$

For physical reasons the energy density T_{44} has to be assumed as a necessarily *positive* quantity at every point of the domain. Due to this property of the energy (or mass) the centre of mass ξ_k is necessarily *inside* the domain τ.

If we integrate (26) over the volume τ, the second term becomes, in view of (19), $-P_k$, while the last term becomes, on account of the definition (27)

$$\frac{d}{dx_4}(E\xi_k) = \frac{d}{icdt}(mc^2\xi_k)$$

$$= -imc\frac{d\xi_k}{dt} + \xi_k\frac{dE}{dx_4} \qquad (8.12.28)$$

We can get rid of the last term by subtracting the equation (21), multiplied by ξ_k, thus obtaining

$$-P_k - imc\frac{d\xi_k}{dt} + \int(x_k - \xi_k)F_4 d\sigma = 0 \qquad (8.12.29)$$

Finally, dividing the equation (29) by $-ic$, we obtain:

$$p_k = m\frac{d\xi_k}{dt} - \frac{1}{ic}\int(x_k - \xi_k)F_4 d\sigma \qquad (8.12.30)$$

In this equation we recognise "Newton's first law of motion", applied to an arbitrary mechanical system: "the total momentum of a mechanical system is equal to the total mass, multiplied by the velocity of the centre of mass of the system". Actually the last term adds a small correction term in the form of an added momentum which is not of kinematic origin but caused by the external field.

We still have three additional conservation laws which correspond to an extension of the "law of moments" of statics. They correspond to the choice (9) for the vector Φ_i, putting $\Phi_4 = 0$, and choosing for i, k a pair of space indices. The conservation law of momentum is thus complemented by the "conservation of angular momentum" which leads to the following extension of the law of moments: "the time rate of change of the angular momentum is equal to the resulting total moment of the forces acting on the system". This is the fundamental dynamical law which governs the motion of rotating bodies.

Here then are the ten fundamental laws of mechanics, which take their origin (in the field theoretical description of matter) in the divergence-free nature of the matter tensor with its associated solutions of the adjoint homogeneous equation, giving rise to ten constraints between the boundary values of the matter tensor. These constraints, expressed in physical terms, give rise to the following ten conservation laws:

the three equations of the conservation of momentum (Newton's second law of motion);

the one equation of the conservation of energy (which is also the conservation of mass);

the three equations of the conservation of angular momentum (Euler's equations for rotating bodies);

and finally the three equations which give a kinematic interpretation of momentum, in accordance with Newton's first law.

The fundamental laws of dynamics played a decisive role in the evolution of physics, starting with Newton's particle mechanics in which the field concept is not yet present, and culminating in Einstein's General Relativity, in which the divergence-free quality of the matter tensor (interpreted in terms of Riemannian geometry, instead of the flat space-time world of Minkowski) is no longer an external postulate but *an inevitable consequence of the space-time structure of the physical universe.*

8.13. Unconventional boundary value problems

In the previous sections we have studied some of the historically interesting boundary value problems and discussed the analytical methods employed for their solution. We shall now return to more fundamental questions and investigate the theory of boundary value problems from a general standpoint. Historically the differential operators of second order were classified into the three types of elliptic, parabolic, and hyperbolic differential operators and parallel with this classification went the prescription that elliptic differential equations required peripheral boundary conditions, while the parabolic and hyperbolic type of equations had to be characterised by initial type of boundary conditions.

We shall now consider three plausible physical situations which seem to allow a unique mathematical answer and yet do not satisfy the customary conditions.

1. The cooling of a bar is observed. By an oversight the temperature distribution of the bar was not recorded at the time moment $t = 0$ but at a somewhat later time $t = T$. We should like to find by calculation what the temperature distribution was at $t = 0$. Since there is a one-to-one correspondence between $v(x, 0)$ and $v(x, T)$, it must be possible to restore the first function by giving the second one. The problem does not fit the conventional pattern since we have the heat-flow equation with an *end-condition* instead of an initial condition.

2. The vibrating string starts its motion at $t = 0$. Instead of giving $v(x, 0)$ and $v_t(x, 0)$, we take two snapshots of the string at the time moments $t = 0$ and $t = T$, obtaining $v(x, 0)$ and $v(x, T)$. If T is sufficiently small, the difference $v(x, T) - v(x, 0)$ cannot be far from $Tv_t(x, 0)$ and our data

must be sufficient to restore the missing quantity $v_t(x, 0)$. But as a boundary value problem we have violated the condition that a hyperbolic differential equation should not be characterised by peripheral data.

3. The values of the potential $V(x, y, z)$ are given on a very flat ellipsoid σ. By calculation we have obtained V in the neighbourhood of the origin $x = y = z = 0$, in the form of an infinite Taylor expansion which, however, does not converge beyond a certain small radius $r = \rho$ at which the sphere $r = \rho$ touches the ellipsoid. By an accident we have lost the original boundary values whose knowledge is very precious to us. We want to restore the original data from the given Taylor expansion. We know that the solution exists and is in fact obtainable by the method of analytical continuation. But considered as a boundary value problem we can say that V and $\partial V/\partial \nu$ are given on the inner boundary $r = \rho$, while no boundary values are given on the outer boundary σ. These are initial type of boundary conditions for an elliptic differential equation, in contradiction to the general rules.

From the standpoint of the general analytical theory we have the right to ask what motivations are behind these prohibitions. The answer was given by J. Hadamard who, in his celebrated "Lectures on the Cauchy Problem" (cf. [5]), introduced the concept of a "well-posed" or "correctly set" problem ("un problème correctement posé"), by postulating certain conditions that a properly formulated boundary value problem should satisfy. The context of his discussions demonstrates that he considers both under-determined and over-determined problems as not-well-posed. In the under-determined case the solution is not unique, while in the over-determined case the given data are not freely choosable but restricted by the necessary compatibility conditions. Hence Hadamard's "well-posed" problem represents in the language of algebra the case of an $n \times n$ linear system with non-vanishing determinant which establishes a one-to-one correspondence between the left side and the right side.

There is, however, a third condition demanded by Hadamard which has no analogy in the algebraic situation. We will call this the *Condition C*: "*an arbitrarily small perturbation of the data should not cause a finite change in the solution*". It is this condition to which we have to pay particular attention when dealing with the general theory of boundary value problems, in which we abandon the restrictions which go with the special class of "well-posed" problems.

8.14. The eigenvalue $\lambda = 0$ as a limit point

In our general dealings with partial differential operators we came to the conclusion that we do injustice to the nature of such an operator if we try to impose on it the $n \times n$ condition. Generally the function on which the operator operates and the result of the operation belong to completely different manifolds and the condition of a one-to-one correspondence between left side and right side is not satisfied. But a deeper analysis revealed that in proper interpretation *every linear operator establishes a one-to-one correspondence between the U-space in which it is activated and the V-space in which it is activated.* If we do not move out of the space of activation of the operator (the "eigenspace" associated with the operator), we do not observe anything that could give rise to something "not-well-posed". The condition is merely that both solution and given right side shall belong to the eigenspace of the operator. If this condition is satisfied, the relation between right and left sides is unique and one-to-one.

In Chapters 4 and 5 we have seen numerous examples for under-determined and over-determined systems and the manner in which these systems subordinate themselves to the general theory. It will thus be of considerable interest to ask: what happens if we depart from the customary type of "well-posed" boundary value problems and assume data which do not harmonise with the traditional prescriptions? From the very beginning it has been our policy to consider the boundary conditions as an *integrating part of the operator.* The actual numerical *values* of the boundary data are of no significance as far as the operator goes—just as the "right side" of a differential equation does not belong to the operator—but the question is: *what kind* of boundary data are given? Hence it is the *left side* of the boundary conditions which are integrating parts of the operator, and changing these left sides also changes our operator profoundly. Hence the same differential operator, once complemented by peripheral and once by the initial type of boundary conditions, represents in fact two completely different operators. Yet so far as the general theory is concerned, we can see no reason why the one operator should be less amenable to the application of the general principles than the other.

Let us recall briefly the main features of this theory, in order to see whether or not it can serve as a sufficiently broad basis if we venture out into the field of non-conventional boundary conditions. Our basic departure point was the "shifted eigenvalue problem" (5.26.1) which led to the following decomposition of the operator D into eigenfunctions:

$$D = \sum_{i=1}^{\infty} u_i(x)\lambda_i v_i(\xi) \qquad (8.14.1)$$

This is a purely symbolic equation which has no direct significance since the right side represents a necessarily divergent infinite sum. But the significance of this sum was that the operation $Dv(x)$ could be obtained with the help of the following integral operation:

$$u(x) = \int \sum_{i=1}^{\infty} u_i(x)\lambda_i v_i(\xi)v(\xi)d\xi \qquad (8.14.2)$$

This is a meaningful operation since on the right side we can integrate term by term. The resulting new sum is no longer meaningless if $v(x)$ belongs to the "permissible" class of functions which are sufficiently regular and differentiable that the operator D can actually operate on them. In that case the smallness of the definite integral

$$c_i = \int v_i(\xi)v(\xi)d\xi \tag{8.14.3}$$

more than compensates for the largeness of λ_i and even $\lambda_i c_i$ converges to zero. The resulting sum

$$\sum_{i=1}^{\infty} c_i u_i(x) \tag{8.14.4}$$

converges and represents the function $u(x)$ which came about as the result of the operation $Dv(x)$.

The sum (1) finds its natural counterpart in another infinite sum which represents the eigenfunction decomposition of the *inverse operator*:

$$D^{-1} = \sum_{i=1}^{\infty} \frac{v_i(x)u_i(\xi)}{\lambda_i} \tag{8.14.5}$$

This too is a sum which need not have an immediate significance. What we mean by it is once more that this sum operates in the sense of a term by term integration on the function $u(\xi)$, in order to obtain $v(x)$:

$$v(x) = \int \frac{v_i(x)u_i(\xi)}{\lambda_i} u(\xi)d\xi \tag{8.14.6}$$

We can thus go from the left to the right, starting with $v(x)$ and obtaining $u(x)$ on the basis of the operation (2), or we can go from the right to the left, obtaining $v(x)$ on the basis of the operation (6). So far as the analytical theory of linear differential operators is concerned, both operations are of equal interest, although usually we consider only the *second* operation, if our task is to "solve" the given differential equation (with the given boundary conditions).

The operator D^{-1} is much nearer to an actual *value* than D itself. In many problems the infinite sum (5) converges and defines a definite function of the two points x and ξ; the "Green's function" of the problem. But even if the sum (5) did not converge in itself, we could arrive at the Green's function by a proper limit process.

This general exposition has to be complemented by the remark that we have omitted from our expansions the eigenvalue $\lambda = 0$. The significance of the zero eigenvalue was that certain dimensions of the function space were not represented in the operator and exactly for this reason the omission of these axes was justified. However, the zero axes of the U-space were not immaterial. We had to check our data concerning their orthogonality with respect to these axes since otherwise our problem was self-contradictory and thus unsolvable.

What would happen now to this theory if we applied it to the case of boundary data which in the customary sense are injudiciously chosen?

Will the fundamental eigenvalue problem (5.26.1) go out of action? We have seen that neither under-determination nor over-determination can interfere with the shifted eigenvalue problem since the fortunate circumstance prevails that the more over-determined the original operator D is, the more under-determined becomes the adjoint operator D—and vice versa—balancing completely in the final system. Hence it seems hardly possible that an injudicious choice of our boundary data could put our eigenvalue problem out of action, and indeed this is not the case. In every one of the "ill-posed" problems mentioned above, the associated eigenvalue problem is solvable, and provides the necessary and sufficient system of eigenfunctions for expansion purposes. And yet it so happens that the Green's function in the ordinary sense of the word does not exist in any one of these problems. For example in the problem of the cooling bar with given end-condition the definition of the Green's function requires a heat source at a certain point $x = \xi, t = \tau$ with the added condition that the temperature distribution shall become zero at a time $t = T$ which is *beyond* the time $t = \tau$. Such a function does not exist. Nor does the Green's function exist in any of the other ill-posed problems. In fact, if the Green's function *did* exist, we could solve our problem in the conventional manner, and there would be no chance of choosing our boundary values injudiciously.

If we examine the sum (5) closer, we observe that here too the eigenvalue spectrum reveals a danger spot. The infinite sum (4) could not converge because the eigenvalues λ_i increase to infinity. Now the same danger that $\lambda = \infty$ represents for the sum (4), is represented by the value $\lambda = 0$ for the sum (5). It is true that division by zero cannot occur since we have excluded the eigenvalue $\lambda = 0$ from our eigenvalue spectrum (being non-existent so far as the given operator is concerned). But we have to envisage the possibility that $\lambda = 0$ may be a *limit point* of the eigenvalue spectrum. This means that although $\lambda = 0$ is excluded as an eigenvalue, *we may have an infinity of λ_i which come to zero as near as we wish*. If this is the case, the sum (5) cannot converge under any circumstances, and the non-existence of the Green's function is explained. For example, if the λ-spectrum contains a set of numbers which follow the law

$$\lambda_k = \frac{C}{k^2} \qquad (8.14.7)$$

where k is an integer which increases to infinity, our eigenvalue spectrum remains discrete and positive, but our spectrum has a "point of condensation" or "limit point" at $\lambda = 0$, although $\lambda = 0$ is never reached *exactly*. We will call a spectrum of this kind—which is the characteristic feature of all "ill-posed" types of boundary value problems which violate the "Condition C" (cf. Section 13) of Hadamard—a "parasitic spectrum". It is characterised by an infinity of discrete eigenvalues which do not collapse into zero but come *arbitrarily near* to zero, and thus "crowd" around zero in a parasitic way.

It is this parasitic spectrum which distinguishes the non-traditional type of boundary value problems from the traditional ones. As far as the eigen*functions* go, they represent once more a complete function system within the activated space of the operator. But the eigen*values* show the peculiarity that they fall in *two categories*. We can start from a certain $\lambda_1 > \epsilon$ where ϵ may be chosen as small as we wish, and now arrange our eigenvalues in a sequence of increasing magnitude:

$$\epsilon < \lambda_1 \leq \lambda_2 \leq \lambda_3 \ldots \tag{8.14.8}$$

In the boundary value problems of the conventional type we exhaust the entire λ-spectrum by this procedure. In a non-conventional type of problem, however, a second infinite set of eigenvalues remains which has to be arranged in *decreasing order*:

$$\epsilon > \lambda'_1 \geq \lambda'_2 \geq \lambda'_3 > \ldots \tag{8.14.9}$$

These eigenvalues—denoted by λ'_i in order to distinguish them from the regular eigenvalues λ_i of the normal spectrum—cause the non-existence of the Green's function because the infinite sum

$$\sum_{i=1}^{\infty} \frac{v'_i(x)u'_i(\xi)}{\lambda'_i} \tag{8.14.10}$$

cannot converge, exactly as the sum (4) could not converge on account of the limit point $\lambda_i = \infty$ of the regular spectrum.

But here again the divergence of the sum (10) does not mean that the solution (6) has to go out of bound. The substitution of a permissible function in (2) had the consequence that the right side of (4) approached a definite limit which was $u(x)$. Now, if we go backward by starting with $u(x)$ as the given function, we shall obtain the right side of (5) as a convergent sum because the expansion coefficients

$$\gamma'_i = \int u'_i(\xi)u(\xi)d\xi \tag{8.14.11}$$

become sufficiently small to compensate for the smallness of the denominator λ'_i and even γ'_i/λ'_i converges to zero.

We see that now the given right side $u(x)$ cannot be chosen freely from the class of sectionally continuous functions of bounded variations but has to be submitted to more stringent constraints, in order to make the solution $v(x)$ possible. This, however, cannot be considered as objectionable since we have accepted the fact that our data may have to satisfy some given constraints. They had to be strictly orthogonal to all the zero-solutions of the adjoint equation and such solutions may be present in infinite number if the eigenvalue $\lambda = 0$ has infinite multiplicity. Now our requirements are *less stringent*. We do not demand that the expansion coefficients of $u(x)$ in the direction of the eigenfunctions $u'_i(x)$ shall *vanish*. It suffices if they are sufficiently *small*, in order to make the sum

$$v'(x) = \sum_{i=1}^{\infty} \frac{\gamma'_i}{\lambda'_i} v'_i(x) \tag{8.14.12}$$

convergent at all points x of our domain.

The following treatment of our problem is then possible. We separate the parasitic spectrum (9) from the regular spectrum (8). As far as the regular spectrum goes, we can obtain our solution in the usual fashion:

$$v(x) = \sum_{i=1}^{\infty} \frac{\gamma_i}{\lambda_i} v_i(x) \tag{8.14.13}$$

with

$$\gamma_i = \int u_i(\xi) u(\xi) d\xi \tag{8.14.14}$$

In fact, for this part of the solution even a Green's function can be constructed and we can put

$$v(x) = \int G(x, \xi) u(\xi) d\xi \tag{8.14.15}$$

because the sum (5), extended only over the eigenfunctions $v_i(x)$, even if it does not converge immediately, will converge after an arbitrarily small smoothing.

We now come to the parasitic spectrum for which a solution in terms of a Green's function is not possible. Here the sum

$$v'(x) = \sum_{i=1}^{\infty} \frac{\gamma'_i}{\lambda'_i} v'_i(x) \tag{8.14.16}$$

has to remain in the form of a sum and we have to require the convergence of this sum at all points x of our domain. This implies—since $v(x)$ must be quadratically integrable—that we should have

$$\sum_{i=1}^{\infty} \left(\frac{\gamma'_i}{\lambda'_i}\right)^2 = \text{finite} \tag{8.14.17}$$

as a necessary condition to which our data have to be submitted. This condition need not be sufficient, however, to insure pointwise convergence of the sum (16) and we may have to ask the fulfilment of the further condition

$$\sum_{i=1}^{\infty} \left|\frac{\gamma'_i}{\lambda'_i}\right| = \text{finite} \tag{8.14.18}$$

But these conditions are much milder than strict orthogonality to the parasitic spectrum which would make the right sides of (17) and (18) not finite but *zero*.

The resulting solution $\bar{v}(x)$ of our problem is now the sum of the contributions of the regular and the parasitic spectrum:

$$\bar{v}(x) = v(x) + v'(x) \tag{8.14.19}$$

Problem 346. Consider the problem of the cooling bar (9.9–10), but replacing the initial condition (9.7) by the end condition

$$v(x, T) = F(x) \tag{8.14.20}$$

Obtain the compatibility condition of the function $F(x)$, on the basis of the Fourier expansion (9.11).

[Answer:

$$\sum c_k{}^2 e^{2k^2 T} < \infty \tag{8.14.21}$$

where

$$c_k = \frac{2}{\pi} \int_0^\pi F(\xi) \sin k\xi \, d\xi \tag{8.14.22}]$$

Problem 347. The Cauchy-Riemann differential equations (4.3) can be written in the form of a single equation for the complex function $u + iv = f(z)$:

$$f_x + i f_y = 0 \tag{8.14.23}$$

Transform this equation to polar coordinates (r, θ) and assume that $f(1, \theta)$ is given as a (complex) function of θ:

$$f(1, \theta) = \varphi(\theta) \tag{8.14.24}$$

Given the further information that $f(z)$ is analytical between the circles $r = 1$ and $r = R$. Find the compatibility condition to be satisfied by $\varphi(\theta)$.

[Answer:

$$r f_r + i f_\theta = 0 \tag{8.14.25}$$

$$\sum_{n=1}^{\infty} c_n c^*{}_n R^{2n} < \infty \tag{8.14.26}$$

where

$$c_n = \frac{1}{2\pi} \int_{-\pi}^{+\pi} \varphi(\theta) e^{-in\theta} d\theta \tag{8.14.27}]$$

Problem 348. Given the values of the two-dimensional potential function $V(r, \theta)$ and its normal derivative on the unit circle:

$$\begin{aligned} V(1, \theta) &= \varphi(\theta) \\ V_r(1, \theta) &= \psi(\theta) \end{aligned} \tag{8.14.28}$$

Find again the compatibility condition of this problem under the same assumptions as those of the previous problem. Interpret the result in terms of the Cauchy-Riemann equations (25).

[Answer:

$$\sum_{n=1}^{\infty} c_n c^*{}_n R^{2n} < \infty \tag{8.14.29}$$

where

$$c_n = \frac{1}{2\pi} \int_{-\pi}^{+\pi} \left(\varphi(\theta) + \frac{\psi(\theta)}{n} \right) d\theta \tag{8.14.30}]$$

8.15. Variational motivation of the parasitic spectrum

The early masters of calculus assumed that the initial values (8.7.26) of the problem of the vibrating string have to be prescribed as *analytical* functions of x. They were led to this assumption by the decomposition of the motion into eigenfunctions which are all analytical and thus any linear combination of them is likewise analytical. Later the exact limit theory of Cauchy revealed the flaw in this argument which comes from the fact that an infinite series, composed of analytical terms, can *in the limit* approach a function which is non-analytical. But exactly the same argument can also be interpreted in the sense that a non-analytical function can be replaced by an analytical function with an error which can be made at each point as small as we wish. Hence we would think that the difference between demanding analytical or non-analytical boundary values cannot be of too great importance. And yet, the decisive difference between the well-posed and ill-posed type of boundary value problems lies exactly in the question whether the nature of the given problem allows non-analytical data, or demands analytical data. An analytical function is characterised by a very high degree of consistency, inasmuch as the knowledge of the function along an arbitrarily small arc uniquely determines the course of the function along the large arc, while a non-analytical function may change its course capriciously any number of times. But then, if an analytical function has such a high degree of predicatability, we recognise at once that a boundary value problem which requires analytical data will be automatically over-determined to an infinite degree, and we can expect that conditions will prevail which deviate radically from the "well-posed" type of problems whose data need not be prescribed with such a high degree of regularity. In this section we shall show that the parasitic spectrum comes into existence automatically in problems of this type.

As a starting point we will consider the first unusual boundary value problem listed in Section 13, the cooling of a bar whose temperature distribution has been observed at the time moment $t = T$, while our aim is to find by calculation what the temperature distribution was at the earlier time moment $t = 0$. We realise, of course, that the function $v(x, T) = F(x)$ is by no means freely at our disposal. But we can assume that $F(x)$ is given to us as the result of *measurements* and we have the right to idealise the physical situation by postulating that our recording instrument provides the course of $F(x)$ free of errors, to any degree of accuracy we want. Hence the compatibility of our data is assured in advance. We have given the function $v(x, T)$ which has developed from an initially given non-analytical but permissible temperature distribution $v(x, 0) = f(x)$. If we can obtain $v(x, T)$ from $v(x, 0)$, we must also be able to obtain $v(x, 0)$ from $v(x, T)$. And in fact the solution (9.11) is reversible. By obtaining the coefficients c_i from the given initial distribution we could obtain $v(x, t)$ at any later time moment. But if we start with $v(x, T)$, the expansion coefficients of this function will give us c_i multiplied by an exponential function and thus the coefficients c_i themselves require a multiplication by the same exponential function, but changing the sign of the exponent to the opposite. The original $f(x)$ thus becomes

$$f(x) = \sum_{k=1}^{\infty} \gamma_k e^{\pi^2 l - 2k^2 t} \sin \frac{k\pi}{l} x, \quad \gamma_k = \frac{2}{\pi} \int_0^l F(\xi) \sin \frac{k\pi}{l} \xi \, d\xi \quad (8.15.1)$$

This sum would diverge, of course, if we had started with the wrong data, but it remains convergent if $F(x)$ has been properly given.

We will now investigate the eigenvalue spectrum associated with our problem. This problem can be conceived as the solution of a *minimum problem*. The shifted eigenvalue problem

$$\begin{aligned} Dv &= \lambda u \\ \tilde{D}u &= \lambda v \end{aligned} \quad (8.15.2)$$

yields for v alone the differential equation

$$\tilde{D}Dv = \lambda^2 v \quad (8.15.3)$$

and for u alone the differential equation

$$D\tilde{D}u = \lambda^2 u \quad (8.15.4)$$

The differential equation (3) can be conceived as the solution of the following minimum problem. Minimise the positive definite variational integral

$$Q = \int (Dv)^2 d\tau \quad (8.15.5)$$

with the auxiliary condition

$$\int v^2 d\tau = 1 \quad (8.15.6)$$

Similarly the differential equation (4) is derivable from the variational integral

$$\tilde{Q} = \int (\tilde{D}u)^2 d\tau \quad (8.15.7)$$

with the auxiliary condition

$$\int u^2 d\tau = 1 \quad (8.15.8)$$

The eigenvalue λ^2 has the following striking significance: it is equal to the value of the variational integral if we substitute in it the solution of the variational problem. The eigenvalue λ^2 is thus the *minimum itself*, obtained by evaluating the integral Q (resp. $\tilde{Q}$) for the actual solution of the given variational problem.

We will ask in particular for the *smallest possible* minimum of Q, respectively $\tilde{Q}$; then we will obtain the *smallest eigenvalue* with which our eigenvalue spectrum starts. Since both problems (5) and (7) lead to the same eigenvalue λ_1^2, we can obtain λ_1^2 in two different ways:

$$Q = \tilde{Q} = \lambda_1^2 \quad (8.15.9)$$

This reasoning would not hold in the case where the minimum is zero because $Dv(x) = 0$, or $\tilde{D}u(x) = 0$ may have non-vanishing solutions, although the *other* equation may have no such solution. We assume, however, that $\lambda = 0$ is *not* included in the eigenvalue spectrum.

This condition is satisfied in our cooling problem. The boundary condition for $v(x, t)$ is

$$v(x, T) = 0 \qquad (8.15.10)$$

and no regular solution of the heat flow equation exists which would give a uniformly vanishing solution at $t = T$, without vanishing identically. The same can be said of the adjoint equation $Du = 0$, under the boundary condition

$$u(x, 0) = 0 \qquad (8.15.11)$$

But the analytical nature of the heat flow equation for any $t > 0$ allows a much more sweeping conclusion. Let us assume that $F(x, T)$ is not given along the entire rod between $x = 0$ and $x = l$, but only on a *part* of the rod. Then the corresponding boundary condition (10) will now involve only the range $x = [0, l - \epsilon]$. Yet even that is enough for the conclusion that the homogeneous equation has no non-vanishing solution, because an analytical function must vanish identically if it vanishes on an arbitrarily small arc. Then our minimum problem requires that we shall minimise the integral (5) (with the auxiliary condition (6)), under the boundary condition

$$v(x, T) = 0 \qquad (x = [0, l - \epsilon]) \qquad (8.15.12)$$

This condition is *less stringent* than the previous condition (10) which required the vanishing of $v(x, T)$ for the *complete* range of x. This greater freedom in choosing our function $v(x, t)$ must give us a *better minimum* than before; that is, the smallest eigenvalue must *decrease*. But let us view exactly the same problem from the standpoint of the *adjoint* equation. Here the shrinking of the boundary for $v(x, t)$ *increases* the boundary value for $u(x, t)$, because now it is not enough to require that $u(x, 0)$ shall be zero. It has to be zero also on that portion of the upper boundary on which $v(x, T)$ remained free. We have thus a *more restricted* minimum problem which must lead to an *increase* of the smallest eigenvalue. And thus we come to the contradictory conclusion that the same eigenvalue must on the one hand decrease, on the other hand increase. We have tacitly assumed in our reasoning that there *exists* a smallest eigenvalue. The contradiction at which we have arrived forces us to renounce this assumption, and this can only mean that the eigenvalue spectrum can become *as small as we wish*, because in that case there is no smallest eigenvalue. And thus we have been able to demonstrate the existence of the parasitic spectrum in the given cooling problem by a purely logical argument, without any explicit calculations.

Quite similar is the situation concerning the third of the problems enumerated in Section 13. Here the potential function was characterised by giving the function and its normal derivative along a certain portion of the boundary surface σ (for example on an inner boundary σ' which, however, can be considered as part of the boundary surface). Here again the existence of the parasitic spectrum follows once more by the same argument that we employed in the case of the parabolic heat flow equation, since the potential function is likewise an analytical function everywhere inside the boundaries. This shows that here again the Green's function in the ordinary sense does not exist, since it has to be complemented by an infinite sum which cannot converge, as discussed in Section 14. The solution of our problem exists, however, if the boundary data are properly given.

Yet in this problem we encounter a situation which is even more surprising. Let us consider the following minimum problem. Minimise the integral

$$Q = \int (\Delta v)^2 d\tau \qquad (8.15.13)$$

under the constraints that the values of v and $\partial v/\partial \nu$ are prescribed on the boundary surface σ. The problem leads to the differential equation

$$\Delta \Delta v = 0 \qquad (8.15.14)$$

The associated eigenvalue problem

$$\Delta \Delta v = \mu v \qquad (8.15.15)$$

is identical with that obtained for the functions v, u of the previous paragraph, if we identify μ with λ^2. Our problem seems "well-posed", and in fact it is, if the given boundary data extend over the *complete* boundary. Then the λ-spectrum starts with a definite finite λ_1 and the parasitic spectrum does not appear. The solution is unique and the data freely choosable. But let us now assume that once more the same minimum problem is given, but with boundary data which *omit* one part of the boundary, be that part ever so small. At this moment the situation changes completely. The smallest eigenvalue falls to an infinitesimal quantity; we get the parasitic spectrum, and the problem becomes unsolvable with boundary data which are not properly given. This means that our minimum problem has *no solution*. We can *approach* a certain minimum as near as we wish but a definite minimum cannot be obtained.

Indeed, the analytical solution demands once more the fulfilment of the differential equation (14), with the boundary conditions

$$v = \frac{\partial v}{\partial \nu} = 0 \qquad (\text{on } \sigma - S) \qquad (8.15.16)$$

to which the variational principle adds the boundary conditions

$$\Delta v = \frac{\partial}{\partial \nu} \Delta v = 0 \qquad (\text{on } S) \qquad (8.15.17)$$

The equation (14) can now be written in the form

$$\begin{aligned} \Delta v &= u \\ \Delta u &= 0 \end{aligned} \qquad (8.15.18)$$

and in the boundary conditions (17) we can replace Δv by u. But we know from the analytical nature of the potential function that u must vanish identically if the boundary conditions (17) hold even along an arbitrarily small portion of the boundary. And thus the analytical solution of our minimum problem demands such boundary data as make the equation

$$\Delta v = 0 \qquad (8.15.19)$$

possible. In that case we get zero for the requested minimum. But for any other choice of boundary data our problem becomes unsolvable. And yet the solution exists immediately if we add further conditions to our problem, for example by requiring that v and $\partial v/\partial \nu$ shall vanish on the remaining portion S of the boundary surface σ.

The "method of least squares" is based on the principle that a function of some parameters which is everywhere positive must have a minimum for some values of the parameters. This theorem is true in algebra, where the number of parameters is finite. It seemed reasonable to assume that the same theorem will hold in the realm of positive definite differential operators. Hence the attempt was made to demonstrate the existence of the solution of the boundary value problems of potential theory on this basis. This principle is called (although with no historical justification) "Dirichlet's principle". In the case of the potential equation this principle is actually applicable, no matter whether the boundary values are prescribed on the total boundary or only on some parts of it. But our result concerning the "biharmonic equation" (14) shows that this principle can have no universal validity. It holds in all cases in which the parasitic spectrum does not exist. But here we have an example of a completely "elliptic" type of differential equation, with apparently well-chosen peripheral boundary conditions, which is in fact "ill-posed" in Hadamard's sense. This peculiarity of our problem is then traceable to the appearance of the parasitic spectrum which again is closely related to the failure of Dirichlet's principle.

8.16. Examples for the parasitic spectrum

An explicit construction of the parasitic spectrum requires in most cases heavy calculations, because in most physical situations we are led to the solution of differential equations of fourth order which are less familiar to us than differential equations of the second order. In the case of heat conduction, however, we are in the fortunate position that the explicit solution is obtainable with the help of simple tools. In the one-dimensional problem of the cooling bar our equation is separable in x and t. The separation in x reduces the problem to an ordinary differential equation of first order in t alone.

For the sake of simplicity we will normalise the length of the bar to

$$l = \pi \qquad (8.16.1)$$

and put

$$v(x, t) = v(t) \sin kx$$
$$u(x, t) = u(t) \sin kx$$
$$(k = 1, 2, 3, \ldots) \qquad (8.16.2)$$

Now the shifted eigenvalue problem appears in the following form:

$$v' + k^2 v = \lambda u$$
$$-u' + k^2 u = \lambda v \qquad (8.16.3)$$

In order to familiarise ourselves with the nature of the problem, we will first treat the case of the "regular" eigenvalue spectrum, in which $v(x, t)$ is given at the initial moment $t = 0$. Then the boundary conditions of the system (3) become:

$$v(0) = u(T) = 0 \qquad (8.16.4)$$

We assume both u and v in exponential form:

$$v = A e^{\alpha t}$$
$$u = B e^{\alpha t} \qquad (8.16.5)$$

Substitution in (3) yields the two conditions

$$(\alpha + k^2)A = \lambda B$$
$$(-\alpha + k^2)B = \lambda A \qquad (8.16.6)$$

from which

$$k^4 - \alpha^2 = \lambda^2$$

and thus

$$\alpha = \pm \sqrt{k^4 - \lambda^2} \qquad (8.16.7)$$

To every given λ *two* exponents are obtained, namely $\pm \alpha$, if we agree that α is defined as the *positive* value of the square root appearing in (7). The full solution now becomes

$$v = A_1 e^{\alpha t} + A_2 e^{-\alpha t}$$
$$u = B_1 e^{\alpha t} + B_2 e^{-\alpha t}$$
$$= \frac{1}{\lambda}[(k^2 + \alpha)A_1 e^{\alpha t} + (k^2 - \alpha)A_2 e^{-\alpha t}] \qquad (8.16.8)$$

The first boundary condition (4) demands $A_2 = -A_1$ and hence we can put:

$$v = C \sinh \alpha t$$
$$u = \frac{C}{\lambda}(k^2 \sinh \alpha t + \alpha \cosh \alpha t) \qquad (8.16.9)$$

Now the relation (7) does not make α necessarily real. For any λ which is larger than k^2, α becomes imaginary. We can show at once that indeed this is the *only* possibility. In the former case we see from (9) that $u(t)$ is a monotonously increasing function of t which cannot vanish for any value of t, in contradiction to the second boundary condition (4). This shows that of necessity

$$\lambda > k^2 \qquad (8.16.10)$$

and the possibility of a parasitic spectrum is excluded.

Let us now assume that we have given $v(x, t)$ at the *end point* $t = T$, instead of the initial point $t = 0$. Then our new boundary conditions become

$$v(T) = u(0) = 0 \qquad (8.16.11)$$

and exactly with the same reasoning as above we obtain the solution

$$u = C \sinh \alpha t$$
$$v = \frac{C}{\lambda} (k^2 \sinh \alpha t - \alpha \cosh \alpha t) \qquad (8.16.12)$$

In this case the possibility of a real α cannot be ruled out. The first boundary condition (11) requires the condition

$$\tanh \alpha T = \frac{\alpha}{k^2} \qquad (8.16.13)$$

Since we are only interested in the possibility of very small λ_i (which we shall denote by λ'_i), we can put, in view of (7):

$$\alpha = k^2 - \frac{\lambda^2}{2k^2} \qquad (8.16.14)$$

and obtain for sufficiently large k the relation

$$1 - 2e^{-2k^2T} = 1 - \frac{\lambda^2}{2k^4} \qquad (8.16.15)$$

and thus

$$\lambda'_k = 2k^2 e^{-k^2T} \qquad (8.16.16)$$

For large k the λ'_k decrease rapidly and come arbitrarily near to zero. We have thus proved the existence of a parasitic spectrum.

The analysis of this solution shows two characteristic features:

1. The parasitic spectrum is a *one-dimensional sequence*; to every k (for large k) only *one* λ'_k can be found, while the regular eigenvalue spectrum is *two-dimensional* (to every k an infinity of periodic solutions can be found).

2. The division by a very small λ'_k makes the data exceedingly vulnerable to small errors and in principle our data have to be given with infinite accuracy, in order to solve the given problem. Hadamard's "Condition C" (see Section 13), is not fulfilled. But a closer examination reveals that the very small λ'_k belong to very high k values. If the time T is sufficiently small, then the dangerously small λ'_k will occur at such large k values that even the complete *omission* of the parasitic spectrum will cause a minor error, provided that the initial temperature distribution is sufficiently smooth. Under such circumstances we can restore from our data $v(x, T)$

the initial temperature distribution $v(x, 0)$, if our data are given with sufficiently high, but not infinitely high accuracy. The time T of the backward extrapolation depends on the accuracy of our data, and it is clear that for large T an excessive (but still not infinite) accuracy is demanded, if we want to obtain $v(x, 0)$ with a given finite accuracy. An absolute accuracy of the data would only be required if we do not tolerate *any* error in the finding of $v(x, 0)$.

In this example the parasitic spectrum came into existence on account of the unconventional type of boundary value problem from which we started. Much more surprising is the appearance of this spectrum in a perfectly regular and "well-posed" problem, namely the Cauchy-problem (initial value problem) associated with the *vibrating string*. The peculiar riddles which we have encountered in the last part of Section 8, find their resolution in the unexpected fact that even in this very well-posed problem the parasitic spectrum cannot be avoided, if we formulate our problem in that "canonical form" which operates solely with first derivatives, the derivatives of higher order being absorbed by the introduction of surplus variables (cf. Chapter 5.11).

We have formulated the canonical system associated with our problem in the equations (8.7). The eigenvalue problem becomes (in view of the self-adjoint character of the differential operator):

$$v_x - p_1 = \lambda q_1 \qquad u_x - q_1 = \lambda p_1$$
$$v_t + p_2 = \lambda q_2 \qquad u_t + q_2 = \lambda p_2 \qquad (8.16.17)$$
$$-p_{1x} - p_{2t} = \lambda u \qquad -q_{1x} - q_{2t} = \lambda v$$

We can first of all separate in the variable x by putting

$$v = v(t) \sin kx \qquad u = u(t) \sin kx$$
$$p_1 = p_1(t) \cos kx \qquad q_1 = q_1(t) \cos kx \qquad (8.16.18)$$
$$p_2 = p_2(t) \sin kx \qquad q_2 = q_2(t) \sin kx$$

(k = integer; the length of the string is normalised to π).

The new system becomes

$$kv - p_1 = \lambda q_1 \qquad ku - q_1 = \lambda p_1$$
$$v' + p_2 = \lambda q_2 \qquad u' + q_2 = \lambda p_2 \qquad (8.16.19)$$
$$kp_1 - p_2 = \lambda u \qquad kq_1 - q'_2 = \lambda v$$

The first two horizontal lines can be solved algebraically for p_1, p_2, q_1, q_2, obtaining

$$(1 - \lambda^2)p_1 = k(v - \lambda u), \qquad (1 - \lambda^2)q_1 = k(u - \lambda v)$$
$$(1 - \lambda^2)p_2 = -(v' + \lambda u'), \qquad (1 - \lambda^2)q_2 = -(u' + \lambda v') \qquad (8.16.20)$$

Substitution in the third line yields the two simultaneous equations:

$$k^2(v - \lambda u) + v'' + \lambda u'' = \lambda(1 - \lambda^2)u$$
$$k^2(u - \lambda v) + u'' + \lambda v'' = \lambda(1 - \lambda^2)v \qquad (8.16.21)$$

Assuming an exponential form of the solution we can put

$$v = Ae^{at}$$
$$u = Be^{at}$$

(8.16.22)

which yields for the constants A and B the relations

$$\lambda(1 - \lambda^2 + k^2 - \alpha^2)B = (k^2 + \alpha^2)A$$
$$\lambda(1 - \lambda^2 + k^2 - \alpha^2)A = (k^2 + \alpha^2)B$$

(8.16.23)

We put

$$\lambda(1 - \lambda^2 + k^2 - \alpha^2) = \mu$$

(8.16.24)

and obtain

$$\mu = \pm (k^2 + \alpha^2)$$
$$B = \pm A$$

(8.16.25)

Accordingly the full solution of our problem becomes

$$v(x) = A_1 e^{\alpha_1 x} + A_2 e^{-\alpha_1 x} + A_3 e^{\alpha_2 x} + A_4 e^{-\alpha_2 x}$$
$$u(x) = A_1 e^{\alpha_1 x} + A_2 e^{-\alpha_1 x} - (A_3 e^{\alpha_2 x} + A_4 e^{-\alpha_2 x})$$

(8.16.26)

where $\alpha_1{}^2$ and $\alpha_2{}^2$ are determined by the two roots of the equation

$$(1 \pm \lambda)\alpha^2 = -k^2 \pm \lambda(1 - \lambda^2 + k^2)$$

(8.16.27)

The free constants of our solution will be determined by the *boundary conditions* which have to be fulfilled. Our original problem demanded the boundary conditions

$$v(0) = v'(0) = 0$$
$$u(T) = u'(T) = 0$$

(8.16.28)

These, however, are *not* the boundary conditions of our canonical problem. The derivative $v'(t)$ was absorbed by the new variable p_2, similarly $u'(t)$ by q_2. The conditions

$$v(0) = p_2(0) = 0, \quad u(T) = q_2(T) = 0$$

(8.16.29)

demand now (in view of (20)), the boundary conditions

$$v(0) = (v' + \lambda u')(0) = 0$$
$$u(T) = (u' + \lambda v')(T) = 0$$

(8.16.30)

We want to find out whether or not these conditions can be met by *very small* λ-values. In that case λ^3 becomes negligible on the right side of (27) and we have to solve the equation

$$(1 \pm \lambda)\alpha^2 = -k^2(1 \mp \lambda)$$

(8.16.31)

Since α must become imaginary, we will put $\alpha = i\beta$

$$(1 \pm \lambda)\beta^2 = k^2(1 \mp \lambda)$$

(8.16.32)

and write (26) in trigonometric rather than exponential form. The first of the boundary conditions (29) reduces the free constants of our solution to only three constants:

$$v(t) = B_1 \sin \beta_1 t + B_2 \sin \beta_2 t + C(\cos \beta_1 t - \cos \beta_2 t)$$
$$u(t) = B_1 \sin \beta_1 t - B_2 \sin \beta_2 t + C(\cos \beta_1 t + \cos \beta_2 t)$$
(8.16.33)

The second boundary condition (30) prescribed at $t = 0$ demands

$$\beta_1 B_1 (1 + \lambda) + \beta_2 B_2 (1 - \lambda) = 0 \tag{8.16.34}$$

which in view of (32) becomes

$$B_1 k \sqrt{1 - \lambda^2} + B_2 k \sqrt{1 - \lambda^2} = 0 \tag{8.16.35}$$

and thus

$$B_1 = -B_2 = B \tag{8.16.36}$$

$$v(t) = B(\sin \beta_1 t - \sin \beta_2 t) + C(\cos \beta_1 t - \cos \beta_2 t)$$
$$u(t) = B(\sin \beta_1 t + \sin \beta_2 t) + C(\cos \beta_1 t + \cos \beta_2 t)$$
(8.16.37)

At this point our problem is reduced to but *two* constants of integration but we still have to satisfy the two conditions (30) at the point $t = T$. The first condition gives directly

$$B(\sin \beta_1 T + \sin \beta_2 T) + C(\cos \beta_1 T + \cos \beta_2 T) = 0 \tag{8.16.38}$$

while the second condition yields, by the same reasoning that led to (34) and (35):

$$B(\cos \beta_1 T + \cos \beta_2 T) - C(\sin \beta_1 T + \sin \beta_2 T) = 0 \tag{8.16.39}$$

The simultaneous fulfilment of these two conditions is only possible if

$$\sin \beta_1 T + \sin \beta_2 T = 0$$
$$\cos \beta_1 T + \cos \beta_2 T = 0$$
(8.16.40)

This means

$$\beta_2 T = \beta_1 T + \pi(2m + 1) \tag{8.16.41}$$

$$2\lambda' k T = \pi(2m + 1)$$
$$\lambda'_{mk} = \frac{\pi}{2kT}(2m + 1)$$
(8.16.42)

(m = integer). We see that *for any choice of the integer m the eigenvalue λ'_{mk} can be made as small as we wish by choosing k sufficiently large.* The existence of a very extended parasitic spectrum is thus demonstrated and we now understand why the solution of the canonical system (8.7) is less smooth than the right side $\beta(x)$, put in the place of zero in the third equation. The propagation of singularities along the characteristics—which is in such strange contradiction to our expectations if we approach the problem from the standpoint of expanding both right side and solution into their respective eigenfunctions—can now be traced to the properties of the parasitic spectrum which emerges unexpectedly in this problem.

Problem 349. Obtain the parasitic spectrum for the following (non-conventional) boundary value problem:

$$\frac{\partial^2 v}{\partial x^2} + \frac{\partial^2 v}{\partial y^2} = 0 \qquad (x = [0, \pi], \quad y = [0, l])$$

$$v(0, y) = v(\pi, y) = 0$$

$$v(x, 0) = f(x), \quad v_y(x, 0) = g(x)$$

(8.16.43)

[Answer:

$$\alpha_1{}^2 = k^2 + \lambda, \quad \alpha_2{}^2 = k^2 - \lambda \tag{8.16.44}$$

$$\left(\frac{\alpha_1}{\alpha_2} + \frac{\alpha_2}{\alpha_1} - 2\right) \sinh \alpha, l \sinh \alpha_2 l = 2[1 + \cosh (\alpha_1 - \alpha_2)l] \tag{8.16.45}$$

for small λ and large k:

$$\lambda'_k = 4k^2 e^{-kl} \tag{8.16.46}]$$

Problem 350. The analytical function $f(z)$ (see equation (14.23)) is known to be analytical in the strip between $y = 0$ and $y = l$. Moreover, it is known to be periodic with the period 2π:

$$f(z + 2\pi) = f(z) \tag{8.16.47}$$

The value of $f(z)$ is given on the line $y = 0$:

$$f(x + i \cdot 0) = \varphi(x) \tag{8.16.48}$$

Find the parasitic spectrum of this problem.

[Answer:

$$\alpha = \sqrt{k^2 - \lambda^2}, \quad \tanh \alpha l = \frac{\alpha}{k} \tag{8.16.49}$$

For large k and small λ:

$$\lambda'_k = 2k e^{-kl} \tag{8.16.50}]$$

8.17. Physical boundary conditions

The solution of a differential equation with data given on the boundary is primarily a *mathematical* problem and the associated shifted eigenvalue problem need not have any direct physical significance. But in the vibration problems of mathematical physics an actual physical situation is encountered which puts the eigenvalue problem in action, not as a mathematical device for the solution of a given problem, but as a *natural phenomenon*, such as the elastic vibrations of solids and fluids, the electromagnetic vibrations of antennae or wave-guides, or the atomic vibrations of wave mechanics. Here one may ask what significance should be attached to the "boundary conditions" which play such a vital role in the mathematical solution of eigenvalue problems.

And here we have first of all to record the fact that from the physical standpoint a "boundary condition" is always a simplified description of an unknown mechanism which acts upon our system from the outside. A completely isolated system would not be subjected to any boundary conditions. The mathematical boundary conditions of a certain vibration problem would follow automatically from the underlying mechanical principles which provide not only the equations of motion but also the "natural boundary conditions" of the given physical problem. Imposed boundary conditions are merely circumscribed interventions from outside which express in simplified language the coupling which in fact exists between the given system and the outer world. The actual forces which act on the system, modify the potential energy of the inner forces and the physical phenomenon is in reality not a modification of the boundary conditions of the isolated system but a modification of its *potential energy*. Hence it is the *differential operator* which in reality should be modified and not the boundary conditions which actually remain the previous "natural boundary conditions".

As a concrete example let us consider the vibrations of a membrane for which the boundary condition

$$v(s) = 0 \tag{8.17.1}$$

is prescribed, where s indicates the points of a certain closed curve along which the membrane is fixed, thus making its displacement zero on the boundary. Such a condition cannot be taken with full rigour. We would need infinitely large forces for the exact fulfilment of this condition. In fact we have large but not infinitely large forces acting on the boundary.

Hence the question arises, how we could take into account more realistically the actual physical situation. For this purpose we will replace the given condition (1) by another condition which is completely equivalent to the original formulation. We will demand the fulfilment of the integral condition

$$\int v^2(s)ds = 0 \tag{8.17.2}$$

where the integration is extended over the entire boundary. Although our condition is now of a *global* character, whereas before we demanded a condition that had to hold at every *point* of the boundary, yet our new condition can only hold if the integrand becomes zero at every point, and thus we are back at the original formulation.

But now we will make use of the "Lagrangian multiplier method" that we have employed so often for the variational treatment of auxiliary conditions. Our original variational integral was given in the form

$$Q = \frac{1}{2} \int \left[\left(\frac{\partial v}{\partial t}\right)^2 - \left(\frac{\partial v}{\partial x}\right)^2 - \left(\frac{\partial v}{\partial y}\right)^2 \right] dxdydt \tag{8.17.3}$$

and required—according to the "principle of least action"—that we should minimise the time integral of the $T - V$, where T is the kinetic, V the potential energy of the system. The method of the Lagrangian multiplier requires that we modify our variational integral in the following sense:

$$Q' = Q - \frac{\mu}{2} \int v^2(s)dsdt \tag{8.17.4}$$

Let us observe that we would obtain the same Lagrangian if the condition (2) were replaced by the less extreme condition

$$\int v^2(s)ds = \epsilon^2 \tag{8.17.5}$$

where ϵ is not zero but small.   It is the magnitude of the constant μ which
will decide what the value of the right side of the condition (5) shall become.
With increasing μ the constant ϵ decreases and would become zero if μ
grows to infinity.
The term that we have added to the Lagrangian $L = T - V$:

$$-V' = -\frac{\mu}{2} \int v^2(s)ds \tag{8.17.6}$$

represents in physical interpretation *the potential energy of the forces which
maintain the constraint* (5).   Hence we cannot let μ go to infinity but must
consider it as a large but finite constant.   The motion law of the membrane
now becomes

$$\frac{\partial^2 v}{\partial t^2} - \Delta v = 0 \tag{8.17.7}$$

which is the same partial differential equation we had before.   The added
term comes in evidence only when we establish the natural boundary
condition of our problem, which now becomes

$$\frac{\partial v}{\partial \nu} + \mu v = 0 \tag{8.17.8}$$

This again shows that the exact condition (1) would come about if μ
became infinite, which is prohibited for physical reasons.   The changed
boundary condition (8) instead of (1) would eventually come into evidence
in the vibrational modes of extremely high frequencies.

However, even so, we cannot be satisfied by the expression (6).   If the
potential energy of the elastic forces require an integration over the two-
dimensional domain of the coordinates (x, y) we cannot assume that the
boundary forces will be concentrated on a line.   Although apparently the
membrane is fixed on the boundary line only, physically there is always a
small but finite *band* along the boundary on which the external forces act.
Accordingly we have to introduce the potential energy of the forces which
maintain the constraint on the boundary in the form

$$V' = \tfrac{1}{2} \int W(x, y)v^2(x, y)dxdy \tag{8.17.9}$$

where the function $W(x, y)$ has the property that it vanishes everywhere
except in a very thin strip of the width ϵ in the immediate vicinity of the
boundary curve s, where $W(x, y)$ assumes a large constant value:

$$W(x, y) = W_0 \tag{8.17.10}$$

In fact, however, we cannot be sure that the force acting on the boundary
has the same strength at all points.   We could have started our discussion
by replacing the boundary condition (1) by the integral condition

$$\int \rho(s)v^2(s)ds = 0 \tag{8.17.11}$$

where the weight factor $\rho(s)$ is everywhere positive but not necessarily constant. Accordingly we cannot claim that the large positive value of $W(x, y)$ in a thin strip along the boundary will be necessarily a constant along the boundary curve s. It may be a function of s, depending on the physical circumstances which prevail on the boundary. For the macroscopic situation this function is of no avail, since *practically* we are entitled to operate with the strict boundary condition (1). But the method we have outlined—and which can be applied to every one of the given boundary conditions, for example to the two conditions $v = 0$ and $\partial v / \partial \nu = 0$ in the case of a clamped plate, which now entails the addition of *two* expressions of the form (9)—has the great advantage that it brings into play the actual physical mechanism which is hidden behind a mathematical boundary condition. We have modified the potential energy of our system and thus the given differential operator with which we have to work. The "imposed boundary conditions" are now gone. They have been absorbed by the modification of the differential operator. The actual boundary conditions follow from the variational problem itself and become the *natural boundary conditions* of the given variational integral.

We can now answer the question whether the parasitic spectrum encountered in our previous discussions might not have been caused by the imposition of artificial boundary conditions, and might not disappear if we operate with the actual physical situation in which only natural boundary conditions occur. The answer is negative: the parasitic spectrum cannot be removed by the replacement of the imposed boundary conditions with the potential energy of forces which maintain that condition. Indeed, the replacement of the given constraint by a potential energy *weakens* the constraint. Hence the chances of a minimum under the given conditions have improved and the eigenvalue must be *lowered* rather than increased. The parasitic spectrum must thus remain, with a very small alteration toward *smaller* λ'_i. And thus we arrive at a strange conclusion. We have seen that the smallest eigenvalue of the shifted eigenvalue problem can always be defined as the minimum of a certain positive definite variational integral— in fact as the *absolute minimum* of that integral. We should think that at least under natural boundary conditions a definite minimum must exist. Now we see that this is not so. In the large class of problems in which the parasitic spectrum makes its appearance (and that includes not only the non-conventional type of boundary value problems, but the well-posed hyperbolic type of problems in which the parasitic spectrum is a natural occurrence if the problem is formulated in its canonical form), we obtain no definite minimum, in spite of the regular nature of the given differential operator, the finiteness of the domain, and the fact that we do not impose any external boundary conditions on the problem. "Dirichlet's principle" fails to hold in this large class of problems. The minimum we wanted to get can only be reached as a *limit*, since we obtain an infinity of stationary values which come to zero as near as we wish without ever attaining the value zero.

An interesting situation arises in connection with the celebrated Schrödinger's wave equation (1.7) for the hydrogen atom. Here

$$V(r) = -\frac{e^2}{r} \qquad (8.17.12)$$

We know that there exists a negative eigenvalue spectrum, given by

$$\lambda_n = -\frac{A}{n^2} \qquad (8.17.13)$$

where A is a universal constant and n an integer. As n goes to infinity, $\lambda = 0$ becomes a limit point. And yet the usual phenomena which accompany a parasitic spectrum, do not come into evidence. The Green's function of the differential equation exists and we do not experience the infinite sensibility of the solution relative to a small perturbation of the inhomogeneous equation.

A closer examination reveals that the parasitic spectrum is not genuine in this instance. It comes into being solely by the infinity of the domain in which we have solved our problem. If we enclose the hydrogen atom in a sphere of a large but finite radius R, we sweep a certain small but finite range around $\lambda = 0$ free of eigenvalues. The negative energy states are now present in a finite number only and the positive energy states form a dense but discrete spectrum which starts with a definite $\epsilon > 0$. Under these circumstances it is clear that the Green's function cannot go out of existence and that the original limit point $\lambda = 0$ has to be conceived as the result of a limit process, by letting R go to infinity.

8.18. A universal approach to the theory of boundary value problems

We have travelled a long way since the beginning of this chapter and encountered many strange phenomena on our journey. In retrospection we will summarise our findings. We have seen that the method of the separation of variables is an eminently useful tool for the solution of some particularly interesting differential equations, if the boundary is of sufficient regularity. For the general understanding of the basic properties of boundary value problems, however, another approach was more powerful which made use of an auxiliary function of considerable generality. This function had to satisfy the given inhomogeneous boundary conditions but was otherwise free of constraints. Hence the differential equation remained unsatisfied, but the function performed the programme of transforming the originally given homogeneous differential equation with inhomogeneous boundary conditions, into an inhomogeneous differential equation with homogeneous boundary conditions. To this problem we could apply directly the usual analysis in eigenfunctions, expanding both right side and solution into their proper eigenfunctions. Hence the eigenfunction analysis remained our basic frame of reference and included within its scope all boundary value problems, irrespective of how judiciously or injudiciously the given boundary data may have been chosen. The basic problem thus remained the solution of the inhomogeneous differential equation

$$Dv(x) = \beta(x) \tag{8.18.1}$$

with the proper homogeneous boundary conditions. The "given boundary data" are thus transformed into the "given right side" $\beta(x)$ of the differential equation (1).

We obtained a unique solution by demanding orthogonality of $v(x)$ to all the zero-axes of the V-space:

$$\int v(\xi)v^i(\xi)d\xi = 0 \tag{8.18.2}$$

Furthermore, the solvability of our problem demanded that the right side $\beta(x)$ should be orthogonal to all the zero-axes of the U-space:

$$\beta^j = \int \beta(\xi)u^j(\xi)d\xi = 0 \tag{8.18.3}$$

These conditions can in fact be replaced by the "completeness relation"

$$\int \beta^2(\xi)d\xi = \sum_{i=1}^{\infty} \beta_i{}^2 \tag{8.18.4}$$

where

$$\beta i = \int \beta(\xi)u_i(\xi)d\xi \tag{8.18.5}$$

and the zero axes are omitted. Indeed, if $u(x)$ had projections in the direction of the u^j-axes, we should have to add to the right side the sum

$$\sum_j (\beta^j)^2 \tag{8.18.6}$$

The omission of this sum is only justified if we have

$$\sum_j (\beta^j)^2 = 0 \tag{8.18.7}$$

which is only possible if *each one* of the β^j defined by (3) vanishes. Hence the compatibility of the data can be replaced by the single scalar condition (4), which makes no reference to the missing axes.

This, however, is generally not enough. In addition to the regular spectrum whose eigenvalues increase to infinity, we may have a "parasitic spectrum", whose eigenvalues λ'_i converge to zero. While the given function $\beta(x)$ need not be orthogonal to the axes $u'_i(x)$ of the parasitic spectrum, it is necessary that the projections in the direction of these axes shall be sufficiently weak to make the following sum convergent:

$$\sum_{i=1}^{\infty} \left(\frac{\beta'_i}{\lambda'_i}\right)^2 < \infty \tag{8.18.8}$$

Beyond this condition—which is necessary but not always sufficient—we have to demand the *pointwise convergence* of the infinite sum

$$v'(x) = \sum_{i=1}^{\infty} \frac{\beta'_i}{\lambda'_i} v'_i(x) \tag{8.18.9}$$

at all points x of the domain.

In this general theory we have moved far from the restricted class of "well-posed" problems which forbid the eigenvalue zero with respect to both V- and U-spaces and forbid also the parasitic spectrum. Hence it is of interest to see that even in the most "ill-posed" problems we are in fact not far from a "well-posed" problem, because a small but finite perturbation transforms all ill-posed problems into the well-posed category. We do that by a device discussed earlier in Chapter 5.29, establishing a weak coupling between the given and the adjoint operator. We consider the given equation (1) as the limit of the following system (letting ϵ go to zero):

$$\begin{aligned} Dv + i\epsilon u &= \beta \\ \tilde{D}u + i\epsilon v &= 0 \end{aligned} \tag{8.18.10}$$

The operator on the left side has exactly the same eigenfunctions as the original one but the eigenvalues are shifted by a small amount. This shift eliminates the zero eigenvalue and also its immediate neighbourhood—that is the parasitic spectrum. We now have a *complete and unconstrained operator* which satisfies all the requirements of a "well-posed" problem: the solution is unique, the right side $\beta(x)$ is not subjected to compatibility conditions and the solution is not infinitely sensitive to small changes of the data. The Green's function exists and we can find the solution in the usual fashion with the help of this function. The eigenfunction analysis is likewise applicable and we need not distinguish between small and large eigenvalues since none of the eigenvalues becomes smaller in absolute value than ϵ.

The solution of the modified problem exists even for data which from the standpoint of our original problem are improperly given. Moreover, the solution is unique. We analyse the given $\beta(x)$ in terms of the eigenfunctions $u_i(x)$:

$$\beta(x) = \sum \beta_i u_i(x), \quad \beta_i = \int \beta(\xi) u_i(\xi) d\xi \tag{8.18.11}$$

Then the differential equation (10) yields:

$$u(x) = -i\epsilon \sum_{i=1}^{\infty} \frac{\beta_i}{\lambda_i^2 + \epsilon^2} u_i(x)$$

$$v(x) = \sum_{i=1}^{\infty} \frac{\beta_i \lambda_i}{\lambda_i^2 + \epsilon^2} v_i(x) \tag{8.18.12}$$

The difference between a well-posed and an ill-posed problem has disappeared in this approach and we have arrived at a *universal basis* for the treatment of arbitrarily over-determined or under-determined systems. The distinction between properly and improperly given data comes into appearance only if we investigate what happens if ϵ converges to zero. The criterion for properly given data becomes that the solution $(v_\epsilon(x), u_\epsilon(x))$ *must approach a definite limit*:

$$\lim_{\epsilon \to 0} v_\epsilon(x) = v(x)$$

$$\lim_{\epsilon \to 0} u_\epsilon(x) = 0$$

(8.18.13)

The fact that $v(x)$ is the limit of $v_\epsilon(x)$ may also be interpreted by saying that, for an ϵ which is sufficiently small, the difference between $v_\epsilon(x)$ and $v(x)$ can be made at all points x *as small as we wish*. We thus come to the following result: "If the data are given adequately, the difference between an arbitrarily ill-posed and a well-posed problem can be reduced at every point of the domain to an arbitrarily small amount."

BIBLIOGRAPHY

[1] Bergman, S., and M. Schiffer, *Kernel Functions and Differential Equations* (Academic Press, New York, 1953)

[2] Churchill, R. V., *Fourier Series and Boundary Value Problems* (McGraw-Hill, 1941)

[3] Courant, R., and D. Hilbert, *Methods of Mathematical Physics*, Vol. II (German Edition, Springer, Berlin, 1937)

[4] Gould, S. H., *Variational Methods for Eigenvalue Problems* (University of Toronto Press, 1957)

[5] Hadamard, J., *Lectures on the Cauchy Problem* (Dover Publications, New York, 1953)

[6] Kellogg, O. D., *Foundations of Potential Theory* (Springer, Berlin, 1929)

[7] McConnell, A. J., *Applications of the Absolute Differential Calculus* (Blackie & Sons, London, 1942)

[8] Smithies, F., *Integral Equations* (Cambridge University Press, 1958)

[9] Spain, B., *Tensor Calculus* (Oliver & Boyd, Edinburgh, 1956)

10] Synge, J. L., and A. Schild, *Tensor Calculus* (University of Toronto Press, 1956)

[11] Tricomi, F. G., *Integral Equations* (Interscience Publishers, 1957)

CHAPTER 9

NUMERICAL SOLUTION OF TRAJECTORY PROBLEMS

Synopsis. In this chapter we deal with the numerical solution of ordinary differential equations, transformed into a first order system. We study the step-by-step procedures which start from a given initial value and advance in equidistant small steps from point to point, on the basis of local Taylor expansions, truncated to a finite number of terms. Although the truncation error may be small at every step, we have no control over the possible accumulation of these errors beyond a danger point. It is thus advisable to complement the local integration method by a global method which considers the entire range of the independent variable as one unified whole and adds a linear correction to the preliminary solution obtained by the step-by-step procedure.

9.1. Introduction

The rapid and spectacular development of the large electronic computers provided a new and powerful tool for the solution of problems which before had to be left unsolved. The physicist and the construction engineer face frequently the situation that they need the numerical solution of a differential equation which is not simple enough to allow a purely analytical solution. The coding for the electronic computer makes many of these problems solvable in purely numerical terms. But purely numerical computations remain a groping in the dark if they are not complemented by the principles of analysis. The adequate translation of a mathematical problem into machine language can only occur under the proper guidance of analysis.

The numerical solution of partial differential equations requires much circumspection and elaborate preparation which goes beyond the framework of our studies and has to be left to the specialised literature dealing with this subject.* The much more circumscribed problem of *ordinary* differential equations, however, is closely related to some of the topics we have encountered in the early phases of our studies, particularly the problems of interpolation and harmonic analysis. It will thus be our aim to give a comprehensive view of the basic analytical principles which lead to an adequate numerical treatment of ordinary differential equations.

* Cf. particularly the comprehensive textbooks [2] and [6] of the Chapter Bibliography.

9.2. Differential equations in normal form

The problems of mechanics are subordinated to a fundamental principle, called the "principle of least action", which demands that the time integral of a certain function, called the "Lagrangian" of the variational principle, shall be made a minimum, or at least a stationary value. This Lagrangian L is a given function of certain variables, called the "generalised co-ordinates" of the mechanical system and usually denoted by $q_1, q_2, \ldots, q_n$. It contains also the "velocities" of the generalised coordinates, i.e., $\dot{q}_1, \dot{q}_2, \ldots, \dot{q}_n$, and it may happen that the independent variable (the time t) is also explicitly present in L:

$$Q = \int L dt$$

$$L = L(q_1, q_2, \ldots, q_n; \dot{q}_1, \dot{q}_2, \ldots, \dot{q}_n; t) \tag{9.2.1}$$

For example in the problem of a missile we may consider the missile as a rigid body whose position in space requires six parameters for its characterisation, namely the three rectangular coordinates of its centre of mass, plus three angles which fix the orientation of the missile relative to the centre of mass, since the missile is capable of *translation* and of *rotation*. We say that the missile has "six degrees of freedom" because its mechanical state requires that six variables shall be given as definite functions of the time t. Here the number of "generalised coordinates" is six and the Lagrangian L (which is defined as the difference between the kinetic and the potential energy of the body) becomes a given function of the six parameters $q_1, q_2, \ldots, q_6$ and their time derivatives $\dot{q}_1, \dot{q}_2, \ldots, \dot{q}_6$.

The direct minimisation of Q by the principles of variational calculus leads to the "Lagrangian equations of motion" which form a system of n (in our case six) differential equations of the second order. We have seen, however, in Chapter 5.10, how by the method of surplus variables we can always avoid the appearance of derivatives of higher than first order. We do that by transforming the equations of motion into the "Hamiltonian form":

$$\dot{q}_i = \frac{\partial H}{\partial p_i}$$
$$\dot{p}_i = -\frac{\partial H}{\partial q_i} \qquad (i = 1, 2, \ldots, n) \tag{9.2.2}$$

The characteristic feature of these equations is that on the right side the function H (the "Hamiltonian function") is only a function of the variables q_i, p_i (and possibly the time t), without any derivatives.

If the basic differential equation—or system of such equations—is not derivable from a variational principle, or if we deal with a mechanical system in which frictional forces are present (which do not allow a variational treatment), we shall nevertheless succeed in reducing our problem to a first order system by the proper introduction of surplus variables. No matter how complicated our original equations have been and what order derivatives appear in them, we can always introduce surplus variables which will finally reduce the system to the following normal form:

$$v'_i = F_i(v_1, v_2, \ldots, v_n; x) \tag{9.2.3}$$

where the F_i are given functions of the dynamical variables v_i and x. The equations (2) can be conceived as special cases of the general system (3), considering the complete set of variables $q_1, \ldots, q_n; p_1, \ldots, p_n$ as our v_i and thus $n = 2m$. Moreover, it is a specific feature of the Hamiltonian system that the right sides F_i can be given in terms of a single scalar function H, while in the general case (3) such a function cannot be found. However, while for the general analytical theory of the Hamiltonian "canonical equations" the existence of H is of greatest importance, for the *numerical* treatment this fact is generally of no particular advantage. Important is only that we shall reduce our system to the normal form (3).

Let us assume for example that the given differential equation is a single equation of the order n. The general form of such an equation is

$$F(y, y', y'', \ldots, y^{(n)}, x) = 0 \tag{9.2.4}$$

We now solve this equation for $y^{(n)}(x)$ and write it in the explicit form

$$y^{(n)}(x) = G(y, y', \ldots, y^{(n-1)}, x) \tag{9.2.5}$$

Then we introduce the derivatives of $y(x)$ as new variables and replace the original single equation of the order n by the following system of n equations of the first order—denoting y by v_1:

$$\begin{aligned}
v'_1 &= v_2 \\
v'_2 &= v_3 \\
&\;\;\vdots \\
v'_{n-1} &= v_n \\
v'_n &= G(v_1, v_2, \ldots, v_n; x)
\end{aligned} \tag{9.2.6}$$

We have thus succeeded in formulating our problem in the normal form (3).

9.3. Trajectory problems

For the numerical solution of the system (2.3) it is necessary to start from a definite *initial position* of the system. This means that the quantities $v_1, v_2, \ldots, v_n$ have to be given at the initial time moment $x = 0$:

$$v_1(0) = a_1, \quad v_2(0) = a_2, \ldots \quad v_n(0) = a_n \tag{9.3.1}$$

In a mechanical problem this condition is usually satisfied. In other cases we may have different conditions and some of the boundary values may be given at the other end point $x = l$. For example in the problem of the elastic bar, the bar may be clamped at both ends which means that two of the four boundary conditions are given at the one end point, and the other two at the other end point. If our problem is linear, we can follow a systematic treatment and make use of the superposition principle of linear operators. We can adjust our solution to arbitrary boundary conditions if we proceed as follows. We make n separate runs, with the initial conditions

$$
\begin{aligned}
(1) \quad & v_1(0) = 1, \quad v_2(0) = 0, \ldots, \quad v_n(0) = 0 \\
(2) \quad & v_1(0) = 0, \quad v_2(0) = 1, \ldots, \quad v_n(0) = 0 \\
& \quad \vdots \\
(n) \quad & v_1(0) = 0, \quad v_2(0) = 0, \ldots, \quad v_n(0) = 1
\end{aligned}
\tag{9.3.2}
$$

By taking an arbitrary linear superposition of these n solution systems we have the n constants $C_1, C_2, \ldots, C_n$ at our disposal which we can adjust to any given boundary conditions. Hence a maximum of n separate runs with a subsequent determination of n constants will solve our problem. In actual fact the number of runs is smaller since the m initial conditions given at $x = 0$ will immediately determine m of the constants and we need only $n - m$ runs with the subsequent determination of $n - m$ constants, obtained by satisfying the $n - m$ boundary conditions at the point $x = l$.

If our problem is non-linear, we cannot make use of the superposition principle and we have to change our parameters by trial and error until finally all the required boundary conditions of the problem are satisfied.

We can give a geometrical interpretation of our system (2.3) by imagining a space of n dimensions in which a general point P has the rectangular coordinates $v_1, v_2, \ldots, v_n$. Then the actual solution

$$
v_i = v_i(x) \tag{9.3.3}
$$

represents a definite *curve* or *trajectory* of this n-dimensional space. The initial conditions (1) express the fact that the trajectory takes its origin from a definite initial point

$$
P_0 = (v_1(0), v_2(0), \ldots, v_n(0)) \tag{9.3.4}
$$

of the n-dimensional space, while the differential equation (3) determines the *tangent* with which the motion starts. The entire mechanical system is thus absorbed by a single point of an imaginary space of n dimensions, the "configuration space" E.

After moving for a very short time $x = h$ along this tangent, the direction of the tangent will change, because now we have to substitute in the equations (2.3) the values of $v_i(h)$, $x = h$ instead of $v_i(0)$, $x = 0$. We continue along the new tangent up to the time moment $x = 2h$ and once more we have to change our course. Continuing in this fashion we obtain a space polygon of small sides which approaches more and more to a *continuous curve* as h recedes to zero. It is this limiting curve that we want to obtain by our calculations.

9.4. Local expansions

The successive construction we have indicated in terms of consecutive tangents, can actually yield a numerical solution but at the expense of much labour and very limited accuracy.   We could change our differential equation to a *difference equation*:

$$\frac{\Delta v_i}{\Delta x} = F_i(v_1, v_2, \ldots, v_n; x) \tag{9.4.1}$$

considering $\Delta x = h$ as a small quantity which converges to zero.   Then we have a simple step-by-step procedure in which the solution at the point x_k:

$$v_1(x_k), v_2(x_k), \ldots, v_n(x_k) (x_k = kh) \tag{9.4.2}$$

yields directly the solution at the point x_{k+1}:

$$v_1(x_{k+1}), v_2(x_{k+1}), \ldots, v_n(x_{k+1}) (x_{k+1} = (k+1)h) \tag{9.4.3}$$

because of the relation

$$v_i(x_{k+1}) = v_i(x_k) + hF_i(v_1(x_k), v_2(x_k), \ldots, v_n(x_k)) \tag{9.4.4}$$

While this procedure seems on the surface appealing on account of its great simplicity, it has the drawback that for an efficient approximation of a curve with the help of polygons we would have to make the step-size h *excessively small*, but then the inevitable numerical rounding errors would swamp our results.

For this reason our aim must be to render our approximating curve *more flexible* by changing it from a straight line to a parabola or a polynomial of third or even fourth order.   In this manner we can increase the step-size h and succeed with a smaller number of intermediate points.

All methods for the numerical integration of ordinary differential equations operate with such approximations.   Instead of representing the immediate neighbourhood of our curve as a straight line:

$$v_i(x_k + \xi) = v_i(x_k) + \alpha_i(x_k)\xi \tag{9.4.5}$$

we prefer to expand every one of the functions $v_i(x_k)$ in a *polynomial* of a certain order (usually 3 to 5) and obtain $v_i(x_{k+1})$ on the basis of this polynomial.   For this purpose we could make use of the Taylor expansion around the point x_k:

$$v_i(x_k + \xi) = v_i(x_k) + v'_i(x_k)\xi + \tfrac{1}{2}v''_i(x_k)\xi^2 + \cdots \tag{9.4.6}$$

The coefficients of this expansion are in principle obtainable with the help of the given differential equation, by successive differentiations.   But in practice this method would be too unwieldy and has to be replaced by more suitable means.   The various methods of step-by-step integration endeavour to obtain the expansion coefficients of the local Taylor series by numerically appropriate methods which combine high accuracy with ease of computation and avoidance of an undue accumulation of rounding errors (these are caused by the limited accuracy of numerical computations and should not be confounded with the truncation errors, which are of purely analytical origin).   A certain compromise is inevitable since the physical universe operates with the continuum as an actuality while for our mental faculties the continuum is accessible only as a *limit* which we may approach but never reach. All our numerical operations are *discrete* operations which can never be fully adequate to the nature of the continuum. And thus we must be reconciled to the fact that every step in the step-by-step procedure is not more than *approximate*. The calculation with a limited number of decimal places involves a definite *rounding error* in all our computations. But even if our computations had absolute accuracy, another error is inevitable because an *infinite* expansion of the form (6) is replaced by a *finite* expansion. We thus speak of a "truncation error" caused by truncating an infinite series to a finite number of terms. Such a truncation error is inevitable in every step of our local integration process. No matter how small this error may be, its effect can accumulate to a large error which is no longer negligible (as we shall see in Section 13). Hence it is not enough to take into account the possible damage caused by the accumulation of numerical rounding errors and devise methods which are "numerically stable", i.e. free from an accumulation of rounding errors. We must reckon with the damaging effect of the accumulation of truncation errors which is beyond our control and which may upset the apparently high *local* accuracy of our step-by-step procedure. The only way of counteracting this danger is not to consider our local procedure as the final answer, but to complement it by a *global* process in which considerations *in the large* come into play, against the purely local expansions which operate with the truncated Taylor series. This we shall do in Section 17.

For the time being we will study the possibilities of local integration from the principal point of view. Our procedure must be based on the method of *interpolation*. We have at our disposal a certain portion of the curve which we can interpolate with sufficient accuracy by a polynomial of not too high order, provided that we choose the step-size $\Delta x = h$ small enough. This polynomial can now be applied for *extrapolating* to the next point x_{k+1}. Then we repeat the procedure by including this new point and dropping the extreme left point of the previous step. In this fashion there is always the same length of curve under the search light, while this light moves slowly forward to newer and newer regions, until the entire range of x is exhausted.

9.5. The method of undetermined coefficients

The first step-by-step integration process which became widely known was the so-called "Runge-Kutta method".* This method requires a great amount of wasted labour since we do not make use of the information available from the previous part of the curve, but start directly at the point x_k and extrapolate to the point x_{k+1}.

The much more efficient "Method of Milne"† is based on Simpson's quadrature formula and makes use of four consecutive equidistant values of $v(x)$, namely $v(x_{k-3})$, $v(x_{k-2})$, $v(x_{k-1})$, and $v(x_k)$, in order to extrapolate to

* Cf. [4], pp. 233, 236; [6], p. 72.
† Cf. [6], p. 66.

$v(x_{k+1})$. A preliminary extrapolation, which is not accurate enough, is followed by a correction. Every new point of the curve is thus the result of two operations: a prediction and a correction.

We will approach our problem from an entirely general angle and take into account all potentialities. We assume that our procedure will be based on local Taylor expansions. We also assume that our scheme is self-perpetuating. These two postulates allow us to develop a definite programme.

Let us assume that by some extrapolation method we have arrived at a certain point $x = kh$. This means that we have now the functional values $v_{ik} = v_i(kh)$ at our disposal. But then, by substituting on the right side of the differential equation

$$v'_1 = F_1(v_1, v_2, \ldots, v_n; x)$$
$$v'_2 = F_2(v_1, v_2, \ldots, v_n; x)$$
$$\vdots \qquad\qquad\qquad\qquad (9.5.1)$$
$$v'_n = F_n(v_1, v_2, \ldots, v_n; x)$$

we can evaluate the derivatives $v'_{ik} = v'_i(kh)$. From now on we will omit the subscript i of $v_i(x)$ since the same interpolation and extrapolation process will be applicable to all our $v_i(x)$. Hence $v(x)$ may mean *any* of the generalised coordinates $v_i(x)$ of our problem.

It is clear that in a local process only a limited number of ordinates can be used. This number cannot become too large without making the step-size h excessively small since we want to stay in the immediate vicinity of the point in question. We will not go beyond three or four or perhaps five successive ordinates. We will leave this number optional, however, and assume that we have at our disposal the following $2m$ data:

$$y_0 = v(x - mh) \qquad\qquad y'_0 = v'(x - mh)$$
$$y_1 = v(x - (m-1)h) \qquad y'_1 = v'(x - (m-1)h)$$
$$\vdots \qquad\qquad\qquad\qquad\qquad \vdots \qquad\qquad\qquad\qquad (9.5.2)$$
$$y_{m-1} = v(x - h) \qquad\qquad y'_{m-1} = v'(x - h)$$

We choose as our point of reference the point x into which we want to extrapolate. Hence we want to determine $y_m = v(x)$ on the basis of our data. For this purpose we expand $v(x)$ in the vicinity of the point x in a Taylor series, leaving the number of terms of the expansion undetermined:

$$y_{m-1} = v(x - h) = v(x) - hv'(x) + \frac{h^2}{2!}v''(x) - \frac{h^3}{3!}v'''(x) + \ldots$$

$$y_{m-2} = v(x - 2h) = v(x) - 2hv'(x) + \frac{4h^2}{2!}v''(x) - \frac{8h^3}{3!}v'''(x) + \ldots$$
$$\vdots \qquad\qquad\qquad\qquad\qquad\qquad\qquad\qquad\qquad (9.5.3)$$
$$y_0 = v(x - mh) = v(x) - mhv'(x) + \frac{m^2h^2}{2!}v''(x) - \frac{m^3h^3}{3!}v'''(x) + \ldots$$

To these expressions we will add the expansions of the first derivatives:

$$-hy'_{m-1} = -hv'(x - h) = -hv'(x) + h^2v''(x) - \frac{h^3}{2!}v'''(x) + \dots$$

$$-hy'_{m-2} = -hv'(x - 2h) = -hv'(x) + 2h^2v''(x) - \frac{4h^3}{2!}v'''(x) + \dots$$

$$\vdots \tag{9.5.4}$$

$$-hy'_0 = -hv'(x - mh) = -hv'(x) + mh^2v''(x) - \frac{m^2h^3}{2!}v'''(x) + \dots$$

Now we multiply all these equations by some undetermined factors α_1, α_2, $\dots$, α_m as far as the first group is concerned and $-\beta_1$, $-\beta_2$, $\dots$, $-\beta_m$ as far as the second group is concerned. Then we add all these equations. On the left side we get the sum

$$\alpha_1y_{m-1} + \alpha_2y_{m-2} + \dots + \alpha_my_0 \tag{9.5.5}$$
$$h(\beta_1y'_{m-1} + \beta_2y'_{m-2} + \dots + \beta_my'_0)$$

On the right side the factor of $v(x)$ becomes

$$\alpha_1 + \alpha_2 + \dots + \alpha_m \tag{9.5.6}$$

This factor we want to make equal to 1 because our aim is to predict $v(x)$ (and that is y_m) in terms of all the previous y_{m-1}:

$$y_m = \sum_{k=1}^{m} \alpha_k y_{m-k} + h \sum_{k=1}^{m} \beta_k y'_{m-k} \tag{9.5.7}$$

Our programme cannot be carried out with absolute accuracy but it can be accomplished with a *high degree of accuracy* if we succeed in obliterating on the right side the factors of h, h^2, and so on. We have $2m$ coefficients at our disposal and thus $2m$ degrees of freedom. One degree of freedom is absorbed by making the sum (6) equal to 1. The remaining $2m - 1$ degrees of freedom can be used to obliterate on the right side the powers of h, up to the order h^{2m-1}. Hence we shall obtain the extrapolated value y_m with an error which is proportional to h^{2m}. An extrapolation will be possible on the basis of m functional values and m derivatives, which is of the order $2m$ if we denote the "order of the approximation" according to the power of h to which the error is proportional. Hence extrapolation on the basis of 1, 2, 3, and 4 points will be of the order 2, 4, 6, and 8.

The linear system of equations obtained for the determination of the coefficients α_i, β_i is given as follows:

$$\alpha_1 + \alpha_2 + \dots + \alpha_m = 1$$
$$\alpha_1 + 2\alpha_2 + \dots + m\alpha_m - (\beta_1 + \beta_2 + \dots + \beta_m) = 0$$
$$\alpha_1 + 4\alpha_2 + \dots + m^2\alpha_m - 2(\beta_1 + 2\beta_2 + \dots + m\beta_m) = 0 \tag{9.5.8}$$
$$\vdots$$
$$\alpha_1 + 2^{2m-1}\alpha_2 + \dots + m^{2m-1}\beta_m$$
$$- (2m - 1)(\beta_1 + 2^{2m-2}\beta_2 + \dots + m^{2m-2}\beta_m) = 0$$

9.6. Lagrangian interpolation in terms of double points

Instead of solving the linear algebraic system (5.8) for the coefficients α_i, β_i we will follow a somewhat different line of approach which sheds new light on the nature of our approximation. We have dealt in Chapter 5.20 with a Lagrangian interpolation problem in which every point of interpolation was used as a *double point*, making use of *function and derivative* at the points of interpolation. This is exactly what we are trying to do in our present problem in which we want to make a prediction on the basis of our data which list the values of y_i and y'_i. The formula (5.20.11) gave the contribution of a double point $x = x_k$ in explicit terms. For this purpose we have to start with the construction of the fundamental polynomial (5.20.7) which in our case becomes

$$F(x) = [(x + 1)(x + 2) \ldots (x + m)]^2 \qquad (9.6.1)$$

Moreover, we want to obtain the value of $f(x)$ at the critical point $x = 0$ to which we want to extrapolate.

The second and third derivatives of $F(x)$ in the formula (5.20.11) come into play because the root factors $(x - x)$ are squared. If we write

$$F(x) = [G(x)]^2 \qquad (9.6.2)$$

with

$$G(x) = (x + 1)(x + 2) \ldots (x + m) = \frac{(x + m)!}{x!} \qquad (9.6.3)$$

we get the simpler formula, immediately applied to $x = 0$:

$$f(0) = \sum_{k=1}^{m} \left[\frac{G(0)}{G'(x_k)} \right]^2 \left[\left(1 + \frac{G''(x_k)}{G'(x_k)} x_k \right) \frac{f(x_k)}{x_k^2} - \frac{f'(x_k)}{x_k} \right] \qquad (9.6.4)$$

If we translate this formula to our present problem, we obtain the following result:

$$v(x) \doteq y_m = \sum_{k=1}^{m} (\alpha_k y_{m-k} + h \beta_k y'_{m-k}) + \frac{(m!)^2}{(2m)!} h^{2m} v^{(2m)}(\bar{x}) \qquad (9.6.5)$$

where

$$\alpha_k = \binom{m}{k}^2 \left[1 - 2k \left\{ \frac{1}{1-k} + \frac{1}{2-k} + \ldots + \frac{1}{m-k} \right\} \right] \qquad (9.6.6)$$

and

$$\beta_k = \binom{m}{k}^2 k \qquad (9.6.7)$$

The expression in braces { } is taken on the understanding that the term $1/(k - k)$ is omitted. The last term of (5) estimates the error of our extrapolation, in accordance with the general Lagrangian remainder formula (1.5.10), here applied to the case of double points; $\bar{x}$ refers to a point which is somewhere between $x - mh$ and x.

9.7. Extrapolations of maximum efficiency

If we identify m with $1, 2, 3, 4, \ldots$, we obtain a sequence of formulae which extrapolate with maximum accuracy, the error being of the order h^{2m}. The resulting formulae up to $m = 4$ follow:

$$\underline{m = 1}$$

$$y_1 = y_0 + hy'_0 + \frac{h^2}{2} v''(\bar{x}) \tag{9.7.1}$$

$$\underline{m = 2}$$

$$y_2 = -4y_1 + 5y_0 + h(4y'_1 + 2y'_0) + \frac{h^4}{6} v''''(\bar{x}) \tag{9.7.2}$$

$$\underline{m = 3}$$

$$y_3 = -18y_2 + 9y_1 + 10y_0$$
$$+ h(9y'_2 + 18y'_1 + 3y'_0) + \frac{h^6}{20} v^{(6)}(\bar{x}) \tag{9.7.3}$$

$$\underline{m = 4}$$

$$y_4 = -\frac{128}{3} y_3 - 36y_2 + 64y_1 + \frac{47}{3} y_0$$
$$+ h(16y'_3 + 72y'_2 + 48y'_1 + 4y'_0) + \frac{h^8}{70} v^{(8)}(\bar{x}) \tag{9.7.4}$$

9.8. Extrapolations of minimum round-off

In algorithms of a repetitive nature the danger exists that small rounding errors may rapidly accumulate. For example in the construction of a difference table the rounding error of half a unit in the last decimal place rapidly increases as we go to higher and higher differences and quickly destroys the reliability of high order differences. In our step-by-step process we repeat the same extrapolation formula again and again, and we must insist that a rounding error in the last decimal place has no cumulative effect as we continue our process. We speak of "numerical stability", if this condition is satisfied. The examination of the formulae of the previous section reveals that this condition is far from being fulfilled. The ordinates y_t are multiplied by numerical factors which are greater than 1 and thus in a few steps the original rounding error of a half unit in the last decimal place would rapidly advance to the lower decimals. Our process would quickly come to a standstill, because of the intolerably large increase of rounding errors. This does not mean that the formulae of Section 7 are necessarily computationally useless. As we shall see later, we can make very good use of them as long as they are not used in a *repetitive* algorithm, but only a few times and with the proper precaution. For our regular algorithm, however, we have to abandon our hopes of gaining a large power of h by the use of double points.

But then we can take advantage of the flexibility of the scheme (5.5) and add conditions which will guarantee numerical stability. The danger does not come from the coefficients β_i which are multiplied by the small factor h and thus can hardly cause any harm from the standpoint of rounding errors. Hence the m degrees of freedom of the β_i are still at our disposal, thus giving us a chance to reduce the error at least to the order of magnitude h^{m+1}. As far as the α_i go, they have to satisfy the condition

$$\alpha_1 + \alpha_2 + \ldots + \alpha_m = 1 \tag{9.8.1}$$

but otherwise they are freely at our disposal. Now it seems reasonable to make all the α_i *uniformly small by choosing them all equal:*

$$\alpha_1 = \alpha_2 = \ldots = \alpha_m = \frac{1}{m} \tag{9.8.2}$$

By taking the arithmetic mean of the ordinates

$$\frac{1}{m}(y_0 + y_1 + \ldots y_{m-1}) \tag{9.8.3}$$

we have minimised the effect of rounding errors since the best statistical averaging of random errors is obtainable by taking the arithmetic mean of the data.

There is still another reason why a small value of the α_i is desirable. It should be our policy to put the centre of gravity of our interpolation formula on the β_i and not on the α_i since the β_i are multiplied by the *derivatives* rather than the ordinates themselves. But these very derivatives are determined by the given differential equation and it is clear that the less we rely on this equation, the more we shall lose in accuracy and vice versa. Hence we should emphasise the role of the β_i as much as possible at the expense of the α_i. This we accomplish by choosing all the α_i as uniformly small.

The question of the propagation of a small numerical error during the iterative algorithm (5.7) is answered as follows. The second term is negligible, in view of the smallness of the step-size h. We have to investigate the roots of the algebraic equation

$$\lambda^m = \sum_{k=1}^{m} \alpha_k \lambda^{m-k} \tag{9.8.4}$$

One of the roots is of necessity equal to 1. The condition of numerical stability demands that all the other roots must remain in absolute value smaller than 1. Now with the choice (2) our equation becomes

$$\lambda^m = \frac{1}{m}(\lambda^{m-1} + \lambda^{m-2} + \ldots + \lambda + 1) \tag{9.8.5}$$

We know that the absolute value of a complex number cannot be greater than the sum of the absolute values.   This yields

$$m|\lambda|^m \le |\lambda|^{m-1} + |\lambda|^{m-2} + \ldots + |\lambda| + 1 \qquad (9.8.6)$$

and thus

$$|\lambda|^{m-1}(1 - |\lambda|) + |\lambda|^{m-2}(1 - |\lambda|^2) + \ldots + (1 - |\lambda|^m) \ge 0 \quad (9.8.7)$$

The assumption $|\lambda| > 1$ would make the left side negative, in contradiction to the inequality (7).   Moreover, the assumption $|\lambda| = 1$ but $\lambda \ne 1$ would exclude the equal sign of (7) and is thus likewise eliminated.   The only remaining chance that $\lambda = 1$ is a double root is disproved by the fact that $F''(\lambda)$ cannot be zero at $\lambda = 1$.   The numerical stability of our process is thus established.

The coefficients β_i of the formula

$$y_m = \frac{1}{m}(y_0 + y_1 + \ldots y_{m-1}) + \sum_{k=1}^{m} \beta_k y'_{m-k} \qquad (9.8.8)$$

can be obtained by solving the algebraic system (5.8) for the β_i; (substituting for the α_i the constant value (2)).   But we can also conceive our problem as an ordinary Lagrangian interpolation problem for $f'(x)$ instead of $f(x)$, with the m points of interpolation $x = -1, -2, \ldots, -m$ (for the sake of convenience we can normalise h to 1).   Then $f(x)$ is obtainable by integration.

Let us consider for example the case $m = 2$.   Then

$$\begin{aligned} f'(x) &= f'(1) + [f'(1) - f'(0)](x + 1) \\ &= 2y'_1 - y'_0 + (y'_1 - y'_0)x \end{aligned} \qquad (9.8.9)$$

Integrating with respect to x we determine the constant of integration by the condition

$$f(-1) + f(-2) = 0 \qquad (9.8.10)$$

This gives

$$f(x) = (2y'_1 - y'_0)\left(x + \frac{3}{2}\right) + (y'_1 - y'_0)\left(\frac{x^2}{2} - \frac{5}{4}\right) \qquad (9.8.11)$$

and extrapolating to $x = 0$:

$$f(0) = \tfrac{1}{4}(7y'_1 - y'_0) \qquad (9.8.12)$$

To this we have to add the term $\frac{1}{2}(y_0 + y_1)$.

In this manner the following formulae can be established for the cases $m = 1$ to $m = 5$:

$$m = 1$$

$$y_1 = y_0 + hy'_0 + \frac{h^2}{2}y''$$

$$m = 2$$

$$y_2 = \frac{1}{2}(y_1 + y_0) + \frac{h}{4}(7y'_1 - y'_0) + \frac{3}{8}h^3 y'''$$

$$m = 3$$

$$y_3 = \frac{1}{3}(y_2 + y_1 + y_0) + \frac{h}{6}(13y'_2 - 4y'_1 + 3y'_0) + \frac{13}{36}h^4 y'''' \qquad (9.8.13)$$

$$m = 4$$

$$y_4 = \frac{1}{4}(y_3 + y_2 + y_1 + y_0)$$

$$+ \frac{h}{48}(123y'_3 - 79y'_2 + 89y'_1 - 13y'_0) + \frac{95}{288}h^5 y^{(5)}$$

$$m = 5$$

$$y_5 = \frac{1}{5}(y_4 + y_3 + y_2 + y_1 + y_0)$$

$$+ \frac{h}{240}(697y'_4 - 686y'_3 + 936y'_2 - 322y'_1 + 95y'_0) + \frac{461}{1440}h^6 y^{(6)}$$

9.9. Estimation of the truncation error

The estimation of the truncation error on the basis of the Lagrangian remainder formula is of not more than theoretical significance; in practice we do not possess the derivatives of higher than first order in explicit form. We have to find some other method for the numerical estimation of the local truncation error in each step of our process. This can be done in the following way. Together with the extrapolation to $y_m = v(x)$ we extrapolate also to the next point $y_{m+1}v(x + h)$. The error of this ordinate will be naturally much larger than that of $v(x)$. We shall not use this y_{m+1} as an actual value of $v(x + h)$ but store it for the next step when we shall get $v(x + h)$ anyway. The difference between the preliminary value—which we will denote by $\bar{y}_{m+1}$—and the later obtained value y_{m+1} can serve for the estimation of the truncation error. The basis of this estimation is the assumption that $y^{(m+1)}(\bar{x})$ has not changed drastically as we proceeded from y_m to y_{m+1}.

The following table gives the evaluation of $\bar{y}_{m+1}$, together with the estimated truncation error η, in terms of Δ, which denotes the difference between the preliminary $\bar{y}_{m+1}$ and the later obtained $y_{m+1} = v(x + h)$:

$$\Delta = \bar{y}_{m+1} - y_{m+1} \qquad (9.9.1)$$

The standard term

$$M_m = \frac{1}{m}(y_0 + y_1 + \ldots + y_{m-1}) \qquad (9.9.2)$$

remains unaltered in all these formulae.

$$\underline{m = 1}$$

$$\bar{y}_2 = M_1 + 2hy'_0 + 2h^2y''(\bar{x}) \qquad\qquad \eta = 0.33\Delta$$

$$\underline{m = 2}$$

$$\bar{y}_3 = M_2 + \frac{h}{4}(17y'_1 - 7y'_0) + \frac{55}{24}h^3y'''(\bar{x}) \qquad\qquad \eta = 0.2\Delta$$

$$\underline{m = 3}$$

$$\bar{y}_4 = M_3 + \frac{h}{12}(79y'_2 - 72y'_1 + 29y'_0) + \frac{95.5}{36}h^4y''''(\bar{x}) \qquad\qquad \eta = 0.16\Delta$$

$$(9.9.3)$$

$$\underline{m = 4}$$

$$\bar{y}_5 = M_4 + \frac{h}{48}(445y'_3 - 665y'_2 + 511y'_1 - 123y'_0)$$

$$+ \frac{4277}{1440}h^5y^{(5)}(\bar{x}) \qquad\qquad \eta = 0.125\Delta$$

$$\underline{m = 5}$$

$$\bar{y}_6 = M_6 + \frac{h}{360}(4411y'_4 - 9226y'_3 + 10272y'_2 - 5110y'_1 + 1093y'_0)$$

$$+ \frac{4738}{1440}h^6y^{(6)}(\bar{x}) \qquad\qquad \eta = 0.11\Delta$$

A different principle for the estimation of the local error can be established by checking up on the accuracy with which we have satisfied the given differential equation. The polynomial by which we have extrapolated y_m, can also extrapolate the derivative y'_m. Let us call this extrapolated value $\bar{y}'_m$. In the absence of errors this $\bar{y}'_m$ would coincide with the y'_m obtained by substituting the y_m values of all the functions $v_k(x)$ into $F_i(v_k, x)$. In view of the discrepancy between the two values we can say that we have not solved the differential equation

$$v'_i - F_i = 0 \qquad\qquad (9.9.4)$$

but the differential equation

$$v'_i - F_i = \beta_i \qquad\qquad (9.9.5)$$

where β_i is the difference

$$\beta = \bar{y}'_m - y'_m \qquad\qquad (9.9.6)$$

applied to the ith function $v_i(x)$. The smallness of β_i is not necessarily an indication for the smallness of the error of $v_i(x)$ itself, as we shall see in Section 13. But the increase of the estimated error beyond reasonable limits can serve as a warning signal that our h has become too large and should be reduced to a smaller amount.

The following table contains the formulae for the calculation of the extrapolated value $\bar{y}'_m$.

$$\underline{m = 2}$$

$$\bar{y}'_2 = -y'_0 + 2y'_1$$

$$\underline{m = 3}$$

$$\bar{y}'_3 = y'_0 - 3y'_1 + 3y'_2 \qquad (9.9.7)$$

$$\underline{m = 4}$$

$$\bar{y}'_4 = -y'_0 + 4y'_1 - 6y'_2 + 4y'_3$$

$$\underline{m = 5}$$

$$\bar{y}'_5 = y'_0 - 5y'_1 + 10y'_2 - 10y'_3 + 5y'_4$$

9.10. End-point extrapolation

Another possible choice of the coefficients α_i is that we make our prediction solely on the basis of the *last ordinate* y_{m-1}. This means the choice

$$\alpha_1 = 1, \quad \alpha_2 = \alpha_3 = \ldots \alpha_m = 0 \qquad (9.10.1)$$

This method is also distinguished by complete numerical stability and is known as the "Method of Adams".*

The formulae of this method are once more deducible on the basis of integrating the interpolation for $f'(x)$ (cf. Section 8), the only difference being that the constant of integration is now determined by the condition

$$f(-1) = 0 \qquad (9.10.2)$$

For example, integrating (9.8.9) for the case $m = 2$ we now obtain

$$f(x) = (2y'_1 - y'_0)(x + 1) + (y'_1 - y'_0)\frac{x^2 - 1}{2} \qquad (9.10.3)$$

and thus the extrapolation to $x = 0$ yields

$$f(0) = \tfrac{1}{2}(3y'_1 - y'_0) \qquad (9.10.4)$$

Carrying through the calculations systematically up to $m = 5$, we obtain the following table which also contains the extrapolation to y_{m+1} for the sake of a numerical estimation of the truncation error, in full analogy to the previous table (9.3). The estimation of the error in the differential equation by comparing $\bar{y}'_m$ with y'_m can once more occur on the basis of the table (9.7) which remains applicable to the present method.

* Cf. [6], pp. 3 and 53.

$$m = 2$$

$$y_2 = y_1 + \frac{h}{2}(3y'_1 - y'_0) + \frac{5}{12}h^3 y^{(3)}$$

$$\bar{y}_3 = y_1 + h(4y'_1 - 2y'_0) + \frac{28}{12}h^3 y^{(3)} \qquad\qquad \eta = 0.22\Delta$$

$$m = 3$$

$$y_3 = y_2 + \frac{h}{12}(23y'_2 - 16y'_1 + 5y'_0) + \frac{9}{24}h^4 y^{(4)}$$

$$\bar{y}_4 = y_2 + \frac{h}{3}(19y'_2 - 20y'_1 + 7y'_0) + \frac{64}{24}h^4 y^{(4)} \qquad\qquad \eta = 0.16\Delta$$

$$m = 4$$

$$y_4 = y_3 + \frac{h}{24}(55y'_3 - 59y'_2 + 37y'_1 - 9y'_0) + \frac{251}{720}h^5 y^{(5)} \qquad (9.10.5)$$

$$\bar{y}_5 = y_3 + \frac{h}{3}(27y'_3 - 44y'_2 + 31y'_1 - 8y'_0) + \frac{2152}{720}h^5 y^{(5)} \qquad \eta = 0.13\Delta$$

$$m = 5$$

$$y_5 = y_4 + \frac{h}{720}(1901y'_4 - 2774y'_3 + 2616y'_2 - 1274y'_1 + 251y'_0)$$

$$+ \frac{2725}{8640}h^6 y^{(6)}$$

$$\bar{y}_6 = y_4 + \frac{h}{90}(1079y'_4 - 1316y'_3 + 2544y'_2 - 2396y'_1 + 269y'_0)$$

$$+ \frac{27504}{8640}h^6 y^{(6)} \qquad\qquad \eta = 0.11\Delta$$

9.11. Mid-point interpolations

The step-by-step algorithm discussed in the previous sections can be described as follows. Let us assume that we have at our disposal m equidistant ordinates $y_0, y_1, \ldots, y_{m-1}$, and the corresponding derivatives $y'_0, y'_1, \ldots, y'_{m-1}$. Then the formulae of Section (8) or those of Section (10) permit us to evaluate the next ordinate y_m as a certain weighted average of the given data y_j and y'_j. We do that for every one of the n functions $v_i(x)$ of the system (5.1). We have now obtained the next system of values $v_i(x_m)$ and substituting in the functions $F_i(v_1, \ldots, v_n; x_m)$ we immediately obtain also the corresponding $v'_i(x_m)$. Now we proceed by the same algorithm to the next point $v_i(x_{m+1})$ by omitting our previous y_0 and y'_0 and operating with the $2m$ new values $(y_1, y_2, \ldots, y_m)$ and $(y'_1, y'_2, \ldots, y'_m)$. Once more we obtain all the $v_i(x_{m+1})$ by the weighted averaging, substitute again in the functions $F_i(v_1, \ldots, v_n; x_{m+1})$, obtain $y'_{m+1} = v'_i(x_{m+1})$, and thus we continue the process in identical terms, all the time estimating also the effect of the truncation error in the solution or in the differential equation or both. Now it may happen that this error estimation indicates that the error begins to increase beyond a danger point which necessitates the choice of a smaller h. We shall then stop and continue our algorithm with an h which has been reduced to half its previous value. This can be done without any difficulty if we use our interpolating polynomial for the evaluation of the functional values half way between the gridpoints. Then we substitute the values thus obtained into the functions F_i of our differential equation (5.1), thus obtaining the mid-point values of the y'_k. The new points combined with the old points yield a consecutive sequence of points of the step-size $h/2$ and we can continue our previous algorithm with the new reduced step-size.

The following table contains in matrix form the evaluation of the mid-point ordinates, using as α-term the arithmetic mean of all the ordinates y_k. The notation y_{01} refers to the mid-point value of y half way between y_0 and y_1, similarly y_{12} to the mid-point value of y half way between y_1 and y_2, and so on. It is tacitly understood that the arithmetic mean M is added to the tabular products. For example the line y_{23}, for $m = 4$, has the following significance. Obtain the ordinate at an x-value which is half way between that of y_2 and y_3 by the following calculation:

$$y_{23} = \frac{1}{4}\,(y_0 + y_1 + y_2 + y_3) + \frac{h}{384}\,(31y'_0 + 163y'_1 + 205y'_2 - 15y'_3)$$

The common denominator is listed on the right side of the table.

$$m = 2$$

	hy'_0	hy'_1	
$y_{01} =$	1	-1	$\div 8$

$$m = 3$$

	hy'_0	hy'_1	hy'_2	
$y_{01} =$	2	-11	-3	$\div 24$
$y_{12} =$	3	11	-2	

$$m = 4$$

	hy'_0	hy'_1	hy'_2	hy'_3	
$y_{01} =$	15	-205	-163	-31	
$y_{12} =$	31	147	-147	-31	$\div 384$
$y_{23} =$	31	163	205	-15	

(9.11.1)

$$m = 5$$

	hy'_0	hy'_1	hy'_2	hy'_3	hy'_4	
$y_{01} =$	348	-7518	-5454	-3930	-726	
$y_{12} =$	794	3178	-5178	-3794	-760	$\div 11520$
$y_{23} =$	760	3794	5178	-3178	-794	
$y_{34} =$	726	3930	5454	7518	-348	

9.12. The problem of starting values

The step-by-step methods of the previous sections are self-perpetuating if we are in possession of m consecutive equidistant ordinates of the mutual distance h. For the sake of starting our scheme we must first of all generate these ordinates. For this purpose we can make use of those "extrapolations of maximum efficiency" which we have studied in Section 7. Although the accumulation of rounding errors prohibits the use of these formulae on a repetitive basis, there is no objection to their use on a small scale if care is taken for added accuracy in this preliminary phase of our work. This can be done without substantial numerical hardships if we take into account that the change of values from one point to the next is small and thus we may take out a constant of our computations and concentrate completely on the *change* of this constant, using the full decimal accuracy in the computation of the change. The unification of the constant with its change occurs only afterward, when we are through with the first three or four stages of our calculations. Under these circumstances we are prepared to make use of unstable formulae in the evaluation of the starting values which precede the application of the regular step-by-step algorithm.

The initial value problem associated with the differential equation (5.1) prescribes the values of $v_i(x)$ at the initial point $x = 0$:

$$v_i(0) = a_i \qquad (i = 1, 2, \ldots, n) \tag{9.12.1}$$

To obtain the next value $v_i(h)$ we have to rely on a local Taylor expansion around the origin $x = 0$. However, it would be generally quite cumbersome to obtain the high order derivatives of the functions $F_i(v_k, x)$ by successive differentiations. We can avoid this difficulty by starting our numerical scheme with a particularly small value of h. For this purpose we transform our independent variable x into a new variable t by putting

$$x = t^2 \tag{9.12.2}$$

Such a transformation may appear artificial and unnecessary at the present stage of our investigation. It so happens, however, that somewhat later, when we come to the discussion of the weaknesses of a purely local integration process, and look around for a "global" process in complementation of the local procedure, we shall automatically encounter the necessity of transforming the independent variable x into a new angle variable θ. This transformation entails around the origin $x = 0$ a change of variable of the type (2).

The transformation (2) has the following beneficial effect. We want to assume that it is not too difficult to obtain the partial derivatives of the F_i functions with respect to the v_k in explicit form:

$$\frac{\partial F_i}{\partial v_k} = A_{ik}(v_1, \ldots, v_n ; x) \tag{9.12.3}$$

This is in fact unavoidable if our aim is to correct a small error in the v_k by an added correction scheme.

The A_{ik} form the components of an $n \times n$ matrix. By implicit differentiation we can now form the second derivative of $v_i(x)$ with respect to x:

$$v''_i(x) = \frac{d}{dx} v'_i = \sum_{k=1}^{n} \frac{\partial F_i}{\partial v_k} v'_k + \frac{\partial F_i}{\partial x}$$

$$= \sum_{k=1}^{n} A_{ik} F_k + \frac{\partial F_i}{\partial x} \qquad (9.12.4)$$

The knowledge of the second derivatives of all the $v_i(x)$ would still not be sufficient, however, to obtain $v_i(h)$ with sufficient accuracy. But the situation is quite different if we abandon the original variable x and consider from now on the functions v_i as functions of the new variable t. Then the expansion around the origin appears as follows:

$$v(t) = v(0) + v'(0)t^2 + \frac{v''(0)}{2} t^4 + \frac{v'''(0)}{6} t^6 + \dots \qquad (9.12.5)$$

(We will agree that we denote differentiation with respect to x in the previous manner by a dash, while derivatives with respect to t shall be denoted by a *dot*.) Hence in the new variable t we obtain

$$v(h) = v(0) + v'(0)h^2 + \frac{v''(0)}{2} h^4 \left[+ \frac{v'''(\bar{x})}{6} h^6 \right] \qquad (9.12.6)$$

We have thus obtained $v(t)$ at the second point $t = h$ with an error which is of sixth order in h, without differentiating with respect to x more than twice. Then, by substituting these $v_k(h)$ into the functions $F_i(v_k(h); h)$ we obtain also

$$\dot{v}_i(h) = 2hv'_i = 2hF_i(v_k; h) \qquad (9.12.7)$$

We now come to the third point $t = 2h$. Here we can make use of the six data $v(0)$, $\dot{v}(0)$, $\ddot{v}(0)$, $\dddot{v}(0)$ and $v(h)$, $\dot{v}(h)$, on the basis of the (unstable) formula:

$$v(2h) = 49v(0) - 48v(h) + h[34\dot{v}(0) + 16\dot{v}(h)]$$

$$+ 10h^2\ddot{v}(0) + \frac{4}{3} h^3\dddot{v}(0) \left[+ \frac{h^6}{45} v^{(6)}(\bar{x}) \right] \qquad (9.12.8)$$

(in our case this formula is simplified on account of $\dot{v}(0) = \ddot{v}(0) = 0$). Then again we evaluate by substitution the quantities $\dot{v}_i(2h)$.

From here we proceed to the fourth point $t = 3h$ on the basis of the (likewise unstable) formula

$$v(3h) = 10v(0) + 9v(h) - 18v(2h)$$

$$+ h[3\dot{v}(0) + 18\dot{v}(h) + 9\dot{v}(2h)] \left[+ \frac{h^6}{20} v^{(6)}(\bar{x}) \right] \qquad (9.12.9)$$

Now we are already in the possession of four ordinates (and their derivatives) and we can start with the regular algorithm of Section 8 or 10, if we are satisfied with $m = 4$, which leads to an error of the order h^5. If, however, we prefer an accuracy of one higher order ($m = 5$), we can repeat the process (9) once more, obtaining $v(4h)$ on the basis of the points h, $2h$, $3h$. Then we arrived at our five points which are demanded for the step-by-step algorithm with $m = 5$.

We add four further formulae which may be of occasional interest, two of them stable, the other two unstable:

Stable

$$f(x + 2h) = \frac{1}{2}[f(x) + f(x + h)] + \frac{h}{4}[7f'(x) - f'(x + h)]$$
$$+ \frac{h^2}{8}[5f''(x) + 11f''(x + h)] + \frac{251}{480}h^5 f^{(5)}(\bar{x})$$

$$f(x + 2h) = f(x + h) + \frac{h}{2}[3f'(x) - f'(x + h)]$$
$$+ \frac{h^2}{2}[7f''(x) + 17f''(x + h)] + \frac{31}{720}h^5 f^{(5)}(\bar{x})$$

Unstable

$$f(x + 2h) = 17f(x) - 16f(x + h) + h[10f'(x) + 8f'(x + h)]$$
$$+ 2h^2 f''(x) + \frac{h^5}{15}f^{(5)}(\bar{x})$$

$$f(x + 2h) = -31f(x) + 32f(x + h) - h[14f'(x) + 16f'(x + h)]$$
$$+ h^2[-2f''(x) + 4f''(x + h)] + \frac{h^6}{90}f^{(6)}(\bar{x})$$

9.13. The accumulation of truncation errors

The accumulation of numerical rounding errors can be avoided by choosing an algorithm which satisfies the condition of numerical stability. The accumulation of truncation errors is quite a different matter. The presence of truncation errors came in evidence by the discrepancy which existed between the extrapolated value $\bar{y}'_m$ and the actual value y'_m obtained by substituting the obtained ordinates in the given differential equation. We could interpret this difference as an error term to be applied on the right side of the given differential equation. What our step-by-step procedure has given, can thus be interpreted as the solution of a differential equation whose right side differs from the correct right side by a very small error at every point of the range. The question is now: what effect will a very small error committed in the differential equation have on the solution?

Since we have constantly used the notation $v_i(x)$ for the function obtained by the local extrapolation process although this function satisfied not the correct differential equation (5.1) but the modified differential equation (9.5), we shall employ the notation $v^*_i(x)$ for the *correct* solution of the given differential equation (5.1). We can assume that the difference between $v^*_i(x)$ and our preliminary $v_i(x)$ is small enough for a linear perturbation method, neglecting the second and higher powers of the difference. We will thus put

$$v^*_i(x) = v_i(x) + \epsilon u_i(x) \tag{9.13.1}$$

and assumes that terms of the order ϵ^2 are negligible. This has the great advantage that we obtain a *linear* differential equation for the correction $u_i(x)$. If we substitute the expression (1) in (5.1) and make use of the notation A_{ik} (cf. (12.3)) for the partial derivatives of F_i with respect to the v_k, we obtain for u_i the following differential equation:

$$u'_i - \sum_{k=1}^{n} A_{ik}(x)u_k = -\frac{\beta_i}{\epsilon} \tag{9.13.2}$$

Originally the A_{ik} are given as functions of the v_i and x. But we can assume that we have substituted for $v_i(x)$ the explicit functions found in our step-by-step process. This makes the A_{ik} mere functions of x.

Now we want to know how a local error, committed in the jth equation at a certain point $x = \xi$, will influence the solution. This means that we want to solve the equation

$$u'_i - \sum_{k=1}^{n} A_{ik}u_k = \delta_j(x, \xi) \tag{9.13.3}$$

($\delta_j(x, \xi)$ denotes the delta function put in the jth equation, while the other equations have zero on their right side). We know from the properties of the Green's function that all $u_i(x)$ will vanish up to the point $x = \xi$, while beyond that point we have to take a certain linear combination of the homogeneous solutions which satisfy the condition that all the $u_i(x)$ vanish at $x = \xi$, except for $u_j(x)$ which becomes 1 at the point $x = \xi$:

$$u_j(\xi) = 1 \tag{9.13.4}$$

Now we cannot solve the homogeneous equation

$$u'_i - \sum_{k=1}^{n} A_{ik}u_k = 0 \tag{9.13.5}$$

in explicit analytical form. But an *approximate* solution can be found by considering the A_{ik} (which are actually functions of x), in the small neighbourhood of a point as *constants*. Then the solution can be given as a linear superposition of exponential functions of the form

$$u_i(x) = B_i e^{\lambda x} \tag{9.13.6}$$

Substitution of (6) in the system (5) yields for λ the characteristic equation

$$\begin{vmatrix} A_{11} - \lambda & A_{12} & \cdots & A_{1n} \\ A_{21} & A_{22} - \lambda & \cdots & A_{2n} \\ \vdots & & & \\ A_{n1} & A_{n2} & \cdots & A_{nn} - \lambda \end{vmatrix} = 0 \qquad (9.13.7)$$

The n solutions of this algebraic equation yield n (generally complex) values for λ and to each $\lambda = \lambda_k$ a system of the form (6) can be found, with a free universal factor C_k remaining in each solution. An arbitrary linear superposition of these n particular solutions yields the general (approximate) solution of (5).

The decisive question is now how the *real parts* $\mathcal{Re}\lambda$ of these n roots λ_k behave. If all $\mathcal{Re}\lambda_k$ are *negative*, then the solutions are exponentially decaying functions and this means that a small error in the differential equation will quickly extinguish itself. If, however, one or more of the λ_k have a *positive* real part, then a small right side of (2) will induce a solution which *increases* exponentially, and will cause a large error of $v_i(x)$. We have no way of preventing this error since the nature of the roots of (7) is entirely dictated by the given differential operator which cannot be altered. Whether this accumulation of local errors is damaging or not, will depend on the nature of the solution. The solution itself may increase exponentially with a speed which is equal or even larger than the largest of the $\mathcal{Re}\lambda_k$. In that case the relative error of $v_i(x)$ does not increase unduly. But if it so happens that the solution locks itself on an exponent which is *smaller* than the largest $\mathcal{Re}\lambda_k$, then the speed with which the errors accumulate in the solution, will eventually cause an intolerably large relative error, in spite of the smallness of local errors observed during the step-by-step process.

Consider for example the situation exemplified by the Bessel functions along the imaginary axis. The transformation of Bessel's differential equation into the normal form yields a pair of first order equations. The characteristic equation (7)—for not too small x—yields a positive and a negative root. Now the given initial value problem may be such that it calls for the exponentially *decreasing* function. Then it will inevitably happen that the truncation errors cause the appearance of the other, exponentially increasing solution which will sooner or later overpower the desired solution.

Under these circumstances we must ask ourselves whether *any* step-by-step procedure can truly solve our integration problem and whether it would not be more adequate to consider the solution thus obtained as a *preliminary* solution which should be corrected by a *global* method. In this global method we abandon the idea of breaking up our domain into small local sections, but consider the entire range of x between $x = 0$ and $x = l$ as one integral unit. But then what procedure will be at our disposal? Shall we replace the approximations by low order polynomials in the neighbourhood of a point by a high order polynomial in the entire range? We have seen that such a procedure cannot succeed in the case of equidistant data because the interpolating polynomial of high order will generally *diverge* between the points of interpolation and cannot be applied for an approximation in the large. We have to abandon the programme of equidistant steps and introduce a properly chosen *unequal* distribution of ordinates. The first outstanding example of such an integration process was discovered by Gauss, who introduced an exceptionally powerful method for the global evaluation of a definite integral. This method is closely related to those "extrapolations of maximum efficiency" which we have studied in Sections 6 and 7. In fact, the entire Gaussian quadrature method can be conceived as a special case of Lagrangian interpolation in terms of double points, but distributing these points in a particularly judicious manner.

9.14. The method of Gaussian quadrature

We consider once more the method of Section 6 in which not only the functional values but also their derivatives are employed for the purpose of interpolation and extrapolation. Let us consider a function $f(x)$ which is the indefinite integral of a given function $g(x)$. Then

$$f'(x) = g(x) \qquad (9.14.1)$$

From the standpoint of $f(x)$ the given functional values represent the *derivatives* $f'(x_k)$ which appeared in the formula (5.20.11). But then the difficulty arises that the ordinates $f(x_k)$ themselves, which enter the first term of the formula, are not known. We can overcome, however, this difficulty by the ingenious device of *putting the points x_k in such positions that their weights automatically vanish.* Then our interpolation (or extrapolation) formula will not contain any other data than those which are given to us.

Let us assume that we want to obtain the definite integral

$$A = \int_{-1}^{+1} g(x)dx = f(1) - f(-1) \qquad (9.14.2)$$

Since in $f(x)$ we have a constant of integration free, we can define

$$f(-1) = 0 \qquad (9.14.3)$$

and consider this condition as additional data, to be added to the n ordinates $f'(x_k) = g(x_k)$. This means that the fundamental polynomial $F(x)$ becomes of the order $2n + 1$ because we have added the single point $x = -1$ to the double points x_k (we consider $x = -1$ as single since we have no reason to demand that $f'(x)$ must be given at this point). Hence

$$F(x) = (x + 1)[(x - x_1)(x - x_2) \ldots (x - x_m)]^2 \qquad (9.14.4)$$

Now the vanishing of the factor of $f(x_k)$ demands the following condition (cf. (5.20.11)):

$$1 - \frac{1}{3} \frac{F'''(x_k)}{F''(x_k)} (x - x_k) = 0 \qquad (9.14.5)$$

We cannot satisfy this condition simultaneously for various x-values. But our aim is—according to (2)—to obtain $f(1)$ and thus we can identify the point x with $x = 1$. Let us put

$$(x - x_1)(x - x_2) \ldots (x - x_n) = G(x) \qquad (9.14.6)$$

Then

$$
\begin{aligned}
F(x) &= (x + 1)G^2(x) \\
F' &= 2(x + 1)GG' + G^2 \\
F'' &= 4GG' + 2(x + 1)(GG'' + G'^2) \\
F''' &= 6GG'' + 6G'^2 + 2(x + 1)(GG''' + 3G'G'')
\end{aligned}
\qquad (9.14.7)
$$

Since $G(x_k) = 0$, we obtain

$$\frac{1}{3}\frac{F'''(x_k)}{F''(x_k)} = \frac{1}{1 + x} + \frac{G''(x_k)}{G'(x_k)} \qquad (9.14.8)$$

and the condition (5) (for $x = 1$), demands

$$\frac{G''(x_k)}{G'(x_k)} = \frac{1}{1 - x_k} - \frac{1}{1 + x_k} = \frac{2x_k}{1 - x_k^2} \qquad (9.14.9)$$

or

$$(1 - x_k^2)G''(x_k) - 2x_kG'(x_k) = 0 \qquad (9.14.10)$$

We may modify this condition to

$$\left[(1 - x^2)\frac{d^2}{dx^2} - 2x\frac{d}{dx} + n(n + 1)\right]G(x) = 0 \quad \text{(at } x = x_k) \qquad (9.14.11)$$

because the addition of the last term does not change anything, since $G(x_k) = 0$. But now the differential operator in front of $G(x)$ has the property that it obliterates the power x^n and thus transforms a polynomial of the order n into a polynomial of lower order. But a polynomial of an order less than n cannot vanish at n points without vanishing *identically*. And thus the differential equation (11)—which is Legendre's differential equation —must hold for $G(x)$ not only at the points x_k but *everywhere*. This identifies $G(x)$ as the nth Legendre polynomial $P_n(x)$, except for an irrelevant factor of proportionality:

$$G(x) = c_nP_n(x) \qquad (9.14.12)$$

The zeros of the Gaussian quadrature are thus identified as the *zeros of the nth Legendre polynomial*.

In this approach to the problem of Gaussian quadrature we obtain the weight factors of the Gaussian quadrature formula in a form which differs from the traditional expression. The traditional weights of the Gaussian quadrature formula

$$A = \sum_{k=1}^{n} w_kg(x_k)$$

are given as follows*:

$$w_k = \frac{1}{P'_n(x_k)} \int_{-1}^{+1} \frac{P_n(x)}{x - x_k} \, dx \tag{9.14.13}$$

while the second term of the formula (5.20.11) gives, without any integration, in view of the formulae (7):

$$w_k = \frac{2}{(1 - x_k^2)[P'_n(x_k)]^2} \tag{9.14.14}$$

The remainder of the Gaussian quadrature is likewise obtainable without any integration, on the basis of the Lagrangian remainder formula associated with the interpolation of $f(x)$:

$$\eta_n = \frac{f^{(2n+1)}(\bar{x})}{(2n + 1)!} F(1) = \frac{g^{(2n)}(\bar{x})}{(2n + 1)!} 2c_n^2 P_n^2(1) \tag{9.14.15}$$

The factor c_n in (12) is determined by the highest power of x which is 1 on the left side, while the highest coefficient of $P_n(x)$ becomes $(2n)!(n!)^{-2}2^{-n}$. Hence

$$c_n = \frac{(n!)^2}{(2n)!} 2^n \tag{9.14.16}$$

and we obtain

$$\eta_n = 2^{2n+1} \frac{(n!)^4}{(2n!)^2} \frac{g^{(2n)}(\bar{x})}{(2n + 1)!} \tag{9.14.17}$$

which agrees with the traditional expression of η_n.

9.15. Global integration by Chebyshev polynomials

The Gaussian quadrature method possesses an exceptionally high degree of accuracy since we accomplish with n ordinates what otherwise would require $2n$ ordinates. This was the original motivation in the Gaussian discovery. In actual fact more is accomplished than a mere saving of ordinates. The Gaussian choice of the zeros yields a *convergent* method of integration since the Legendre polynomials form an *orthogonal* set of functions. The more ordinates we take into account, the nearer we come to the true value of A. This is by no means so if equidistant ordinates are employed, even if we are willing to go through the double amount of computational labour. Equidistant ordinates do not have the tendency to converge—as we have seen in Chapter 1—except if $f(x)$ belongs to a very limited class of functions. The Gaussian quadrature, on the other hand, converges even if applied to non-analytical functions. If we want to operate with polynomials of high order, we must dispense with the use of equidistant ordinates and replace them by ordinates which are related to some orthogonal set of functions.

The Gaussian method works only for the evaluation of a *definite* integral. If an indefinite integral is in question, we must follow a somewhat different approach. Here we cannot expect the very high degree of accuracy that

---

* Cf. e.g. A. A. (6–10.4) and (8), p. 398.

characterises the Gaussian quadrature. But we can save the other outstanding feature of the Gaussian method, namely that it operates with *orthogonal* polynomials and thus provides a global approximation which converges better and better as the degree of the polynomial increases.

We obtain a perfect counterpart of the Gaussian quadrature for the case of an indefinite integral if we replace the Legendre polynomials by another outstanding set of polynomials, called "Chebyshev polynomials".* The operation with these polynomials is equivalent to the transformation

$$x = l \sin^2 \frac{\theta}{2} \tag{9.15.1}$$

which transforms the range $[0, l]$ of x into the range $[0, \pi]$ of θ. The originally given function $g(x)$, if viewed from the variable θ, becomes a *periodic* function of θ of the period 2π, which can be expanded in a Fourier series. Moreover, it is an *even* function of θ which requires only *cosine* functions for its representation. We assume that $g(x)$ is given at n points which are equidistant in the variable θ but not equidistant in the variable x. Two distributions are in particular of interest:

$$\theta_k = \frac{\pi}{n} k \qquad (k = 0, 1, 2, \ldots, n - 1) \tag{9.15.2}$$

and

$$\theta_k = \frac{\pi}{2n} (2k - 1) \qquad (k = 1, 2, \ldots, n) \tag{9.15.3}$$

The irrational distribution of points from the standpoint of the variable x causes a certain inconvenience in the case of tabulated functions but is not objectionable if the integration of a differential equation is involved where no tabulated functions occur, and the unknown functions are to be determined on the basis of the differential equation itself. Here we have no difficulty in abandoning the variable x from the beginning and introducing immediately the new variable θ.

We will first solve the integration problem associated with the Fourier series, paying no attention to the specific conditions which prevail in our problem on account of the even character of the function involved. We start with the trigonometric identity

$$\tfrac{1}{2} + \cos \theta + \ldots + \cos (n - 1)\theta + \tfrac{1}{2} \cos n\theta = \tfrac{1}{2} \sin n\theta \cot \frac{\theta}{2} \tag{9.15.4}$$

and observe that a function $\varphi(\theta)$, given in any $2n$ equidistant points θ_k of the mutual distance π/n, allows the following trigonometric interpolation:

$$\varphi_n(\theta) = \sum_{k=1}^{2n} \varphi(\theta_k) C(\theta - \theta_k) \tag{9.15.5}$$

* Cf. A. A., p. 245.

where

$$C(t) = \frac{1}{n} \left(\tfrac{1}{2} + \cos t + \ldots + \tfrac{1}{2} \cos nt \right) \tag{9.15.6}$$

Indeed, the form of the function (6) shows that the sum (5) represents a Fourier series of sine and cosine terms up to the order n. Moreover, if we put $\theta = \theta_k$, we obtain

$$\varphi_n(\theta_k) = \varphi(\theta_k) \tag{9.15.7}$$

and thus we have actually interpolated the given $2n$ data with the help of a Fourier series of lowest order which fits these data.

Now we want to *integrate* the function $\varphi(\theta)$, denoting the indefinite integral by $\Phi(\theta)$:

$$\Phi'(\theta) = \varphi(\theta) \tag{9.15.8}$$

For this purpose we replace the exact $\varphi(\theta)$ by its approximation on the basis of trigonometric interpolation, that is $\varphi_n(\theta)$. Then the formula (5) yields

$$\Phi(\theta) = \sum_{k=1}^{2n} \varphi(\theta_k) B(\theta - \theta_k) + C \tag{9.15.9}$$

where

$$B(t) = \frac{1}{n} \left(\frac{t}{2} + \sin t + \frac{\sin 2t}{2} + \ldots + \frac{1}{2} \frac{\sin nt}{n} \right)$$

$$= \frac{1}{n} \int_0^t \frac{\sin n\xi}{\xi} \left(\frac{\xi}{2} \cot \frac{\xi}{2} \right) d\xi \tag{9.15.10}$$

The integral in the last line is for large n very nearly equal to Si (nt) where Si x is the sine integral, defined by (2.3.11). It is preferable to introduce a slightly different function that we want to denote by $K(t)$:

$$K(t) = \frac{1}{\pi} \left(\frac{t}{2} + \sin t + \ldots + \frac{1}{2} \frac{\sin nt}{n} \right) \tag{9.15.11}$$

and put

$$\Phi(\theta) = \frac{\pi}{n} \sum_{k=1}^{n} \varphi(\theta_k) K(\theta - \theta_k) + C \tag{9.15.12}$$

The function $K(t)$ has the following two fundamental properties:

$$K(t + 2\pi) = K(t) \tag{9.15.13}$$

$$K(-t) = -K(t) \tag{9.15.14}$$

We now return to our original global integration problem in the variable x. We want to obtain

$$\int_0^x g(\xi) d\xi = \frac{l}{2} \int_0^\theta g(x(t)) \sin t \, dt \tag{9.15.15}$$

The problem of integration can then be solved as follows.   We first define our fundamental data by putting

$$\gamma_k = \frac{l\pi}{2n} g(x_k) \sin \theta_k \qquad \left(x_k = l \sin^2 \frac{\theta_k}{2}\right) \tag{9.15.16}$$

The subscript n runs from 0 to n in the case of the distribution (2) and from 1 to n in the case of the distribution (3) (in the first case $\gamma_0 = \gamma_n = 0$; this means that we lose our two end-data $g(0)$ and $g(l)$ which enter all our calculations with the weight zero.   The loss is not serious, however, if n is sufficiently large).   We must extend our domain of θ_k-values toward negative k, in order to have a full cycle.   Hence we define

$$\gamma_{-k} = -\gamma_k \tag{9.15.17}$$

The full range of k extends now from $-n$ to $+n$ (including $k = 0$ in the first distribution and excluding it in the second).

Now the indefinite integral of $g(x)$, expressed in the variable becomes:

$$G(\theta) = \sum_{k=-n}^{n} \gamma_k K(\theta - \theta_k) + \text{const.} \tag{9.15.18}$$

Although the data (16) have to be weighted for every value of θ separately, yet the formula (18) shows that *it suffices to give a one-dimensional sequence of weight factors for every* n.   Let us assume that we want to obtain $G(\theta)$ at the data points.   Then

$$G(\theta_m) = \sum_{k=-n}^{n} \gamma_k K(\theta_m - \theta_k) + \text{const.} \tag{9.15.19}$$

and it suffices to evaluate the n coefficients

$$W_s = -K\left(\frac{\pi s}{n}\right) = -\frac{s}{2n} - \frac{1}{\pi} \sum_{k=1}^{n-1} \frac{\sin k \frac{\pi}{n} s}{k} \qquad (s = 0, 1, 2, \ldots, n-1) \tag{9.15.20}$$

(with the added condition $W_{-s} = -W_s$), in order to obtain

$$G(\theta_m) = \sum_{k=-n}^{n} \gamma_k W_{k-m} \tag{9.15.21}$$

We may prefer to obtain the integral not at the data points but *half way between* these points.   This is advisable in order to minimise the errors. The error oscillations of trigonometric interpolation follow the law

$$A(\theta) \sin (n\theta - \varphi) \tag{9.15.22}$$

where φ is a constant phase angle (which is zero for the first and $\pi/2$ for the second distribution of data points), while the amplitude $A(\theta)$ changes slowly, compared with the rapid oscillations of the second factor. Hence the approximate law of the error oscillations of the integral becomes

$$- \frac{1}{n} A(\theta) \cos (n\theta - \varphi) \qquad (9.15.23)$$

which is zero half-way between the data points. At these points we can expect a particularly great accuracy, gaining the factor n compared with the average amplitude of the error oscillations. For these points the coefficients W_s have to be defined as follows:

$$W_s = -K \left(\frac{\pi}{2n} (2s - 1) \right) = \frac{1 - 2s}{2n} - \frac{1}{\pi} \sum_{k=1}^{n} {}' \frac{\sin k \dfrac{\pi}{2n} (2s - 1)}{k} \qquad (9.15.24)$$

(the prime attached to sigma shall indicate that the *last term* is to be taken with half weight). The symmetry pattern of this weighting is somewhat different from the previous pattern, due to the exclusion of the subscript $s = 0$.

9.16. Numerical aspects of the method of global integration

The global method of integration is based on the great efficiency of the Chebyshev polynomials for the global representation of functions, defined in a finite range. However, the actual expansion of $g(x)$ (and also $G(x)$) into Chebyshev polynomials does not appear in explicit form but remains latent in the technique of trigonometric interpolation. The range of applicability of this method is wider than that of the usual "point to point integration techniques" which are based on Simpson's formula or some formula of a similar type. Such techniques assume the existence of derivatives up to a certain order. Here the analytical nature of the function $g(x)$ is not required. Functions of the type $\log x$, $\sqrt{x}$, and other similar functions which are integrable but not differentiable, are included in the validity domain of the method of global integration, in full analogy to the Gaussian quadrature method which, however, is restricted to the evaluation of a *definite* integral. Compared with the Gaussian method we lose the advantage of halving the number of ordinates; that is, the accuracy obtained is comparable to the Gaussian accuracy with half as many ordinates. However, the saving of ordinates is not as decisive as the advantage of a global technique which avoids the accumulation of local truncation errors.

We will study the numerical aspects of the method for $n = 12$. First of all we will generate the sequence of weight factors W_s, defined by (15.20) and (15.24). Since all the weights are near to $\pm \frac{1}{2}$, we will omit the constant $\frac{1}{2}$ from our W_s. Hence we will put (for positive s):

$$w_s = W_s + \tfrac{1}{2} \qquad (9.16.1)$$

and retain the definition

$$w_{-s} = -w_s \qquad (9.16.2)$$

For the set (15.24) we obtain the following sequence:

s	w_s	s	w_s	
1	0.06385709	7	0.00096689	
2	−0.01214645	8	−0.00079306	
3	0.00495169	9	0.00068299	(9.16.3)
4	−0.00271080	10	−0.00061348	
5	0.00174643	11	0.00057229	
6	−0.00125080	12	−0.00055307	

These values represent very nearly the oscillations of the function

$$-\frac{1}{\pi}\left(\text{Si}\,(\theta) - \frac{\pi}{2}\right) \tag{9.16.4}$$

and the amplitudes are particularly small since we are near to the *nodal points* of these oscillations, namely half-way between the consecutive minima and maxima which occur at the points

$$\theta = k\pi \tag{9.16.5}$$

Hence we obtain only a *small correction* of the principal contributions of the weights W_s which is $-\frac{1}{2}$ for positive s and $+\frac{1}{2}$ for negative s. This contribution can be taken into account separately. Apart from an additive constant which is irrelevant—since the indefinite integral contains a free additive constant anyway—it amounts to the *summing of the ordinates* $\gamma_1, \gamma_2, \ldots$, up to γ_m.

For a more concrete elucidation of the numerical technique we will normalise the range of x to [0, 1] and we will assume that the data points are given as the n points

$$x_k = \sin^2 \frac{\pi}{4\pi}\,(2k - 1) \qquad (k = 1, 2, \ldots, n) \tag{9.16.6}$$

which corresponds to the *second* distribution (9.15.3). We wish to obtain the indefinite integral at the *midpoints* between the data points, that is at the values

$$x_m = \sin^2 \frac{\pi}{2n}\,m \tag{9.16.7}$$

In order to normalise the constant of integration, we assume that the initial value $G(0)$ of the integral is given.

Now the formula (15.21) expresses a numerical method known as the "movable strip technique".* Let us assume that the data γ_k are put in proper succession on a horizontal strip which is fixed. The weights w_k

* Cf. A. A., p. 13.

are put on a parallel strip which gradually moves to the right. The initial position of these two strips is as follows:

(9.16.8)

We multiply corresponding elements and form the sum. Let the result be G_0;

$$G_0 = \sum_{k=-n}^{n} \gamma_k w_k \qquad (9.16.9)$$

We now move the strip *one place to the right*, obtaining the following picture:

(9.16.10)

In view of the cyclic nature of our method we should rather think of a movable "band" which has no beginning and no end. The number w_n which disappeared at the right end, comes back at the left end. The movable strip is thus all the time filled up with the same set of numbers; it is only the left starting point which gradually moves from $-n$ to n, then to $n - 1, n - 2, \ldots$, as the strip moves forward.

Once more we form the products of corresponding elements and their sum, noting down the result as G_1. Then the strip moves forward again, giving rise to G_2, and so on (this kind of procedure is particularly well adapted to the task of coding for the electronic computer). Generally

$$G_m = \sum_{k=-n}^{n} \gamma_k w_{k-m} \qquad (9.16.11)$$

The integral of the given function $g(x)$ is now evaluated as follows:

$$G(x_m) - G(0) = G_m - G_0 + \gamma_1 + \gamma_2 + \ldots \gamma_m \quad \left(x_m = \sin^2 \frac{\pi}{2n} m \right) \qquad (9.16.12)$$

As a concrete numerical example we will consider the following function of the range [0, 1]:

$$g(x) = \frac{1}{1 + (2x - 1)^2} = \frac{1}{1 + \cos^2 \theta} \qquad (9.16.13)$$

First of all we need our data γ_k and for this purpose we have to form the products (15.16) for $n = 12$, $\theta_k = 7.5°$, $22.5°$, $37.5°$, $52.5°$, $67.5°$, $82.5°$.

Since $g(x)$ is symmetric with respect to the centre point $x = \frac{1}{2}$ (or $\theta = 90°$), the second set of data $k = 7$ to 12 merely repeats the first set in reverse order. Hence we will list the γ_k only up to $k = 6$.

k	γ_k
1	0.00861632
2	0.02702547
3	0.04890525
4	0.07577005
5	0.10548729
6	0.12760580

$$(9.16.14)$$

If now we carry out the calculations according to the movable strip technique, with the data γ_k on the fixed strip and the weights (3) on the movable strip, we obtain the following results. The exact integral is available in our example:

$$G^*(x) = \tfrac{1}{2} \text{ arc tan } (2x - 1) \qquad (9.16.15)$$

which is a tabulated function.* Hence we have no difficulties in checking our numerical results. The following table contains the exact values (given to 8 decimal places) of the function

$$G^*(x_m) - G^*(0) = \tfrac{1}{2}[\text{arc tan } 1 - \text{arc tan } (1 - 2x_m)]$$
$$= \frac{\pi}{8} - \frac{1}{2} \text{ arc tan } \left(\cos \frac{\pi}{12} m \right) \qquad (9.16.16)$$

(evaluated at points which are in the variable θ half-way between the data points), and the corresponding values obtained by global integration. The sum of the ordinates is listed separately, in order to show the effect of the correction.

m	$G^*(x_m) - G^*(0)$	$\sum\limits_{i=1}^{m} \gamma_i$	$G_m - G_0$	$G(x_m) - G(0)$
1	0.00866532	0.00861632	0.00004918	0.00866550
2	0.03583689	0.03564178	0.00019590	0.03583747
3	0.08495923	0.08454704	0.00041242	0.08495945
4	0.16087528	0.16031709	0.00055811	0.16087520
5	0.26606830	0.26580438	0.00026414	0.26606852
6	0.39269908	0.39341018	-0.00071101	0.39269917

$$(9.16.17)$$

The "sum of the ordinates" $\sum \gamma_i$ corresponds to the simple "trapezoidal rule" of obtaining an area. Since the correction is small, the global integration method can only be effective if the ordinates are sufficiently close to allow a satisfactory application of the trapezoidal rule (although it is of interest to remember that in our case we would get exact results for any $g(x)$ which is given as an arbitrary polynomial not exceeding the order 11).

---

* "Tables of the Arc tan x", NBS, Applied Mathematical Series 26 (U.S. Government Printing Office, Washington, D.C., 1953).

The correction, although very small, is highly effective and extends the accuracy to the seventh decimal place.

We will, however, add another distribution of our data points which corresponds to (15.2) and which is more suitable in view of the application to the solution of differential equations. We will now assume that our data are given at multiples of the angle π/n and that the evaluation of the integral shall occur at the same points. Then we need the weights which correspond to the definition (15.20). These weights are contained in the following table.

s	w_s	s	w_s	
0	0	7	-0.01012223	
1	-0.08891091	8	0.00762260	
2	0.04742642	9	-0.00547171	
3	-0.03134027	10	0.00354076	(9.16.18)
4	0.02267382	11	-0.00174001	
5	-0.01712979	12	0	
6	0.01317377			

The values in this table oscillate with much larger amplitudes than those of the previous table (3) since at present we are at the points of the minima and maxima of the function (4). Special attention has to be given to the value w_0 which according to the general rule should be listed as $\frac{1}{2}$ instead of zero. But again we will take into account the effect of this large constant separately. It amounts to the following modification of the previous law of the ordinates. Instead of adding up the γ_i according to (12), we have to modify the sum by *taking the two limiting ordinates with half weight*:

$$G\left(m\,\frac{\pi}{n}\right) - G(0) = G_m - G_0 + \tfrac{1}{2}\gamma_0 + \gamma_1 + \cdots \gamma_{m+1} + \tfrac{1}{2}\gamma_m \quad (9.16.19)$$

Once more we have to construct our data according to the formula (15.16) but now associated with the angles $\theta_k = 15°,\ 30°,\ 45°,\ 75°,\ 90°$. Once more we need not list our data beyond $k = 6$ because they return in reversed order (but the symmetry point now is at $k = 6$ which is *not* repeated; $\gamma_7 = \gamma_5,\ \gamma_8 = \gamma_4$, etc.).

k	γ_k	
0		
1	0.01752670	
2	0.03739991	
3	0.06170671	(9.16.20)
4	0.09068997	
5	0.11850131	
6	0.13089969	

The movable strip (8) comes again into operation but with a slightly modified symmetry pattern since $k = 0$ is now included in our subscripts:

$$\gamma_k$$ FIXED STRIP

(9.16.21)

$$w_k$$ MOVABLE STRIP

We list the results once more in tabular form, in full analogy to the previous table (17).

m	$G^*(x_m) - G^*(0)$	$\sum_{i=0}^{m}{}' \gamma_i$	$G_m - G_0$	$G(x_m) - G(0)$
1	0.00866532	0.00876336	-0.00010103	0.00866232
2	0.03583689	0.03622666	-0.00038977	0.00358369
3	0.08495923	0.08577996	-0.00082367	0.08495629
4	0.16087528	0.16197830	-0.00110330	0.16087500
5	0.26606830	0.26657394	-0.00050791	0.26606603
6	0.39269908	0.39127445	$+0.00142313$	0.39269758

(9.16.22)

The error has now moved up to the sixth decimal place which is a considerable increase compared with the much smaller errors of Table (17); this is what we have expected on the basis of the error behaviour of the Fourier series.

Actually the two sets of weights (3) and (18), if applied to the same sets of data, give redundant results inasmuch as they both define the same function $G(\theta)$, computed at two sets of points. Since this function can be generated in terms of trigonometric interpolation by one set of data, we can predict the results of the weighting by the other set of coefficients if we apply half-way interpolation to the computed points, on the basis of the formula (15.5). Hence the much less accurate ordinates which have been obtained by the weights (18), are nevertheless able to restore the much more accurate ordinates half-way between (which are also available by applying the weights (3) to the data), through the medium of trigonometric interpolation.

The objection can be raised that the chosen function (13) is too smooth for a true testing of the method since it belongs to that class of functions which are amenable even to equidistant interpolation. A more characteristic choice would have been a similar function, encountered in Chapter 1:

$$g(x) = \frac{1}{a^2 + (2x - 1)^2}$$ (9.16.23)

where a had the value 0.2 instead of 1. However, the more or less smooth character of the function merely changes the number of ordinates needed for a certain accuracy. In the present example we are in the possession of an exact error analysis and we can show that the accuracy of the tables (17), respectively (22), could have been matched with the much less smooth function (23) (with $a = 0.2$), but at the expense of a much larger number of ordinates since the present number $n = 12$ would have to be raised to

$n = 50$. Our aim was merely to demonstrate the numerical technique and for that purpose a simpler example seemed more adequate.

9.17. The method of global correction

In the previous methods of solving trajectory problems by inching forward step by step on the basis of local Taylor expansions, we did not succeed in exhibiting an explicit solution in the form of a set of continuous functions $v_i(x)$. The point of reference was constantly shifting and we could not arrive at a solution which could be truly tested whether or not it satisfied the given differential equation. Even if we did find that the differential equation is actually satisfied at every point with a high degree of accuracy, we would still have to convince ourselves that these small local errors will not accumulate and possibly cause a large error in the solution. But the principal objection to the method of shifting centres of expansion is that we have obtained a set of discrete values $v_i(x_m)$ instead of true functions of x: $v_i(x)$. We usually get round the difficulty in a purely empirical fashion by reducing the step-size h and making repeated trials, until we come to the point where the y_m-values become "stabilised", by approaching more and more definite limits. But a real "solution" in the sense of testable functions has not been achieved.

This situation is quite different, however, if we change from the variable x to the angle variable θ, obtaining once more the solution by the previous step-by-step process. Although we have once more only a discrete sequence of ordinates, we can now combine all these ordinates into a continuous and differentiable function by the method of trigonometric interpolation. Now we have actually obtained our $v_i(\theta)$ and can test explicitly to what extent the given differential equation has been satisfied.

It is more satisfactory to start with the *derivatives* y'_m, which we possess in all the data points. By trigonometric interpolation we can combine these data to a true function $v'(\theta)$ and then, by the technique of integration discussed in the sections 15 and 16 we also obtain $v(\theta)$. In contradistinction to the previous problem our "data" are now more simply constructed. In the previous case (15.16) the factor $l/2 \sin \theta_k$ appeared because we had to obtain $dg/d\theta$ in terms of the given dg/dx. In the present problem we have abandoned from the very beginning the operation with the original variable x and changed over to the variable θ. Hence we possess $dg/d\theta$ without any transformation and our γ_k now become simply

$$\gamma_k = \frac{\pi}{n} v'(\theta_k) \qquad (9.17.1)$$

We apply the "movable strip technique" and arrive at the functional values $v_i(\theta_m)$. These $v_i(\theta_m)$ will generally not agree, however, with the previous y_m-values obtained in the course of the step-by-step process. We have thus to distinguish between three functions: the function $v_i(\theta)$, found by the step-by-step procedure at the data points θ_m (and combinable to a true function $v_i(\theta)$ by the process of trigonometric interpolation): then the function $\bar{v}_i(\theta)$ found by integrating the derivative data y'_m, and finally the true function $v^*_i(\theta)$ which actually satisfies the given differential equation (5.1). The difference

$$\bar{v}_i(\theta) - v_i(\theta) = \rho_i(\theta) \tag{9.17.2}$$

can be explicitly obtained at all the data points because the previous process gave us $v_i(\theta_m)$ and now we have found the new $\bar{v}_i(\theta_m)$. We assume that the difference $\rho_i(\theta_m)$ is small. Then we have a good indication that the function $\bar{v}_i(\theta)$ will need only a small correction in order to obtain the true function $v^*(\theta)$. Hence we will put

$$v^*_i(\theta) = \bar{v}_i(\theta) + u_i(\theta) \tag{9.17.3}$$

The substitution of this expression in our differential equation (9.5.1) yields:

$$u'_i - F_i(v_j + \rho_j + u_j, \theta) + F_i(v_j, \theta) = 0 \tag{9.17.4}$$

Then, in view of the smallness of u_i and ρ_i, we can be satisfied with the solution of the linear perturbation problem

$$u'_i - \sum_{k=1}^{n} A_{ik}u_k = \sum_{k=1}^{n} A_{ik}\rho_k \tag{9.17.5}$$

If the second term were absent, we could immediately integrate this equation by the previous global integration technique. The second term has the effect that instead of an explicit solution $u_i(\theta_m)$ at the data points we obtain a large scale linear system of algebraic equations for the determination of the $u_i(\theta_m)$. Since the matrix of this system is nearly triangular, we have no difficulty in solving our system by the usual successive approximations.

We have then obtained the global correction which has to be added to the preliminary step-by-step solution and we have the added advantage that now we really possess the functions $v^*_i(\theta)$ at all points, instead of a discrete set of ordinates which exist only in a selected sequence of isolated points.

Great difficulties arise, however, if it so happens that the linear algebraic system associated with the linear differential equation (5) is "badly conditioned". Then the smallness of the right side cannot guarantee the smallness of $u_i(\theta)$. Not only is our perturbation method then in danger of being put out of action, but the further danger exists that the solution of the given problem (5.1) is exceedingly sensitive to very small changes of the initial data. The solution of such problems is very difficult and we may have to resort to the remedy of *sectionalising* our range and applying the step-by-step method in combination with the global correction technique separately in every section, thus reducing the damaging influence of explosive error accumulations.

Numerical Example. The following numerical example characterises the intrinsic properties of the local and global integration procedures. We choose a differential equation whose solution is known in analytical form and which is tabulated with great accuracy.* Bessel's differential equation for imaginary arguments of x and the order $p = 0$ can be written in the normal form (2.3) as a pair of first order equations:

$$u' = v$$
$$v' = u - \frac{v}{x}$$

We choose the interval $x = [1, 5]$ and give at $x = 1$ the initial conditions for the Hankel function $iH^{(1)}(ix)$, which has the property that it decreases exponentially to zero as x increases to infinity, while the second solution $iH^{(2)}(ix)$—which will enter in consequence of the truncation and rounding errors—increases exponentially.

Our aim is to study the error pattern of the local and the global procedure, excluding the accumulation of the purely numerical rounding errors. The results are thus tabulated to 6 decimal places, while the computations were made to 10 decimal places. In the local procedure the interval $h = 0.2$ was chosen which means that the total interval was covered by 21 points. It it convenient to change the independent variable x to the new variable $x_1 = 1.2x$. In the new variable h becomes 0.24 and the formula for the step-by-step integration, with an error of the order h^5, becomes, if we use Adams' method (cf. 10.5):

$$y_4 = y_3 + 0.55y'_3 - 0.59y'_2 + 0.37y'_1 - 0.09y'_0$$

Since our aim is to study the gradual accumulation of truncation errors, we assume optimum starting conditions and give the first 4 values of $u(x)$ and $v(x)$ with 10 place accuracy, taken from the NBS tables.

In the global procedure we will not use 20 sections since this would make the analytical errors so small that the only remaining errors would be the numerical rounding errors. We will employ the *half* number of points by dividing the interval in 10 sections only, satisfying the given differential equation in the 10 equidistant points

$$\theta_k = 0, \frac{\pi}{10}, 2\frac{\pi}{10}, \ldots, 9\frac{\pi}{10}$$

where

$$x = 3 - 2\cos\theta$$

While the local procedure uses parabolas of fourth order with a constantly shifting origin, the global procedure uses the *same* trigonometric cosine polynomial of tenth order. It is this *sameness* which prevents the gradual increase of the truncation errors. The trigonometric interpolation is characterised by *periodic* rather than exponentially increasing errors.

* Tables of the Bessel Functions $Y_0(Z)$ and $Y_1(Z)$ for Complex Arguments (Computation Laboratory, National Bureau of Standards; Columbia University Press, New York, 1950).

In the following table u_l denotes the successive $u(x)$ values, obtained by the step-by-step procedure, while u_g denotes the values obtained by the global method. The correct values $u(x)$ (taken from the NBS tables), are listed under u^*. The function $u(x)$ corresponds to the negative real part of $Y_0(ix)$ of the tables (on p. 364). The same notations hold for the function $v(x)$ which corresponds to the negative imaginary part of $Y_1(ix)$.

x	u_l	u_g	u^*	$-v_l$	$-v_g$	$-v^*$
1.0	0.268032	0.268032	0.268032	0.383186	0.383186	0.383186
1.2	0.202769	0.202779	0.202769	0.276670	0.276687	0.276670
1.4	0.155116	0.155117	0.155116	0.204250	0.204280	0.204250
1.6	0.119656	0.119642	0.119656	0.153192	0.153181	0.153192
1.8	0.093298	0.092887	0.092903	0.117695	0.116220	0.116261
2.0	0.072447	0.072502	0.072507	0.090559	0.089029	0.089041
2.2	0.056892	0.056831	0.056830	0.070781	0.068732	0.068689
2.4	0.044176	0.044693	0.044701	0.055069	0.053363	0.053301
2.6	0.034641	0.035237	0.035268	0.043698	0.041585	0.041561
2.8	0.026656	0.027850	0.027897	0.034486	0.032506	0.032539
3.0	0.020664	0.022071	0.022116	0.027903	0.025510	0.025564
3.2	0.015487	0.017543	0.017567	0.022502	0.020131	0.020144
3.4	0.011537	0.013947	0.013979	0.018755	0.015881	0.015915
3.6	0.007987	0.011133	0.011140	0.015695	0.012704	0.012602
3.8	0.005170	0.008850	0.008891	0.013724	0.010068	0.009999
4.0	0.002501	0.007020	0.007104	0.012201	0.007944	0.007947
4.2	0.000224	0.005580	0.005683	0.011446	0.006305	0.006327
4.4	−0.002078	0.004467	0.004551	0.011042	0.005118	0.005044
4.6	−0.004244	0.003574	0.003648	0.011241	0.004208	0.004027
4.8	−0.006663	0.002787	0.002927	0.011785	0.003315	0.003218
5.0	−0.008996	0.002265	0.002350	0.012917	0.002780	0.002575

We observe that the errors of u_l and v_l become gradually worse as we approach the end of the range and eventually cease to constitute even a rough approximation, although the trend of the successive values remains smooth and a purely numerical examination would not lead us to suspect that something had gone out of order. Actually, around the end of range even the sign of $u_l(x)$ changes from positive to negative. Moreover, the negative values would increase more and more if we had continued our calculations for larger values of x, although the actual function goes to zero. This is inevitable since the truncation errors invoke the exponentially increasing solution with greater and greater force. We observe the same phenomenon in $-v_l(x)$ where a certain minimum is reached and then the values increase again.

However, the global approximation behaves quite differently. We observe the presence of small periodic errors from the beginning, but later a secular error of exponentially increasing strength becomes manifest. This seems puzzling since on the basis of the properties of the finite Fourier series we have been expecting solely *periodic* errors. And yet even here the unwanted exponentially increasing solution seems to make itself felt around the end of the range.

The reason of this phenomenon is that at the start $x = 1$ we should make allowance for the existence of a small periodic error. We have suppressed this error by satisfying the given initial conditions *exactly*. The consequence is that the undesired solution is excited with a small constant factor, thus causing a gradually increasing secular error which is imposed on the small periodic errors. There is, however, a fundamental difference between this error and the error of the step-by-step procedure. In the latter case the truncation errors provide a constant energy source which feeds the exponentially increasing solution all the time. Hence the undesired solution comes into play in constantly increasing strength, while in the global method this strength remains constant. Moreover, this constant strength recedes rapidly as the number of points increases. Had we used 15 instead of 10 points of the given interval, the secular error would have disappeared in the first six decimal places.

BIBLIOGRAPHY

[1] Bennett, A. A., W. E. Milne, and H. Bateman, *Numerical Integration of Differential Equations* (Dover Publications, New York, 1956)

[2] Collatz, L., *The Numerical Treatment of Differential Equations* (Springer, Berlin, 1960)

[3] Fox, L., *The Numerical Solution of Two-Point Problems in Ordinary Differential Equations* (Clarendon Press, Oxford, 1957)

[4] Hildebrand, F. B., *Introduction to Numerical Analysis* (McGraw-Hill, 1956)

[5] Levy, H., and E. A. Baggot, *Numerical Studies in Differential Equations* (Watts, London, 1934)

[6] Milne, W. E., *Numerical Solution of Differential Equations* (Wiley, New York, 1953)

[7] Morris, M., and O. E. Brown, *Differential Equations* (Prentice-Hall, New York, 1952)

APPENDIX

TABLE I: Smoothing of the Gibbs oscillations of a Fourier series; 9 decimal place table of the sigma factors

$$\sigma_k = \frac{\sin k \dfrac{\pi}{n}}{k \dfrac{\pi}{n}}$$

for $n = 2$, to $n = 20$; cf. Chapter 2.14.

TABLE II: Double smoothing of the Gibbs oscillations of a Fourier series; 9 decimal place table of the square of the sigma factors, for $n = 2$ to $n = 20$.

k	σ_k	$\sigma_k{}^2$	k	σ_k	$\sigma_k{}^2$
	$n = 2$	$n = 2$		$n = 7$	$n = 7$
1	0.636619772	0.405284735	1	0.966766385	0.934637243
			2	0.871026416	0.758687017
	$n = 3$	$n = 3$	3	0.724101450	0.524322909
			4	0.543076087	0.294931636
1	0.826993343	0.683917990	5	0.348410566	0.121189923
2	0.413496672	0.170979497	6	0.161127731	0.025962146
	$n = 4$	$n = 4$		$n = 8$	$n = 8$
1	0.900316316	0.810569469	1	0.974495358	0.949641203
2	0.636619772	0.405284735	2	0.900316316	0.810569469
3	0.300105439	0.090063274	3	0.784213304	0.614990506
			4	0.636619772	0.405284735
	$n = 5$	$n = 5$	5	0.470527982	0.221396582
			6	0.300105439	0.090063274
1	0.935489284	0.875140200	7	0.139213623	0.019380433
2	0.756826729	0.572786697			
3	0.504551153	0.254571865		$n = 9$	$n = 9$
4	0.233872321	0.054696263	1	0.979815536	0.960038485
			2	0.920725429	0.847735315
	$n = 6$	$n = 6$	3	0.826993343	0.683917990
			4	0.705316598	0.497471504
1	0.954929659	0.911890653	5	0.564253279	0.318381763
2	0.826993343	0.683917990	6	0.413496672	0.170979497
3	0.636619772	0.405284735	7	0.263064408	0.069202883
4	0.413496672	0.170979497	8	0.122476942	0.015000601
5	0.190985932	0.036475626			

k	σ_k	$\sigma_k{}^2$	k	σ_k	$\sigma_k{}^2$
	$n = 10$	$n = 10$		$n = 14$	$n = 14$
1	0.983631643	0.967531210	1	0.991628584	0.983327250
2	0.935489284	0.875140200	2	0.966766385	0.934637243
3	0.858393691	0.736839729	3	0.926160517	0.857773303
4	0.756826729	0.572786697	4	0.871026416	0.758687017
5	0.636619772	0.405284735	5	0.803004434	0.644816121
6	0.504551153	0.254571865	6	0.724101450	0.524322909
7	0.367883011	0.135337909	7	0.636619772	0.405284735
8	0.233872321	0.054696263	8	0.543076087	0.294931636
9	0.109292405	0.011944830	9	0.446113574	0.199017321
			10	0.348410566	0.121389923
	$n = 11$	$n = 11$	11	0.252589232	0.063801320
			12	0.161127731	0.025962146
1	0.986463589	0.973110413	13	0.076279122	0.005818504
2	0.946502244	0.895866498			
3	0.882062724	0.778034649		$n = 15$	$n = 15$
4	0.796248357	0.634011446			
5	0.693153891	0.480462317	1	0.992705200	0.985463613
6	0.577628242	0.333654387	2	0.971012209	0.942864710
7	0.454999061	0.207024145	3	0.935489284	0.875140200
8	0.330773521	0.109411122	4	0.887063793	0.786882173
9	0.210333832	0.044240321	5	0.826993343	0.683917990
10	0.098646084	0.009731050	6	0.756826729	0.572786697
			7	0.678356039	0.460166915
	$n = 12$	$n = 12$	8	0.593561534	0.352315294
			9	0.504551153	0.254571865
1	0.988615929	0.977361456	10	0.413496672	0.170979497
2	0.954929659	0.911890653	11	0.322568652	0.104050535
3	0.900316316	0.810569469	12	0.233872321	0.054696263
4	0.826993343	0.683917990	13	0.149386494	0.022316325
5	0.737912976	0.544515560	14	0.070907514	0.005027876
6	0.636619772	0.405284735			
7	0.527080697	0.277814061		$n = 16$	$n = 16$
8	0.413496672	0.170979497			
9	0.300105439	0.090063274	1	0.993586851	0.987214831
10	0.190985932	0.036475626	2	0.974495358	0.949641203
11	0.089874175	0.008077367	3	0.943165321	0.889560823
			4	0.900316316	0.810569469
	$n = 13$	$n = 13$	5	0.846927993	0.717287025
			6	0.784213304	0.614990506
1	0.990295044	0.980684275	7	0.713585488	0.509204249
2	0.961518870	0.924518537	8	0.636619772	0.405284735
3	0.914673491	0.836627595	9	0.555010935	0.308037138
4	0.851382677	0.724852462	10	0.470527982	0.221396582
5	0.773824776	0.598804783	11	0.384967270	0.148199799
6	0.684642939	0.468735954	12	0.300105439	0.090063274
7	0.586836804	0.344377435	13	0.217653536	0.047373062
8	0.483640485	0.233908118	14	0.139213623	0.019380433
9	0.378392301	0.143180734	15	0.066239124	0.004387621
10	0.274402047	0.075296483			
11	0.174821613	0.030562596			
12	0.082524587	0.006810307			

k	σ_k	σ_k^2	k	σ_k	σ_k^2
	$n = 17$	$n = 17$		$n = 19$	$n = 19$
1	0.994317898	0.988668081	1	0.995449621	0.990919947
2	0.977387746	0.955286806	2	0.981872985	0.964074559
3	0.949555184	0.901655048	3	0.959492150	0.920625185
4	0.911386930	0.830626137	4	0.928672398	0.862432423
5	0.863657027	0.745903461	5	0.889915138	0.791948953
6	0.807328089	0.651778643	6	0.843848160	0.712079718
7	0.743528055	0.552833969	7	0.791213480	0.626018771
8	0.673523068	0.453633324	8	0.732853009	0.537073533
9	0.598687172	0.358426330	9	0.669692359	0.448487856
10	0.520469639	0.270888645	10	0.602723123	0.363275163
11	0.440360776	0.193917613	11	0.532984007	0.284071952
12	0.359857095	0.129497129	12	0.461541197	0.213020277
13	0.280426748	0.078639161	13	0.389468382	0.151685620
14	0.203476111	0.041402528	14	0.317826835	0.101013897
15	0.130318366	0.016982876	15	0.247645973	0.061328528
16	0.062144869	0.003861985	16	0.179904778	0.032365729
			17	0.115514469	0.013343593
	$n = 18$	$n = 18$	18	0.055302757	0.003058395
1	0.994930770	0.989887237		$n = 20$	$n = 20$
2	0.979815536	0.960038485			
3	0.954929659	0.911890653	1	0.995892735	0.991802340
4	0.920725429	0.847735315	2	0.983631643	0.967531210
5	0.877822270	0.770571938	3	0.963397762	0.928135248
6	0.826993343	0.683917990	4	0.935489284	0.875140200
7	0.769148875	0.591589991	5	0.900316316	0.810569469
8	0.705316598	0.497471504	6	0.858393691	0.736839729
9	0.636619772	0.405284735	7	0.810331958	0.656637882
10	0.564253279	0.318381763	8	0.756826728	0.572786697
11	0.489458375	0.239569501	9	0.698646585	0.488107151
12	0.413496672	0.170979497	10	0.636619772	0.405284735
13	0.337623950	0.113989932	11	0.571619933	0.326749348
14	0.263064408	0.069202883	12	0.504551152	0.254571865
15	0.190985932	0.036475626	13	0.436332593	0.190386132
16	0.122476942	0.015000601	14	0.367883011	0.135337909
17	0.058525340	0.003425215	15	0.300105439	0.090063274
			16	0.233872321	0.054696263
			17	0.170011370	0.028903866
			18	0.109292405	0.011944830
			19	0.052415407	0.002747375

TABLE III: The four transition functions from the exponential to the periodic domain and vice versa; KWB method; cf. Chapter 7.23, 24.

x	$\varphi_1{}^e(x)$	$\varphi_2{}^e(x)$	x	$\varphi_1{}^p(x)$	$\varphi_2{}^p(x)$
0	1.258542	1.089930	0	1.215658	−0.325735
−0.1	1.166995	1.169574	0.1	1.191711	−0.237065
−0.2	1.079780	1.250403	0.2	1.166576	−0.148173
−0.3	0.988343	1.333746	0.3	1.139111	−0.058999
−0.4	0.903038	1.421106	0.4	1.108239	0.030336
−0.5	0.821332	1.514167	0.5	1.072938	0.119535
−0.6	0.743722	1.614818	0.6	1.032281	0.208122
−0.7	0.670563	1.725180	0.7	0.985441	0.295447
−0.8	0.602089	1.847645	0.8	0.931716	0.380690
−0.9	0.538425	1.984923	0.9	0.870551	0.462875
−1.0	0.479599	2.140103	1.0	0.801567	0.540885
−1.1	0.425564	2.316729	1.1	0.724587	0.613477
−1.2	0.376206	2.518894	1.2	0.639658	0.679316
−1.3	0.331359	2.751352	1.3	0.547075	0.737001
−1.4	0.290817	3.019658	1.4	0.447403	0.785106
−1.5	0.254345	3.330337	1.5	0.341493	0.822226
−1.6	0.221688	3.691089	1.6	0.230484	0.847022
−1.7	0.192576	4.111043	1.7	0.115812	0.858282
−1.8	0.166739	4.601059	1.8	−0.000807	0.854970
−1.9	0.143903	5.174106	1.9	−0.117393	0.836294
−2.0	0.123803	5.845721	2.0	−0.231732	0.801758
−2.1	0.106179	6.634567	2.1	−0.341425	0.751220
−2.2	0.090786	7.563125	2.2	−0.443943	0.684944
−2.3	0.077392	8.658546	2.3	−0.536697	0.603639
−2.4	0.065780	9.953694	2.4	−0.617115	0.508492
−2.5	0.055747	11.488444	2.5	−0.682740	0.401183
−2.6	0.047109	13.311284	2.6	−0.731322	0.283883
−2.7	0.039698	15.482199	2.7	−0.760925	0.159239
−2.8	0.033359	18.070646	2.8	−0.770025	0.030326
−2.9	0.027956	21.167626	2.9	−0.757616	−0.099413
−3.0	0.023365	24.880518	3.0	−0.723292	−0.226254

INDEX

A CATALOG OF SELECTED
DOVER BOOKS
IN SCIENCE AND MATHEMATICS

A CATALOG OF SELECTED
DOVER BOOKS
IN SCIENCE AND MATHEMATICS

QUALITATIVE THEORY OF DIFFERENTIAL EQUATIONS, V.V. Nemytskii and V.V. Stepanov. Classic graduate-level text by two prominent Soviet mathematicians covers classical differential equations as well as topological dynamics and ergodic theory. Bibliographies. 523pp. 5⅜ x 8½. 65954-2 Pa. $14.95

MATRICES AND LINEAR ALGEBRA, Hans Schneider and George Phillip Barker. Basic textbook covers theory of matrices and its applications to systems of linear equations and related topics such as determinants, eigenvalues and differential equations. Numerous exercises. 432pp. 5⅜ x 8½. 66014-1 Pa. $10.95

QUANTUM THEORY, David Bohm. This advanced undergraduate-level text presents the quantum theory in terms of qualitative and imaginative concepts, followed by specific applications worked out in mathematical detail. Preface. Index. 655pp. 5⅜ x 8½. 65969-0 Pa. $14.95

ATOMIC PHYSICS (8th edition), Max Born. Nobel laureate's lucid treatment of kinetic theory of gases, elementary particles, nuclear atom, wave-corpuscles, atomic structure and spectral lines, much more. Over 40 appendices, bibliography. 495pp. 5⅜ x 8½. 65984-4 Pa. $13.95

ELECTRONIC STRUCTURE AND THE PROPERTIES OF SOLIDS: The Physics of the Chemical Bond, Walter A. Harrison. Innovative text offers basic understanding of the electronic structure of covalent and ionic solids, simple metals, transition metals and their compounds. Problems. 1980 edition. 582pp. 6⅛ x 9¼. 66021-4 Pa. $16.95

BOUNDARY VALUE PROBLEMS OF HEAT CONDUCTION, M. Necati Özisik. Systematic, comprehensive treatment of modern mathematical methods of solving problems in heat conduction and diffusion. Numerous examples and problems. Selected references. Appendices. 505pp. 5⅜ x 8½. 65990-9 Pa. $12.95

A SHORT HISTORY OF CHEMISTRY (3rd edition), J.R. Partington. Classic exposition explores origins of chemistry, alchemy, early medical chemistry, nature of atmosphere, theory of valency, laws and structure of atomic theory, much more. 428pp. 5⅜ x 8½. (Available in U.S. only) 65977-1 Pa. $11.95

A HISTORY OF ASTRONOMY, A. Pannekoek. Well-balanced, carefully reasoned study covers such topics as Ptolemaic theory, work of Copernicus, Kepler, Newton, Eddington's work on stars, much more. Illustrated. References. 521pp. 5⅜ x 8½. 65994-1 Pa. $12.95

PRINCIPLES OF METEOROLOGICAL ANALYSIS, Walter J. Saucier. Highly respected, abundantly illustrated classic reviews atmospheric variables, hydrostatics, static stability, various analyses (scalar, cross-section, isobaric, isentropic, more). For intermediate meteorology students. 454pp. 6⅛ x 9¼. 65979-8 Pa. $14.95

RELATIVITY, THERMODYNAMICS AND COSMOLOGY, Richard C.
Tolman. Landmark study extends thermodynamics to special, general relativity; also
applications of relativistic mechanics, thermodynamics to cosmological models.
501pp. 5⅜ x 8½. 65383-8 Pa. $13.95

APPLIED ANALYSIS, Cornelius Lanczos. Classic work on analysis and design of
finite processes for approximating solution of analytical problems. Algebraic equations, matrices, harmonic analysis, quadrature methods, much more. 559pp. 5⅜ x 8½.
65656-X Pa. $13.95

INTRODUCTION TO ANALYSIS, Maxwell Rosenlicht. Unusually clear, accessi-
ble coverage of set theory, real number system, metric spaces, continuous functions,
Riemann integration, multiple integrals, more. Wide range of problems.
Undergraduate level. Bibliography. 254pp. 5⅜ x 8½. 65038-3 Pa. $8.95

INTRODUCTION TO QUANTUM MECHANICS With Applications to
Chemistry, Linus Pauling & E. Bright Wilson, Jr. Classic undergraduate text by Nobel
Prize winner applies quantum mechanics to chemical and physical problems.
Numerous tables and figures enhance the text. Chapter bibliographies. Appendices.
Index. 468pp. 5⅜ x 8½. 64871-0 Pa. $12.95

ASYMPTOTIC EXPANSIONS OF INTEGRALS, Norman Bleistein & Richard A.
Handelsman. Best introduction to important field with applications in a variety of scientific disciplines. New preface. Problems. Diagrams. Tables. Bibliography. Index.
448pp. 5⅜ x 8½. 65082-0 Pa. $12.95

MATHEMATICS APPLIED TO CONTINUUM MECHANICS, Lee A. Segel.
Analyzes models of fluid flow and solid deformation. For upper-level math, science
and engineering students. 608pp. 5⅜ x 8½. 65369-2 Pa. $14.95

ELEMENTS OF REAL ANALYSIS, David A. Sprecher. Classic text covers fundamental concepts, real number system, point sets, functions of a real variable, Fourier
series, much more. Over 500 exercises. 352pp. 5⅜ x 8½. 65385-4 Pa. $11.95

PHYSICAL PRINCIPLES OF THE QUANTUM THEORY, Werner Heisenberg.
Nobel Laureate discusses quantum theory, uncertainty, wave mechanics, work of
Dirac, Schroedinger, Compton, Wilson, Einstein, etc. 184pp. 5⅜ x 8½.
60113-7 Pa. $6.95

INTRODUCTORY REAL ANALYSIS, A.N. Kolmogorov, S.V. Fomin. Translated
by Richard A. Silverman. Self-contained, evenly paced introduction to real and functional analysis. Some 350 problems. 403pp. 5⅜ x 8½. 61226-0 Pa. $10.95

PROBLEMS AND SOLUTIONS IN QUANTUM CHEMISTRY AND
PHYSICS, Charles S. Johnson, Jr. and Lee G. Pedersen. Unusually varied problems,
detailed solutions in coverage of quantum mechanics, wave mechanics, angular
momentum, molecular spectroscopy, scattering theory, more. 280 problems plus 139
supplementary exercises. 430pp. 6½ x 9¼. 65236-X Pa. $13.95

CATALOG OF DOVER BOOKS

ASYMPTOTIC METHODS IN ANALYSIS, N.G. de Bruijn. An inexpensive, comprehensive guide to asymptotic methods–the pioneering work that teaches by explaining worked examples in detail. Index. 224pp. 5⅜ x 8½.   64221-6 Pa. $7.95

OPTICAL RESONANCE AND TWO-LEVEL ATOMS, L. Allen and J. H. Eberly. Clear, comprehensive introduction to basic principles behind all quantum optical resonance phenomena. 53 illustrations. Preface. Index. 256pp. 5⅜ x 8½.
65533-4 Pa. $8.95

COMPLEX VARIABLES, Francis J. Flanigan. Unusual approach, delaying complex algebra till harmonic functions have been analyzed from real variable viewpoint. Includes problems with answers. 364pp. 5⅜ x 8½.   61388-7 Pa. $9.95

ATOMIC SPECTRA AND ATOMIC STRUCTURE, Gerhard Herzberg. One of best introductions; especially for specialist in other fields. Treatment is physical rather than mathematical. 80 illustrations. 257pp. 5⅜ x 8½.   60115-3 Pa. $7.95

APPLIED COMPLEX VARIABLES, John W. Dettman. Step-by-step coverage of fundamentals of analytic function theory–plus lucid exposition of five important applications: Potential Theory; Ordinary Differential Equations; Fourier Transforms; Laplace Transforms; Asymptotic Expansions. 66 figures. Exercises at chapter ends. 512pp. 5⅜ x 8½.   64670-X Pa. $12.95

ULTRASONIC ABSORPTION: An Introduction to the Theory of Sound Absorption and Dispersion in Gases, Liquids and Solids, A.B. Bhatia. Standard reference in the field provides a clear, systematically organized introductory review of fundamental concepts for advanced graduate students, research workers. Numerous diagrams. Bibliography. 440pp. 5⅜ x 8½.   64917-2 Pa. $11.95

UNBOUNDED LINEAR OPERATORS: Theory and Applications, Seymour Goldberg. Classic presents systematic treatment of the theory of unbounded linear operators in normed linear spaces with applications to differential equations. Bibliography. 199pp. 5⅜ x 8½.   64830-3 Pa. $7.95

LIGHT SCATTERING BY SMALL PARTICLES, H.C. van de Hulst. Comprehensive treatment including full range of useful approximation methods for researchers in chemistry, meteorology and astronomy. 44 illustrations. 470pp. 5⅜ x 8½.
64228-3 Pa. $12.95

CONFORMAL MAPPING ON RIEMANN SURFACES, Harvey Cohn. Lucid, insightful book presents ideal coverage of subject. 334 exercises make book perfect for self-study. 55 figures. 352pp. 5⅜ x 8¼.   64025-6 Pa. $11.95

OPTICKS, Sir Isaac Newton. Newton's own experiments with spectroscopy, colors, lenses, reflection, refraction, etc., in language the layman can follow. Foreword by Albert Einstein. 532pp. 5⅜ x 8½.   60205-2 Pa. $12.95

GENERALIZED INTEGRAL TRANSFORMATIONS, A.H. Zemanian. Graduate-level study of recent generalizations of the Laplace, Mellin, Hankel, K. Weierstrass, convolution and other simple transformations. Bibliography. 320pp. 5⅜ x 8½.
65375-7 Pa. $8.95

THE ELECTROMAGNETIC FIELD, Albert Shadowitz. Comprehensive undergraduate text covers basics of electric and magnetic fields, builds up to electromagnetic theory. Also related topics, including relativity. Over 900 problems. 768pp. 5⅜ x 8¼. 65660-8 Pa. $18.95

FOURIER SERIES, Georgi P. Tolstov. Translated by Richard A. Silverman. A valuable addition to the literature on the subject, moving clearly from subject to subject and theorem to theorem. 107 problems, answers. 336pp. 5⅜ x 8½. 63317-9 Pa. $9.95

THEORY OF ELECTROMAGNETIC WAVE PROPAGATION, Charles Herach Papas. Graduate-level study discusses the Maxwell field equations, radiation from wire antennas, the Doppler effect and more. xiii + 244pp. 5⅜ x 8½. 65678-0 Pa. $6.95

DISTRIBUTION THEORY AND TRANSFORM ANALYSIS: An Introduction to Generalized Functions, with Applications, A.H. Zemanian. Provides basics of distribution theory, describes generalized Fourier and Laplace transformations. Numerous problems. 384pp. 5⅜ x 8½. 65479-6 Pa. $11.95

THE PHYSICS OF WAVES, William C. Elmore and Mark A. Heald. Unique overview of classical wave theory. Acoustics, optics, electromagnetic radiation, more. Ideal as classroom text or for self-study. Problems. 477pp. 5⅜ x 8½.
64926-1 Pa. $13.95

CALCULUS OF VARIATIONS WITH APPLICATIONS, George M. Ewing. Applications-oriented introduction to variational theory develops insight and promotes understanding of specialized books, research papers. Suitable for advanced undergraduate/graduate students as primary, supplementary text. 352pp. 5⅜ x 8½.
64856-7 Pa. $9.95

A TREATISE ON ELECTRICITY AND MAGNETISM, James Clerk Maxwell. Important foundation work of modern physics. Brings to final form Maxwell's theory of electromagnetism and rigorously derives his general equations of field theory. 1,084pp. 5⅜ x 8½. 60636-8, 60637-6 Pa., Two-vol. set $25.90

AN INTRODUCTION TO THE CALCULUS OF VARIATIONS, Charles Fox. Graduate-level text covers variations of an integral, isoperimetrical problems, least action, special relativity, approximations, more. References. 279pp. 5⅜ x 8½.
65499-0 Pa. $8.95

HYDRODYNAMIC AND HYDROMAGNETIC STABILITY, S. Chandrasekhar. Lucid examination of the Rayleigh-Benard problem; clear coverage of the theory of instabilities causing convection. 704pp. 5⅜ x 8¼. 64071-X Pa. $14.95

CALCULUS OF VARIATIONS, Robert Weinstock. Basic introduction covering isoperimetric problems, theory of elasticity, quantum mechanics, electrostatics, etc. Exercises throughout. 326pp. 5⅜ x 8½. 63069-2 Pa. $9.95

DYNAMICS OF FLUIDS IN POROUS MEDIA, Jacob Bear. For advanced students of ground water hydrology, soil mechanics and physics, drainage and irrigation engineering and more. 335 illustrations. Exercises, with answers. 784pp. 6⅛ x 9¼.
65675-6 Pa. $19.95

NUMERICAL METHODS FOR SCIENTISTS AND ENGINEERS, Richard Hamming. Classic text stresses frequency approach in coverage of algorithms, polynomial approximation, Fourier approximation, exponential approximation, other topics. Revised and enlarged 2nd edition. 721pp. 5⅜ x 8½. 65241-6 Pa. $15.95

THEORETICAL SOLID STATE PHYSICS, Vol. 1: Perfect Lattices in Equilibrium; Vol. II: Non-Equilibrium and Disorder, William Jones and Norman H. March. Monumental reference work covers fundamental theory of equilibrium properties of perfect crystalline solids, non-equilibrium properties, defects and disordered systems. Appendices. Problems. Preface. Diagrams. Index. Bibliography. Total of 1,301pp. 5⅜ x 8½. Two volumes. Vol. I: 65015-4 Pa. $16.95
Vol. II: 65016-2 Pa. $16.95

OPTIMIZATION THEORY WITH APPLICATIONS, Donald A. Pierre. Broad spectrum approach to important topic. Classical theory of minima and maxima, calculus of variations, simplex technique and linear programming, more. Many problems, examples. 640pp. 5⅜ x 8½. 65205-X Pa. $16.95

THE CONTINUUM: A Critical Examination of the Foundation of Analysis, Hermann Weyl. Classic of 20th-century foundational research deals with the conceptual problem posed by the continuum. 156pp. 5⅜ x 8½. 67982-9 Pa. $6.95

ESSAYS ON THE THEORY OF NUMBERS, Richard Dedekind. Two classic essays by great German mathematician: on the theory of irrational numbers; and on transfinite numbers and properties of natural numbers. 115pp. 5⅜ x 8½.
21010-3 Pa. $5.95

THE FUNCTIONS OF MATHEMATICAL PHYSICS, Harry Hochstadt. Comprehensive treatment of orthogonal polynomials, hypergeometric functions, Hill's equation, much more. Bibliography. Index. 322pp. 5⅜ x 8½. 65214-9 Pa. $9.95

NUMBER THEORY AND ITS HISTORY, Oystein Ore. Unusually clear, accessible introduction covers counting, properties of numbers, prime numbers, much more. Bibliography. 380pp. 5⅜ x 8½. 65620-9 Pa. $10.95

THE VARIATIONAL PRINCIPLES OF MECHANICS, Cornelius Lanczos. Graduate level coverage of calculus of variations, equations of motion, relativistic mechanics, more. First inexpensive paperbound edition of classic treatise. Index. Bibliography. 418pp. 5⅜ x 8½. 65067-7 Pa. $12.95

MATHEMATICAL TABLES AND FORMULAS, Robert D. Carmichael and Edwin R. Smith. Logarithms, sines, tangents, trig functions, powers, roots, reciprocals, exponential and hyperbolic functions, formulas and theorems. 269pp. 5⅜ x 8½.
60111-0 Pa. $6.95

THEORETICAL PHYSICS, Georg Joos, with Ira M. Freeman. Classic overview covers essential math, mechanics, electromagnetic theory, thermodynamics, quantum mechanics, nuclear physics, other topics. First paperback edition. xxiii + 885pp. 5⅜ x 8½. 65227-0 Pa. $21.95

HANDBOOK OF MATHEMATICAL FUNCTIONS WITH FORMULAS, GRAPHS, AND MATHEMATICAL TABLES, edited by Milton Abramowitz and Irene A. Stegun. Vast compendium: 29 sets of tables, some to as high as 20 places. 1,046pp. 8 x 10½. 61272-4 Pa. $26.95

MATHEMATICAL METHODS IN PHYSICS AND ENGINEERING, John W. Dettman. Algebraically based approach to vectors, mapping, diffraction, other topics in applied math. Also generalized functions, analytic function theory, more. Exercises. 448pp. 5⅜ x 8¼. 65649-7 Pa. $10.95

A SURVEY OF NUMERICAL MATHEMATICS, David M. Young and Robert Todd Gregory. Broad self-contained coverage of computer-oriented numerical algorithms for solving various types of mathematical problems in linear algebra, ordinary and partial, differential equations, much more. Exercises. Total of 1,248pp. 5⅜ x 8½. Two volumes. Vol. I: 65691-8 Pa. $16.95
Vol. II: 65692-6 Pa. $16.95

TENSOR ANALYSIS FOR PHYSICISTS, J.A. Schouten. Concise exposition of the mathematical basis of tensor analysis, integrated with well-chosen physical examples of the theory. Exercises. Index. Bibliography. 289pp. 5⅜ x 8½. 65582-2 Pa. $8.95

INTRODUCTION TO NUMERICAL ANALYSIS (2nd Edition), F.B. Hildebrand. Classic, fundamental treatment covers computation, approximation, interpolation, numerical differentiation and integration, other topics. 150 new problems. 669pp. 5⅜ x 8½. 65363-3 Pa. $16.95

INVESTIGATIONS ON THE THEORY OF THE BROWNIAN MOVEMENT, Albert Einstein. Five papers (1905–8) investigating dynamics of Brownian motion and evolving elementary theory. Notes by R. Fürth. 122pp. 5⅜ x 8½. 60304-0 Pa. $5.95

CATASTROPHE THEORY FOR SCIENTISTS AND ENGINEERS, Robert Gilmore. Advanced-level treatment describes mathematics of theory grounded in the work of Poincaré, R. Thom, other mathematicians. Also important applications to problems in mathematics, physics, chemistry and engineering. 1981 edition. References. 28 tables. 397 black-and-white illustrations. xvii + 666pp. 6⅛ x 9¼. 67539-4 Pa. $17.95

AN INTRODUCTION TO STATISTICAL THERMODYNAMICS, Terrell L. Hill. Excellent basic text offers wide-ranging coverage of quantum statistical mechanics, systems of interacting molecules, quantum statistics, more. 523pp. 5⅜ x 8½. 65242-4 Pa. $12.95

STATISTICAL PHYSICS, Gregory H. Wannier. Classic text combines thermodynamics, statistical mechanics and kinetic theory in one unified presentation of thermal physics. Problems with solutions. Bibliography. 532pp. 5⅜ x 8½. 65401-X Pa. $12.95

ORDINARY DIFFERENTIAL EQUATIONS, Morris Tenenbaum and Harry Pollard. Exhaustive survey of ordinary differential equations for undergraduates in mathematics, engineering, science. Thorough analysis of theorems. Diagrams. Bibliography. Index. 818pp. 5⅜ x 8½. 64940-7 Pa. $18.95

STATISTICAL MECHANICS: Principles and Applications, Terrell L. Hill. Standard text covers fundamentals of statistical mechanics, applications to fluctuation theory, imperfect gases, distribution functions, more. 448pp. 5⅜ x 8½. 65390-0 Pa. $11.95

ORDINARY DIFFERENTIAL EQUATIONS AND STABILITY THEORY: An Introduction, David A. Sánchez. Brief, modern treatment. Linear equation, stability theory for autonomous and nonautonomous systems, etc. 164pp. 5⅝ x 8¼. 63828-6 Pa. $6.95

THIRTY YEARS THAT SHOOK PHYSICS: The Story of Quantum Theory, George Gamow. Lucid, accessible introduction to influential theory of energy and matter. Careful explanations of Dirac's anti-particles, Bohr's model of the atom, much more. 12 plates. Numerous drawings. 240pp. 5⅜ x 8½. 24895-X Pa. $7.95

THEORY OF MATRICES, Sam Perlis. Outstanding text covering rank, nonsingularity and inverses in connection with the development of canonical matrices under the relation of equivalence, and without the intervention of determinants. Includes exercises. 237pp. 5⅜ x 8½. 66810-X Pa. $8.95

GREAT EXPERIMENTS IN PHYSICS: Firsthand Accounts from Galileo to Einstein, edited by Morris H. Shamos. 25 crucial discoveries: Newton's laws of motion, Chadwick's study of the neutron, Hertz on electromagnetic waves, more. Original accounts clearly annotated. 370pp. 5⅜ x 8½. 25346-5 Pa. $10.95

INTRODUCTION TO PARTIAL DIFFERENTIAL EQUATIONS WITH APPLICATIONS, E.C. Zachmanoglou and Dale W. Thoe. Essentials of partial differential equations applied to common problems in engineering and the physical sciences. Problems and answers. 416pp. 5⅜ x 8½. 65251-3 Pa. $11.95

BURNHAM'S CELESTIAL HANDBOOK, Robert Burnham, Jr. Thorough guide to the stars beyond our solar system. Exhaustive treatment. Alphabetical by constellation: Andromeda to Cetus in Vol. 1; Chamaeleon to Orion in Vol. 2; and Pavo to Vulpecula in Vol. 3. Hundreds of illustrations. Index in Vol. 3. 2,000pp. 6⅛ x 9¼. 23567-X, 23568-8, 23673-0 Pa., Three-vol. set $44.85

CHEMICAL MAGIC, Leonard A. Ford. Second Edition, Revised by E. Winston Grundmeier. Over 100 unusual stunts demonstrating cold fire, dust explosions, much more. Text explains scientific principles and stresses safety precautions. 128pp. 5⅜ x 8½. 67628-5 Pa. $5.95

AMATEUR ASTRONOMER'S HANDBOOK, J.B. Sidgwick. Timeless, comprehensive coverage of telescopes, mirrors, lenses, mountings, telescope drives, micrometers, spectroscopes, more. 189 illustrations. 576pp. 5⅝ x 8¼. (Available in U.S. only) 24034-7 Pa. $11.95

SPECIAL FUNCTIONS, N.N. Lebedev. Translated by Richard Silverman. Famous Russian work treating more important special functions, with applications to specific problems of physics and engineering. 38 figures. 308pp. 5⅜ x 8½. 60624-4 Pa. $9.95

OBSERVATIONAL ASTRONOMY FOR AMATEURS, J.B. Sidgwick. Mine of useful data for observation of sun, moon, planets, asteroids, aurorae, meteors, comets, variables, binaries, etc. 39 illustrations. 384pp. 5⅝ x 8¼. (Available in U.S. only) 24033-9 Pa. $8.95

INTEGRAL EQUATIONS, F.G. Tricomi. Authoritative, well-written treatment of extremely useful mathematical tool with wide applications. Volterra Equations, Fredholm Equations, much more. Advanced undergraduate to graduate level. Exercises. Bibliography. 238pp. 5⅜ x 8½. 64828-1 Pa. $8.95

POPULAR LECTURES ON MATHEMATICAL LOGIC, Hao Wang. Noted logician's lucid treatment of historical developments, set theory, model theory, recursion theory and constructivism, proof theory, more. 3 appendixes. Bibliography. 1981 edition. ix + 283pp. 5⅜ x 8½. 67632-3 Pa. $8.95

MODERN NONLINEAR EQUATIONS, Thomas L. Saaty. Emphasizes practical solution of problems; covers seven types of equations. ". . . a welcome contribution to the existing literature...."–*Math Reviews*. 490pp. 5⅜ x 8½. 64232-1 Pa. $13.95

FUNDAMENTALS OF ASTRODYNAMICS, Roger Bate et al. Modern approach developed by U.S. Air Force Academy. Designed as a first course. Problems, exercises. Numerous illustrations. 455pp. 5⅜ x 8½. 60061-0 Pa. $10.95

INTRODUCTION TO LINEAR ALGEBRA AND DIFFERENTIAL EQUATIONS, John W. Dettman. Excellent text covers complex numbers, determinants, orthonormal bases, Laplace transforms, much more. Exercises with solutions. Undergraduate level. 416pp. 5⅜ x 8½. 65191-6 Pa. $11.95

INCOMPRESSIBLE AERODYNAMICS, edited by Bryan Thwaites. Covers theoretical and experimental treatment of the uniform flow of air and viscous fluids past two-dimensional aerofoils and three-dimensional wings; many other topics. 654pp. 5⅜ x 8½. 65465-6 Pa. $16.95

INTRODUCTION TO DIFFERENCE EQUATIONS, Samuel Goldberg. Exceptionally clear exposition of important discipline with applications to sociology, psychology, economics. Many illustrative examples; over 250 problems. 260pp. 5⅜ x 8½. 65084-7 Pa. $8.95

LAMINAR BOUNDARY LAYERS, edited by L. Rosenhead. Engineering classic covers steady boundary layers in two- and three- dimensional flow, unsteady boundary layers, stability, observational techniques, much more. 708pp. 5⅜ x 8½. 65646-2 Pa. $18 95

LECTURES ON CLASSICAL DIFFERENTIAL GEOMETRY, Second Edition, Dirk J. Struik. Excellent brief introduction covers curves, theory of surfaces, fundamental equations, geometry on a surface, conformal mapping, other topics. Problems. 240pp. 5⅜ x 8½. 65609-8 Pa. $8.95

CATALOG OF DOVER BOOKS

ROTARY-WING AERODYNAMICS, W.Z. Stepniewski. Clear, concise text covers aerodynamic phenomena of the rotor and offers guidelines for helicopter performance evaluation. Originally prepared for NASA. 537 figures. 640pp. 6⅛ x 9¼.
64647-5 Pa. $16.95

DIFFERENTIAL GEOMETRY, Heinrich W. Guggenheimer. Local differential geometry as an application of advanced calculus and linear algebra. Curvature, transformation groups, surfaces, more. Exercises. 62 figures. 378pp. 5⅜ x 8½.
63433-7 Pa. $9.95

INTRODUCTION TO SPACE DYNAMICS, William Tyrrell Thomson. Comprehensive, classic introduction to space-flight engineering for advanced undergraduate and graduate students. Includes vector algebra, kinematics, transformation of coordinates. Bibliography. Index. 352pp. 5⅜ x 8½.
65113-4 Pa. $9.95

A SURVEY OF MINIMAL SURFACES, Robert Osserman. Up-to-date, in-depth discussion of the field for advanced students. Corrected and enlarged edition covers new developments. Includes numerous problems. 192pp. 5⅜ x 8½.
64998-9 Pa. $8.95

ANALYTICAL MECHANICS OF GEARS, Earle Buckingham. Indispensable reference for modern gear manufacture covers conjugate gear-tooth action, gear-tooth profiles of various gears, many other topics. 263 figures. 102 tables. 546pp. 5⅜ x 8½.
65712-4 Pa. $14.95

SET THEORY AND LOGIC, Robert R. Stoll. Lucid introduction to unified theory of mathematical concepts. Set theory and logic seen as tools for conceptual understanding of real number system. 496pp. 5⅜ x 8¼.
63829-4 Pa. $12.95

A HISTORY OF MECHANICS, René Dugas. Monumental study of mechanical principles from antiquity to quantum mechanics. Contributions of ancient Greeks, Galileo, Leonardo, Kepler, Lagrange, many others. 671pp. 5⅜ x 8½.
65632-2 Pa. $14.95

FAMOUS PROBLEMS OF GEOMETRY AND HOW TO SOLVE THEM, Benjamin Bold. Squaring the circle, trisecting the angle, duplicating the cube: learn their history, why they are impossible to solve, then solve them yourself. 128pp. 5⅜ x 8½.
24297-8 Pa. $4.95

MECHANICAL VIBRATIONS, J.P. Den Hartog. Classic textbook offers lucid explanations and illustrative models, applying theories of vibrations to a variety of practical industrial engineering problems. Numerous figures. 233 problems, solutions. Appendix. Index. Preface. 436pp. 5⅜ x 8½.
64785-4 Pa. $11.95

CURVATURE AND HOMOLOGY, Samuel I. Goldberg. Thorough treatment of specialized branch of differential geometry. Covers Riemannian manifolds, topology of differentiable manifolds, compact Lie groups, other topics. Exercises. 315pp. 5⅜ x 8½.
64314-X Pa. $9.95

HISTORY OF STRENGTH OF MATERIALS, Stephen P. Timoshenko. Excellent historical survey of the strength of materials with many references to the theories of elasticity and structure. 245 figures. 452pp. 5⅜ x 8½.
61187-6 Pa. $12.95

CATALOG OF DOVER BOOKS

CHALLENGING MATHEMATICAL PROBLEMS WITH ELEMENTARY SOLUTIONS, A.M. Yaglom and I.M. Yaglom. Over 170 challenging problems on probability theory, combinatorial analysis, points and lines, topology, convex polygons, many other topics. Solutions. Total of 445pp. 5⅜ x 8½. Two-vol. set.

Vol. I: 65536-9 Pa. $7.95
Vol. II: 65537-7 Pa. $7.95

FIFTY CHALLENGING PROBLEMS IN PROBABILITY WITH SOLUTIONS, Frederick Mosteller. Remarkable puzzlers, graded in difficulty, illustrate elementary and advanced aspects of probability. Detailed solutions. 88pp. 5⅜ x 8½.

65355-2 Pa. $4.95

EXPERIMENTS IN TOPOLOGY, Stephen Barr. Classic, lively explanation of one of the byways of mathematics. Klein bottles, Moebius strips, projective planes, map coloring, problem of the Koenigsberg bridges, much more, described with clarity and wit. 43 figures. 210pp. 5⅜ x 8½. 25933-1 Pa. $6.95

RELATIVITY IN ILLUSTRATIONS, Jacob T. Schwartz. Clear nontechnical treatment makes relativity more accessible than ever before. Over 60 drawings illustrate concepts more clearly than text alone. Only high school geometry needed. Bibliography. 128pp. 6⅛ x 9¼. 25965-X Pa. $7.95

AN INTRODUCTION TO ORDINARY DIFFERENTIAL EQUATIONS, Earl A. Coddington. A thorough and systematic first course in elementary differential equations for undergraduates in mathematics and science, with many exercises and problems (with answers). Index. 304pp. 5⅜ x 8½. 65942-9 Pa. $8.95

FOURIER SERIES AND ORTHOGONAL FUNCTIONS, Harry F. Davis. An incisive text combining theory and practical example to introduce Fourier series, orthogonal functions and applications of the Fourier method to boundary-value problems. 570 exercises. Answers and notes. 416pp. 5⅜ x 8½. 65973-9 Pa. $11.95

AN INTRODUCTION TO ALGEBRAIC STRUCTURES, Joseph Landin. Superb self-contained text covers "abstract algebra": sets and numbers, theory of groups, theory of rings, much more. Numerous well-chosen examples, exercises. 247pp. 5⅜ x 8½. 65940-2 Pa. $8.95

STARS AND RELATIVITY, Ya. B. Zel'dovich and I. D. Novikov. Vol. 1 of *Relativistic Astrophysics* by famed Russian scientists. General relativity, properties of matter under astrophysical conditions, stars and stellar systems. Deep physical insights, clear presentation. 1971 edition. References. 544pp. 5⅜ x 8½. 69424-0 Pa. $14.95

Prices subject to change without notice.

Available at your book dealer or write for free Mathematics and Science Catalog to Dept. GI, Dover Publications, Inc., 31 East 2nd St., Mineola, N.Y. 11501. Dover publishes more than 250 books each year on science, elementary and advanced mathematics, biology, music, art, literature, history, social sciences and other areas.